BURNHAM'S CELESTIAL HANDBOOK

An Observer's Guide to the Universe
Beyond the Solar System

ROBERT BURNHAM, JR.

Staff Member, Lowell Observatory, 1958–1979

IN THREE VOLUMES
Volume One, Andromeda–Cetus
REVISED AND ENLARGED EDITION

DOVER PUBLICATIONS, INC.
NEW YORK

The author takes great pleasure in offering his special thanks and appreciation to Herbert A. Luft, whose unflagging interest and support has helped immeasurably to make the *Celestial Handbook* a reality.

FRONTISPIECE: The Great Nebula NGC 6611, some 8000 light years distant in the constellation Serpens. New stars are born in clouds of dust and gas such as this one, which measures about 20,000 times the diameter of our Solar System. Palomar Observatory photograph with the 200-inch telescope.

Copyright © 1966, 1978 by Robert Burnham, Jr.
All rights reserved under Pan American and International Copyright Conventions.

Published in Canada by General Publishing Company, Ltd., 30 Lesmill Road, Don Mills, Toronto, Ontario.
Published in the United Kingdom by Constable and Company, Ltd., 10 Orange Street, London WC2H 7EG.

This Dover edition, first published in 1978, is an expanded and updated republication of the work originally published by Celestial Handbook Publications, Flagstaff, Arizona, in 1966.

INTERNATIONAL STANDARD BOOK NUMBERS:
paperbound edition: 0-486-23567-X
clothbound edition: 0-486-24063-0
Library of Congress Catalog Card Number: 77-082888

Manufactured in the United States of America
Dover Publications, Inc.
180 Varick Street
New York, N. Y. 10014

The *CELESTIAL HANDBOOK*

*is affectionately dedicated
to all the young friends
who have traveled with me
to the far reaches
of the Universe:*

*Bill James Holmes
Cleo Ray Green
Thomas A. Lyman
Gary Lyman
Terry Charles Goyette
Billy Bruce Piland
Rick Dale Piland
Andrew Niero
Bruce Thomas
Kevin Boyce
Colin Boyce
Lawrence Jones
Nicholas Franz*

MAY THERE ALWAYS BE STARLIGHT ON THE PATH

Midnight.....
There is no sound in the forest –
only the phantom murmur
of the far wind
and the wind's shadow drifting
as smoke
through ebon branches; there a single star
glistens in the heart of night....
A star!
Look skyward now...
and see above...INFINITY
Vast and dark and deep
and endless....your heritage:
Silent clouds of stars,
Other worlds uncountable and other suns
beyond numbering
and realms of fire-mist and star-cities
as grains of sand....
drifting...
Across the void....
Across the gulf of night:...
Across the endless rain of years....
Across the ages.
Listen!
Were you the star-born you should hear
That silent music of which the ancient sages spoke
Though in silent words...
Here then is our quest
and our world
and our Home.
Come with me now, Pilgrim of the stars,
For our time is upon us and our eyes
shall see the far country
and the shining cities of Infinity
which the wise men knew
in ages past, and shall know again
in ages yet to be.
Look to the east....there shines
the Morning Star...soon shall the sunrise come...
We await the Dawn,
Rise, oh eternal light;
Awaken the World!
With trumpets and cymbals and harp and the sound
of glad song!
And now...
The clouds of night are rolled away;
Sing welcome to the Dawn
Of the bright new day!

THE CELESTIAL HANDBOOK
TABLE OF CONTENTS

THE CONE NEBULA

" .... a picture of such stranegeness and splendor that it scarcely seems natural............"

REGION OF NGC 2264   PALOMAR   OBSERVATORY   200-INCH TELESCOPE

AMATEUR ASTRONOMY - A PERSONAL VIEW
THE IDEA BEHIND THE CELESTIAL HANDBOOK

If astronomy is the oldest of the sciences, surely amateur astronomy may rightfully claim to be the oldest of the scientific hobbies. No one can date that remote epoch when astronomy "began" - we can say only that the fascination of the heavens is as old as man's ability to think; as ancient as his capacity to wonder and to dream. And in company with most of the special enchantments of human life, the unique appeal of astronomy is incommunicable, easily understood through direct experience, but not to be precisely defined or explained. Nor should any explanation be thought necessary. The appeal of astronomy is both intellectual and aesthetic; it combines the thrill of exploration and discovery, the fun of sight-seeing, and the sheer pleasure of firsthand acquaintance with incredibly wonderful and beautiful things. But it also offers the privilege, not to be taken lightly, of adding something to the knowledge and understanding of man.

There is one other factor which I think deserves comment. An amateur, in the true and original meaning of the word, is one who pursues a study or interest for sheer love of the subject; and in this respect the division between professionals and amateurs is indeed indefinite. We are all impelled by the same wonder and curiosity, we are all exploring the same Universe, and we all have the enviable opportunity of contributing something to the store of human knowledge.

Now I should like to phrase one of these considerations in a somewhat less conventional manner, at the risk of being accused of undue whimsicality by the sternly serious minded. Considered as a collector of rare and precious things, the amateur astronomer has a great advantage over amateurs in all other fields, who must usually content themselves with second and third-rate specimens. For example, only a few of the world's mineralogists could hope to own such a specimen as the Hope diamond, and I have yet to meet the amateur fossil collector who displays a complete tyrannosaurus skeleton in his cabinet. In contrast, the amateur astronomer has access at all times to the original objects of his study; the masterworks of the heavens belong to him as much as to the great observatories of the world. And there is no privilege like that of being allowed to stand in the presence of the original.

THE IDEA BEHIND THE CELESTIAL HANDBOOK

Yet it sometimes happens, perhaps because of the very real aesthetic appeal of astronomy and the almost incomprehensible vastness of the Universe, that the more solidly practical and duller mentalities tend to see the study as an "escape from reality" - surely one of the most thoroughly lop-sided views ever propounded. The knowledge obtained from astronomy has always been, and will continue to be, of the greatest practical value. But, this apart, only the most myopic minds could identify "reality" solely with the doings of man on this planet. Contemporary civilization, whatever its advantages and achievements, is characterized by many features which are, to put it very mildly, disquieting; to turn from this increasingly artificial and strangely alien world is to escape from unreality; to return to the timeless world of the mountains, the sea, the forest, and the stars is to return to sanity and truth.

ABOUT THIS BOOK

There is no lack of astronomical literature today; on the contrary, the flood of new material is so great that the compiler of a good small astronomical library faces a serious task. The chief problem, in fact, is to choose those items which will quickly become dog-eared under continual use, as contrasted to those which will quietly disappear under an accumulation of dust on one's bookshelf. Astronomical literature may be broadly grouped into four classes: popular works for the general reader, textbooks for the student, technical reports and bulletins for the professional worker, and guidebooks for the observer. A book of the first type can perform a valuable service if well and accurately written, though the majority appear to be oriented toward the casual reader who is content to study astronomy from his armchair. Textbooks also deal very sketchily with observational matters, and the more technical publications are suitable only for the advanced worker. The Celestial Handbook belongs in the fourth category, and is being offered in the hope that it will fill a very real gap in astronomical literature for the observer.

It is, briefly, intended to be a standard catalog and detailed descriptive handbook of the many thousands of objects available to observers with telescopes in the 2-inch to 12-inch range. Its realm is the entire Universe beyond our own Solar System, and it deals with those

THE IDEA BEHIND THE CELESTIAL HANDBOOK

celestial objects which are now popularly known as "deep-sky wonders". I can claim no originality for this idea, of course. Other such books have been produced in the past, of which the most complete and successful was T.W.Webb's *CELESTIAL OBJECTS FOR COMMON TELESCOPES*. This remarkable work, since 1962, has been available in a revised edition from Dover Publications in New York. Nothing else of comparable value has been produced since, to meet the needs of the modern observer.

Aside from the obvious fact that all the older books are now very much out of date, there are a number of other reasons why a complete new Celestial Handbook is needed. To begin with, the earlier observing guides were written for the possessor of the standard telescope of about 1900, the classic 3" refractor. Today's average amateur telescope is a 6" to 12" instrument, and the increasing availability of good quality large reflectors has opened up a vast new world of deep-sky objects for the modern observer.

Secondly, the vast increase of astronomical knowledge has resulted in an enormous shift of interest in the last 50 years. Older books concentrated heavily on such relatively local objects as double stars and the brighter variables. The more spectacular star clusters and nebulae were included, but descriptions were often limited to visual appearances because of the scarcity of facts. Galaxies as such were not mentioned at all, since nothing was known of the true nature of the "spiral nebulae". The situation is now radically different. If nothing else had happened in this century, the final identification of the spirals as external galaxies was sufficient to alter our whole conception of the large-scale features of the Universe. We can now speak with reasonable accuracy not only of the distances, masses, temperatures, etc., of the celestial objects, but through the growth of astrophysics we can analyse some of the physical processes at work and begin to understand what these things mean from the viewpoint of the evolutionary history of the Universe. Surely there has never been any intellectual adventure to equal this, and even the most casual observer is entitled to share something of its wonder.

The existence of the *Celestial Handbook* is a result of this gradual widening of our horizons, and reflects,

THE IDEA BEHIND THE CELESTIAL HANDBOOK

I believe, the interests and needs of today's amateur. The book had its beginning about 2 dozen years ago, and started as an attempt to keep facts in order for my own use, to bring together into one place the data from many different sources, to bridge the gap between the elementary beginner's star books and the more technical publications, and to maintain a permanently up-to-date guidebook by constant addition, correction, and revision. In 1958 I joined the Lowell Observatory staff, and the resources of the Lowell library were thus made available, as well as the superb collection of photographic plates made with the 13-inch wide angle camera which discovered Pluto in 1930. In the years between 1958 and 1965 the Handbook more than doubled in size, and eventually grew to occupy four thick loose-leaf volumes, totalling nearly 2000 pages.

The work now includes virtually all the objects of interest which appear on the present-day star atlases such as Norton's and the new Skalnate Pleso "Atlas Coeli". But no simple catalog listing can convey much of the real interest and importance of many of the celestial wonders. Detailed descriptions are necessary, and a very simple and direct policy was adopted: if an object was considered worthy of a detailed description, it was given one. Also, a number of objects have been included which are usually regarded as being beyond the range of amateur telescopes, but which appear to me to be of exceptional interest. Such decisions, of course, depend very much upon one's personal interests, and each observer would undoubtedly make a somewhat different selection. My own choices are based upon more than thirty years of actual observing with instruments ranging from field glasses up to large observatory reflectors. I think that none of the famous old favorites have been neglected in this book, and I hope that some objects will be introduced to many observers for the first time.

The number of objects listed is well over 7000, of which many hundreds are given additional detailed descriptions. The book is illustrated by more than 250 photographic plates, collected from many different observatories, and a fine selection of the work of some amateur astro-photographers has been included. There are also several hundred finder charts, orbit diagrams, graphs and tables of various types .

THE IDEA BEHIND THE CELESTIAL HANDBOOK

Although intended primarily for the serious observer and advanced amateur, there is no absolute reason why this book should not prove useful even to a beginner who is willing and able to learn. Chapters 2 and 3 have been prepared to introduce such a beginner first to the Universe itself, and then to the world of the astronomical observer with its special terminology and symbols. There may be little in these two sections which will be of any real use to the experienced amateur, except possibly to assist in the instruction of the novice. But, after much thought, I have decided to let them stand as they are - if they help only a few of the users of this Handbook, their inclusion will have been justified. And, to shift to the other end of the spectrum of potential users, every attempt has been made to maintain a standard of accuracy which will assure the value of this book as a quick reference even for the professional astronomer.

In such a large compilation the question of errors and discrepancies deserves some mention. Typographical errors and other definite mistakes should be reported to the author so that corrections can be made in any possible future editions. The question of discrepancies between various authorities is not so easily handled, and raises serious problems. If one standard catalogue gives a star a spectral class of K0, another is sure to classify it as G7 and a third will offer K2. The same galaxy may be classed by three different authorities as "irregular ?", "late-type spiral" and "elliptical peculiar". Published values for the distances of objects often show very large discrepancies. In addition to these typical uncertainties, there are numerous cases involving errors which were corrected long ago, but which still exist in books that are in wide use today. The galaxy M74, for example, is called a globular star cluster in the NGC listing, while the planetary nebula NGC 6026 is included as a galaxy in the Shapley-Ames catalogue. Similarly, the small galaxy NGC 2283 is marked as a diffuse nebulosity on the Skalnate Pleso Atlas. All such cases which have come to my attention have been corrected in the Handbook, but there must undoubtedly be others which have so far escaped notice. The user of this book should therefore not be distressed to find four-way discrepancies (and worse!) between this book and authors

INTRODUCTION

THE IDEA BEHIND THE CELESTIAL HANDBOOK

A, B, and C. This is a reflection on the present state of knowledge. Many astronomical facts depend upon very precise and difficult measurements of very tiny quantities, such as the parallax of a star, and there is always a good margin of error. Other astronomical questions remain frankly controversial, and the best authorities differ in their interpretations. It would be a rash astronomer indeed who would claim absolute exactness in such matters as the precise distance of Polaris or the Ring Nebula, the exact size or luminosity of Antares, or the exact orbital period of Zeta Aquarii. And in such enigmatic objects as Epsilon Aurigae, SS Cygni, and 3C273, it is best to admit at once that we do not know exactly <u>what</u> is going on. We have just reached the point where we are beginning to find out what the questions are, and what methods may be used to study the problems. The Universe remains – as it probably always will – an awesome mystery, and Newton's great Ocean of Truth still lies undiscovered before us. But let us make no apology for this. Much of the fascination of astronomy lies in the fact that there are still so many unknowns, so many puzzles and mysteries yet to be solved. May it always be so!

ACKNOWLEDGEMENTS

The Celestial Handbook is a collection of data from a great number of separate sources. No one person could, in a dozen lifetimes, accomplish more than a small fraction of the research which is represented by the information contained in this book. Any compiler of such a work is, of necessity, forced to rely upon the studies and investigations of literally hundreds of other observers, both of the past and present. In nearly thirty years of observing I have actually seen, at one time or another, possibly half of the celestial objects listed herein, and the visual descriptions are largely the results of my own records. Aside from this, my chief task has been the collection, checking, and inter-comparison of data.

The chief source-books of information are listed in the classified bibliography. In addition, the publications of many different observatories were checked each month for the presence of any new information relevant to the project, while both the Astronomical Journal and the Astro-

THE IDEA BEHIND THE CELESTIAL HANDBOOK

physical Journal were periodically searched in the same way. Information obtained from any of these current sources is identified in this book by the name of the author and the date. The source of each photograph is given in the caption. To all the astronomical research workers of the world, who have made my task so engrossing, so rewarding, and so endless (!) I acknowledge my deep indebtedness. At the same time, it should be evident that the present spectacular rate of increase of astronomical information tends to render my whole project a basic impossibility. No modern astronomical handbook can possibly remain "up-to-date" for even as long as a year; the best of modern measurements, data, theories, and interpretations will be totally superseded within a decade. In preparing the revised manuscript for the Dover edition of this work, the author has taken the opportunity to correct some minor errors and update some of the information. But it should be obvious that a work of this nature could be expanded, corrected, and updated forever! One must call a halt somewhere! "Were I to await perfection," wrote the 13th Century Chinese historian Tai T'ung, "my book would never be finished."

　　　Finally, it is a very great pleasure to acknowledge the primary source of information for this book - the fine astronomical library of Lowell Observatory. To Dr. John S. Hall, the Director, and to Mr. Henry Giclas of the Observatory staff, I must express my deep gratitude for generously allowing the use of the Lowell telescopes and the Observatory plate collection. Without the use of these and other Observatory facilities, the Celestial Handbook could never have reached anything comparable to its present degree of completeness. In addition, many of my astronomical friends, both amateur and professional, have expressed their interest in the project and have offered valuable encouragement, assistance, and advice. It is through their efforts, as well as mine, that the idea behind this book has become a concrete reality.

Robert Burnham, Jr.

Flagstaff, Arizona
October, 1976

SPLENDOR OF THE HEAVENS

In the vast reaches of the Universe modern telescopes reveal many vistas of unearthly beauty and wonder.......

REGION OF STAR CLUSTER NGC 6611

LICK OBSERVATORY

12

A CELESTIAL SURVEY
THE DISTANCE SCALE OF THE UNIVERSE

We are beginning a journey.

It will be a journey both strange and wonderful.   In our tour of the Universe we shall travel the   vast empty pathways of   limitless space   and   explore   the uncharted wilderness of creation.   Here, in the dark unknown immensity   of   the heavens, we shall meet with glories beyond description and witness scenes of inexpressible splendor. In the great black gulfs of space and in the realm of the innumerable stars, we shall find mysteries   and   wonders undreamed of.   And when we return to Earth, we shall try to remember something of what we have learned about the incredible Universe which is our home.

Let us now prepare for our journey.   We shall need, first of all, an imaginary spacecraft which can travel at any desired speed—no matter how incredible (or impossible). The reason for this will soon be seen.   If we limit ourselves to speeds which are "possible," our journey to even relatively near regions   of space will require a period of untold ages.   And, we simply haven't the time. We must also equip ourselves with abnormally sensitive eyes which   can instantly see such details as would ordinarily be revealed only by many hours' exposure of the photographic plate.   We shall also take along one of the world's largest telescopes so that we may still better view the passing scenery.

Finally, we shall require some briefing   on   the distance scale used on our journey.   If we were to travel to New York, for instance, we would want to know   that   the distance to be covered was 600   or 800 or 1200 miles; but this information would scarcely be of any value if we had no idea of the meaning of the term "mile."   Now,   on   our journey, we shall find that the celestial distances are so incredibly vast that to express them in miles would be like giving the distance to Hong Kong in millionths of an inch. So, we shall require a new distance unit—the <u>light</u> <u>year</u>.

There is nothing particularly complicated about the light year.   All radiant energy, as we may remember from high school physics, travels at the same velocity, commonly called the "speed of light."   This velocity is the greatest known in the Universe and appears from all   known facts to be the ultimate velocity possible.   The proofs of this lie in the abstruse realm of relativistic   mathematics   which few of us are prepared to tackle. For the moment, we shall

A CELESTIAL SURVEY

accept as a fact the principle that the velocity of light cannot be exceeded, except, of course, in the imagination. The speed of light is close to 186,300 miles per second. At this speed we could travel seven times around the Earth in one second. We could reach the Moon in less than two seconds, the Sun in eight minutes, and all the planets in a few hours. Traveling for a year at this speed, we would cover a distance of slightly less than six trillion miles, and we would find ourselves about a quarter of the way to the nearest star.

A light year, then, is simply the distance that light travels in a year—slightly less than six trillion miles. It is a unit of distance—not a unit of time.

At this point, we must pause briefly to make some attempt, however inadequate, to understand the implication of such an enormous distance unit as the light year. If we cannot grasp—in some degree—the chilling vastness expressed by such a concept, we shall never have more than the haziest notion of the scale of the Universe. Our minds may be jolted into some degree of comprehension by such statements as "It would require over 150,000 years to count the number of miles in one light year," but the final impression is still one of incomprehensibly large numbers. Let us try instead the old and classic "scale model" method.

The Earth is a planet, one of nine relatively small bodies revolving about a typical sort of star we call the Sun. These objects, together with a number of smaller bodies, make up the "Solar System," our own familiar corner of the Universe. The distance of the Earth from the Sun is approximately 93 million miles, or about eight "light minutes," a distance which we will dignify by a special title, the "astronomical unit" (AU). The most remote planet, Pluto, is 40 times more distant from the sun than we are, and so we may say that the distance of Pluto is about 40 AU. The diameter of the planetary system is thus some 80 AU.

One light year is equal to 63,000 AU. The nearest of the stars is 4.3 light years distant; this is about 270,000 AU, or some 7,000 times the distance of Pluto. The light from Pluto reaches the Earth in 5½ hours; from the nearest star it requires 4.3 years.

A CELESTIAL SURVEY

It would be advisable to read that paragraph again.

By a fortunate circumstance, the number of inches in a mile is very nearly equal to the number of AU in one light year, a perfect arrangement for the purpose of constructing mental scale models. Let us then imagine a scale model of the Solar System, with the Earth represented as a speck one inch away from the pinpoint Sun.   Pluto is then about $3\frac{1}{2}$ feet from the Sun. The nearest star on this model will be nearly $4\frac{1}{2}$ miles away.And all the stars are, on the average, as far from each other as the nearest ones are from us.

Imagine, then, several hundred billion stars scattered throughout space, each one another Sun, each one separated by a distance of several light years from its nearest neighbors. Comprehend, if you can, the almost terrifying isolation of any one star in space. How many of these distant Suns are surrounded by planetary systems, and how many other inhabited worlds may exist somewhere?   Have we any hope of ever knowing? For a planet the size of the Earth would be completely beyond the range of any telescope in the world, even at the distance of the very nearest of the stars.

Finally, let us imagine our several hundred billion stars, with their enormous separations, arranged in space in the form of a great "star city", a round flattened cloud like a vast wheel, fully 100,000 light years in diameter, and surrounded by a few hundred thousand light years of complete nothingness.

This wheel is our Galaxy. Far beyond it lie millions of other similar galaxies—the nearest ones a good part of a million light years across the void.

This is the Universe which is being explored by the great telescopes of the Earth, and by the amateur with his homemade telescope. This is our home in space.

****

Armed with this knowledge, let us now enter our imaginary spacecraft and launch ourselves out approximately one million light years from the Earth. In reality even a beam of light would require more than ten thousand lifetimes to traverse such a distance. The aiming of our spacecraft does not especially matter, for the general features of our surroundings would appear the same, regardless of the

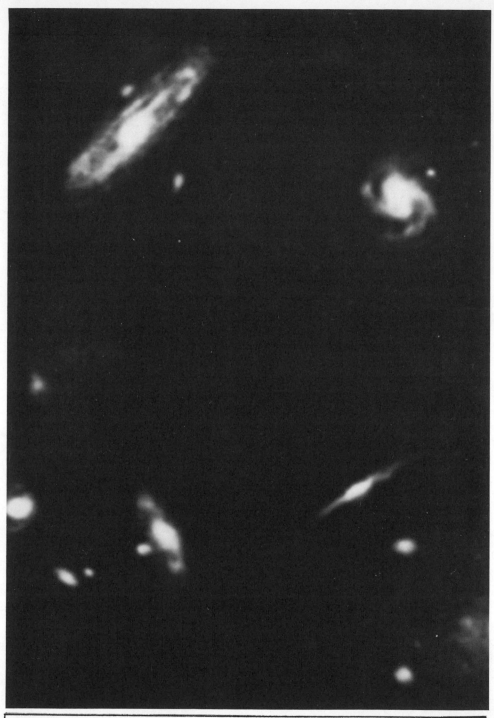

GALAXIES OF THE UNIVERSE WOULD APPEAR SOMETHING LIKE THIS PICTURE TO AN OBSERVER ONE MILLION LIGHT YEARS FROM THE EARTH. EACH OF THESE OBJECTS IS AN ENTIRE "STAR CITY", CONTAINING SEVERAL HUNDRED BILLION SUNS. THESE ARE THE MAJOR UNITS OF CREATION.

A CELESTIAL SURVEY

direction taken. Now let us examine the view.

A strange view it is! We appear to be suspended in a blank nothingness. The Earth is gone, of course, and the Sun and Moon; even the stars we knew at home have vanished. They are all so far away now that not even the greatest telescopes could detect them. Instead of our familiar skies, we seem to be gazing into the blackness of eternal night, and the view is the same in any direction we look.

A few objects, however, are visible, floating in the blackness of space. One of them appears as a softly glowing oval with a luminous center. In the opposite part of the sky we can see another similar object. Around the rest of the sky we can barely make out a few more luminous spots of the same type. All else is blackness. If we employ our telescope, however, we can see hundreds of these glowing islands of light. The general view we get will resemble the picture opposite.

These objects are the galaxies—the major units of creation.

If we now set our spacecraft (and our imagination) in motion, we can travel through the Universe at a few hundred million times the speed of light, discovering and counting galaxies by the hundreds, by the thousands, by the millions. If we photograph the whole sky bit by bit with our giant telescope—a project which will take us several centuries— we will find that perhaps a billion galaxies are within range. They extend throughout space as far as the limit of telescopic exploration.

The galaxies, then, are the primary units of the Universe—we may sometimes hear them referred to as "island universes" in themselves. They are immensely large objects, as you may imagine, after our recent attempt to construct mental scale models. Their diameters range from a few thousand light years up to about 100,000 light years. The average separation between any two galaxies is on the order of a million light years, though in some regions we will find them grouped in more closely associated clusters. Some of the largest galaxies appear as great flat discs, often showing a distinct "pinwheel" or spiral appearance; others are irregular, and many are globular or elliptical and show no structural features.

Let us now choose one galaxy—a large spiral type—for

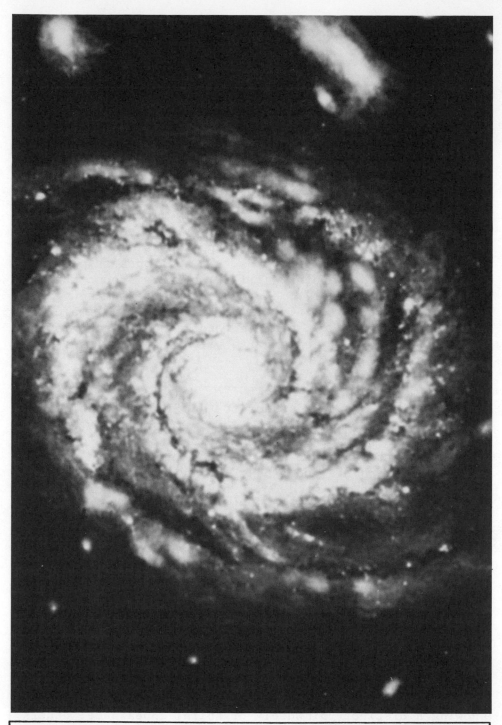

LARGE SPIRAL GALAXY APPEARS LIKE THIS TO AN OBSERVER APPROACHING FROM INTERGALACTIC SPACE. IN THIS IMAGINARY VIEW, BASED ON PHOTOGRAPHS OF THE GALAXY M51, WE ARE A FEW HUNDRED THOUSAND LIGHT YEARS DISTANT. TWO SMALL IRREGULAR GALAXIES APPEAR NEAR THE TOP OF THE PICTURE.

A CELESTIAL SURVEY

detailed exploration. This galaxy, like many others, is a member of a loosely associated group containing twenty or so members. We note that several of the smaller members— irregular galaxies—are near enough to the big spiral to be regarded as possible satellite systems. As we approach this group, the view we get is something like the picture opposite.

The spiral shape suggests rotation, and, indeed, if we wait a few million years, we shall be able to watch the process ourselves. Any shorter time will scarcely suffice, for a typical large galaxy may require more than 100 million years for one complete turn. The galaxy does not rotate as a rigid body, however. The inner regions, because of the greater concentration of mass, rotate more rapidly than the outer portions; and there is thus a continual "shear" effect across the body of the galaxy. The spiral pattern may be attributed in part to this differential rotation, but it also appears to be connected in some way with electric and magnetic fields in the galaxy. The full explanation is not yet clear.

The particular galaxy we have chosen is about 100,000 light years in diameter, and when seen edge-on is about 10,000 light years thick. From our position several hundred thousand light years out, the great galaxy resembles a softly glowing pinwheel with a bright center and spiral arms. Its actual constitution, however, remains as much of a mystery as ever from this distance. So we must now turn our telescope upon it to discover that the glow comes from uncountable billions of separate points of light. We thus answer our next question—What is a galaxy? It is a colossal aggregation of stars—stars by the billions. In many places we can see great clouds of dust and gas, many light years in extent, some luminous and some dark, scattered among the stars. Most of this material lies in the regions of the spiral arms. The nuclear region of the galaxy seems relatively free from it.

Before beginning our detailed exploration of this galaxy, let us pause for a few moments in another attempt to understand the scale of things. The picture (opposite) shows the galaxy as it appears from our vantage point a few hundred thousand light years out. We have chosen this particular galaxy because it is our own—somewhere among all those billions of stars   lies   one rather ordinary sort of star—our Sun, with its attendant planets.

How much will we have to enlarge this picture in order to find the Sun, its planetary family, and our home, the Earth?

The answer, as we   expected, is literally numbing to the mind, but once again serves to shock us into some degree of comprehension of the vastness of space. An enlargement of a few times will not help us—or of a few hundred times. But suppose we can enlarge the picture <u>until it covers all of North America</u>.   Then the billions of individual stars will appear as pinpoint specks averaging about 600 feet apart. The Solar System—if we can locate the exact spot to look for it—will be about two inches in diameter, and the Sun and Earth will appear as two pinpoint dots about 1/30 inch apart. The Earth, in fact, will be totally invisible to the naked eye on this scale, and we shall have to examine our super-enlarged picture with a super-microscope to eventually locate the Earth as a sub-microscopic dot a few millionths of an inch in diameter.

Keeping this scale in mind, let us now begin a tour through the Galaxy.

Approaching the nearest star cloud, we watch for some time in awed silence as the cloud resolves first into a haze of millions of glimmering flecks of light, each becoming eventually a brilliant star and ultimately spreading out to spangle the heavens all about us. We are not surprised to find that the individual stars are separated from each other by relatively enormous distances, even though our telescopic view made them appear to be crowded together in incredibly thick masses. This was an illusion due to their unimaginable numbers and the great distance from which we viewed them. It was something like looking at a great city at night from a distance of many miles; the countless lights seem to melt together into a single glowing haze. The average distance between any two stars is on the order of three or four light years, and since the diameter of an average star is less than one light minute, we can see that the relative separation is vast indeed.

There are many exceptions to this general rule, though, as in many places we find two or more stars quite close together and revolving about each other. Some of these pairs,

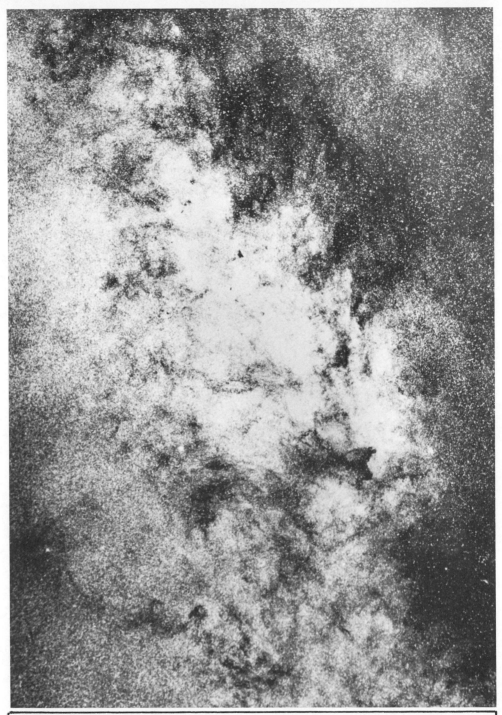

SPIRAL ARM OF THE GALAXY IS REVEALED AS A VAST COMPLEX OF STAR CLOUDS. ABOUT 400 BILLION INDIVIDUAL STARS POPULATE THIS STELLAR SYSTEM, WHICH IS ONE OF THE LARGEST GALAXIES KNOWN.

PHOTOGRAPH by ALAN McCLURE

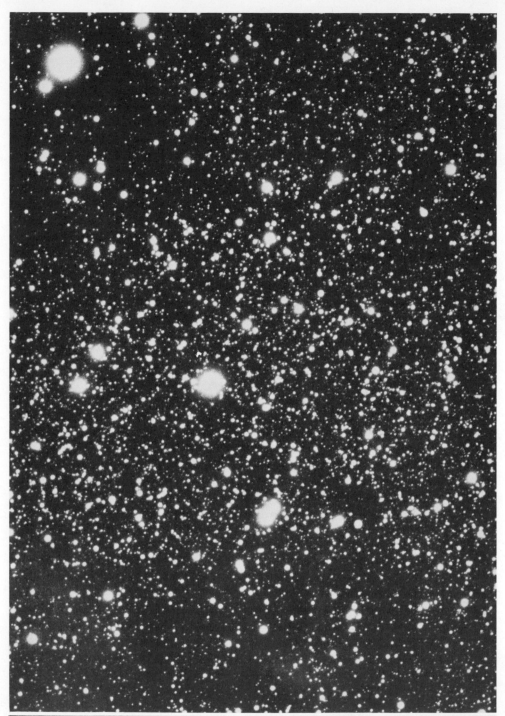

AN INFINITE PROFUSION OF SHINING SUNS. THIS STAR FIELD IS A SMALL PORTION OF THE GREAT CYGNUS STAR CLOUD; THESE ARE A FEW OF THE ESTIMATED 400 BILLION STARS WHICH COMPRISE OUR GALAXY.

LOWELL OBSERVATORY

A CELESTIAL SURVEY

known as "binary stars," are nearly in contact and complete one revolution in less than a day, while others are so widely separated as to require centuries for a single revolution. As we continue to study the stars of the Galaxy, we learn that such double, triple, and multiple stars are by no means rare. Possibly about half of all the stars are double or multiple systems.

In addition, we frequently come upon great clusters of stars containing hundreds or even thousands of members, gravitationally connected and moving together through space. As we survey the fantastic blazing beauty of one of these clusters, we realize it is time to answer another question: What is a star?

To answer this, we shall take a close look at one, not from a distance of several light years, but from a mere light minute (10 million miles or so). What we see now is not a tiny point of light, but an immense blazing globe of such intense brilliance that we cannot gaze upon it without protection for our eyes. We are looking at a typical, normal star.

The particular star we have chosen is about average in size; its diameter is slightly less than one million miles, or about six light seconds. After our experience with distances of thousands of light years, a distance of six light seconds seems absurdly small. Yet, a star of this size could easily contain a million globes the size of the Earth.

As we examine the dazzling surface of this star, we can find nothing solid about any part of it; every part appears to be boiling, erupting, and exploding in various ways. Great flame-like tongues of glowing matter are continually leaping up and falling back, huge jets of blazing gas shoot up like gigantic fountains, and immense turbulent whirlpools move like fiery clouds across the boiling surface. In only a few hours such clouds may rise to heights of several hundred thousand miles above the star.

The entire star is a colossal globe of superheated gas. By "gas" we do not mean to convey the impression of an airy nothing, like the inside of an inflated balloon. This star has enough mass to make over 300,000 planets like the earth, and its density is greater than many normally liquid or solid substances. In some stars, in fact, the density exceeds to an astonishing degree that of our

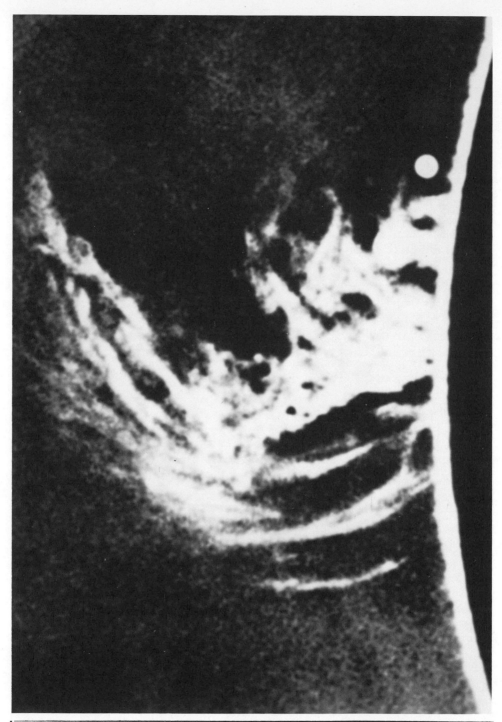

ENORMOUS CLOUD OF GLOWING GAS RISES FROM THE SURFACE OF THE SUN LIKE A TOWERING FOUNTAIN OF FLAME. SUCH ERUPTIONS, CALLED "PROMINENCES," ILLUSTRATE SOME OF THE TYPICAL ACTIVITY ON THE SURFACE OF A STAR. THE SMALL WHITE DISC SHOWS THE EARTH TO THE SAME SCALE.

JULY 9, 1917 —   MT. WILSON OBSERVATORY

A CELESTIAL SURVEY

heaviest known metals. This material is gaseous here because nothing solid or liquid can exist at the temperature
found on a star.

The surface temperature of this particular star is
about 6000° C, and the interior temperature is nearly 20
million degrees. This temperature allows a nuclear reaction—the conversion of hydrogen into helium—to proceed,
which explains the energy source of the star. A typical
average star such as this contains sufficient hydrogen for
20 or 30 billion years of energy production. So, as we can
see, a star is a natural atomic furnace. We might think of
it as a natural and perpetual nuclear reactor .

"It looks just like the Sun," someone says. And, in
this simple and direct observation we have the whole key to
understanding the stars. They are other suns. Or, to put
the statement in its correct form, our Sun is merely one of
the stars.

With this knowledge, let us continue our trip through
the galaxy. We are now surrounded by stars; we pass them
by the millions. Our realization that each one is in reality a blazing sun capable of swallowing up a million earths
gives us an incredible and even frightening picture of the
unimaginable vastness of the Universe. Now we turn our
attention to the individual stars and see what else can be
learned from them.

The first thing we notice is a definite variety of
color. We find that some stars are white or bluish; others
are yellowish, and some show an orange or reddish tint. If
we know our physics, we will deduce that this is due to a
difference in temperature. White and blue stars are the
hottest, red stars are the coolest. The coolest stars may
have a surface temperature of a mere 2000° C, while a great
blue giant may be ten or fifteen times hotter.

There is also a great difference in stellar diameters
and densities. Some of the great red giants are hundreds
of times bigger than our Sun, but their temperatures are
low and their substance is highly rarified. At the other
extreme we find the astonishing "white dwarf" stars which
are only a fraction of the size of the Sun but weigh tons
to the cubic inch. The difference in actual luminosity is
another subject for study in our celestial survey. We come
across some great stars which outshine the Sun by more than

A CELESTIAL SURVEY

50,000 times, but such supergiants are exceedingly rare. Much more numerous are the tiny "red dwarf" stars whose feeble illumination may be a mere 1/100,000 the luminosity of the Sun. For every supergiant we find thousands of stars like our Sun, and for every star of the solar type we find thousands of red dwarfs, by far the commonest citizens in the stellar community. Our Sun thus appears as a fairly respectable member of the stellar population.

Among the more unusual and peculiar members are the "variable stars." These are suns which do not remain constant in their outpourings of light and heat, but fluctuate in a peculiar manner, some periodically, others erratically. The commonest types are red giants, which appear to pulsate, expanding and contracting like a beating heart. If we watch a few typical examples, we shall soon see that their average period is about a year and their light range is very great, usually more than 100 times and, in some examples, over 1000 times. These stars are called "long-period variables." Another similar type of pulsating star has a period of a few days or weeks and a light range of only two or three times. These stars are white or yellow giants and are technically known as "cepheids."

The long-period variables and the cepheids appear to follow certain rules which allow us to predict their future behavior with fair accuracy. Not so obliging, and therefore more intriguing to the curious astronomer, are the more erratic variable stars, some of which fluctuate in a completely unpredictable manner. There are small dense white stars, for example, that flare up suddenly in the course of a day or so, increasing forty or sixty times in brightness, thereafter slowly fading back to normal, only to repeat the process again and again at intervals of a few months. Some of the tiny red dwarfs show a similar effect, flaring up to double their normal light in a few minutes, then fading just as suddenly. We also find stars whose variations are completely irregular; they may fade or brighten at any time from some still unknown cause. Our steady, dependable Sun seems comfortably unexciting when compared to these erratic, unstable stars.

The most spectacular of all the variable stars are the novae. If we travel around the galaxy for a year or so, we have a fair chance to see one. A nova is a star that

GALACTIC STAR CLUSTER IS A MASSED GROUP OF SEVERAL HUNDRED SUNS, OFFERING A SPLENDID SIGHT FOR SMALL TELESCOPES. THIS EXAMPLE IS M37 IN THE CONSTELLATION OF AURIGA.

13" TELESCOPE — LOWELL OBSERVATORY

GREAT DIFFUSE NEBULA KNOWN AS M8 LIES IN THE SUMMER SKIES IN THE CONSTELLATION OF SAGITTARIUS. THIS VAST CLOUD OF RARIFIED GASES SHINES BY THE LIGHT OF AN INVOLVED CLUSTER OF HIGHLY LUMINOUS STARS.

42" TELESCOPE – LOWELL OBSERVATORY

A CELESTIAL SURVEY

literally explodes, blasting its outer layers into space with titanic violence and rising to unheard-of brilliance for a period of a few days or weeks.   A <u>supernova</u> is a similar phenomenon on a vastly greater scale—it results in the more-or-less complete destruction of a giant star, with an explosion that makes an ordinary nova look pale by comparison. A first-rate supernova may equal the total luminosity of all the other billions of stars in a galaxy put together. Unfortunately, the supernovae are very rare, and we might have to cruise around the Galaxy for several centuries before seeing one.

To the question, "Why do certain stars fluctuate, pulsate, or explode?" it must be admitted that we have no lack of theories, but also no definite proven answers. All these spectacular happenings are symptoms of various kinds of stellar instability, generally thought to indicate that the star is approaching the point of exhausting its nuclear "fuel" supply. Such stars, it seems, may first expand into red giants, then later collapse into the super-dense "white dwarf" state, where their life history ends.

The stars of the Universe are thus seen to be in various stages of evolution, from the virtually newborn to the dying cinders. We have seen some of the dying cinders—let us now study a region where new stars are still being formed.

You will remember that, when we looked at this Galaxy from our telescopic vantage point far out in space, we noticed that much dust and gaseous material was distributed among the stars, especially in the region of the spiral arms. We are now entering one of these regions. The great glowing cloud of gases and dust which we see floating ahead of us is called a "<u>nebula</u>." To be precise, we should call it a "diffuse nebula" or "gaseous nebula," so as to avoid confusion with the so-called "spiral nebulae," which are actually galaxies. Galaxies should no longer be referred to as "nebulae," of course; but the term was applied in the early days, when the true nature of the galaxies was not known, and the term is often used even today.

The particular diffuse nebula we have chosen is a fair example of the type, measuring about twenty light years in extent and consisting of highly rarified gases, chiefly hydrogen, helium, nitrogen, and oxygen (picture opposite). The Galaxy contains a tremendous amount of this matter; in

A CELESTIAL SURVEY

fact, it is estimated that over a third of the mass of the Galaxy is in the form of dust and gas clouds. This material, however, remains invisible unless illuminated by nearby stars. Thus the nebula at which we are now looking is visible only because of the hot stars embedded in it, which cause the gases to shine. There are two processes by which the nebula may be illuminated. If the star is of a sufficiently high temperature (20,000°K or above), the gases may be excited to luminescence by the star's ultraviolet radiation, a process known as "fluorescence." If the temperature of the star is too low, the nebula will merely shine by reflected starlight. Fluorescence produces the more brilliant illumination and also gives the glowing gases a peculiar ghostly greenish color. It happens that the spiral arms of the Galaxy are extraordinarily rich in great blue-hot supergiant stars, and the involved nebulosity glows in vivid splendor. The presence of these stars is no accident. We find them in the regions where they were born, and the great diffuse nebulae are the "star factories" of the galaxy. We are witnessing one of the scenes in the drama of creation.

In studying these vast chaotic masses of glowing gas, we often find them associated with nearby clouds of dark dust, producing strange and spectacular patterns against the starry background. Some of them appear in great foggy masses of indistinct form; others seem like slender wisps and streamers of light reaching from star to star; and others, lacking illumination, appear as great dark blots against the bright star clouds of the Galaxy. Some of this matter may have come from exploding stars or other celestial cataclysms, but it seems more likely that most of it has never been part of any star but is rather a portion of the original "building material" from which the galaxies were formed and will therefore supply the material for many stars in ages to come. The Universe is still young, as we can see. Although the galaxies have probably existed for 10 billion years or so, this is but a fraction of their total "life expectancy."

A typical diffuse nebula, as we said, is something like 20 light years in diameter, and we might pause again for a moment of contemplation concerning the scale of things. With the Earth and Sun one inch apart, our scale

BRIGHT AND DARK NEBULOSITY PRODUCES STRANGE PATTERNS IN THE REGION OF THE SPECTACULAR "HORSEHEAD NEBULA" IN THE CONSTELLATION OF ORION. THIS PHOTOGRAPH WAS MADE WITH AN 8" REFLECTING TELESCOPE.

KENT DE GROFF, LOWELL OBSERVATORY

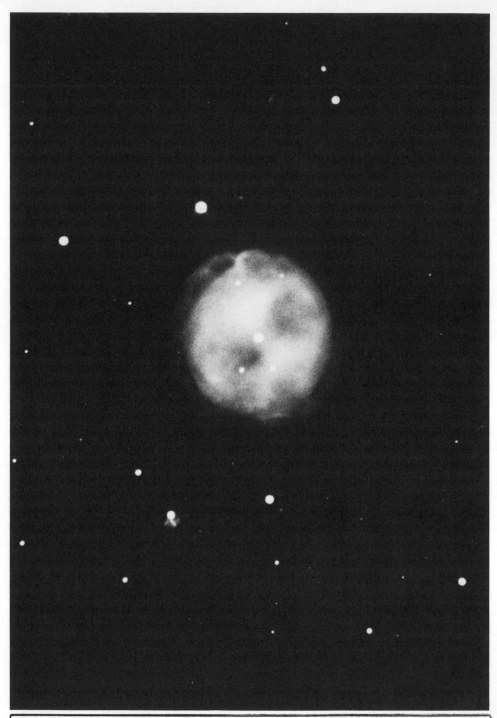

TYPICAL PLANETARY NEBULA CONSISTS OF A HOLLOW SHELL, BUBBLE, OR RING OF GAS SURROUNDING A HOT BLUE DWARF STAR. THIS EXAMPLE IS THE "OWL NEBULA" M97 IN URSA MAJOR.

42" TELESCOPE — LOWELL OBSERVATORY

A CELESTIAL SURVEY

model of this nebula would spread over 20 miles of country-
side.   There is enough material in one of these clouds to
make many hundreds of stars like our Sun.

Somewhat smaller is another type of nebula which we
shall now briefly visit (picture opposite). This type, as
you will notice, rather resembles a hazy globe of smoke.
This vaporous sphere is composed of the same sort of rarified gases which we met in the diffuse nebulae.   In the
center of the sphere is a small but extremely hot blue-
dwarf star which supplies the illumination for the entire
nebula.   These objects are not as large as the diffuse
nebulae, but are often remarkably bright for their size.
The particular example we are studying has a diameter of
somewhat less than one light year. An object of this type
is called a "planetary nebula." The term does not imply any
connection with planets, but merely indicates that such
nebulae rather resemble the pale disc of a distant planet
as seen through a small telescope.

The actual significance of the planetary nebulae
remains obscure. They are undoubtedly produced by some sort
of outbursts which have occurred on the central stars, but
the outbursts appear to have been much less violent than
the explosions we call "novae." Perhaps we should classify
a planetary nebula as an unusually "lazy" type of nova.

Continuing our voyage through the spiral arms of the
Galaxy, we are repeatedly astonished by the extraordinary
richness of these regions.   Not only do the vast nebulae
present countless vistas of great interest and beauty, but
we also notice enormous numbers of star groups, clusters,
and associations, some of them measuring over 100 light
years in extent and containing thousands of stars of different sizes and color. Many of them are embedded in glowing nebulosity or smothered in great dark clouds of cosmic
dust.   All of these star clusters are examples of one type,
the "galactic" or "open" cluster, so called from their
scattered appearance and the fact that they are found inside the Galaxy, chiefly along the spiral arms.   There is
one other type of star cluster; and, in order to see it, we
must travel out about 30,000 light years to the outer rim
of the Galaxy.

From our new position we can now see a number of fuzzy-
looking objects distributed around the edges of the Galaxy

FINEST GLOBULAR STAR CLUSTER IN THE SKY IS THIS MAGNIFICENT
SWARM OF COUNTLESS STARS KNOWN AS OMEGA CENTAURI OR NGC 5139, ONE OF
THE NEAREST CLUSTERS OF THIS TYPE TO THE EARTH.

HARVARD COLLEGE OBSERVATORY

A CELESTIAL SURVEY

on all sides. As we approach one, it gradually grows into a great silvery ball spangled by thousands of star-points, and finally, from a few hundred light years away, we get the view shown in the picture opposite.

This magnificent object, resembling a vast celestial swarm of bees, is called a "globular star cluster"; it is incomparably richer and more compact than the galactic type, as you can see.   It is   also   noted for its beautifully spherical outline.   About a hundred of these clusters are known; the particular one at which we are looking is about 200 light years in diameter and contains a million stars. Its total luminosity is more than 500,000 times the light of the Sun.

When we viewed some of the rich star clouds of the galaxy, we received the impression that the stars were packed so thickly as to be almost in actual contact.   The globular cluster now gives us the same impression, which again turns out to be an illusion. As we travel through the cluster, we find that the real separation of the stars is on the order of half a light year or so, pinpoints averaging 3000 feet apart on our much-used scale model! This cannot be called crowding, although it is much more "crowded" than most regions of the Galaxy. Only the Galactic Center, in fact, is so thickly populated.

And the Galactic Center, incidentally, is where we are now headed.

Having looked at the Galaxy from far out in space, you will remember the bright central nucleus which appeared to dominate the general glow of the great spiral. That nucleus was the Galactic Center, the great concentration of mass about which the rest of the system revolves. The brightness of the Center is not due to any greater luminosity of its stars; on the contrary, the great hot supergiant stars are quite absent from this region. It is instead the unbelievably vast numbers of the stars which give the Center a brilliance exceeding even the rich star clouds of the spiral arms.

No one can guess how many stars are contained in the Galactic Center—certainly hundreds of millions, perhaps billions. We are at the very hub of the Galaxy, and in this region of awesome splendor we have reached the climax of our celestial journey. Let us think for a moment of the

A CELESTIAL SURVEY

billions of other Suns, the nebulae and clusters, and the countless other wonders we have seen. Let us think also of the millions of other galaxies beyond this one, each another star city containing a few hundred billion Suns. If we can even begin to grasp these concepts, however hazily, we have made a good start at understanding the astronomer's Universe, and we are also in a good position to increase our knowledge by asking the right questions:

QUESTION: If we had chosen some other galaxy at random instead of this one, would we have seen the same things?

ANSWER: Yes. Any large spiral galaxy would be a fair duplicate of the one we have just explored. Some of the smaller galaxies, however, particularly the "elliptical" types, contain no dust and gas and would therefore contain no diffuse nebulae. Their stellar population also appears to be limited, as they contain none of the blue supergiants which we found in the spiral arms of our own Galaxy.

QUESTION: What is the reason for this difference?

ANSWER: The full explanation is not yet known. There are two widely contrasting types of stellar population. Some regions are occupied by gas and dust clouds and blue giant stars; their presence identifies "Population I." In other regions we find no dust or gas, and the brightest stars are red giants of only moderate luminosity; these are the earmarks of "Population II." From this we can see that the spiral arms of our Galaxy, and also the irregular galaxies, are Population I systems. The Galactic Center, the globular star clusters, and the spherical and elliptical galaxies are all Population II. There is no real doubt that this difference is due primarily to a matter of age. In a Population I system we still see star formation in progress, while in a Population II system the stars appear to be very ancient and are in an advanced state of stellar evolution. But it is not known why some galaxies appear to be so much older than others. These facts must be considered in any attempt to formulate a general theory of the history of the Universe.

QUESTION: Isn't our Galaxy sometimes referred to as the "Milky Way"?

ANSWER: Yes. The galaxy is shaped like a wheel, you remember. The stars, therefore, appear more thickly massed when we look in one direction—down the longer diameter of the wheel. In this direction we see stars by the billions. But, because of the vast size of the Galaxy, most of these stars are too distant to be separately visible, and only their combined light reaches us in the form of a misty luminescence which appears to circle the heavens in a great glowing band. This appearance is easily noticed from Earth, and it is called the "Milky Way." The Milky Way, then, is merely the result of viewing our Galaxy from the inside. When we look toward this glowing band, we are looking across the long dimension of the Galaxy. And the brightest portion, which we see from Earth in the southern sky in summer, marks the direction to the Galactic Center.

QUESTION: Where is our own Solar System located in our Galaxy?

ANSWER: It is approximately 30,000 light years out from the center of the Galaxy, or a little more than halfway from the center to the outer edge. It is not located in any definite cluster, but is a member of one of the loose star clouds which compose part of a spiral arm of the Galaxy.

From a distance of 50 light years, the Sun appears as a faint yellowish star barely visible to the naked eye and completely indistinguishable from the myriads of other faint stars scattered over the heavens. From about four light years out it has become one of the brightest stars in the sky. And, finally, from a distance of less than one light hour (about 670 million miles), it begins to show as a disc instead of a point of light, and its brilliance has become dazzling.

The Sun, from anywhere else in the galaxy, would be regarded as a very typical sort of star like billions of others, but for us it seems to have a certain uniqueness. As we already know, it is not a lone wanderer in space, but is accompanied by a "family" of objects known as the "Solar System," all the members of which are within a few light hours of the dominant central star. The chief objects of interest in this system are nine small solid worlds known as the "planets," which revolve about the Sun at different distances and various speeds. The nearest planet to the Sun is 36 million miles away and revolves about it in 88 days; the farthest is 100 times more distant and requires 248 years for one revolution. The planets also differ

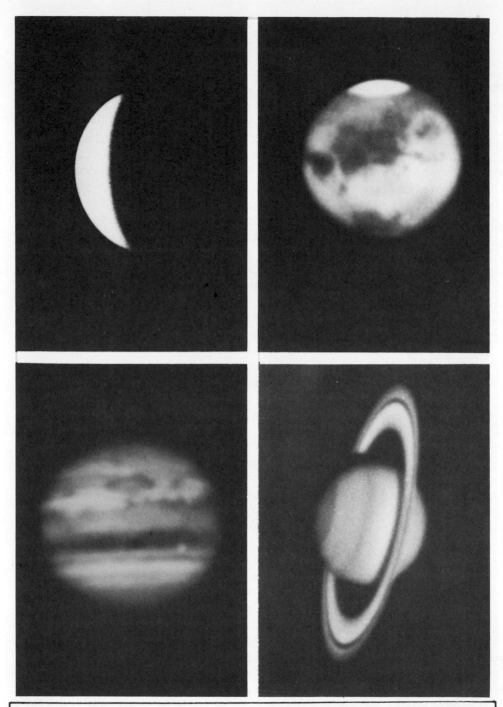

PLANETS OF THE SOLAR SYSTEM ARE THE OTHER WORLDS WHICH ORBIT THE SUN IN COMPANY WITH OUR OWN EARTH. SHOWN HERE ARE VENUS, MARS, JUPITER, AND SATURN.

LOWELL OBSERVATORY PHOTOGRAPHS

THE MOON IS THE ONLY KNOWN NATURAL SATELLITE OF THE EARTH AND THE NEAREST OBJECT IN SPACE. THOUSANDS OF GIANT CRATERS POCK—MARK THE SURFACE; THE HUGE FORMATION SHOWN HERE IS CLAVIUS, 135 MILES IN DIAMETER.

LOWELL OBSERVATORY PHOTOGRAPH

A CELESTIAL SURVEY

greatly in size, having diameters ranging from 3200 to 88,000 miles. The roll call of the planets—going outward from the Sun—reads as follows: Mercury, Venus, Earth, Mars, Jupiter, Saturn, Uranus, Neptune, and Pluto.

Some of the planets have smaller bodies called "satellites" or "moons" revolving about them. There are 34 such objects in the solar system. Our Earth, as we all know, has only one natural satellite.

In addition to the planets and their satellites, we also find a host of "asteroids," sometimes called "planetoids" or "minor planets," revolving about the Sun, mainly between the orbits of Mars and Jupiter. Possibly fragments of a shattered world, they must be numbered in the thousands, although the majority are mere hunks of rock less than 5 miles in diameter. Possibly 4 dozen or so have diameters exceeding 100 miles; the two largest, Ceres and Pallas, have diameters of about 620 and 375 miles.

Meteorites are perhaps similar to asteroids, except in the matter of size. They must be numbered in billions, but are usually no larger than tiny pebbles or sand grains. The Earth collides with millions of them each day, their fiery passage through our atmosphere producing the phenomenon called a "shooting star." On rare occasions a meteorite may be large enough to reach the surface of the Earth before being completely incinerated by friction against the air. Meteorites are naturally of the greatest scientific interest and value since they are the only samples of material from beyond the Earth-Moon System which can be studied and ana-lysed in the laboratory. The largest single specimen known today is the "Hoba" in South Africa, a solid mass of nickel-iron weighing some 60 tons. The great majority of meteor-ites are composed of stony material, however, and a few consist of metallic iron and stony material mixed together. Giant meteorites have occasionally struck the Earth; the mile-wide Meteor Crater in Arizona was produced by such a collision some 30,000 years ago.

The final members of the Solar System are the comets, large chunks of frozen gases and ices mixed together with dust and meteoritic material. Such an object is generally no more than 10 or 20 miles in diameter, but when near the Sun becomes surrounded by a huge glowing cloud of vapor which may be several times the size of the Earth. Comets

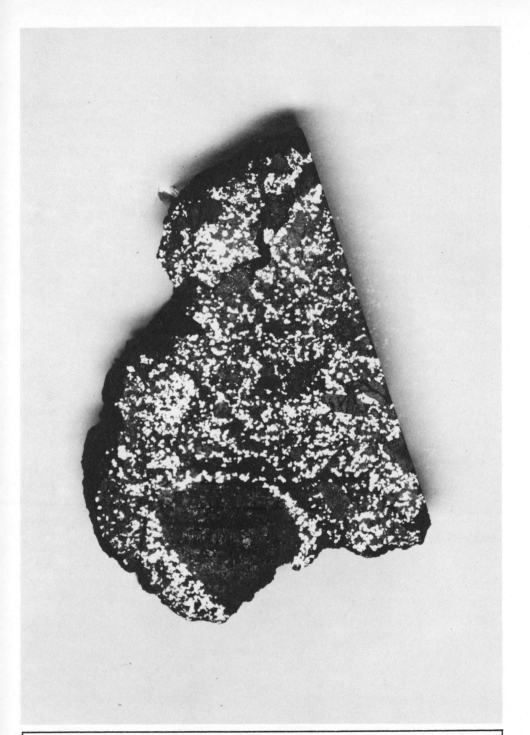

STRUCTURE OF A STONY-IRON METEORITE. A POLISHED FRAGMENT OF THE 85-POUND METEORITE WHICH FELL AT PATWAR, INDIA, JULY 29, 1935; SHOWING BRIGHT METALLIC NICKEL-IRON GRAINS IN A DARK MATRIX OF SILICATE MINERALS.

2 X ENLARGEMENT   BURNHAM COLLECTION

COMET BURNHAM 1959k, DISCOVERED BY THE AUTHOR AT LOWELL OBSERVATORY IN
DECEMBER 1959. THIS PHOTOGRAPH SHOWS THE COMET ON APRIL 22, 1960, SHORTLY
BEFORE ITS CLOSEST APPROACH TO THE EARTH.

PHOTOGRAPH BY ALAN MC CLURE

A CELESTIAL SURVEY

revolve about the Sun in periods which may range from a few years up to many centuries. A comet orbit is usually a long ellipse, bringing the comet very near the Sun and the Earth at the closest approach, but afterwards carrying it out beyond the farthest planets at the other end of its journey. Comets are visible only when relatively near the Sun and appear as hazy "stars" often accompanied by a long glowing "tail" of fine dust and gas. Tail activity is connected in some way with the radiation from the Sun and with the streams of electrically charged particles being emitted by the Sun; the tail vanishes as the comet recedes from the Sun and heads out into interplanetary space. The return of certain comets may be predicted when the period is known; the most famous example is undoubtedly Halley's Comet, which has a period of about 76 years and will return again in 1986.

To review briefly, then, the Solar System contains:

One Star (The Sun)
Nine Planets) (worlds)
Thirty-four Satellites (moons)
Thousands of Asteroids (minor planets)
Thousands of Comets
Billions of Meteorites

One final question is certain to occur to us—Is the Sun's family unique, or are other stars surrounded by similar "solar systems"? The answer is quickly given. We do not know. Among the billions of stars composing our Galaxy, which is itself only one among millions, it appears certain that other solar systems must exist—somewhere. There may, in fact, be millions of other planetary systems. But, because of the vastness of space, we shall never be able to observe them directly. Even at the distance of the nearest star—a mere four light years—a planetary system such as ours would be completely beyond the range of the greatest telescope in the world; which is in the nature of a comment on both the incredible immensity of our Universe and the almost microscopic insignificance of our home in space.

And, having returned once again to Earth, let us leave our travelers alone with their thoughts.

CELESTIAL PHOTOGRAPHY is one of the many exciting fields open to the amateur astronomer who has gained a "working knowledge of the heavens". This is the galaxy M101, photographed with a 12½-inch reflector, by Evered Kreimer.

In the preceding chapter we toured the Universe with the speed of imagination and learned something of the various heavenly bodies and their general arrangement in space. Now we must begin a different sort of survey—an examination of the sky as it appears from the Earth. We want a "working knowledge" of the heavens; we want to be able to find our way about among the various celestial wonders; we want to be able to locate specific objects and find them easily at any time. We want to be able to find the Great Orion Nebula, the Hercules Cluster, the Andromeda Galaxy, and eventually such esoteric objects as Omicron Eridani, NGC 7789, and UX Ursa Majoris. We are dealing now with the big three fundamentals of actual observing: How do we find an object? When will it be visible? Where do we look? These are the items which will concern us in this chapter.

Following this, we shall briefly introduce some of the terms, definitions and symbols used constantly by observers, among them the magnitude system, the celestial coordinate system, and the use of angular measurement in astronomy. These things often seem wonderfully mystical to a beginner, but soon become as comfortably familiar as the hours and minutes on a clock face or the divisions on a foot rule. In astronomy, as in many other things, we grow in knowledge automatically, through experience.

How then do we begin? Let us first consider our position on Earth. We live on a rotating globe which is a satellite of one star—the Sun—some 93 million miles away. As the globe turns, we find ourselves facing sunward for a period of approximately 12 hours; then we are carried across the dark half of our planet for another 12 hours. This continual cycle of day and night is one of the outstanding features of our physical world.

For some reason, perhaps explainable by the psychology of primitive peoples, it seems natural to regard the daytime appearance of things as the normal situation and to consider the night as strange, perhaps even somewhat fearful. Nothing can destroy this false impression more quickly and thoroughly than astronomy. When we face toward the Sun, our atmosphere is so brilliantly illuminated by scattered light that the feeble rays of the stars and planets are overpowered by the glare. It is only when we are on the "night side" of the Earth that we can see the sky as it

THE CONSTELLATIONS

really is and as it would appear at all times if there were
no atmosphere surrounding the Earth.

Obviously, the best introduction to our subject would
be an actual look at the real sky itself. So let us choose
a clear moonless night in some secluded spot with a clear
view in all directions, away from the lights and glare of
cities and highways. Here the heavens are revealed in
unforgettable grandeur seldom, if ever, glimpsed by the
unfortunate city dweller. Here we are face to face with the
beauty, the wonder, and the mystery of creation.

"One might think," wrote Emerson, "that the atmosphere
was made transparent with this design; to give man, in the
heavenly bodies, the perpetual presence of the sublime..."

THE CONSTELLATIONS. Using our imaginations, we can trace
out many seeming patterns in the stars: squares, crosses,
circlets, and more elaborate figures such as the outlines
of men, animals, and other objects. Several thousand years
ago, the men of ancient times—the world's first amateur
astronomers—invented a number of such star patterns and
named them after the heroes, beasts, and demi-gods of their
own rich mythology. In more recent times, a number of additional groups were introduced by more modern astronomers,
and there are now 88 standard configurations recognized—
they are called the "constellations." Among them we find
such groups as the familiar Ursa Major (Great Bear or Big
Dipper) in the north; Orion, the Hunter, in winter skies,
Sagittarius, the Archer, in the summer, and Cetus, the sea
monster, and Pegasus, the flying horse, in the autumn heavens. Since 1930, the constellations all have definite
official boundary lines defined by the International
Astronomical Union, so there is never any doubt where one
constellation ends and another begins. Every star in the
sky has a definite constellation "address."

Although a beginner usually thinks of a constellation
as an imaginary outline figure, the serious observer soon
begins to regard it as a definite area of the sky, rather
like a state or country on Earth. It should be scarcely
necessary to point out here that the constellations are
merely artificial star groups defined by arbitrary boundaries, in much the same way as the countries of the world
are areas artificially divided and marked off by arbitrary

boundary lines. The stars of a constellation do not necessarily have any connection with each other; they simply lie
in the same direction as seen from the Earth.   The stars in
any constellation are at varying distances, some of them
hundreds of times more distant than others, so that the
whole constellation would appear radically different if
seen from some other point in space.   If we suddenly took
up residence in some other solar system many light years
away, we would have to invent an entirely new set of constellations to fill our skies.

Nevertheless, for us on Earth the constellations form
a definite and, for all practical purposes, permanent system
of recording the apparent positions of the celestial objects.

We say "for all practical purposes" because it is evident that these star patterns must gradually change in the
course of time.   The Galaxy is rotating, the Solar System
is in motion, and all of the stars are moving in various
directions   at   velocities averaging many miles per second.
It may seem surprising, therefore, that the constellation
figures are not noticeably changing from year to year, or
even from day to day.   The answer to this puzzle lies in
the vast distances separating the stars.   The nearest star
to our own Solar System is more than four light years distant, which is quite typical of the average separation of
stars throughout space.   Now a star may be moving at 30
miles per second, for example; but if we must observe it at
a distance of 30 trillion miles, we shall not be able to
detect any direct evidence of motion, except with highly
accurate measuring instruments used over a period of years.
The star's naked-eye position will not be noticeably altered,
even after centuries.   Thus, it is not difficult to see why
the constellation figures that we see tonight are the same
ones seen by the ancient Egyptians, Chaldeans, and the pre-
Homeric Greeks, three thousand years ago.

LEARNING THE CONSTELLATIONS. The problem of finding our way
about the sky is thus substantially reduced to the problem
of learning the constellations.   Perhaps the simplest aid
for the beginner is the device known as the "planisphere"
or "star finder," a revolving chart which may be set for
any date and time and shows all the star patterns in their
relation to the horizon.   An inexpensive and very useful

CONSTELLATION IDENTIFICATION

form of planisphere called the "Star Explorer" may be obtained at small cost from the Hayden Planetarium in New York City.   A very similar model called the "Star and Satellite Pathfinder" is available for about $0.75 from the Edmund Scientific Company in Barrington, New Jersey. More elaborate and expensive models are obtainable from various dealers in astronomical books and supplies.

There are a number of relatively inexpensive books dealing with constellation identification.   A long-time favorite has been the Field Book of the Skies by W. T. Olcott and E. W. Putnam. Another was McKready's Beginner's Star Book, now unfortunately out of print, but still obtainable frequently from dealers in secondhand books. In using any of the older books, the reader must remember that many of the astronomical facts given will often be badly out-of-date; the constellation charts, however, will remain usable indefinitely.

In using any of these charts, the observer must choose the correct chart for the time and date and then face the proper direction. Sectional charts will usually be labeled as to north, south, etc.   A chart showing the entire sky will have the compass directions printed around the outer circumference, and the observer must hold the chart so that the direction he is facing appears at the bottom of the chart. The same rule holds true with the revolving type of chart. The lower edge of the chart will then correspond more or less to the horizon in that direction.

For the beginner attempting constellation identification for the first time, a few simple rules may be of some help: Start with the brighter stars and more conspicuous groups.   Begin near the center of the field and work outward. Choose a clear moonless night and use a dim red light when referring to charts and maps. Such an observing light may be made by mounting a piece of red paper over a flashlight lens. Never attempt to accomplish any constellation identification with one of the artistic 18th Century types of star chart—the variety which depicts the heavens as a confusing clutter of Greek heroes and mythical monsters. These imaginary figures have no connection at all with real astronomy. Confine your attention to the charts which show the actual star patterns as they appear in the sky.

Finally, it may be well to point out that a certain

THE APPARENT MOTION OF THE CONSTELLATIONS

amount of distortion is unavoidable in making any chart of the sky. The heavens appear to us as the inside of a great globe, and no globe—not even an imaginary one—can be truthfully represented on a flat surface such as a sheet of paper. The distortion is negligible on charts of small areas of the sky, but becomes noticeable on any map which attempts to show the entire visible heavens—or the greater part of it—on a single chart. Thus, on the rotating planisphere the constellations near the southern horizon always appear somewhat squashed from top to bottom and elongated left to right. Individual monthly charts of the sky, such as those in Sky and Telescope magazine, are considerably less affected by this sort of distortion.

APPARENT MOTION OF THE CONSTELLATIONS. A study of the sky (or of the planisphere) will soon show that the constellation groups do not remain fixed in the heavens, but appear to move across the sky hour by hour, in the same manner as the Sun and Moon. This apparent motion is due to the 24-hour rotation of the Earth. To verify this motion, observe the night sky at intervals of several hours, choosing some fairly conspicuous stars and comparing their positions against some earthly landmarks such as nearby trees or telephone poles. Stars in the east are rising, those in the south are moving from left to right, and those in the west are setting. The motion of stars in the northern part of the sky is more peculiar; they appear to be moving in concentric circles around a fairly bright star not quite halfway up the sky. This star is the well known Polaris, or North Star, toward which the Earth's axis happens very nearly to point. As a result of this circumstance, Polaris does not appear to share noticeably in the apparent turning of the sky; it seems rather to remain motionless in the north while the heavens slowly revolve about it.

SEASONAL CHANGES OF THE CONSTELLATIONS. We must now consider a second apparent motion of the entire constellation pattern, a seasonal shift caused by the Earth's motion around the Sun. The effect of this is to make every star reach the same position four minutes earlier each night. Thus a star which rises at 10 p.m. on January 1 will rise at 8 p.m. on February 1 and at 6 p.m. on March 1. In this

THE CELESTIAL SPHERE

way the constellations gradually drift westward and are replaced by new groups as the year progresses. The process is a yearly cycle and makes it possible for us to classify the constellations by seasons. Orion, Lepus, Canis Major and Taurus are typical winter constellations, for example. In the spring skies we have such groups as Leo, Virgo, Coma Bernices and Hydra. In the summer skies we have Lyra, Cygnus, Sagittarius and Scorpius; while the heavens of autumn present to our view such time-honored groups as Perseus, Andromeda, Cetus and Pegasus.

Constellations near enough to the North Star will remain continuously in view, as their apparent daily motion around Polaris will never carry them below the horizon. Such groups are called the "north circumpolar constellations." Their number obviously depends upon the observer's position on Earth. At the North Pole, Polaris would appear directly overhead, and all visible constellations would be circumpolar. At the Equator, Polaris would appear to lie on the horizon, and no constellations would be circumpolar. From the latitude of the United States, the constellations classified as circumpolar include the Big and Little Dippers (Ursa Major and Ursa Minor), Draco, Cepheus, Cassiopeia, and Camelopardus.

There are likewise a group of constellations classified as "south circumpolar," of little interest to North Americans, as they never rise above the southern horizon as seen from the United States.

THE CELESTIAL SPHERE. Before introducing the next topic, let us pause briefly to review some of the facts given in Chapter 2. When we look toward the sky, we know that we are looking into limitless space; we have learned that the stars are actually great blazing suns many trillions of miles away. We are familiar with the idea that beyond these stars lie billions more, all together forming a vast disc-shaped aggregation called the "Milky Way Galaxy." We know that this galaxy is, in turn, only one among hundreds of millions. All these things we know. Yet, for our present purpose, we must temporarily ignore such facts and return to a concept which may seem quite primitive, but which was once regarded as literally true. For purposes of observing convenience, we are going to pretend that the sky

THE CELESTIAL SPHERE
THE ECLIPTIC AND THE ZODIAC

is a great hollow sphere with the Earth at its center. We must imagine the stars as being fixed to the inside surface of this "Celestial Sphere." We can see only half the sphere at any one time, of course, the other half being out of sight beneath the horizon. As we have already learned, this fictitious sphere is in slow rotation, making one turn in four minutes less than a day. The north and south celestial poles, upon which the sphere rotates, are located directly above the Earth's north and south poles. The Celestial Sphere likewise has an equator, which is a great circle drawn around the sphere midway between the poles. It passes directly overhead as seen from any point on the Earth's equator.

The North Celestial Pole is thus located directly above the true north point of the horizon, and its altitude is equal to the observer's latitude on Earth. The South Celestial Pole, of course, is always below the southern horizon for observers in the Earth's northern hemisphere.

Another great circle on the Celestial Sphere is the Ecliptic, the apparent yearly path of the Sun and the approximate path of the Moon and planets. It is also the path of the Earth as seen from the Sun, Moon, and planets. The ecliptic is tilted at an angle of 23½° from the celestial equator and intersects the equator at two points called the "equinoxes." The Sun passes the equinoxes at March 21 and September 21. The Spring or "Vernal Equinox" is also called the "First Point of Aries," although the point is now actually located in the constellation of Pisces.

The twelve constellations lying along the ecliptic form a band around the sky called the "Zodiac." All the bright planets will always be found somewhere within this band. The twelve zodiacal constellations are Aries, Taurus, Gemini, Cancer, Leo, Virgo, Libra, Scorpius, Sagittarius, Capricornus, Aquarius, and Pisces.

We may now begin to see why the imaginary "Celestial Sphere" will prove useful for observational purposes. Since the sphere has both poles and an equator and since the stars are (for all ordinary purposes) fixed in their positions upon it, it is evident that we can set up a system of "celestial coordinates" for the location of objects in the sky.

CELESTIAL COORDINATES. The system will be analogous to the latitude and longitude system used on Earth. We shall be using the terms "declination" and "right ascension" instead of latitude and longitude, however. A star's declination is its distance north (+) or south (-) of the celestial equator, measured in degrees from 0° to 90°. For accurate positions, degrees are subdivided into sixty parts called minutes of arc ('), and each minute is divided into sixty parts called seconds of arc ("). These same units are used in giving apparent sizes and apparent distances between celestial objects, as described later in this chapter.

Right ascension is measured in hours from 0 to 24, beginning at the First Point of Aries and going eastward. The hour lines of right ascension are spaced 15° apart at the equator and converge like spokes going toward a hub at the north and south celestial poles. Hours (h) of right ascension (RA) are further divided into sixty minutes (m), and each minute is divided into sixty seconds (s).

The celestial coordinate system makes it possible to state the position of any object or to locate any object whose coordinates are known. The practical aspects of this system will now be briefly discussed.

Let us suppose that we are using one of the standard star atlases such as Norton's. On each of the star charts we find a superimposed grid of coordinate lines. The lines traversing the chart from top to bottom are the lines of right ascension. We will find them labeled 0h, 1h, 2h, 3h, etc., at the top and bottom of each map, the numbers increasing as we go toward the left. Smaller marks between the hour lines indicate subdivisions of each hour. These are usually given at intervals of five minutes, though on some of the simpler atlases they may be given at ten-minute intervals instead. With some care we can plot (or estimate) the RA of an object to better than one-minute accuracy.

The horizontal lines traversing the chart left to right are the parallels of declination, analogous to the parallels of latitude on Earth. The lines are usually drawn for every 10° of declination, and the smaller marks between generally indicate divisions of one degree. In reading or reporting a declination we must always remember to specify whether it is north (+) or south (-). The letters "n" and "s" may also be used, as they are in this book. In all the lists of

USING CELESTIAL COORDINATES

objects in this Handbook, the celestial coordinates are given in the column headed "RA & DEC." They appear in a contracted form as follows:

Sample position		As it appears in this book
RA	Dec	RA & Dec
12h 22.7m	+17° 31'	12227n1731
8h 7m 34s	-6° 43.6'	08076s0644

In the second example, the 34s has been rounded to 0.6 minute in the RA, and the 43.6' has been rounded to 44' in the Dec. All positions are thus given to the nearest tenth of a minute RA and to the nearest minute of arc in Dec. This degree of accuracy is more than sufficient for all ordinary purposes.

Celestial coordinates may be used in two chief ways. First, they may be used to identify objects plotted on the star charts or to add objects which are not already shown. For example, suppose we wish to observe the very interesting double star called "61 Cygni." The catalog position is: 21044n3828. As we should know by now, this decodes to read: RA 21h 04.4m, Dec +38°28'. Plotting this position in Norton's Atlas, we find that it falls within the borders of the constellation Cygnus and that there is a small star shown at that position labeled "61." The identification is thus certain, and it remains only to compare the chart with the actual sky and find the object, perhaps with the aid of a pair of binoculars.

Suppose, however, that our chosen object was the famous explosive variable star called "SS Cygni." When we plot the catalog position (21407n4321), we find nothing shown on the atlas at that exact spot, and there is no star labeled "SS" in the region. In this case we are dealing with an object which is too faint to have been included in the atlas we are using. We can, however, still plot the position as accurately as possible on the map and then turn the telescope toward that precise spot in the sky, using a fairly wide field eyepiece. We shall now see a multitude of faint stars in the field, perhaps more than 100. Which one is SS Cygni? To identify it, we need a detailed chart of the field. For

PRECESSION

this star, and most of the other variables chosen for detailed description, finder charts are given in this book. After some problems with the orientation of the chart and perhaps after rechecking the atlas position and re-aiming the telescope several times, we eventually identify the brighter stars in the field and locate SS Cygni.

The second way in which RA and Dec may be used involves the installation of a pair of calibrated dials called "setting circles" on the telescope. These read the celestial coordinates directly and enable the observer to point the telescope toward any object of known position. To do this, however, the telescope must be equatorially mounted and accurately aligned on the celestial pole. Preferably, it should also be equipped with a good clock drive. The opinion has occasionally been expressed that setting circles are an absolute necessity if the observer wishes to locate the thousands of interesting objects in the skies above. This is, of course, utter nonsense and is somewhat equivalent to saying that it is impossible to locate New York without knowing its latitude and longitude. Finding objects is a simple matter of becoming familiar with the star patterns so that star charts may be used intelligently. The most elaborate telescopic equipment is no substitute for the observer's self-achieved knowledge of the celestial geography—it may in fact act as a barrier to the attainment of that knowledge. Plan to know the sky. Keep mechanical devices in their proper place—as servants and aids to learning, but do not let them rob you of the joy and pride of firsthand knowledge.

PRECESSION.   It is obvious that the celestial coordinates of an object are not absolutely permanent values. All the stars are moving in space in various directions, and these motions will gradually change every star's position on the celestial sphere. As we have seen, such changes are extremely slow. Much more rapid, however, is another phenomenon known as "Precession of the Equinoxes," or simply "precession," a gradual shifting of the direction of the Earth's axis in space. Precession does not alter the constellation outlines or change their boundaries with respect to the stars, but causes a slow drifting of the entire celestial coordinate system on the celestial sphere. Thus

PRECESSION

the right ascensions and declinations of all the stars are
gradually changing, not because of the actual motion of the
stars, but because of the drifting of the coordinate lines
past them.

Precession is caused chiefly by the pull of the Moon
on the Earth's equatorial bulge and, to a lesser extent, by
the pull of the Sun and the planets. The total effect is
to make the Earth's axis perform a circular "wobble" resembling the motion of a "dying top," with a period of about
25,800 years. In the course of one such wobble, the Earth's
axis traces out a circle some 47° in diameter in the sky.
Obviously, this means that the position of the celestial
poles is constantly changing with respect to the star patterns. About 12,000 years ago the Earth's axis pointed in
the general direction of the star Vega, which was then the
Pole Star.   When the Pyramids were built, the star Thuban
occupied the position of honor. Now the star we call
"Polaris" marks the point.   In 1960 it was slightly less
than one degree distant from the true Pole, and the distance will continue to decrease during the next century.

Because of precession, a star chart or a star position
is absolutely accurate only at the time or "epoch" for
which it is intended. It is customary to adopt certain
years as "standard epochs" for star charts and position
catalogs, that of 1950 being the current standard. The
well known Norton's Star Atlas and the new Skalnate-Pleso
atlas are both "epoch 1950," as are all the positions given
in this Handbook.   All objects in this book can therefore
be plotted directly on either of these standard atlases .
If, however, we have an "epoch 1920" position of an object
from an older catalog and wish to plot it on an "epoch 1950"
chart, we must first apply a slight correction to bring the
position up to date. For this we refer to the "Precession
Tables" in Norton's Atlas.

The effect of precession over a few years is not great
and may be safely ignored by the beginner or average amateur.
In the course of time, however, the effect adds up and becomes noticeable. A few pages ago we mentioned the fact
that the so-called "First Point of Aries" (the intersection
of the celestial equator and the ecliptic at 0 hours RA) is
now actually in the constellation of Pisces.   In the last
2,000 years the shift caused by precession has amounted to

DIRECTIONS IN THE SKY

As seen from Latitude 35° to 40° North

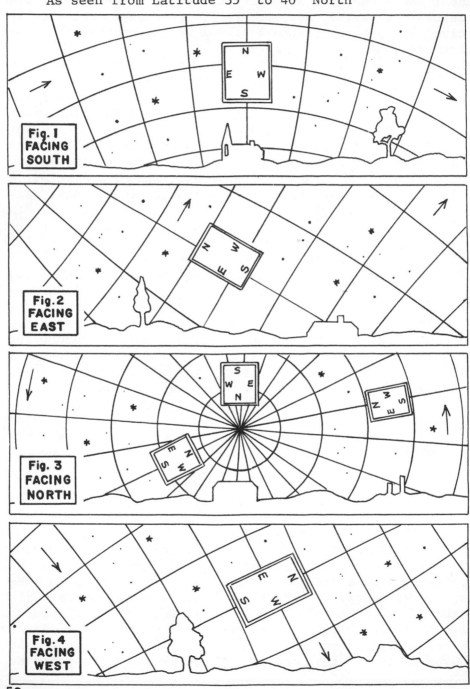

Fig. 1
FACING
SOUTH

Fig. 2
FACING
EAST

Fig. 3
FACING
NORTH

Fig. 4
FACING
WEST

DIRECTIONS IN THE SKY

nearly 30°, sufficient to move the point into the next constellation. In the course of a century the RA of a star may
change by as much as a degree or more, depending on the
position of the individual star.   During the next 25,800
years, most of the stars will pass through every hour of RA,
from 0h to 24h.

Although precession may be of no actual practical
interest to the beginning observer, it has been thought
advisable to include a short discussion of the effect in
this chapter, to demonstrate that the RA and Dec of celestial objects are not absolutely unchanging values, and also
to help the earthbound observer to sense his situation on
a moving, oscillating platform.

DIRECTIONS IN THE SKY are defined by the celestial coordinate system. An object is said to be directly east or west
of another if both lie on the same declination; an object
is directly north or south of another if both lie on the
same line of RA. For the simplest case, let us imagine we
are facing directly south and looking at a bright star, as
shown in Figure 1. Then "north" will be straight up, "south"
will be straight down, "east" will be toward the left, and
"west" toward the right. The seeming reversal of east and
west may confuse beginners, but it is the natural and obvious result of looking outward from the Earth instead of down
upon it. When we face south, east IS toward our left, and
west is to our right.

In other parts of the sky the directions become more
confusing because our guiding lines of RA and Dec may appear tilted in various directions.   In Figures 2, 3, and
4, we attempt to clarify this situation by showing the orientation of the lines as they would appear if "printed" on
the sky, along with a properly oriented star chart.

Another way of giving directions in the sky is by the
use of "position angles" (PA). A position angle of 0° indicates exact north; 90° is east; 180° is due south; and 270°
is west. This system is used in giving the direction of a
faint companion from its primary in double stars and in
other applications, such as the direction of a moving object.

The question of directions in the sky has been briefly
discussed here in order to discourage the use of such inadequate terms as "toward the upper left" or "a little below

THE ZENITH, THE MERIDIAN
SIDEREAL TIME

and to the right" in astronomy.   These orientations are
suitable only for the immediate moment but will change continually as the Earth rotates, and therefore cannot be used
in any permanent definition of a direction in the sky.   It
is far better that the serious observer should attempt to
familiarize himself with the "framework" of the celestial
sphere and the coordinate system.   In the learning of such
things, the use of star charts under the actual sky will
soon accomplish what no amount of written instruction can
do.   The facts given in this chapter are not to be memor-
ized or "learned"; they are to be used.   Firsthand knowledge is our goal, and the sky itself is our teacher.

OTHER TERMS. We must now introduce a few astronomical terms
which may not be completely unfamiliar, even to the average
beginner. The ZENITH of a place is the point directly overhead; the corresponding point below the feet is called the
"nadir." As astronomers, we shall have little use for this.
A line drawn across the sky from the north to south, passing through the zenith, is called the "MERIDIAN." As the
celestial sphere rotates, all objects must cross or "transit" the meridian once a day; such a passage is called a
"culmination." At culmination, a celestial object is midway between rising and setting and is also at its highest
altitude above the horizon. As a familiar example, the Sun
culminates at local noon.   As we have already seen, every
star culminates four minutes earlier each day.
     "Sidereal time" or "star time" is measured by the pas-
sage of the RA hour lines across the meridian. The sidereal
time at any instant is simply the RA which coincides with
the meridian at that moment.   The bright star Aldebaran,
for example, has an RA of 4h 33m. When this star is crossing the meridian, the sidereal time is therefore 4h 33m. For
the beginning observer, sidereal time is chiefly useful as
an indication of the constellations which are well placed
for observation at any given time. For example, suppose we
are planning to observe at 10 p.m. on August 5. From the
accompanying table (Table I), we find that the approximate
sidereal time will be 19h. Referring to our star atlas, we
now find the 19-hour line of RA on our star chart and note
that it runs through the constellations of Sagittarius,
Aquila, Lyra, and Draco. These will be the groups that will

TABLE I — APPROXIMATE SIDEREAL TIME

DATE AND TIME	5 pm	6 pm	7 pm	8 pm	9 pm	10 pm	11 pm	12 m	1 am	2 am	3 am	4 am	5 am	6 am
Jan 5	0	1	2	3	4	5	6	7	8	9	10	11	12	13
13	0½	1½	2½	3½	4½	5½	6½	7½	8½	9½	10½	11½	12½	13½
21	1	2	3	4	5	6	7	8	9	10	11	12	13	14
28	1½	2½	3½	4½	5½	6½	7½	8½	9½	10½	11½	12½	13½	14½
Feb 5	2	3	4	5	6	7	8	9	10	11	12	13	14	15
13	2½	3½	4½	5½	6½	7½	8½	9½	10½	11½	12½	13½	14½	15½
20	3	4	5	6	7	8	9	10	11	12	13	14	15	16
28	3½	4½	5½	6½	7½	8½	9½	10½	11½	12½	13½	14½	15½	16½
Mar 7	4	5	6	7	8	9	10	11	12	13	14	15	16	17
15	4½	5½	6½	7½	8½	9½	10½	11½	12½	13½	14½	15½	16½	17½
22	5	6	7	8	9	10	11	12	13	14	15	16	17	18
29	5½	6½	7½	8½	9½	10½	11½	12½	13½	14½	15½	16½	17½	18½
Apr 6	6	7	8	9	10	11	12	13	14	15	16	17	18	19
14	6½	7½	8½	9½	10½	11½	12½	13½	14½	15½	16½	17½	18½	19½
22	7	8	9	10	11	12	13	14	15	16	17	18	19	20
29	7½	8½	9½	10½	11½	12½	13½	14½	15½	16½	17½	18½	19½	20½
May 7	8	9	10	11	12	13	14	15	16	17	18	19	20	21
15	8½	9½	10½	11½	12½	13½	14½	15½	16½	17½	18½	19½	20½	21½
22	9	10	11	12	13	14	15	16	17	18	19	20	21	22
30	9½	10½	11½	12½	13½	14½	15½	16½	17½	18½	19½	20½	21½	22½
Jun 6	10	11	12	13	14	15	16	17	18	19	20	21	22	23
14	10½	11½	12½	13½	14½	15½	16½	17½	18½	19½	20½	21½	22½	23½
22	11	12	13	14	15	16	17	18	19	20	21	22	23	0
29	11½	12½	13½	14½	15½	16½	17½	18½	19½	20½	21½	22½	23½	0½
Jul 7	12	13	14	15	16	17	18	19	20	21	22	23	0	1
15	12½	13½	14½	15½	16½	17½	18½	19½	20½	21½	22½	23½	0½	1½
22	13	14	15	16	17	18	19	20	21	22	23	0	1	2
30	13½	14½	15½	16½	17½	18½	19½	20½	21½	22½	23½	0½	1½	2½
Aug 6	14	15	16	17	18	19	20	21	22	23	0	1	2	3
14	14½	15½	16½	17½	18½	19½	20½	21½	22½	23½	0½	1½	2½	3½
22	15	16	17	18	19	20	21	22	23	0	1	2	3	4
29	15½	16½	17½	18½	19½	20½	21½	22½	23½	0½	1½	2½	3½	4½
Sep 6	16	17	18	19	20	21	22	23	0	1	2	3	4	5
13	16½	17½	18½	19½	20½	21½	22½	23½	0½	1½	2½	3½	4½	5½
21	17	18	19	20	21	22	23	0	1	2	3	4	5	6
29	17½	18½	19½	20½	21½	22½	23½	0½	1½	2½	3½	4½	5½	6½.
Oct 6	18	19	20	21	22	23	0	1	2	3	4	5	6	7
14	18½	19½	20½	21½	22½	23½	0½	1½	2½	3½	4½	5½	6½	7½
21	19	20	21	22	23	0	1	2	3	4	5	6	7	8
29	19½	20½	21½	22½	23½	0½	1½	2½	3½	4½	5½	6½	7½	8½
Nov 6	20	21	22	23	0	1	2	3	4	5	6	7	8	9
13	20½	21½	22½	23½	0½	1½	2½	3½	4½	5½	6½	7½	8½	9½
21	21	22	23	0	1	2	3	4	5	6	7	8	9	10
28	21½	22½	23½	0½	1½	2½	3½	4½	5½	6½	7½	8½	9½	10½
Dec 6	22	23	0	1	2	3	4	5	6	7	8	9	10	11
14	22½	23½	0½	1½	2½	3½	4½	5½	6½	7½	8½	9½	10½	11½
21	23	0	1	2	3	4	5	6	7	8	9	10	11	12
29	23½	0½	1½	2½	3½	4½	5½	6½	7½	8½	9½	10½	11½	12½

be well placed for study at our chosen time.   Of course, many other constellations, though not actually on the meridian, will still be in conveniently observable positions.

ANGULAR MEASUREMENT.   This   is   a method of expressing the apparent sizes and distances between   objects   in the sky. Uninformed persons often say that the Moon looks "about a foot across" or that one star is located "about two inches" below another. In place of this unscientific and essentially meaningless way of describing apparent   sizes and distances, the astronomer uses a system of "angular measurement" based on the divisions of the circle.   As we know, a circle contains 360 degrees, and thus any line   drawn   from the true horizon to the zenith would be one-quarter of a circle and would therefore contain 90 degrees.   Any object apparently covering 1/9 of this distance would have an angular size of 10 degrees.   The separation of the two bright stars at the front of the Great Dipper's bowl is approximately 5 degrees. The Moon has a diameter of about half a degree; so does the Sun, incidentally,   which   is   one way of pointing out that the angular size or apparent size   has   nothing   to do with actual diameter.   The Sun is actually 400 times larger than the Moon, but as it is also 400 times farther   away,   both have the same angular size of half a degree, as would   a one-inch disc placed 9½ feet away.
    A degree   is   the   apparent   size of any object whose distance is 57.3 times its diameter.
    The symbol for the degree is (°).   For the measuring of objects of small apparent size, it   is   divided into 60 minutes of arc ('), and each minute is further divided into 60 seconds of arc (").   These symbols, unfortunately, are identical   to   those commonly used for feet and inches; so the beginner must be on his guard and remember   their   new meaning.
    A minute of arc (1') is   about   the   smallest angular quantity which can be perceived by the best human eyes without optical aid.   The planet Venus attains an approximate size of 1' when it is nearest   to the Earth; at such times its disc is almost detectable with the naked eye.   A second of arc (1") is   an exceedingly tiny quantity, as you may imagine—it is the apparent size of   a   25-cent piece seen from a distance of slightly over three miles.   The orbit of

THE MAGNITUDE SYSTEM

the Earth, if seen from the distance of the nearest star (4.3 light years), would subtend an angle of about $1\frac{1}{2}''$.

The observer will, of course, wish to know what a second or minute of arc implies in terms of actual observing at the telescope. How does a 10" double star, for example, appear through a typical six-inch telescope? Will it appear widely separated, just barely divided, or single? The answer, naturally, depends upon the size and quality of the instrument and, to a lesser extent, upon the atmospheric conditions, the relative brightness of the two stars, etc. But to answer the question purely in terms of the apparent separation, a 10" pair should be quite easily divided in any good two-inch glass with a medium power, and pairs of 5" should be found not at all difficult with some experience. The theoretical limit for a good three-inch telescope is about $1\frac{1}{2}''$, and for a six-inch about 0.75". Under the best conditions, even closer pairs may be detected by an elongation of the image, although true separation is not achieved.

It should be obvious that there is a direct relation between the apparent or angular size of an object, the actual diameter, and the distance. If any two of these quantities are known, the third may be determined by the principles of geometry. For objects of stellar and galactic distances, the accompanying Table II has been prepared. We give here an example of its use: A certain galaxy has an apparent diameter of 10'. The distance is known to be 20 million light years. What is the actual diameter? From the table the answer is slightly under 60,000 light years. For a second example: A double star has an apparent separation of 2". The distance is known to be 75 light years. Assuming that the two stars are actually being seen side by side, what is the real separation of the pair? In this case we must interpolate between values in the table. The answer is about 46 astronomical units (46 times the Earth-Sun separation).

MAGNITUDE SYSTEM. This is a method of expressing the apparent brightness of a celestial object. In the original system, initiated by Hipparchus and Ptolemy some 2000 years ago, the 20 brightest stars in the sky were collectively grouped together as stars of the first brightness class or "first magnitude." Stars about $2\frac{1}{2}$ times fainter were

TABLE II

1"	5"	10"	20"	45"	60"	Distance-light years
THE ACTUAL SIZE OF AN OBJECT, CORRESPONDING TO APPARENT ANGULAR DIAMETER AND KNOWN DISTANCE						
Entries below in Astronomical units						
.31	1.5	3.1	6.1	13.8	18.4	1
.61	3.1	6.1	12.2	27.6	36.8	2
.92	4.6	9.2	18.4	41.4	55.2	3
1.2	6.1	12.3	24.5	55.2	73.6	4
1.5	7.7	15.3	30.7	69.0	92.0	5
3.1	15	31	61	138	184	10
9	46	92	184	414	552	30
15	77	153	307	690	920	50
31	153	307	613	1380	1840	100
77	383	767	1533	3449	4600	250
123	613	1226	2455	5520	7360	400
153	767	1533	3065	6900	9200	500
230	1150	2300	4600	10350	13800	750
276	1380	2760	5520	12420	16560	900
307	1533	3065	6130	13800	18400	1000
1533	7670	15330	30660	68990	91990	5000
Entries below in light years						
.024	.12	.24	.48	1.09	1.45	5000
.048	.24	.48	.97	2.18	2.91	10,000
.097	.48	.97	1.94	4.36	5.8	20,000
.122	.61	1.22	2.43	5.5	7.3	25,000
.145	.73	1.45	2.9	6.5	8.7	30,000
.194	.97	1.94	3.9	8.7	11.6	40,000
.24	1.2	2.4	4.8	10.9	14.5	50,000
.48	2.4	4.8	9.7	21.8	29.1	100,000
2.4	12	24	48	109	145	500,000
3.6	18	36	73	164	218	750,000
4.8	24	48	97	218	291	1 million
9.7	48	97	194	436	582	2 million
14.5	73	145	291	655	873	3 "
24	121	242	485	1090	1455	5 "
34	170	339	679	1527	2036	7 "
48	242	485	970	2180	2910	10 "
97	485	970	1940	4365	5820	20 "

TABLE II

1'	5'	10'	20'	45'	60'	Distance- light years
THE ACTUAL SIZE OF AN OBJECT, CORRESPONDING TO APPARENT ANGULAR DIAMETER AND KNOWN DISTANCE						
Entries below in astronomical units						
18	92	184	368	828	1104	1
37	184	368	736	1656	2207	2
55	276	552	1104	2483	3311	3
74	368	736	1472	3311	4415	4
92	460	920	1840	4139	5520	5
184	920	1840	3680	8280	11040	10
552	2760	5520	11040	24835	33115	30
920	4600	9200	18400	41390	55190	50
Entries below in light years						
.0145	.073	.145	.29	.65	.87	50
.029	.145	.29	.58	1.31	1.75	100
.073	.36	.73	1.45	3.27	4.36	250
.116	.58	1.16	2.33	5.24	6.98	400
.145	.73	1.45	2.91	6.55	8.73	500
.218	1.09	2.18	4.36	9.82	13.09	750
.262	1.31	2.62	5.24	11.77	15.71	900
.291	1.45	2.91	5.8	13.1	17.5	1000
1.45	7.27	14.5	29.1	65.5	87.3	5000
2.91	14.5	29.1	58.2	131	175	10,000
5.8	29.1	58.2	116	262	349	20,000
7.3	36.4	72.7	145	327	436	25,000
8.7	43.6	87.3	175	393	524	30,000
11.6	58.2	116	233	524	698	40,000
14.5	72.7	145	291	655	873	50,000
29.1	145	291	582	1310	1746	100,000
145	727	1455	2910	6545	8725	500,000
218	1090	2180	4365	9820	13090	750,000
291	1455	2910	5820	13090	17455	1 million
582	2910	5820	11635	26175	34910	2 "
873	4365	8725	17455	39275	52365	3 "
1455	7275	14545	29090	65455	87275	5 "
2035	10180	20365	40730	91640	122185	7 "
2910	14545	29090	58185	130915	174550	10 "
5820	29090	58185	116365	261760	349100	20 "

The MAGNITUDE SYSTEM

classified as stars of the "second magnitude", those still fainter were classified as "third magnitude," and so on. Stars of the sixth magnitude were at the limit of naked-eye visibility.

This system, virtually unaltered, is still in use today. The exact ratio between magnitudes has been set at 2.512, this having the advantage of making a difference of 100 times in brightness for a difference of five magnitudes. Mathematically-minded persons will thus perceive immediately that the magnitude scale is a logarithmic one and that the number 2.512 is the fifth root of 100. A star of the first magnitude is thus 100 times brighter than a star of the sixth magnitude.

The difference in brightness for any difference in magnitude is given in the short table opposite (Table III).

Some writers have spoken of the magnitude system as confusing, since the magnitude number grows larger as the star grows fainter. The confusion will disappear immediately if the word "class" is substituted for "magnitude." Obviously, we would expect a "first class star" to be brighter than a second or third class star. And the term "fourth class" already begins to suggest faintness and unimportance.

The theoretical magnitude limits of various size telescopes are given in tables by various authors, but can never be rigidly defined because of such variable factors as sky conditions, the condition of the telescope, the sensitivity of the observer's eyes, etc. It is commonly said that the limit of a three-inch telescope is magnitude 11.5, of a six-inch about 13.0, and of a ten-inch about 14.1. These estimates are undoubtedly on the conservative side and may possibly be extended up to 1.5 magnitudes under excellent conditions. In April 1965, observations of Pluto were made by the author with a ten-inch reflector at Lowell Observatory. The planet was then magnitude 14.1 and was seen with no difficulty whatever; nearby field stars of magnitude 15.3 were glimpsed frequently. Seeing conditions and sky transparency were good, but not exceptionally so.

Similarly, we may question the supposed naked-eye limit of sixth magnitude. Experienced observers often reach seventh, and in laboratory tests artifical stars of magnitude $8\frac{1}{2}$ have been detected against a completely dark

TABLE III

RATIO OF BRIGHTNESS FOR A KNOWN DIFFERENCE IN MAGNITUDE

Mag.Diff.	Ratio	Mag.Diff.	Ratio	Mag.Diff.	Ratio	Mag.Diff.	Ratio
0.1	1.10	1.6	4.37	3.2	19.0	8.5	2511.9
0.2	1.20	1.7	4.79	3.3	20.9	9.0	3981.1
0.25	1.26	1.8	5.25	3.4	23.0	9.5	6309.6
0.3	1.32	1.9	5.75	3.5	25.1	10.0	10,000
0.4	1.45	2.0	6.31	3.6	27.6	10.5	15,849
0.5	1.58	2.1	6.94	3.7	30.1	11.0	25,119
0.6	1.74	2.2	7.57	3.8	33.2	11.5	39,811
0.7	1.91	2.3	8.33	3.9	36.4	12.0	63,096
0.8	2.09	2.4	9.15	4.0	39.8	12.5	100,000
0.9	2.29	2.5	10.00	4.5	63.1	13.0	158,490
1.0	2.51	2.6	11.00	5.0	100.0	13.5	251,190
1.1	2.75	2.7	12.0	5.5	158.5	14.0	398,110
1.2	3.02	2.8	13.2	6.0	251.2	14.5	630,960
1.3	3.31	2.9	14.5	6.5	398.1	15.0	1,000,000
1.4	3.63	3.0	15.85	7.0	630.96	15.5	1,580,000
1.5	3.98	3.1	17.44	7.5	1000.00	16.0	2,511,900
1.6	4.37	3.2	19.02	8.0	1584.90	16.5	3,980,000

Ratios may be multiplied together to obtain figures
not given in the table. For example: For a magnitude
difference of 4.7 mags, we have the result shown here:

$$4.5^m = \text{ratio } 63.1$$
$$0.2^m = \text{ratio } 1.2$$
$$\text{Then } 63.1 \times 1.2 = 75.7$$

APPARENT AND ABSOLUTE MAGNITUDES

background.   The present limit for any existing telescope appears to be about magnitude 23½, reached by the 200-inch reflector at Mount Palomar. This is about two million times fainter than the faintest stars visible to the unaided eye.

The exact magnitude of a star on the standard scale can be accurately determined by modern photoelectric devices, and magnitudes are usually given to the nearest tenth— often to the nearest hundredth. Polaris, for example, has a catalog magnitude of 1.99, meaning that it may be regarded for all practical purposes as a standard second magnitude star. The magnitudes of the stars forming the Great Dipper, beginning at the front of the bowl and proceeding to the end of the handle, are as follows: 1.81, 2.37, 2.44, 3.30, 1.79, 2.40, and 1.87.   In ordinary astronomical conversation, we would refer to all of these as "second magnitude stars," with the exception of the fourth in the list, which is obviously third magnitude.   In the bowl of the Little Dipper we find an interesting opportunity to familiarize ourselves with the appearance of various magnitudes; the four stars are second, third, fourth and fifth magnitudes.

The 20 brightest stars in the sky are still referred to as "first magnitude stars."   However, they are not all equally brilliant, and only one (Spica) has a magnitude of exactly 1.00.   On the present scale, there are 22 stars brighter than magnitude 1.5.

It is evident that a star 2½ times brighter than magnitude 1.00 will have to be assigned a magnitude of 0.00, while one still brighter will actually have a negative or minus value.   Thus Vega is magnitude 0.04; Sirius, the brightest star in the sky, has a magnitude of -1.42; while the planet Jupiter is usually brighter than -2.0, and Venus reaches -4.4 on occasion.

Here it may be well to point out that the term "magnitude," when used alone, is understood to mean "apparent magnitude," the apparent brightness of a celestial object as seen by us.   This has nothing to do with the actual or real luminosity of the object. The "absolute magnitude" of a star is the magnitude that the star would have if it were brought to a standard distance from the Earth, the distance agreed upon being 10 parsecs, or about 32½ light years.   In astronomy, the symbol for apparent magnitude is usually a small letter "m," while for absolute magnitude a capital

TABLE   IV

DISTANCE MODULI & CORRESPONDING DISTANCES IN LIGHT YEARS						
Mod	= Dist	Mod	= Dist	Mod	= Dist	Mod = Dist
0.0	32.6	7.4	984	14.8	29600	22.2	902,000
0.2	35.7	7.6	1080	15.0	32590	22.4	984,000
0.4	39.1	7.8	1180	15.2	35700	22.6	1,079,000
0.6	42.9	8.0	1297	15.4	39100	22.8	1,180,000
0.8	47.1	8.2	1430	15.6	42900	23.0	1,297,000
1.0	51.8	8.4	1560	15.8	47100	23.2	1,430,000
1.2	56.8	8.6	1710	16.0	51650	23.4	1,560,000
1.4	62.2	8.8	1880	16.2	56800	23.6	1,710,000
1.6	68.1	9.0	2056	16.4	62200	23.8	1,880,000
1.8	74.4	9.2	2250	16.6	68100	24.0	2,056,000
2.0	81.8	9.4	2460	16.8	74400	24.2	2,250,000
2.2	90.2	9.6	2700	17.0	81870	24.4	2,460,000
2.4	98.4	9.8	2960	17.2	90200	24.6	2,700,000
2.6	108	10.0	3259	17.4	98400	24.8	2,960,000
2.8	119	10.2	3570	17.6	107900	25.0	3,259,000
3.0	130	10.4	3910	17.8	118000	25.2	3,570,000
3.2	142	10.6	4290	18.0	129700	25.4	3,910,000
3.4	155	10.8	4710	18.2	143000	25.6	4,290,000
3.6	170	11.0	5165	18.4	156000	25.8	4,710,000
3.8	187	11.2	5680	18.6	171000	26.0	5,165,000
4.0	206	11.4	6220	18.8	188000	26.2	5,680,000
4.2	225	11.6	6810	19.0	205600	26.4	6,220,000
4.4	246	11.8	7440	19.2	225000	26.6	6,810,000
4.6	270	12.0	8187	19.4	246000	26.8	7,440,000
4.8	296	12.2	9020	19.6	270000	27.0	8,187,000
5.0	326	12.4	9840	19.8	296000	27.2	9,020,000
5.2	357	12.6	10790	20.0	325900	27.4	9,840,000
5.4	391	12.8	11800	20.2	357000	27.6	10,790,000
5.6	429	13.0	12970	20.4	391000	27.8	11,800,000
5.8	471	13.2	14300	20.6	429000	28.0	12,970,000
6.0	517	13.4	15600	20.8	471000	28.2	14,300,000
6.2	568	13.6	17100	21.0	516500	28.4	15,600,000
6.4	622	13.8	18800	21.2	568000	28.6	17,100,000
6.6	681	14.0	20560	21.4	622000	28.8	18,800,000
6.8	744	14.2	22500	21.6	681000	29.0	20,560,000
7.0	819	14.4	24600	21.8	744000	29.2	22,500,000
7.2	902	14.6	27000	22.0	818700	29.4	24,600,000

TABLE V

ABSOLUTE MAGNITUDES CONVERTED TO ACTUAL LUMINOSITIES IN TERMS OF THE SUN

Abs.Mag.	=	Lum.	Abs.Mag.	=	Lum.	Abs.Mag.	=	Lum.	Abs.Mag.	=	Lum.
+19.0		0.0000021	+9.5		0.013	0.0		83	-9.5		525,000
+18.5		0.0000033	+9.0		0.021	-0.5		132	-10.0		833,000
+18.0		0.0000052	+8.5		0.033	-1.0		209	-10.5		1,320,000
+17.5		0.0000083	+8.0		0.052	-1.5		332	-11.0		2,090,000
+17.0		0.0000132	+7.5		0.083	-2.0		525	-11.5		3,316,000
+16.5		0.0000209	+7.0		0.132	-2.5		833	-12.0		5,250,000
+16.0		0.0000332	+6.5		0.209	-3.0		1320	-12.5		8,330,000
+15.5		0.000052	+6.0		0.331	-3.5		2090	-13.0		13,170,000
+15.0		0.000083	+5.5		0.52	-4.0		3320	-13.5		20,900,000
+14.5		0.00013	+5.0		0.83	-4.5		5250	-14.0		33 million
+14.0		0.00021	+4.5		1.32	-5.0		8330	-14.5		52 "
+13.5		0.00033	+4.0		2.09	-5.5		13200	-15.0		83 "
+13.0		0.00052	+3.5		3.31	-6.0		20900	-15.5		132 "
+12.5		0.00083	+3.0		5.25	-6.5		33200	-16.0		209 "
+12.0		0.0013	+2.5		8.33	-7.0		52500	-16.5		332 "
+11.5		0.0021	+2.0		13.2	-7.5		83300	-17.0		525 "
+11.0		0.0033	+1.5		20.9	-8.0		132,000	-17.5		833 "
+10.5		0.0052	+1.0		33.2	-8.5		209,000	-18.0		1.32 billion
+10.0		0.0083	+0.5		52.5	-9.0		332,000	-18.5		2.09 "

"DISTANCE MODULI"
COLOR INDICES

"M" is used.   The difference between apparent and absolute magnitudes is the quantity called the "distance modulus" or (m-M), which converts directly to an actual distance (Table IV), assuming that none of the light has been lost through absorption by interstellar dust or nebulosity. In many regions of space, a correction for such a light loss must be made; otherwise the derived distance will be over-estimated.

For an example, the star Rigel is magnitude 0.14 and it is about 900 light years away. If brought to a distance of only 32½ light years, its magnitude would rise to -7.1. Thus its apparent magnitude is 0.14, and its absolute magnitude is -7.1.

For another example, our Sun, if removed to the 32½ light year standard distance, would appear as a star of magnitude 4.8.   This then is its absolute magnitude. The figure for Rigel, as we have just seen, is -7.1. The difference is 11.9 magnitudes, which tells us (Table III) that Rigel is about 57,000 times more luminous than the Sun.

This example is given to dramatize the obvious fact that there is a definite relation between the apparent magnitude, absolute magnitude, and distance of an object. If any two of these quantities are known, the third can be determined.   Absolute magnitude can also be converted directly into actual luminosity in terms of the Sun. (Refer to Table V.)

Before ending this brief description of the magnitude system, we must mention one additional point: we have been speaking of "visual magnitudes," star brightness as seen by the normal eye.   Photographic plates and photoelectric instruments "see" the stars differently, and there are thus a variety of magnitude scales—photographic, photoelectric, bolometric, etc.   The differences between the various measurements are chiefly a function of the color or temperature of the star; and thus the various scales are inter-related by quantities known as "color indices."   In its simplest and original meaning, the color index of a star is the difference between its visual magnitude and its photographic magnitude.   The hotter and bluer stars appear brighter on the ordinary photographic plate than they do to the eye; whereas the redder and cooler stars appear fainter.   By agreement, the two scales were set to give a color index of zero for stars of spectral type A0. A negative color index implies a hotter and bluer star, and a positive index indicates a cooler and redder one, as given in the brief table below:

Spectral type	B0	A0	F0	G5	K5	M5
Color index	-0.31	0.0	+0.30	+0.71	+1.11	+1.61

Color indices greater than two magnitudes are rare; the extremes are reached in the cases of the N-type red stars (S Cephei, R Leporis), where the color index exceeds five magnitudes. In reporting a color index, it is always necessary to specify whether it is "plus" or "minus." It is well to adopt the same policy with regard to absolute magnitudes, even though a figure is understood to be "plus" unless it is actually labeled with a minus sign.

Throughout this "Celestial Handbook" all magnitudes given are visual, unless otherwise noted.

STAR NAMES AND DESIGNATIONS. Most of the brighter stars in the sky have their own proper names which have been used since remote antiquity. We have already mentioned such examples as Vega, Aldebaran, Polaris, and Rigel. About thirty such names are in common use. It would be obviously impossible, however, to assign proper names to all of the 6000 naked-eye stars, not to mention the millions of others visible only through the telescope. A more practical system was devised by Bayer in 1603; he assigned each star in a constellation a letter of the Greek alphabet, beginning usually with Alpha for the brightest, Beta for the second brightest, Gamma for the third, and so on. In a few cases, however, as in the Great Dipper, order of position was used instead of order of brightness. The Greek letter was followed by the name of the constellation written in the possessive or "genitive" form. Thus the brightest star in Lyra became "Alpha Lyrae." The second brightest star in Cepheus is designated "Beta Cephei." It also has a proper name, Alfirk, but such names are rarely used for stars below the first magnitude.

Those interested in the lore of star names are referred to R. H. Allen's classic work Star Names and Their Meanings, now available in a paperback reprint from Dover Publications,

STAR NAMES AND DESIGNATIONS
THE GREEK ALPHABET

Inc., in New York.

The entire Greek alphabet is given here for the bene-
fit of beginners who may be unfamiliar with Greek.   To
encourage those who may despair at the thought of learning
such strange symbols, it should be stated here that no
attempt to memorize the list is necessary. Any amateur who
uses star charts and atlases will soon gain a knowledge of
the letters without even trying.

THE GREEK ALPHABET					
α	Alpha	ι	Iota	ρ	Rho
β	Beta	κ	Kappa	σ	Sigma
γ	Gamma	λ	Lambda	τ	Tau
δ	Delta	μ	Mu	υ	Upsilon
ϵ	Epsilon	ν	Nu	ϕ	Phi
ζ	Zeta	ξ	Xi	χ	Chi
η	Eta	o	Omicron	ψ	Psi
θ	Theta	π	Pi	ω	Omega

When gaining familiarity with the letters, remember that
different star maps may show them in slightly different
styles, just as no two persons make their English letters
exactly the same.   After a little experience, this should
confuse no one.

After the supply of Greek letters is exhausted, the
remaining stars of a constellation are given ordinary numbers according to a system devised by Flamsteed. The numbering begins at the western border of a constellation and
proceeds eastward. Thus we have such star designations as
"61 Cygni," "23 Orionis," and "89 Virginis." To find such
stars, we must consult our charts of Cygnus, Orion, and
Virgo and look for the small stars labeled 61, 23, and 89.
Of course, if we have the celestial coordinates (RA and Dec)
of these stars, we can locate them on the chart whether or
not they are identified by number.

The Greek letters and the Flamsteed numbers take care
of most of the brighter stars of a constellation. Fainter

STAR NAMES AND DESIGNATIONS

objects are usually identified by their numbers in one of the standard catalogs used by professional astronomers. Examples are:

HD 20685	BD +14° 613	LFT 1023
GC 12651	CD -39° 14192	Wolf 359
ADS 3247	Groombridge 1830	Ross 451

The "HD," "GC," "BD," and "CD" are comprehensive lists of thousands of stars by position, magnitude and (in the case of the first two) spectral type. These catalogs would be of very little use to the observing amateur and are mentioned here only because such numbers appear often in astronomical literature and appear occasionally in this book. The other five examples are lists of stars of special interest (doubles and stars of large proper motion) identified in each case by the astronomer who compiled the work. Although we may refer to these numbers from time to time in this Handbook, it is again evident that the catalogs themselves would be of very little use to the amateur observer; they are intended mainly as reference works for the professional astronomer.

For double and variable stars, additional systems of nomenclature are in common use. A faint double star may be designated by its number in one of the standard lists such as those of Aitken (ADS), Struve (Σ or $O\Sigma$), Dunlop ($\triangle$), Sir John Herschel (h), S. W. Burnham (β), T. E. Espin (Es), W. J. Hussey (Hu), etc. Hence we find stars marked h4848, Σ2051, $\triangle$49, β328, etc.

A variable star is usually designated by a single or double Roman capital letter, such as R Virginis, T Ceti, VV Cephei, and SX Persei. This system allows 334 variable stars to be designated in any constellation; any further discoveries are designated V335, V336, etc. However, if a bright variable star already has a Greek letter designation, it is retained, and no additional designations are given.

Novae are designated in the same manner as other variable stars, though frequently a bright nova will be referred to merely by the constellation and date—thus "Nova Persei 1901" and "Nova Aquilae 1918." In the standard variable star catalogs these appear as "GK Persei" and "V603 Aquilae." Both designations are in common use, and both are correct.

STAR NAMES AND DESIGNATIONS

Star clusters, nebulae, and galaxies are generally designated by their numbers in the New General Catalogue (NGC) of John Dreyer, published in 1888 or in the older lists of Charles Messier (M) which date from the last half of the 18th Century. Thus we have such numbers as "NGC 6205" or "M13." There is also a supplement to the NGC called the "Index Catalogue" (IC). Another system, devised by Sir William Herschel, is also in use. In his catalog we find such numbers as "HV 24" and "HIII 830." The "H" stands for Herschel, of course. The "V" and "III" are Roman numerals indicating the class of object. On star atlases these numbers are shortened for practical reasons; the "H" is omitted, and the rest of the number is written in a contracted form: "24^5" and "830^3". Herschel numbers are little-used today except for their appearance in the standard Norton's Atlas, and for this reason alone they are included in the lists in the present Handbook.

Herschel's classification system is given here, though the modern observer must regard these classes with caution. They were based upon the visual appearance of the object in Herschel's telescopes and frequently tell us nothing about the true nature of the object. The distinction between true nebulae and external galaxies had not, of course, been made in Herschel's time; and the great majority of his first five classes are now known to be external galaxies. Some of his class IV objects, "planetary nebulae," are in fact external galaxies. The Herschel classes are:

      I. Bright nebulae
     II. Faint nebulae
    III. Very faint nebulae
     IV. Planetary nebulae
      V. Very large nebulae
     VI. Very compressed rich clusters of stars
    VII. Compressed clusters of small and large stars
   VIII. Coarsely scattered clusters of stars

Practically all the brighter clusters, nebulae, and galaxies have NGC numbers, regardless of whatever other designations they may have. Thus NGC 1976 is also known as M42 and has a further popular name—the "Great Nebula in Orion." The Messier (M) list contains 103 of the brighter objects (some

THE SMALL TELESCOPE
STAR ATLASES

modern lists add several more) and is a good reference for the beginner. Messier never had access to a large telescope, and all the "M" objects are fairly conspicuous in small instruments. The complete list, brought up to date with modern identifications, appears in the final "Index and Tables" section of this Handbook.

THE USE AND CARE OF THE SMALL TELESCOPE is a topic not directly covered in this book. The author assumes that the typical user of this Handbook is the owner of a telescope in the 2-inch to 12-inch range, who has learned by direct experience (the best way!) how to use the instrument effectively. For those who wish advice in the matter of purchasing a good instrument, I recommend a new book by Dr. Henry E. Paul, Telescopes for Skygazing. And for those who need a comprehensive guide to observing techniques, covering all such practical matters as light grasp, resolution, vision, seeing effects, choice of oculars, accessories, etc., no better guide exists than the Amateur Astronomer's Handbook by J. B. Sidgwick.
     The vast majority of the objects listed in the Celestial Handbook can be studied in a good 6-inch reflector, probably the best telescope for a serious start in observing. But I would make one recommendation—whatever the size of the telescope, it should be used in conjunction with a good pair of 7x50 binoculars or other comparable low-power wide field instrument. Some observers, in fact, strongly advise that the beginner spend at least a few months exploring with binoculars, getting to know the sky, before graduating to the telescope. There is much to be said for this viewpoint. And, finally, for the enthusiast who wants the incomparable thrill of viewing the heavens through a telescope of his own making, I recommend A. J. Thompson's Making Your Own Telescope.

USING A STAR ATLAS. Earlier in this chapter we mentioned the use of the planisphere or simple star charts for the purpose of learning the constellations. For the location of many of the specific objects of interest within each constellation, however, we require more detailed charts drawn on a larger scale, naming and identifying various objects of interest. A collection of such charts is called a

STAR ATLASES

"star atlas" and is an absolute necessity for the serious observer. Possibly the best star atlas for the average amateur, and even for the beginner, is Norton's, available from most dealers in scientific and astronomical books. Eight double page maps cover the entire celestial sphere, showing all the naked-eye stars, hundreds of star clusters, nebulae, variable stars, and other objects of interest. Some 60 pages of compact information-packed notes precede the charts, making this one of the most useful reference books the amateur can own. Norton's Atlas measures 11" by 8½". The current price of the 16th edition is $12.50.

On an even larger scale is the new Skalnate Pleso Atlas of the Heavens compiled by Antonin Becvar and his associates at the Skalnate Pleso (Rocky Lake) Observatory in Czechoslovakia. This work is the most complete atlas available for all-purpose observing and may be ordered from the Sky Publishing Corporation, Cambridge, Massachusetts. It is a set of 16 charts, each measuring 22" by 16", bound in the form of a large portfolio-size book and covering the entire heavens down to magnitude 7.75. Double stars, variables, clusters, nebulae, and galaxies are all identified by appropriate symbols printed in various colors. Star clusters are shown in yellow, diffuse nebulae in green, the Milky Way in blue, dark nebulae in grey, and galaxies in red. The Skalnate Pleso Atlas has recently been re-issued in a revised edition in a more compact format, and in 1976 was being offered at $14.00. A simpler "field edition" with stars in white on a black background, and on a somewhat reduced scale is also available at $5.00.

The present Celestial Handbook is intended for use with either Norton's or the Skalnate Pleso Atlas, and the serious observer would do well to have both of them.

* * * * * * * *

The preceding sections of this chapter have been focused on the theme "Gaining a Working Knowledge of the Heavens." We now turn to the celestial objects themselves, briefly reviewing some of the standard terms, definitions, and symbols used.

THE STARS are other Suns, located at immense distances, and

THE SPECTROSCOPE

existing in a wide variety of types and sizes. The Sun itself is so "typical" a star that it may be used as a standard by which to compare the others.   Thus, in this book, the masses, diameters, luminosities, and densities of the stars are given in terms of the Sun; for typical examples of this usage, refer to the descriptions of such objects as Eta Cassiopeiae, Xi Bootis, 61 Cygni, etc. The astronomical symbol for the Sun is ⊙ .

THE SPECTRAL CLASSES of the stars are mentioned constantly throughout this book, requiring even the most casual observer to have some understanding of their meaning. The spectroscope is an instrument which analyzes light. The ultimate nature of light may remain one of nature's mysteries; but as a useful analogy we may imagine that a light beam consists of a series of wave-like "ripples" moving outward at tremendous velocity from the light source, much as the ripples on a pond spread out when a stone is dropped into the water. The color of the light is a function of the crest-to-crest distance or "wavelength" of these ripples, calibrated in "angstrom units" of one ten-millionth of a millimeter and identified by the symbol (λ) preceding the number. The longest visible wavelengths (about $\lambda7600$) produce the sensation of red, and the shortest (about $\lambda3900$) produce the sensation of violet. All the other visible colors lie between these two extremes.

The spectroscope analyzes light by sorting out all the wavelengths from longest to shortest and presenting them in systematic order in the form of a long, colored band called a "spectrum" which may be seen visually or be photographed. The most familiar example of a natural spectrum is the rainbow. In this case, the "dispersion" or spreading out of the light is accomplished by water droplets in the atmosphere, each acting as a tiny prism. In the spectroscope, the light is passed first through a fine, narrow slit, generally less than 1/500 inch in width, and then through a prism or series of prisms. In some models, a "diffraction grating" may take the place of a prism. The resulting spectrum may be directed into an eye-lens for actual viewing or projected onto a photographic plate. Professional spectroscopes are often equipped with a calibrated scale so that the wavelength of any spectral feature may be deter-

STELLAR SPECTRAL CLASSES

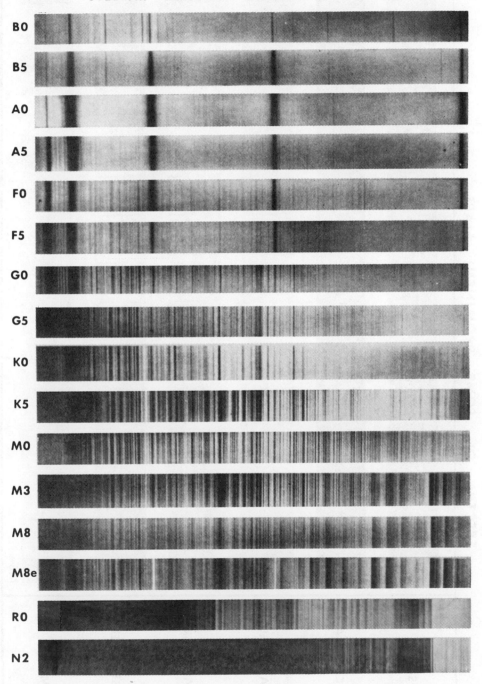

SPECTRA OF SOME PROMINENT STARS

SIRIUS — A1

CAPELLA — G8 + F

ARCTURUS — K2

ALDEBARAN — K5

BETELGEUSE — M2

MIRA — M6e

TYPES OF SPECTRA

mined; in astronomical photography of spectra the calibration is obtained by photographing a "comparison spectrum" from an artificial light source on the same plate.

Since the colors of the spectrum are arranged according to wavelength, they always appear in the same order, with red at one end and violet at the other. Beyond both ends of the visible spectrum are other wavelengths—infrared and ultraviolet—which cannot be seen by the eye, but which may be recorded with special photographic equipment and also with certain photoelectric devices.

There are three basic types of spectra:

1. A glowing solid, liquid, or gas under high pressure emits the full range of all wavelengths, producing a complete band of color. This is a "continuous spectrum." An ordinary electric light shows a spectrum of this type since the light source is an incandescent solid.

2. A glowing gas under low pressure radiates only in certain frequencies, producing an "emission spectrum," a pattern of bright lines at certain definite positions in the spectrum. The reason for this is found in the actual structure of each type of atom; thus the pattern of lines produced by each element or compound is as unique as a fingerprint and positively identifies the type of atom which is emitting the light. Some elements have a very simple pattern consisting of a few lines, while others produce patterns containing many dozens of lines.

3. A rarified gas, when at a lower temperature than the light source, absorbs the same frequencies which it would emit if it were hot and glowing. Thus when light passes through such gas, various wavelengths are absorbed and appear as dark lines or bands in the spectrum. This is called an "absorption spectrum."

A typical star spectrum is of this third type, because the dense glowing mass of the star produces a continuous spectrum, but various wavelengths are absorbed by the thinner gases of the star's atmosphere. The spectrum of our nearest star—the Sun—contains thousands of dark lines, and in this way most of the elements known on Earth have been identified in the Sun. The Sun is not, however, one of the hottest stars or one of the coolest; it is a rather average type called "class G2" by the astronomers. Stars can be arranged in various classes by their spectral

TABLE VI — THE CHIEF SPECTRAL CLASSES

O	Blue-white; high temperatures (35,000°K), large masses, high luminosities; lines of ionized helium, nitrogen, oxygen, in addition to hydrogen. Typical examples = Zeta Puppis, Lambda Orionis, 15 Monocerotis.
B	Blue-white; high luminosities; temperature 20,000°K, large masses. Strong helium lines with greatest intensity at B2, vanishing at A0. Sometimes called "Orion stars". Typical examples = Rigel, Spica, Regulus, Alpha Eridani.
A	White- "Sirian" or "hydrogen" stars; temperature 10,000°K; luminosities average 50 to 100 times Sun. Strong hydrogen lines, helium absent. Examples = Sirius, Vega, Altair.
F	Yellow-white; temperature 7000°K; weaker hydrogen lines, strong lines of calcium with other metallic lines increasing. Examples = Canopus, Procyon, Alpha Persei.
G	Yellow- "Solar type" stars; temperature 6000°K; weaker hydrogen lines, prominent lines of many metals. Examples = the Sun, Capella, Alpha Centauri.
K	Orange- "Arcturian" stars; temperature 4000° to 4700°K; complex spectra with many strong lines of metals, faint hydrogen lines, hydrocarbon bands appear. Examples= Arcturus, Pollux, Alpha Ursa Majoris.
M	Red stars- temperature 2500° to 3000°K; rich spectra showing many strong metallic lines with wide bands produced by titanium oxide. Many M-type variables show bright hydrogen lines, indicated by spectrum "Me". Examples = Antares, Betelgeuse, Mira.
N	Deep red= cool giants of temperature 2500°K; peculiar banded spectra showing carbon compounds; mostly variable stars. Examples = S Cephei, R Leporis, Y Canum Venaticorum.
R	Orange-red; similar to type N, somewhat higher temperature, carbon bands weaker. May form connecting link between classes G and N. Examples = S Camelopardi, RU Virginis.
S	Red; resembles type M, but titanium oxide bands are replaced by zirconium oxide. Complex spectra, usually variable, with hydrogen emission lines. Example = R Cygni.
W	Wolf-Rayet Stars; hot blue giants, high temperatures and luminosities, resemble O-type, but show broad emission features caused by expanding gaseous shell, extremely turbulent atmospheres. Temperature 50,000°K and higher. Example = Gamma Velorum.

STELLAR SPECTRAL CLASSES

characteristics, and this classification system is the central subject of this discussion.

The vast majority of stellar types may be arranged in a logical sequence, each spectral class gradually merging into the next. The chief classes now recognized are identified by the letters O, B, A, F, G, K, and M. Each class contains ten subdivisions numbered from 0 to 9. Thus a "B5" star is approximately midway between B0 and A0. The classes define a temperature sequence—or color sequence—which amounts to the same thing. Stars of type O and B are blue-white; A stars are white; F and G stars are yellowish; K stars are orange, and M stars are red. Three additional classes, R, N, and S, are used for stars which resemble type M but show certain spectral differences, as described in the short summary on the opposite page. Stars of type N are the reddest known.

Prefixes and suffixes are often used to further define the status of a star. Some typical examples are:

dM2   Prefix "d" indicates ordinary dwarf star.
gM5   Prefix "g" indicates giant star.
DA   Prefix capital "D" indicates white dwarf star (degenerate star).
B2e   Suffix "e" indicates emission spectrum, bright lines replacing certain dark absorption lines.
A5p   Suffix "p" indicates spectral peculiarities.

As we study the spectral series from type O to type M, we find a gradual increase in the number and complexity of the spectral lines and bands, the cooler stars showing by far the richest spectra. This does not necessarily imply any fundamental difference in chemical composition, but is largely the effect of temperature. In the cool N-type stars, for example, we find rich banded spectra produced by carbon compounds in the atmosphere of the star; these would be destroyed by significantly higher temperatures. The spectral features depend not only upon the elements present, but also upon the temperature. Thus the interpretation of stellar spectra is a complex study, and many stars display spectral peculiarities which still defy explanation.

The terms "early" and "late" are often used in referring to spectral classes; O and B stars are "early" types,

STAR MOTIONS, TEMPERATURES

and M stars are "late" types. The terms simply refer to the position of the class in the standard sequence of letters O-B-A-F-G-K-M and have no connection with the star's evolution or history.

The "Doppler Effect" is the displacement of the whole pattern of spectral lines due to the motion of the star (or of the observer). If a star is approaching, the lines are shifted toward the violet; if it is receding, they are shifted toward the red. The amount of displacement may be converted to actual velocities in miles per second in the line of sight, a measurement known as the star's "radial velocity." The term "red-shift" has gained great fame in astronomical literature since it is a phenomenon displayed by all the external galaxies, with the exception of the very nearest ones. This "cosmological red-shift," interpreted as a Doppler effect, seems to indicate that the entire Universe is expanding.

The Doppler principle may be used also in detecting and measuring stellar rotation and in detecting orbiting double stars which are too close to be separately seen by any telescope.

STAR MOTIONS. In addition to radial velocity, astronomers deal with several other varieties of star motion. The "proper motion" is the actual displacement of a star over a period of time, compared to the background of extremely distant objects which can be regarded as fixed in position. For example, in comparing photographs made just one year apart, we find that the star Omicron Eridani is changing its position by more than 4" per year. This is a direct indication that the star is relatively close. The greatest known proper motion is shown by Barnard's Star in Ophiuchus, which is moving about 10.29" per year.

Once the distance is known, the proper motion may be converted to actual speed in miles per second across the line of sight; this figure is called the "tangential velocity." Finally, if both the radial and tangential velocities are known, the true "space velocity" is easily computed from simple geometrical principles.

TEMPERATURES in astronomy are usually given on the absolute or "Kelvin" scale; the divisions in degrees are equal to those of the more familiar Centigrade scale, but the starting point is much lower, in fact, at absolute zero. Thus 0°K = -273°C, the temperature at which all molecular motion theoretically ceases. Similarly, 100°C, the boiling point of water, is 373°K. In dealing with very great temperatures such as those found on stars, the two scales may be regarded as virtually identical. In any case, no conversion table is required since the difference is always exactly 273 degrees.

THE H-R DIAGRAM or color-magnitude diagram is a graph upon which stars are plotted by spectral type and actual luminosity. It is named for the two scientists Russell and Hertzsprung who first used it in 1913 in one of the early attempts to arrange the many types of stars into a meaningful system. A typical H-R Diagram is shown on the following page. The vertical coordinate represents absolute magnitude or actual luminosity in terms of the Sun, with the most luminous stars near the top and the faintest near the bottom. The horizontal coordinate represents color, temperature, or spectral type, all of which are naturally interrelated. The hottest stars are at the left, and the coolest toward the right.

The stars plotted on this diagram represent a fairly typical selection of the various stellar types. They include the majority of the stars within a few hundred light years for which the necessary information has been obtained. The most evident fact is that the plotted points are not distributed randomly over the graph, but appear to be restricted to certain definite areas. The main feature is a long band running across the graph from upper left to lower right, demonstrating the existence of a large "family" of stars which range from blue, hot, and bright, down to red, cool, and faint. This band is called the "Main Sequence" and includes all the stars which are operating primarily on the hydrogen-to-helium nuclear reaction. Thus the position of a Main Sequence star depends upon its mass, the more massive stars being naturally hotter and more luminous. Stars in the upper left portion of the graph are thus rather massive objects, in some cases ranging up to 60 or more solar masses. In contrast, any star which falls near the lower right end of the Main Sequence is probably

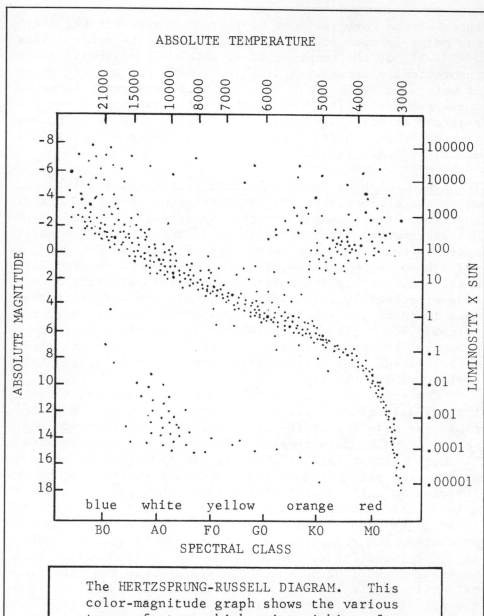

The HERTZSPRUNG-RUSSELL DIAGRAM. This
color-magnitude graph shows the various
types of stars which exist within a few
hundred light years of the Sun. (For
explanation, refer to the text.)

THE H-R DIAGRAM
STELLAR LUMINOSITY CLASSES

an extreme lightweight among stars, containing less than 10 percent the mass of the Sun.

In addition to the normal main sequence stars, we find a few other concentrations of star points at other locations on the H-R Diagram. Across the top is a thin scattering of points representing the extreme supergiants which may attain luminosities of over 50,000 suns. Near the upper right corner are a number of stars which are red and cool, but still have very high luminosities and must therefore have very great actual dimensions. These are the "red giants" which may often be 300 or 400 times the size of our Sun. Finally, in the lower left portion of the graph we find a few peculiar objects which have rather high temperatures but very low luminosities; these are the shrunken "white dwarfs" or "degenerate stars" which have contracted to a state of incredible density, evidently after exhausting their nuclear "fuel" supplies. In addition to these major groups, there are a few stars which fall somewhat above the main sequence and are known as "subgiants," while a few stars falling somewhat below the main sequence are termed "subdwarfs."

Since the spectral type alone does not fully define the status of a star, astronomers often use "MK luminosity classes" (defined by W. Morgan and P. Keenan) in the form of Roman numerals which may be appended to the spectral type. The classes are:

    Ia   Most luminous supergiants
    Ib   Less luminous supergiants
    II   Bright giants
    III   Normal giants
    IV   Subgiants
    V   Main sequence
    VI   Subdwarfs

With this system, the full classification of Betelgeuse would be "M2 Ia," and that of the Sun would be "G2 V." Betelgeuse is an extreme sort of red supergiant, not only vastly larger and more brilliant than our Sun; but also, apparently, operating on a different nuclear energy source. Such differences raise many questions about the life histories of the stars and the evolutionary paths they follow.

DOUBLE STARS

In the article on Betelgeuse (Alpha Orionis) in this Handbook, the general outline of red-giant evolution is presented. For a further discussion of the use of H-R diagrams in the study of stellar evolution, the reader is referred to the article on the star cluster M13 in Hercules.

DOUBLE STARS appear to the unaided eye as single stars, except in the case of unusually wide pairs, but through the telescope may be "split" or resolved into two or more components. If the two stars are known to be actually near each other in space and bound together by gravitational attraction, they are called "physical" doubles. When actual orbital motion has been detected, these are termed "binaries." On the other hand, some apparent doubles are known to be due to the chance alignment of two totally unrelated stars, one being far beyond the other; these are called "optical" doubles.

   The apparent separation of a double at any time is measured in seconds of arc. The apparent orientation of the pair is defined by the "position angle" (PA), always measured from the bright star toward the fainter. Due north is a PA of 0°; due east is 90°, etc. In the case of a binary system, the motion of the companion is said to be direct when the PA is increasing and retrograde when it is decreasing. Periastron is the point in the true orbit where the actual separation between the stars is at a minimum. The semi-major axis is one-half the distance across the longest dimension of the true orbit; it may also be defined as the mean separation of the two stars. It is usually given in seconds of arc, but may be converted directly into an actual separation in miles or astronomical units (Table II) once the distance of the star is known. In the binary stars we are given our only opportunity of the direct determination of stellar masses. By Newton's laws, there is a direct relationship between the separation, period, and the total mass of a binary pair; when any two of these quantities are known, the third may be computed.

   In the case of spectroscopic binaries, the duplicity is revealed by the spectroscope, although the two stars may be too close to be resolved by any telescope. The term "astrometric binary" designates a system where an unseen companion is detected through periodic deviations in the

STAR DISTANCES

motion of a visible star.

STAR DISTANCES may be determined by direct trigonometrical
measurements only in the case of fairly near stars. As the
Earth revolves around the Sun, the nearer stars show a small
yearly shift against the background of more distant stars.
The radius of the Earth's orbit (one astronomical unit) is
the base-line used in measuring the amount of this shift,
termed the star's "parallax."   A distance unit often used
by professional astronomers is the "parsec," the distance
of an object having a parallax of 1" or, to phrase it
another way, the distance at which one AU would subtend an
angle of 1".   The parsec is equivalent to 3.26 light years.
It is a convenient unit to use, since the distance in parsecs is merely the reciprocal of the parallax.   However,
the light year seems firmly entrenched in popular astronomical literature and is used throughout this book. A short
conversion table for the standard distance units is given
below:

1 astronomical unit (AU)	= 93 million miles
1 light year	= 63,240 AU
1 light year	= 5.88 trillion miles
1 light year	= 0.307 parsecs
1 parsec	= 206,000 AU
1 parsec	= 19.17 trillion miles
1 parsec	= 3.26 light years
1 kiloparsec	= 1000 parsecs
1 megaparsec	= 1 million parsecs

There is, of course, no star as near as one parsec. The
nearest star, the triple system of Alpha Centauri, is 1.33
parsecs or 4.3 light years distant and shows a parallax of
about 0.75". Distances less than about 30 light years may
be determined with good accuracy by the direct parallax
method and with fair reliability out to about 100 light
years. At 300 light years, the probable error of the measurement nearly equals the size of the parallax; and at
greater distances the method becomes virtually useless.
More indirect methods must then be used.   The distance of
very remote objects may be determined if, in some way,
their actual luminosities can be learned.   The comparison

VARIABLE STARS

of the apparent and absolute magnitudes (the distance
modulus) then gives the distance. This is the "spectroscopic parallax" method, in which the intrinsic luminosity of the star is determined from various spectroscopic
features. For extremely remote objects, various stars of
known high luminosity may be used as distance indicators,
supergiant stars, novae, etc. Finally, there are the
remarkable pulsating variable stars called "cepheids" whose
periods are proportional to their luminosities; these are
often called the "measuring sticks of the Universe." The
principle of distance calculation by the cepheid rule seems
direct and straightforward, but in practice matters are
complicated by the presence of light-absorbing material in
interstellar space. Objects thus seem fainter than they
are, and distances are overestimated. This effect must be
allowed for in any attempt to measure distances based upon
apparent and absolute magnitudes. It is not surprising that
the distances of many celestial objects are very roughly
known and that modern catalogs still contain numerous discrepancies. (For a summary of the cepheid method of distance determination, refer to the article on Delta Cephei.)

VARIABLE STARS are stars which change in brightness, either
periodically, irregularly, or explosively. The simplest
way to illustrate the typical behavior of any specific example is to plot the magnitude (vertical scale) against the
time (horizontal scale) on a graph which is then known as a
"light curve." Numerous examples will be found throughout
this book. Known periods of variable stars range from a
few hours up to a number of years. The light variations
range from a tiny fraction of a magnitude up to 9 or 10
magnitudes in some extreme cases. The novae, of course, may
exceed even these limits.
     There are many different types of variable stars, but
it is possible to group them into three broad classes:
(1) Pulsating variables, (2) Eruptive or cataclysmic variables, (3) Eclipsing variables. The various sub-types will
now be briefly reviewed.

(1) PULSATING VARIABLES are stars which appear to be periodically oscillating, expanding and contracting. The chief
types are:

TYPES OF VARIABLE STARS

Cepheids—very luminous giants, periods from one to fifty days or so, but usually about one week. Light range up to two magnitudes. Spectral types F, G, and K. These stars show clockwork precision in their pulsations and also display the noted "period-luminosity relation" which makes them valuable as distance indicators. Typical examples are Delta Cephei, Eta Aquilae, RT Aurigae.

Long-period Variables—pulsating red giant stars with periods from about 75 days up to two years or more. The average light range is five or six magnitudes. Spectral class usually M, sometimes R, N, or S. Typical examples are Omicron Ceti, Chi Cygni, R Leonis, R Hydrae. Periods do not show the absolute precision of the cepheids, but may vary by a number of days.(indicated by abbreviation LPV in the lists in this book).

Semi-regular Variables—mostly red giant stars with periods often poorly defined; subject to unpredictable variations. Typical examples are Alpha Herculis, W Cygni, and Rho Persei.

Irregular Variables—giants of various spectral types with no definite period. Betelgeuse and Mu Cephei are placed in this class by some authorities, but classed with the semi-regular types by others.

Cluster Variables or "RR Lyrae" Stars—precise periods generally under one day, range about one magnitude, spectral type A or F, light curve of the "cepheid" type with rapid rise and slower decline (identified "Cl.Var." in lists in this book). Typical example—RR Lyrae.

RV Tauri Stars—pulsating giants with multiple light curves, alternate maxima and minima in a period of 30 to 150 days, superimposed on a longer wave of three or four years. Total range may be about three magnitudes, spectral types G or K. Typical examples—RV Tauri, R Scuti.

Beta Canis Majoris Stars—brilliant stars of types B1-B3, magnitude range very slight, periodic oscillations of spectral lines in period of about 0.2 day. Typical examples are Beta Cephei and Beta Canis Majoris.

Dwarf Cepheids—resemble the RR Lyrae stars but show smaller amplitudes and shorter periods. Spectral types A and F, period less than 0.25 day. Typical examples — CY Aquarii, SX Phoenicis. Another sub-class, Delta Scuti stars, have very slight light changes and periods of under 0.2 day.

TYPES OF VARIABLE STARS

(2) <u>ERUPTIVE VARIABLES</u> consist chiefly of the novae and the nova-like stars. The major classes are:

<u>Novae</u> are hot subdwarfs which brighten explosively by 7 to 15 magnitudes or so in a period of a few days, thereafter fading back to normal in a few years. Typical examples are V603 Aquilae (1918), GK Persei (1901) and DQ Herculis (1934).

<u>Recurrent Novae</u> are those which have shown two or more outbursts. They differ also from the standard novae in showing smaller amplitudes, shorter maxima, and a more rapid return to normal brightness. Typical examples are T Corona Borealis, RS Ophiuchi, and WZ Sagittae.

<u>Supernovae</u> are exploding stars which brighten by 20 or more magnitudes, attaining a brilliance of several hundred million suns. These are thought to be massive stars which do not "return to normal," being largely destroyed in the explosion. Typical examples are: the Nova of 1572 in Cassiopeia and the Nova of 1604 in Ophiuchus.

<u>Dwarf Novae</u> or "SS Cygni Type Stars" are hot dwarfs which show sudden outbursts of up to five magnitudes, returning to normal faintness in a week or so, but repeating the phenomenon again and again at intervals of a few months. Typical examples are U Geminorum and SS Cygni. A small subclass, typified by Z Camelopardi, act in a similar manner, but occasionally show long periods of constant light at some intermediate magnitude between maximum and minimum.

<u>Flare Stars</u> are faint red dwarfs which show extremely sudden outbursts of up to several magnitudes in a time of one or two minutes. Typical examples are: UV Ceti, DO Cephei, and Alpha Centauri C.

<u>R Corona Borealis Stars</u> have light curves resembling "reverse novae." The star remains normally bright, but may fade by eight magnitudes or so at unpredictable intervals, returning to normal in a period of many months. These stars are giants of various spectral types. The standard star of the class is R Corona Borealis.

<u>Nova-like Stars</u> form an uncertain group of irregular variables with erratic behavior, some with composite spectra, as R Aquarii, Z Andromedae, and BF Cygni. Some of these are the so-called "symbiotic stars" in which the spectral features of a red giant and blue dwarf both appear.

TYPES OF VARIABLE STARS
CLASSIFICATION OF NEBULAE AND GALAXIES

(3) ECLIPSING VARIABLES are binary systems in which the two stars occult each other periodically as they revolve in their orbits.

Algol Systems are relatively widely separated so that the light curve remains fairly flat between the large dips representing the eclipses. Typical examples: Beta Persei, U Cephei, U Sagittae.

Lyrid Systems, or Beta Lyrae Stars, are usually giant systems revolving in close proximity, both stars being distorted into ellipsoids by tidal effects and rapid rotation. The light curve is a continuously varying sinusoidal wave with alternate maxima and double minima. Typical examples: Beta Lyrae, 68 Herculis.

Dwarf Eclipsing Systems, or "W Ursa Majoris Stars," are rapidly rotating dwarf binaries with the components nearly in contact; the periods are less than one day. Typical examples: W Ursa Majoris, U Pegasi.

Ellipsoidal Variables are binaries which do not eclipse, but vary in light as they revolve due to the changing amount of luminous surface seen from the Earth. Typical examples: Zeta Andromedae, b^1 Persei.

In addition to the chief classes, some authorities would add a fourth major group—nebular variables—whose changes may be due in some way to surrounding gas and dust clouds. Typical examples: T Tauri, R Monocerotis, RW Aurigae, and T Orionis. There are other stars which cannot conveniently be classed in any of the major groups.

NEBULAE are the clouds of rarified gas and dust found in space, often involving whole groups of stars and distributed chiefly along the spiral arms of the Galaxy. The two chief types—diffuse nebulae and planetary nebulae—were introduced in Chapter 2, and little else concerning them need be said here. For a summary of planetary nebulae, refer to M57 in Lyra; for information on some of the better known diffuse nebulae, refer to M42 in Orion, M8 in Sagittarius, M1 in Taurus, etc.

GALAXIES are the other "Island Universes" beyond our own Milky Way System. They are classified in three main groups—spirals, ellipticals, and irregulars. As with many

CLASSIFICATION OF GALAXIES

classificational systems, there is a certain degree of over-lapping between classes, and more elaborate systems have been devised. Since no two galaxies are exactly alike, it would be possible to refine such systems indefinitely and invent more and more detailed subclasses. For most purposes the original classificational system is quite adequate and is used throughout this book. For researchers specializing in more technical studies of the galaxies, de Vaucouleurs' Reference Catalog of Bright Galaxies is now a standard source of data (University of Texas Press, 1964).

The three chief classes of galaxies are:

(1) Spirals—divided into normal spirals (S) and barred spirals (SB). In a normal spiral the arms curve out directly from the rounded nuclear mass; in a barred spiral they begin at opposite ends of a flattened oblong central bar.

To the primary capital letters "S" or "SB" may be appended three smaller letters, a, b, or c, which further describe the general structure of the spiral. In an "a" system the nuclear bulge dominates the galaxy, and the spiral arms are narrow and tightly wound. In a "b" system the nuclear mass and the spiral arms are about equally prominent. In a "c" system the nucleus is small, and the spiral arms are widely opened and well resolved into clumps and clouds of stars. The Andromeda Galaxy is an Sb spiral, for example, and the great M101 in Ursa Major is a typical type Sc. Our own Galaxy, difficult to study from our position within, is usually considered to be type Sb, but some recent studies suggest a type closer to Sc.

A fourth subclass, Sd, is sometimes used to designate systems in which the nucleus is reduced to a tiny condensation of nearly stellar appearance, and the spiral pattern may be nearly lost in a chaotic mass of star clouds. A typical example is NGC 7793 in Sculptor.

The terms "early" and "late" are sometimes used in referring to the sequence Sa, Sb, Sc; for example, an Sa spiral is said to be "earlier" than type Sb. The use of these terms has no connection with the life history of a galaxy or the direction of its evolutionary development. Indeed, it appears from present evidence that the Sd and Sc stages are actually the earliest in the life of a spiral and that the Sa stage is the last.

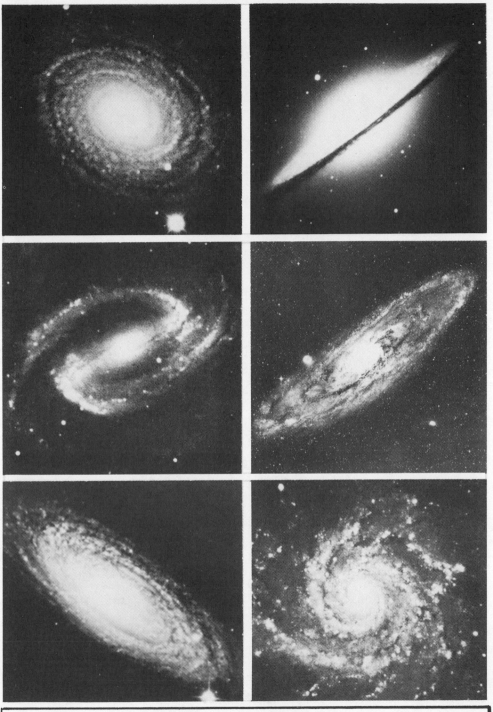

A VARIETY OF GALAXIES

NGC 488, Pisc = Sb	NGC 4594, Virg = Sa/Sb
NGC 1300, Erid = SBb	NGC 224, Andr = Sb
NGC 2841, UMaj = Sb	NGC 628, Pisc = Sc

A VARIETY OF GALAXIES

NGC 4486, Virg = E0 p	NGC 205, Andr = E6
NGC 1365, Forn = SBb	NGC 5457, UMaj = Sc
NGC 1156, Arie = Irr	NGC 5128, Cent = E0/S0 p

CLASSIFICATION OF GALAXIES

(2) Elliptical Galaxies display no spiral pattern or other structure, but are simply spheroidal or elliptical swarms of stars, resembling super-size globular clusters. Usually there is a brighter central condensation. These galaxies are classed by their apparent degree of oblateness from E0 (perfectly round) to E7 (elongated lens-shaped systems).

The transitional type "S0" defines a galaxy which resembles an elliptical type, but in which a flattened central plane or disc may just be detected. If any true spiral pattern is discernible, the galaxy is classed as an "Sa".

(3) Irregular Galaxies, identified by the letter "I", are systems which appear as coarse masses of star clouds, lacking any symmetry or smooth distribution of stars. These galaxies, may, however, grade imperceptibly into the coarser spirals, as one of the typical examples - the Large Magellanic Cloud in Dorado - appears to show an incipient spiral pattern.

In all classes of galaxies, a small letter "p" may be used to indicate various peculiarities such as outer rings or haloes, distorted structure, tidal filaments, and other unusual features. Some of the most interesting cases of this type (as M82 in Ursa Major, M87 in Virgo, and NGC 4038 in Corvus) receive special attention in this Handbook.

THE HUBBLE CONSTANT is the relationship between the red-shift of a galaxy and its distance, a relationship of great importance since it defines not only the scale, but also the **age**, of the Universe. This relationship, which appears to be linear, has been drastically revised since the time of Hubble, and is now (1975) thought to be about 53 km/sec per megaparsec, or roughly 10 miles per second for each million light years of distance.

QUASARS or "Quasi-stellar Radio Sources" are strange and enigmatic objects which have only recently come to the attention of astronomers. Stellar in appearance, they show enormous red shifts which suggest that they are the most distant objects yet identified in the Universe. If the Hubble relationship is assumed to hold true for these objects, their actual luminosities must be higher than any

COSMOLOGY

other type of object known, exceeding even the largest galaxies, and the source of this huge energy output has been a subject for much current speculation. All known quasars seem to be extremely distant objects, which of course means that we are seeing them as they were, far back in the remote past. There is a growing belief, therefore, that the quasar phenomenon represents a stage in the early history of a galaxy. Only one of these strange objects can be considered a subject for the typical amateur telescope, 3C 273 in Virgo, which appears as a star of about the 12th magnitude.

COSMOLOGY is that branch of astronomy which sums up and includes everything else; it deals with the over-all features of the Universe, the structure and distribution in space of the celestial bodies, the great questions of matter, energy, space, time, relativity, cosmic evolution, the mysteries of beginnings and endings, and ultimate destinies. For the last half century, the central theme of cosmology has been the concept of the Expanding Universe, which is still accepted (1976) as the only feasible explanation of the "red-shift". If all the galaxies are receding from each other, then in the past they were much closer together than we now find them, and some 16 or 18 billion years ago must have been very close indeed. The present Universe seems to have had its beginning in a tremendously dense and extremely hot "primordial atom" or "cosmic fireball" at that very distant time. This primeval epoch, however, can be called the "time of creation" only in a very limited sense, since the primordial fireball was obviously preceded by some condition which existed earlier, presumably the gravitational collapse of the universe which existed before the present one. The "Big Bang" hypothesis is thus not necessarily in conflict with the concept of some sort of "steady state" universe, or at least a cyclic universe which undergoes alternate periods of expansion and contraction.

Although lacking final confirmation from astronomical evidence, the oscillating universe seems philosophically more appealing than a "linear" or one-way universe which ends with dead galaxies of burned-out stars expanding forever into limbo. Also, of course, it does not present the knotty and probably artificial problem of explaining an original "creation". The invention of one or more deities does not furnish any real solution; the existence of such beings would constitute but another mystery which must also be explained. Oriental philosophers speak of the *"Tao"*, the all-pervading intelligence of the Universe, never personified or regarded as a "being" of any sort; such a concept seems vastly more appropriate to the Universe we actually live in than do the grossly anthropomorphic and marvelously tortuous theologies of Western thinkers. *"The Tao that can be expressed in words"*, begins the Book of Lao Tzu, *"is not the absolute Tao. The names that can be given are not absolute names."* But...*"from the days of old until now, its manifested forms have never ceased....."*

To search for an object the size of the Earth in our Galaxy is comparable to searching for an object smaller than a dust grain, lost somewhere in North America. It is in the last degree unlikely that the acquired knowledge and ideas developed by the inhabitants of this cosmic speck should constitute the final word on the subject. Every man of honesty recognizes that we are in the earliest stage of knowledge of the Universe; we have barely begun to learn how to learn. The uncritical acceptance of any dogmatic philosophy at this stage constitutes intellectual suicide, since it closes the mind to any new evidence, or to any new vision of the world. *"If we are open only to those discoveries which will accord with what we already know"*, said the English-American philosopher Alan Watts, *"we might as well stay shut."* The study of astronomy has opened our eyes to a Universe whose unimaginably vast extent, both in space and time, surpasses anything of which Man could have dreamed. And yet, the celestial horizons continue to widen, and our present knowledge is just the beginning.....

"Only that day dawns to which we are awake", wrote Henry David Thoreau in the final paragraph of **Walden**. *"There is more day to dawn. The Sun is but a morning star."*

The arrangement of the Celestial Handbook is alphabetical, by constellation. Each constellation is divided into four subject sections, as follows:

LIST OF DOUBLE AND MULTIPLE STARS
LIST OF VARIABLE STARS
LIST OF STAR CLUSTERS, NEBULAE, AND GALAXIES
DESCRIPTIVE NOTES

The lists are arranged in the following manner:

LIST OF DOUBLE AND MULTIPLE STARS

<u>Name-</u> The most commonly used designation of the star. Greek letters and Flamsteed numbers are given preference. Fainter doubles are identified by their numbers in such catalogs as those of:

F.G.W.Struve	= Σ	T.E.Espin	= Es
O.Struve	= OΣ	W.J.Hussey	= Hu
J.Herschel	= h	R.T.Innes	= I
J.Dunlop	= $\triangle$	T.J.J.See	= λ
S.W.Burnham	= β	C.Rumker	= Rmk
R.G.Aitken	= A	G.W.Hough	= Ho
Alvan Clark	= AC	W.H.van den Bos	= B
R.A.Rossiter	= Rst	E.S.Holden	= Hn, Hld
F.Argelander	= Arg	G.P.Kuiper	= Kui
H.A.Howe	= Hwe	J.South	= S
Cordoba Obsvt	= Cor		

<u>Dist-</u> The angular separation of the two stars in seconds of arc. The distances of third or fourth components are given from the primary (A) star unless otherwise noted.

<u>PA-</u> The position angle of the pair in degrees in the usual sense, measured from the brighter to the fainter component.

<u>Yr-</u> The year in which the preceding measurements were made. The last two digits only are given; the first two are understood to be "19". The few measurements made before 1900 are identified by the entry "00".

<u>Mags-</u> The visual magnitudes of the two stars on the standard scale, to the nearest half magnitude.

TERMS, SYMBOLS, AND ABBREVIATIONS USED

Notes- Various information of interest. Abbreviations
and terms are:
relfix= relatively fixed pair; no definite change
in separation or angle in at least 50 years.
PA inc = the position angle is increasing.
Dist dec = the apparent separation is decreasing.
Spect = spectral type; a single entry refers to
the primary star, two entries to the A & B
pair unless otherwise noted.
cpm = common proper motion; the two stars are
moving through space together.
optical = the two stars are not a physical pair.
A,B,C -- These letters are used to identify the
components of multiple systems, in the order
in which they appear in the list. Remarks in
the "Notes" column always refer to the A-B
pair, unless one of the other components is
specifically mentioned by letter "C", "D",
etc.
(*) This sign indicates a more detailed description, following the catalog lists.
RA & DEC - The celestial coordinates (1950 epoch). They
are given in a contracted form, as in the two
following examples:
22115s2119 = RA 22h 11.5m; Dec -21° 19'.
06078n4844 = RA 6h 07.8m; Dec +48° 44'.

LIST OF VARIABLE STARS
An attempt has been made to list all known variables which reach 9.5 magnitude or brighter at
maximum.
Name- The standard designation of the star. Stars with
Greek letter designations are listed first, other
stars in the usual order: R,S,T....Z, RS, RT, etc.
MagVar- The visual magnitudes of the star at maximum and
minimum. Only approximate mean values can be given
for many long-period variables which do not repeat
their cycles exactly.
Per- The period in days. For long-period variables and
other pulsating stars this is the interval between
maxima. In the case of eclipsing variables it is
the interval between minima. Irregular variables

TERMS, SYMBOLS, AND ABBREVIATIONS USED

are identified by the abbreviation "Irr". In the case of periods of less than one day, the entry begins with a decimal point; the "0" being omitted to save space.

Notes - Various information of interest, usually beginning with the class of variable. Some abbreviations used are:

LPV. = Long period variable.
Cl.Var. = cluster variable (RR Lyrae type star)
Ecl.Bin. = eclipsing binary
Semi-reg = Semi-regular variable
Spect = the spectral type
(*) This sign indicates a more detailed description following the catalog lists.

RA & DEC - The celestial coordinates for 1950, as previously described.

LIST OF STAR CLUSTERS, NEBULAE, AND GALAXIES

NGC = The standard number from the New General Catalogue of John Dreyer.

OTH = Other designations, as in the following examples:

M35 - The number in the catalog of Charles Messier.
33^7 The number assigned by Sir William Herschel.
I.405- The number in the "Index Catalogue", a supplement to the NGC.
△309 - The number in Dunlop's list.

TYPE -The class of object. The symbols are:

Galactic star cluster

Globular star cluster

Diffuse nebula

Planetary nebula

Dark Nebula

Galaxy

SUMMARY DESCRIPTION- The visual appearance and chief facts about the object. In the case of galaxies, the first line gives the type, visual magnitude, and the apparent size. For the majority of other objects the

TERMS, SYMBOLS, AND ABBREVIATIONS USED

apparent diameter ("diam") and magnitude ("mag") are given. The remainder of the description employs a simplified code, based on the system used in the NGC:

B =	Bright	1C =	little compressed
b =	brighter	Ri =	rich
L =	Large	M =	middle
pS =	pretty small	N =	nucleus
F =	Faint	mag =	magnitude
vF =	very faint	diam =	diameter
R =	round	g =	gradually
C =	compressed or	s =	suddenly
	condensed	m =	much
S =	small	v =	very
rrr=	well resolved	14^m =	14th magnitude
E =	elongated	st =	star or stars
e =	extremely	irr =	irregular
c =	considerably	neby =	nebulosity
P =	poor	9....	9th mag and fainter
np =	north preceding	nf =	north following
sp =	south preceding	sf =	south following

Class C ---- G. Class numbers for galactic clusters, defining richness and degree of concentration:
C = loose and irregular clusters
D = loose clusters
E = moderately concentrated
F = fairly well compressed, compact clusters
G = very rich compact clusters

Class I ---- XII. Class numbers for globular clusters, defining the degree of concentration:
I = extremely rich and highly compressed
XII = very loose, sparse clusters

(*) This sign indicates a more detailed description or a photograph, or both, following the catalog lists.

RA & DEC - The celestial coordinates for 1950, as previously described.

The standard type symbols for galaxies (Sa, Sb, Sc, E,) etc., are defined on page 92.

TERMS, SYMBOLS, AND ABBREVIATIONS USED

DESCRIPTIVE NOTES

Detailed descriptions for the following objects:

1. All stars in the constellation brighter than magnitude 3.50.
2. All objects in the lists of doubles, variables, clusters, nebulae, and galaxies which were identified by the sign (*)
3. Other objects of special interest which fall into neither of the above groups; for example - 3C273, Barnard's Star, Van Maanen's Star, etc.

Objects given detailed notes are listed in the following order:

1. Stars with Greek letter designations, in alphabetical order. (As Alpha Andromedae)
2. Stars with Flamsteed numbers. (As 36 Andromedae)
3. Stars with standard double-star designations (as Σ 215)
4. Stars with standard variable star designations (as R Andromedae)
5. Stars with miscellaneous designations. (As Wolf 359, or Groombridge 34)
6. Star clusters, nebulae or galaxies with Messier numbers. (As M31)   In numerical order.
7. Star clusters, nebulae or galaxies with NGC numbers. (As NGC 891)   In numerical order.
8. Star clusters, nebulae or galaxies with other designations. (As IC 405)

FINDER CHARTS are provided for the majority of variable stars chosen for a detailed description. These were made directly from plates obtained with the 13-inch telescope at Lowell Observatory, and show stars to about 15th magnitude. The large circle on each chart represents a field of one-degree diameter except when specifically labeled otherwise. North is always at the top. In using these, or any other photographic charts, the observer must remember that the relative brightnesses of stars may differ somewhat from the visual appearance; red stars appearing brighter to the eye than they do on the print, and blue stars just the opposite.

ANDROMEDA

LIST OF DOUBLE AND MULTIPLE STARS

NAME	DIST	PA	YR	MAGS	NOTES	RA& DEC
Ho622	23.6	87	02	7 - 12	spect A0	00002n3532
O Σ514	5.3	168	58	7 - 9½	relfix, spect A2	00020n4149
Σ3056	0.6	144	67	8 - 8	PA dec, spect K0	00021n3359
	23.1	360	26	- 9	$9\frac{1}{2}^m$ star at 96"	
β997	4.0	338	59	8 - 9	relfix, spect F8	00024n4524
Es1293a	12.9	120	25	6½-13½	spect A0	00026n4457
	21.6	235	34	- 9½		
O Σ547	5.7	166	62	8½- 8½	AB binary, about	00028n4532
	84.1	183	25	- 11	360 yrs; PA inc, spect dK2, dM0	
Σ3064	24.8	3	55	7 - 10	optical, PA & dist inc; spect G5	00050n3952
β1014	1.5	335	55	7 -12½	relfix, spect G5	00050n3123
Σ 1	9.8	288	34	8½- 10	no change, spect A5	00063n3656
β483	2.8	43	32	7 -11½	cpm; spect G5; 8^m star at 144"	00064n4034
Σ 3	5.0	83	62	7½- 8½	relfix, spect A3	00074n4607
O Σ2	0.1	332	62	7 - 8	all cpm; AB binary	00108n2643
	17.8	225	44	- 9½	about 700 yrs, PA dec; spect F5	
A1256	0.2	42	57	7 - 7½	PA inc; spect B9	00127n4356
	18.8	344	45	- 13		
h1947	9.0	76	58	6½- 9	relfix; cpm; spect A0	00137n4319
Σ 17	26.9	29	29	8 - 9	spect K0	00139n2901
Σ 17b	2.5	264	15	9 - 12	(β487) relfix	
OΣ 4	0.5	196	61	8 - 9	binary, 112 yrs; PA dec, spect F7	00141n3613
Σ 19	2.3	139	62	7 - 9½	slight PA inc; spect A0	00141n3620
Grb34	39.1	56	18	8 -10½	cpm; red dwarf binary system (*)	00155n4344
Σ 24	5.2	249	62	7 - 8	cpm, relfix, spect A2	00159n2552
26	6.2	238	58	6½- 10	relfix, cpm, spect B9	00160n4331
S384	76.4	20	26	7 - 9½	optical, dist inc, spect G0	00174n3757
A647	0.5	171	58	7 - 9	PA dec, spect F5	00179n4514

LIST OF DOUBLE AND MULTIPLE STARS (Cont'd)

NAME	DIST	PA	YR	MAGS	NOTES	RA & DEC
AC 1	1.7	288	62	7 - 7½	binary, PA & dist slow inc, spect F5	00183n3240
28	2.4	6	48	5½- 13	(β1095) PA slow inc, cpm, spect F0	00275n2929
Σ 33	2.6	210	62	8 - 8	relfix, spect F5	00283n3349
β1310	3.7	212	58	7 - 12	AB cpm, relfix,	00300n2255
	17.7	294	25	- 13	AC dist inc, spect	
	95.2	147	26	- 9	G0	
OΣ14	8.6	160	55	6½- 10	relfix, spect K0	00302n2800
	60.0	84	12	- 12		
Σ40	11.9	312	38	7 - 9	relfix, cpm;	00325n3633
	24.9	123	14	-12½	spect K0	
Ho623	8.3	152	22	7 - 12	spect G5	00335n2345
π	36.1	174	37	4½- 9	AB cpm; star C is	00342n3327
	55.5	357	24	-11½	optical (*)	
Σ 44	11.1	271	41	8½- 9	dist inc, spect K, PA inc.	00357n4043
δ	28.7	298	34	3 -12½	(β491) cpm (*)	00366n3035
Σ47	16.6	205	34	7 - 8½	AB cpm, spect A3	00377n2347
	43.6	244	12	-10½		
Σ 52	1.3	8	62	8 - 9	PA dec, spect F5	00414n4558
Σ 55	2.3	330	62	8 - 9	dist inc, spect A3	00417n3321
Σ I 1	46.8	50	33	7 - 7	optical, spectra K1, K2	00437n3040
Σ 72	23.9	178	25	8 - 9	slight PA dec, spect M	00519n3854
36	0.5	208	68	6 - 6½	binary (Σ73) (*)	00523n2321
A1511	1.3	36	29	7 -11½	spect A2	00527n4007
μ	34.2	126	25	4 - 11	(h1057) dist dec,	00540n3814
	42.2	307	25	- 13	both optical	
Es155	6.2	69	29	8½- 9½		00567n3730
Σ79	7.8	193	58	6 - 7	cpm, relfix, both spectra B9	00572n4427
h2010	10.0	270	03	8 - 9½	spect A0	00599n4726
39	20.0	5	25	6 - 12	(h1064) spect A7	01001n4105
OΣ21	0.7	173	67	7 - 8	binary, 120 yrs; dist inc, spect A3	01001n4706
Ho213	0.2	259	66	7 - 7	PA inc, spect A3	01012n3512
A1516	0.2	83	54	8 - 8	binary, 33½ yrs, PA inc, spect F8	01043n3823

LIST OF DOUBLE AND MULTIPLE STARS (Cont'd)

NAME	DIST	PA	YR	MAGS	NOTES	RA& DEC
β397	8.7	142	55	$7\frac{1}{2}$- 10	relfix, spect K2	01049n4635
	16.6	64	00	- 13		
AC 13	0.5	257	57	8 - 8	(h2018) spect A0	01060n4457
	15.7	2	37	- 11		
ϕ	0.5	154	61	5 - $6\frac{1}{2}$	binary, about 370 yrs, PA dec, spect B7	01066n4659
Ho214	2.9	246	44	8 - 12	relfix, spect F0	01069n3752
A655	0.2	277	62	$8\frac{1}{2}$- 9	binary, 108 yrs; PA inc, spect G5	01084n4057
β398	1.8	47	62	9 - 9	relfix; spect A2	01089n4732
Σ102	0.5	288	60	7 - 8	PA dec, spect A0	01148n4845
	10.0	224	34	- $8\frac{1}{2}$	ABC cpm	
	26.8	62	34	- 11		
OΣ29	21.2	264	25	7 - 11	slight dist inc; spect G5	01160n3942
Σ108	6.0	62	58	7 - 10	relfix, spect A3	01160n3707
Σ112	21.5	334	28	$8\frac{1}{2}$- 9	PA slow inc,	01178n4605
Σ112^b	4.7	187	13	- 14	primary spect G0	
Ho 7	15.3	165	59	6 - 13	dist inc, spect A3	01245n4050
ω	1.9	122	62	5 - 12	(48)(β999) spect F4, PA inc (*)	01246n4509
AC 14	0.8	95	60	8 - 9	relfix, spect G5	01253n4231
Σ133	3.1	186	45	7 -$10\frac{1}{2}$	AB no certain	01300n3535
	24.1	195	25	-$10\frac{1}{2}$	change, AC dist	
	22.9	195	35	-$10\frac{1}{2}$	inc; spect K0	
β1166	2.7	347	28	8 - 11	spect G5	01358n3824
	24.8	9	00	- 13		
Σ140	3.2	177	46	$8\frac{1}{2}$- 9	relfix; spect F8	01361n4049
Σ141	1.6	301	62	8 - $8\frac{1}{2}$	relfix; spect F5	01371n3843
τ	52.5	329	25	5 - 10	(53) cpm, spect B8	01376n4019
Σ154	5.3	126	41	8 - 8	relfix, spect F0	01420n4327
A952	2.2	70	27	$7\frac{1}{2}$- 13	spect G5	01483n4650
Σ179	3.6	160	55	7 - 8	relfix, spect F5	01502n3705
Σ3113	0.8	273	67	$8\frac{1}{2}$- $8\frac{1}{2}$	dist dec, spect G0	01504n4423
56	18.4	79	34	6 -$11\frac{1}{2}$	AB cpm; AC optical,	01532n3701
	190	300	28	- 6	AC spectra K0, K2	
S404	25.0	78	25	8 - 10	optical, dist & PA inc, spect G5	01550n4109
Σ195	2.9	193	46	$8\frac{1}{2}$- 9	relfix, spect A0	01571n4413

LIST OF DOUBLE AND MULTIPLE STARS (Cont'd)

NAME	DIST	PA	YR	MAGS	NOTES	RA & DEC
γ	10.0	63	62	$2\frac{1}{2}$ - 5	(Σ205) gold, blue superb object (*)	02008n4206
γ^b	0.3	121	61	$5\frac{1}{2}$ - 6	binary (*)	
Σ 215	20.3	60	20	$8\frac{1}{2}$- 10	relfix, spect G0 (BX Andr) (*)	02060n4034
Es 48	8.8	208	25	7 - 11	PA inc, spect F5	02065n4238
59	16.6	35	23	$6\frac{1}{2}$ - 7	(Σ 222) relfix, Spect A0, A2	02078n3848
Σ 228	0.7	248	61	$6\frac{1}{2}$ -$7\frac{1}{2}$	binary, 144 yrs; PA inc, spect F5	02108n4715
Σ 245	11.2	294	43	7 - 8	relfix, spect F2, yellow & blue	02155n4003
Σ 248	0.9	125	59	9 - 9	PA & dist dec	02179n4233
Σ 250	3.2	137	49	$8\frac{1}{2}$- 9	relfix, spect A3	02182n3712
Σ 249	2.2	196	33	7 - 9	relfix, spect A2	02184n4422
Σ 251	2.5	264	37	8 - 9	relfix, spect G5	02186n3909
σ 70	56.3	2	17	$6\frac{1}{2}$ -10	spect F2	02198n4110
A1815	1.9	135	31	7 - 11	spect F0	02251n3837
A1816	1.6	250	29	$6\frac{1}{2}$ -11	spect A5	02260n3706
A967	3.8	220	27	$7\frac{1}{2}$ -13	spect G5	02272n4513
A660	0.4	315	60	8 - 8	PA inc, spect A0	02283n4221
β 304	19.9	283	25	$7\frac{1}{2}$-$11\frac{1}{2}$	slight dist inc, spect F0	02284n3714
Es---	19.9	330	23	7 -$11\frac{1}{2}$	spect K0	02287n3754
h1120	16.3	83	59	7 -$11\frac{1}{2}$	dist dec, spect B9	02323n3927
	41.5	321	59	-11		
Σ 279	17.9	67	33	6 - 11	relfix, spect K4	02326n3706
	22.9	206	09	- 12		
β 305	8.5	262	34	$6\frac{1}{2}$ -11	spect F6; AC cpm; AB optical	02352n3731
	20.8	205	25	-$10\frac{1}{2}$		
2	0.5	25	58	5 - $8\frac{1}{2}$	(β1147) PA inc, spect A2 ; B=var?	23003n4229
Σ 2973	7.2	38	58	$7\frac{1}{2}$- 10	relfix, spect B3	23004n4347
Ho194	0.4	60	45	7 - $9\frac{1}{2}$	relfix, spect G5, A3	23048n4131
Σ 2979	3.1	224	45	8 - 10	PA slow inc, spect F2	23054n3931
Σ 2985	15.3	254	55	7 - 8	relfix, spect both G5	23076n4741

LIST OF DOUBLE AND MULTIPLE STARS (Cont'd)

NAME	DIST	PA	YR	MAGS	NOTES	RA & DEC
Σ2987	4.1	156	50	7½- 10	PA slow dec, spect G0	23080n4844
Ho197	0.2	3	58	8 -8½	PA & dist dec; spect F5	23090n3756
	40.3	327	25	-8½		
	50.5	281	25	-8½		
Σ2992	14.2	286	28	7½- 9	relfix, spect A3	23107n3943
8	7.7	162	34	5 - 13	(β717) cpm; spect M2	23154n4845
A202	2.5	257	32	8 - 10	spect B9	23160n4659
OΣ493	8.4	25	11	7½-10½	relfix, spect A5	23166n4812
Σ3004	13.2	178	58	6½ -10	relfix, spect A3	23183n4351
Σ3010	25.7	132	51	8 - 9	spect K0	23210n4530
Σ3010[b]	28.1	105	10	9 - 11		
Σ3024	4.9	307	41	8 - 9	slight PA dec, spect A0	23296n4334
OΣ500	0.5	348	61	6 - 7	PA inc, spect B9	23351n4409
β722	7.4	348	17	7 -12½	spect B9	23360n4214
	38.4	219	16	-11		
	45.1	247	16	-11		
Σ3028	16.0	201	57	7 - 9½	PA & dist dec, spect A2	23361n3446
OΣ501	14.9	163	58	7 - 10	relfix, spect F0	23376n3723
K	46.8	194	23	4 - 11	(h1898) cpm; spect B8	23379n4403
Σ3034	5.4	103	34	8 - 10	relfix, spect A0	23421n4606
β995	0.7	242	62	6½- 8½	PA inc, spect B3	23451n4633
OΣ506	0.3	346	60	7½ -8½	spect G0, AC dist slow inc	23461n3601
	19.5	81	58	-10½		
A796	0.6	14	56	7½ -10	PA dec, spect B9	23489n4729
OΣ510	0.4	312	59	7½- 7½	(h1911) PA dec, spect A5	23490n4148
	21.3	345	25	- 9		
Σ3042	5.3	88	62	7 - 7	slight dist inc, spect F5	23493n3737
β728	1.2	9	62	8½- 8½	PA inc, spect F8	23496n4314
Σ3043	0.4	165	46	8 - 10	(A1496) relfix	23503n3825
	15.6	250	46	- 9½		
Ho205	4.8	181	06	6½- 12	spect F8	23515n3901
OΣ513	3.6	22	39	7 - 9½	relfix, spect A3	23558n3445
Σ3050	1.3	288	67	6 - 6	binary, dist dec, about 800 yrs,	23569n3327

LIST OF DOUBLE AND MULTIPLE STARS (Cont'd)

NAME	DIST	PA	YR	MAGS	NOTES	RA & DEC
β 860	6.5	107	33	7 -11½	PA inc, spectra both G0 relfix, spect B9	23574n3836
Hn 60	0.8	197	61	9 - 9½	binary, 150 yrs; PA dec, spect K1	23598n3922

LIST OF VARIABLE STARS

NAME	MagVar	PER	NOTES	RA & DEC
ζ	4.06--4.20	17.77	ellipsoidal variable; spect K1 (*)	00447n2400
λ	3.7--4.1	55.82	class uncertain; spect G8; spect binary (*)	23351n4611
O	3.6--3.8	1.60	class uncertain; spect possibly composite (*)	22596n4203
R	5.3--15.1	409	LPV. Spect S6e (*)	00214n3817
S	6.0-----	---	Supernova of 1885 in the Galaxy M31 (NGC 224)	00400n4100
T	7.6--14..	281	LPV. Spect M4e	00198n2643
U	9.0--14.8	347	LPV. Spect M6e	01126n4027
V	8.3--14.9	261	LPV. Spect M2e	00474n3523
W	6.7--14.1	397	LPV. Spect M8e	02144n4405
X	8.1--15.0	346	LPV. Spect S3e	00135n4644
Y	8.2--14.3	220	LPV. Spect M3e	01367n3905
Z	8.0--11.5	---	Semi-reg; peculiar type; "symbiotic star" (*)	23312n4832
RR	8.4--15.2	330	LPV. Spect M5e	00486n3406
RS	9.0--9.7	234	Semi-reg; spect M6	23528n4822
RT	9.0--10.1	.629	Ecl.bin.; Spect G0, K1	23089n5245
RV	8.4--11.5	167	Semi-reg; spect M4e	02078n4842
RW	7.7--15.0	431	LPV. Spect M6e	00446n3225

ANDROMEDA

LIST OF VARIABLE STARS (Cont'd)

NAME	MagVar	PER	NOTES	RA & DEC
RX	10.3--13.6	Irr	"cataclysmic variable" or dwarf nova type (*)	01017n4102
RZ	8.8--9.5?		variability uncertain	23073n5246
SS	8.9--9.6	153	Semi-reg; spect M6	23092n5237
ST	8.3--11.5	337	LPV. Spect R3e	23363n3530
SU	7.6--8.5	Irr	Spect N	00020n4316
SV	7.8--13.9	316	LPV. Spect M7e	00017n3950
SW	8.8--10.2	.442	Cl.Var.; spect A3 --F8	00211n2907
SX	8.9--13.0	336	LPV. Spect M7e	01306n4616
SZ	9.0--13.0	341	LPV. Spect M2	22573n4234
TU	7.8--13.1	317	LPV. Spect M6e	00297n2545
TV	8.4--11.0	114	Semi-reg; spect M4e	22558n4228
TW	8.8--11.1	4.123	Ecl.bin.; spect F0, G6	00007n3234
TY	7.9--10.3	135	Semi-reg; spect M6e	23124n4031
TZ	8.2--9.6	974	Semi-reg; spect M5	23483n4714
UX	8.2--9.9	414	Semi-reg; spect M6	02302n4526
UZ	9.0--15.2	314	LPV. Spect M7e	01133n4129
VX	8.0--9.5	367	Semi-reg; spect N7; exceptionally red star	00172n4426
VY	9.6--11.4	Irr	Spect R8	22595n4537
WY	8.7--10.2	109	Semi-reg; spect G6e	23390n4719
AN	6.0---6.2	3.22	Ecl.bin.; (9 Andr) spect A7, A3	23160n4130
AQ	6.9---8.2	332	Semi-reg; spect N	00249n3519
AU	8.5---9.4	Irr	Spect M3	01444n3933
BX	8.6---9.5	.610	(Σ 215) Ecl.bin.; spect G0; 10^m star at 20"	02060n4034
BZ	8.1---9.0	Irr	Spect M5	00349n4520
CC	8.4---8.7	.125	Delta Scuti type; spect F2	00410n4201
CF	8.4---9.7	--	Semi-reg or Irr; spect M7	23014n3734
CG	6.2--6.24	3.74	Alpha Canes Venatici type Spect A0p	23581n4458
EG	7.4---7.6	40	Semi-reg? Spect gM2e	00419n4024
ET	6.3--6.32	.723	Alpha Canes Venatici type Spect B9p	23156n4513
GG	8.4---8.9		Semi-reg? Spect M5	23349n4650

LIST OF STAR CLUSTERS, NEBULAE, AND GALAXIES

NGC	OTH	TYPE	SUMMARY DESCRIPTION	RA & DEC
205	18^5		E6; 10.8; 8.0' x 3.0' B,L,mE,vmbM. Companion to Great Galaxy M31	00376n4125
214	209^2		Sb; 12.8; 1.4' x 0.8' pF,pS,lE, gvlbM	00387n2514
224	M31		Sb; 5.0; 160' x 40' !!! eeB,eL,vmE,sbM,vBSN. Great Andromeda Galaxy (*) Triple system with NGC 205 and NGC 221	00400n4100
221	M32		E2; 9.5; 3.6' x 3.1' vvB,L,R,psmbMN. Companion galaxy to M31	00400n4036
404	224^2		E0 or S0; 11.9; 1.3' x 1.3' pB,cL,R,gbM; np β Andr 6'	01066n3527
708	565^3		E1; 13.5; 0.5' x 0.4' vF,vS,R; brightest member of small group including NGC 703, 704, 705	01498n3555
753			Sc; 12.9; 1.8' x 1.5' pB,S,lE	01546n3541
752	32^7		vvL,pRi; diam 45'; irregular scattered group of 70 stars; mags 8... class D (*)	01548n3726
891	19^5		Sb; 12.2; 12.0' x 1.0' B,vL,vmE; edge-on spiral (*)	02193n4207
7640	600^2		Sb or SBb; 12.5; 9.0' x 1.0' cF,vL,vmE,lbM; nearly edge-on spiral	23197n4035
7662	18^4		vB,pS,R; mag 8½; diam 30"; bright bluish-green disc with central 14^m star (*)	23234n4212
7686	69^8		P,1C,Irr; diam 12'; about 12 stars mags 8....13	23277n4851

DESCRIPTIVE NOTES

ALPHA   Name- ALPHERATZ;   sometimes called "Sirrah".
Mag 2.06; spectrum given by various authorities
as B8, B9, A0, or A1, but peculiar for the unusual strength
of the lines of manganese. Position 00058n2849. Direct
parallaxes obtained at Allegheny and Yerkes agree in giving a distance of about 120 light years; the resulting
luminosity is about 160 times that of the Sun and the absolute magnitude about -0.7. Slightly different results
are obtained from the spectroscopic characteristics which
suggest an absolute magnitude of -0.1; this would reduce
the distance to about 90 light years.
        The annual proper motion is 0.20" in PA 140°, and the
radial velocity is about 7 miles per second in approach.
        Alpheratz is a spectroscopic binary with a period of
96.697 days. The two stars are of unequal brightness and
the companion has not been detected spectroscopically.
According to J.A.Pearce (1937) the orbit of the visible
star has an eccentricity of 0.53; the mean radius of the
orbit may be about 20 million miles.
        In addition, there is a distant optical companion of
the 11th magnitude, discovered by Sir William Herschel and
first measured by F.G.W.Struve in 1836 when the distance
was 64.9". This star is not a true physical companion to
Alpha and the separation in 1954 had increased to 81.5" in
PA 280° from the proper motion of the primary.
        Alpheratz marks the northeast corner of the familiar
Great Square of Pegasus, and is identified on some of the
older atlases as "Delta Pegasi". It is now officially
assigned to Andromeda.

BETA   Name- MIRACH. Mag 2.03; spectrum M0 III. Position
01069n3521. The distance is about 75 light years
according to parallaxes obtained at Mt.Wilson, Allegheny,
and McCormick; the resulting luminosity is about 75 times
that of the Sun, and the absolute magnitude about +0.2.
The star shows an annual proper motion of 0.21" in PA 122°
and the radial velocity is about 0.2 miles per second in
recession.
        Mirach has a companion of the 14th magnitude at 28"
in PA 202°, discovered by E.Barnard at Yerkes in 1898. It
apparently shares the proper motion of the primary, and is
a dwarf star some 800 times fainter than the Sun.   There

BETA ANDROMEDAE. The star is the bright central image with the diffraction spikes; the galaxy NGC 404 is at the upper right.   (42-inch reflector, Lowell Observatory)

are two other stars of the 12th magnitude at 85" and 90",
but these are merely optical companions. Mirach itself,
like many of the red giant stars, has been suspected of
slight variability.

Observers of this star should attempt to find the
12th magnitude galaxy NGC 404 in the same field, a good
test for the light-gathering ability of the telescope. It
is located 6.4' from the star toward the northwest.

GAMMA   Name- ALMACH. Mag 2.12; spectrum K2 II or K3.
Position 02008n4206. This is a beautiful double
star, one of the finest within range of a small telescope.
According to T.W.Webb it was probably discovered by J.T.
Mayer in 1788 (R.H.Allen gives the date as 1778) but the
first recorded measurements appear to be those of F.G.W.
Struve in 1830. The brighter star is golden yellow or
slightly orange, and the companion (mag 5.08) appears a
definite greenish-blue. The color contrast is unusually
fine, and often seems more striking with the eyepiece very
slightly displaced from the position of sharpest focus.
There has been no definite change in separation or angle
in the pair in the last 130 years. In 1962 the measurement
made at Lowell was: 10" in PA 63°.

In 1842, Struve discovered that the companion is
itself a close double. It is a binary with a period of 61
years according to recent computations by P.Muller (1957).
The star was at periastron in 1891 and again in 1952; the
greatest separation of the components is about 0.55" and

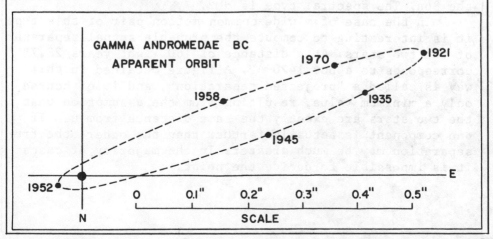

GAMMA ANDROMEDAE BC
APPARENT ORBIT

will be reached about 1982. The apparent orbit is a much-elongated ellipse extending toward PA 110°; the semi-major axis is 0.3" and the eccentricity is 0.93. Both of the stars are late B or early A type; the individual magnitudes are 5.5 and 6.3.

The brightest member of the close pair is itself a spectroscopic binary with a period of 2.67 days and two identical spectra (about B9) visible. Gamma Andromedae is thus a quadruple system. The luminosity of the K-star is about 650 times that of the Sun; the B-C-D system totals about 50 times the light of the Sun. The actual separation of the A-B pair may be about 800 AU, and the B-C separation averages about 30 AU.

The distance is approximately 260 light years; the annual proper motion is 0.07"; the radial velocity is about 7 miles per second in approach. The total absolute magnitude is about -2.4.

DELTA   Mag 3.25; spectrum K3 III. Position 00366n3035.
The computed distance is about 160 light years, the actual luminosity about 100 times that of the Sun. The absolute magnitude is -0.2. The star shows an annual proper motion of 0.16" in PA 125°; the radial velocity is $4\frac{1}{2}$ miles per second in approach.

The 12th magnitude companion at 28.7" was discovered by S.W.Burnham with the 26-inch refractor at the U.S.Naval Observatory in 1878. It shares the proper motion of the primary and is a red dwarf of about 1/40 the luminosity of the Sun. The spectral type is dM2.

In the case of a wide common motion pair of this type it is interesting to compute the probable actual separation of the two stars. At a distance of 160 light years 28.7" corresponds to about 1420 AU. A figure obtained in this way is called a "projected separation", and is of course only a minimum value, resulting from the assumption that the two stars are exactly the same distance from us. If one component is actually farther than the other, the true separation may be much greater. In the majority of cases it is impossible to decide the point.

DESCRIPTIVE NOTES (Cont'd)

ZETA   Mag 4.06 (slightly variable); spectrum K1 III.
Position 00447n2400. This star is a spectroscopic binary with a period of 17.7673 days, and the typical
example of an "ellipsoidal variable" in which the light
changes are due to the fact that both stars are oval in
shape and present varying amounts of luminous surfaces as
they revolve in their orbits. Very small partial eclipses
may also add to the effect. According to S.Jones, the
spectroscopic orbit is nearly circular, with the slight
eccentricity of 0.017, and the brighter star is about 3.9
million miles from the center of gravity of the system.
The light variations were first measured photoelectrically
by J.Stebbins in 1928, and suggest that the two components
are revolving nearly in contact. The larger star may be
about 8 or 10 times the diameter of the Sun.

Direct parallaxes obtained at Allegheny and McCormick give the distance as about 100 light years; the total
luminosity is then about 18 suns. Different results seem to
be obtained, however, from the spectroscopic features;
these suggest a luminosity class of II or III. If the star
is actually a KI III giant, the absolute magnitude should
be about +0.8 (luminosity = 40 suns) and the distance must
then be about 150 light years. An attribution to luminosity class II would further increase this discrepancy.

Zeta Andromedae also has a faint visual companion of
the 13th magnitude at 96" in PA 230°; it apparently shares
the annual proper motion of the primary (0.13") and was
first detected by S.W.Burnham in 1910. It is a red dwarf
at least 400 times fainter than the Sun. The radial velocity of both stars is 14½ miles per second in approach.

LAMBDA   Mag 3.88; (slightly variable); spectrum G8 IV.
Position 23351n4611. A peculiar spectroscopic
binary star, discovered by W.W.Campbell in 1899, and displaying the unusual feature of bright (emission) lines of
calcium in its spectrum. According to J.L.Greenstein (1952)
the stellar absorption lines are quite sharp but the emission lines are fairly broad, raising some interesting
questions concerning the structure of the star's atmosphere
and the possible presence of large prominences. There is
also the peculiar fact that the slight variations in light

show no correlation with the revolution period of the
system. The visual range is about 0.4 magnitude.

The period of the binary pair is 20.5212 days, and
the orbit has the small eccentricity of 0.04 according to
J.A.Pearce and E.C.Walker (1944). The bright star is less
than 1 million miles from the center of gravity of the
system, but the actual separation of the two stars is uncertain. Only one component is detected spectroscopically.
The primary is a subgiant with a computed diameter of 6
times that of the Sun, and a luminosity of about 16 suns.
The absolute magnitude is about +1.9. Parallaxes obtained
at Allegheny and Sproul agree in giving the distance as
about 80 light years.

The annual proper motion of Lambda Andromedae is
0.45" in PA 159°; the mean radial velocity is about 4
miles per second in recession.

OMICRON   Mag 3.63 (slightly variable); spectrum given
as composite (B6 + A1) by some authorities,
simply "B6p" by others. Position 22596n4203. This star is
a peculiar variable of uncertain class, perhaps combining
the features of several different classes. The variations
were first suspected by Guthnick and Prager in 1915, and
confirmed by R.M.Emberson (1939) who found a range of
about 0.5 magnitude. An examination of Harvard patrol
plates showed variations of about one magnitude. Spectra
have been obtained since 1890, often showing the features
of a normal B6 type star, but at other times showing the
presence of a gaseous shell or ring. The shell was apparently present in 1890 but absent in 1893 and 1928, developing again about 1937 and very evident in 1946 and 1952.
There is some evidence that short-period variability in
the star is connected in some way with the presence of the
surrounding shell. The star is remarkable for its extremely high rotational velocity of 215 miles per second (at
the equator), one of the most rapidly rotating stars known.

S.Archer (1959) suggests that there may be some interaction between the rotation period and the pulsation of
the star, when a shell is present. He found in 1958 that
the short-period variations appear to resemble those of the
cluster variables (RR Lyrae stars), and derived a period of
0.7882 days with an amplitude of about 0.5 magnitude.

DESCRIPTIVE NOTES (Cont'd)

At other times, however, the light changes appear to be of an entirely different nature. The observations of H.Schmidt (1959) indicated a period of 1.59984 days, very close to double the period found by Archer, and the light curve strongly resembles that of an eclipsing binary of the lyrid type. Primary minimum has a depth of about 0.15 magnitude. The evidence seems clear that the star is a close binary, but with the added complications of occasional shell activity and short-period pulsations in at least one of the components.

A.Slettebak (1952) calls attention to the interesting discovery that the shell of the star is stratified. The spectrum lines of helium show the greatest rotational broadening and evidently originate in the main body of the star. The lines of magnesium and silicon are sharper, and apparently originate at higher levels in the shell. The iron lines seem to be produced at various levels. This same effect has been found in another famous shell star, 48 Librae.

Parallax measurements of Omicron Andromedae have been inconclusive, but suggest that the distance cannot be less than 450 or 500 light years. The actual luminosity would appear to be in the range of 500 to 800 times that of the Sun, and the absolute magnitude near -2.0. The estimated diameter of the B-star is 4 to 6 times that of the Sun. The spectral peculiarities make it unsafe to attempt to define the luminosity class, but Slettebak (1952) states that the broad "wings" of the hydrogen lines suggests an object which is near the main sequence.

The annual proper motion of Omicron Andromedae is only 0.02"; the radial velocity is about 8½ miles per second in approach.

Pl   Mag 4.43; spectrum B5 V. Position 00342n3327. Pi Andromedae has two visual companions for the telescope; the brighter one at 36" was first measured by Sir William Herschel late in the 18th century, and shares the proper motion of the primary. In addition, the chief star is a spectroscopic binary with a period of 143.606 days; two spectra of nearly identical type are visible. For the orbit of the brighter star J.A.Pearce (1936) found an eccentricity of 0.56; the mean separation of the two stars is in the neighborhood of 150 million miles, but the exact figure depends upon the value accepted for the inclination of the orbit, which is unknown.

The distance, from parallax measurements obtained at Sproul Observatory, may be about 350 light years, and the resulting absolute magnitude about -0.8 (luminosity about 170 suns.) The annual proper motion is very slight, about 0.015"; the radial velocity averages 5½ miles per second in recession.

For the common proper motion companion at 36" we give once again the "projected separation" of the two stars, approximately 3860 AU.

OMEGA   Mag 4.84; spectrum F4 IV. Position 01246n4509.
This is a close and difficult double star, discovered by S.W.Burnham with the 12-inch refractor at Lick Observatory in 1881. The two stars form a binary of long period with a gradual increase in the PA, from 92° at the time of discovery to 122° in 1962. The system shows a fairly large annual proper motion of 0.36" in PA 107°. The McCormick and Allegheny parallaxes agree in giving the distance as about 135 light years; the actual luminosities of the two stars are then 17 and 0.025 suns. The small companion is a red dwarf. According to the Yale "Catalogue of Bright Stars" (1964) Omega Andromedae is an outlying member of the Hyades moving group in Taurus, moving toward the same convergent. The radial velocity is about 6½ miles per second in recession. The Hyades cluster is so close to us that outlying members may be found in almost any part of the sky; they may be identified by their space motions but this can be done only when both the proper motion and radial velocity are accurately known.

In the same field with Omega is the faint double star β382, noted by S.W.Burnham in 1872 when it was 135" from Omega in PA 110°. This star does not share the large motion of Omega itself, and the separation has been steadily decreasing from the proper motion of the bright star; in 1965 it was slightly under 2'. β382 itself has shown no definite change in separation or PA since discovery. According to the Lick "Index Catalogue of Visual Double Stars" (1961) the separation in 1943 was 4.9" in PA 138°, both stars being of magnitude 10.4.

36   Mag 5.45; spectrum Kl IV. Position 00523n2321. The
star is a close but interesting binary, discovered
by F.G.W.Struve in 1836. According to a computation by P.
Muller (1957) the period is about 165 years with periastron occurring in 1957. The components are magnitudes 6.2
and 6.7, and their apparent separation varies from 0.6" to
1.4". The closest approach lasts for a number of years, as
in the interval from 1930 to 1980, and the star is then an
excellent test object for larger amateur telescopes. The
computed orbit gives the semi-major axis as 1.0"; and the
eccentricity is 0.31.

The primary star is a subgiant of class Kl with about
7 times the luminosity of the Sun. The companion is probably similar in type since there is no noticeable color
contrast between the components. "A beautiful strong yellow" says T.W.Webb.

Parallax measurements of 36 Andromedae give the distance as about 160 light years. On this basis the true
separation of the stars averages about 50 AU, somewhat
greater than the distance of Pluto from the Sun. The star
has an annual proper motion of 0.13"; the radial velocity
is about 1.2 miles per second in recession.

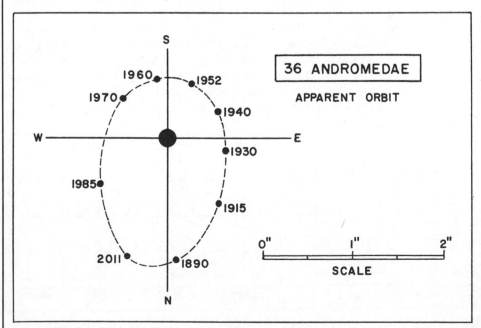

36 ANDROMEDAE

APPARENT ORBIT

Σ 215   Mag 8.6 (variable); spectrum G0. The position is 02060n4034. A rather faint but easy pair, first measured by F.G.W.Struve in 1831. The annual proper motion is given in the Lick "Index Catalogue" as 0.03"; this may apply to both stars since no definite relative change has been noted in more than a century. The present separation is slightly over 20". The distance of this star is not definitely known.

The brighter component is the short-period eclipsing binary BX Andromedae, with a range of 8.6 to 9.5 and a period of 0.6101123 days (14h 38.6m). According to the Moscow "General Catalogue of Variable Stars" (1958) the light curve is of the lyrid type, with a secondary minimum of magnitude 9.0. The shortness of the period suggests that this star is a dwarf system of the W Ursa Major type.

R   Variable; spectrum S6e. Position 00214n3817. This is the brightest of the long-period variables in Andromeda, discovered at Bonn, Germany, in 1858. It is easily located near the bright triangle of stars formed by Theta, Rho, and Sigma Andromedae, about 4° southwest of the Great Galaxy M31. R Andromedae is noted for its exceptionally large range which at times has exceeded nine magnitudes. The star at maximum is visible in binoculars, and on occasion has attained naked-eye visibility. At minimum it is sometimes almost impossible to detect in a good 8-inch telescope. The period averages 409 days, but may vary by a number of days from one cycle to the next.

The star is a pulsating red giant of the general type

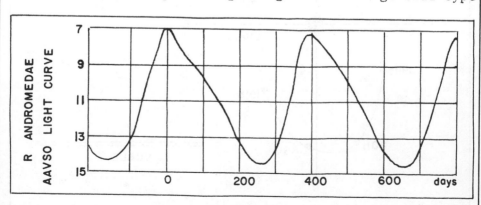

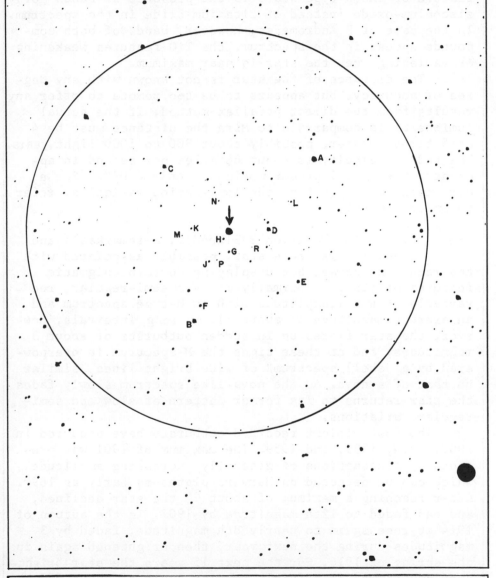

COMPARISON MAGNITUDES (AAVSO) A= 6.9; B= 9.4; C= 9.5;
D= 9.7; E= 10.6; F= 11.0; G=11.6; H= 11.8; J= 12.4;
K= 12.6; L= 12.9; M= 13.5; N= 13.8; P=13.9; R=14.0.

R ANDROMEDAE. Finder chart made from a 13-inch telescope
plate at Lowell Observatory. Circle diameter = 1 degree.
North is at the top. Limiting magnitude about 15. Bright
star at lower right is Rho Andromedae, magnitude 5.1.

to which the famous Mira (Omicron Ceti) belongs, but the
spectral class in this case is type S. The distinguishing
feature of the S-type stars is the presence of bands of
zirconium-oxide instead of titanium-oxide in the spectrum.
In the case of R Andromedae, however, bands of both compounds appear in the spectrum, the TiO features weakening
or vanishing when the star is near maximum.

The distance of the star is not known with any degree of accuracy, but appears to be too remote to offer any
results from the direct parallax method. If the actual
luminosity is comparable to Mira the distance must be 4
or 5 times greater, possibly about 800 to 1000 light years.
The radial velocity is about $6\frac{1}{2}$ miles per second in approach; the annual proper motion is only 0.02". (For a
more detailed account of the long-period variables, refer
to Omicron Ceti.)

Z   Variable. Position 23312n4832. A remarkable and
         peculiar variable star, possibly associated with
the recurrent novae, but displaying certain enigmatic
features of its own. Normally it is a semi-regular red
variable of small amplitude with an M-type spectrum and
an average magnitude of about 11. At long intervals, however, the star flares up in sudden outbursts of about 3
magnitudes, and at these times the M-spectrum is overpowered by a "shell spectrum" of wide bright lines, similar
to that of a nova. As the nova-like spectrum slowly fades
the star returns to its former pattern of slow and semi-
regular variations.

The most violent recorded outbursts have occurred in
1901, 1914, 1939, and 1959. The maximum of 1901 was preceded by fluctuations of gradually increasing amplitude
which can be detected on Harvard plates as early as 1890.
After reaching a maximum of about $9\frac{1}{2}$ the star declined,
and had faded to 12th magnitude by 1907. In the autumn of
1914 it rose again to nearly 8th magnitude, faded by 3
magnitudes during the next year, then brightened again in
the spring of 1916. For the next 15 years the star brightened and dimmed in a fairly regular cycle of about 695
days, but with steadily decreasing amplitude. By 1931 a
normal minimum was reached, and the star remained faint

DESCRIPTIVE NOTES (Cont'd)

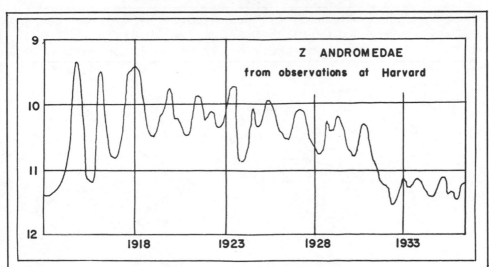

Z ANDROMEDAE
from observations at Harvard

until the outburst of 1939. In the summer of that year it
began to brighten, and reached magnitude 7.9 in November,
probably the greatest brightness yet recorded. In 1959 it
rose to 10th magnitude, and in 1961 brightened to 9.2.

From spectroscopic observations it now appears certain that Z Andromedae is actually a close binary star.
The spectrum is composite, and combines the features of a
low temperature red giant and a hot bluish B-star which
is probably a subdwarf. P.W.Merrill has applied the term
"symbiotic stars" to objects of this type. Z Andromedae
and R Aquarii are perhaps the most typical examples. The
bright outbursts are attributed to the blue stars, but the
red components appear to be variable also, with a range of
one or two magnitudes. In addition, the spectrum shows the
lines which are characteristic of the gaseous nebulae, and
it seems certain that both of the components are enveloped
in a gaseous cloud. In the case of R Aquarii a faint diffuse nebula can actually be seen surrounding the star, and
it is in a state of slow expansion.

A significant fact about Z Andromedae is that the
blue and red components both vary in nearly the same cycle
of about 700 days, and that the radial velocities again
show approximately the same period. The variations of both
stars may be connected in some way with the orbital motion
of the system. The bright outbursts may be attributed to
some process of interaction between the components; the

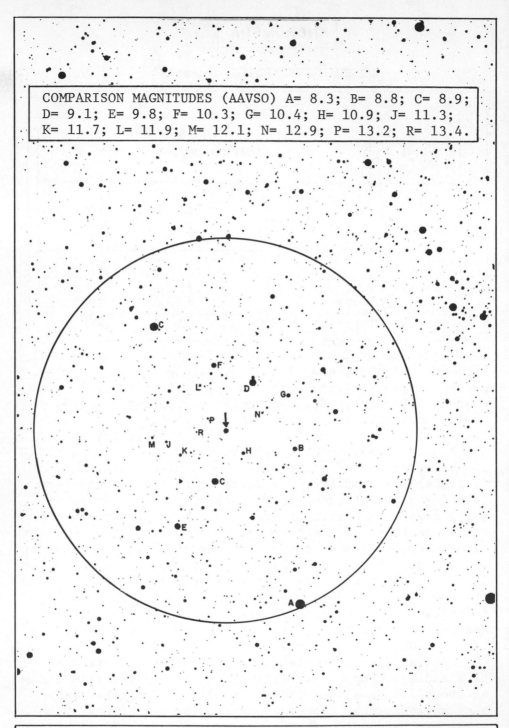

COMPARISON MAGNITUDES (AAVSO) A= 8.3; B= 8.8; C= 8.9; D= 9.1; E= 9.8; F= 10.3; G= 10.4; H= 10.9; J= 11.3; K= 11.7; L= 11.9; M= 12.1; N= 12.9; P= 13.2; R= 13.4.

Z ANDROMEDAE. Finder chart made from a 13-inch telescope plate at Lowell Observatory. Circle diameter = 1 degree. North is at the top. Limiting magnitude about 15.

DESCRIPTIVE   NOTES   (Cont'd)

same mechanism has been proposed also for the   recurrent
novae. But the exact details are quite uncertain. In the
case of the eruptive "dwarf novae" of the U Geminorum
type there is some evidence that the red component is the
seat of the outbursts, rather than the blue subdwarf as
has been generally assumed. At the present time, the whole
subject is well supplied with fascinating uncertainties.
(Refer also to R Aquarii, AG Pegasi, and T Corona Borealis.)

RX   Variable. Position 01017n4102. An erratic and very
         unpredictable variable star, belonging to the rare
class of which Z Camelopardi is the prototype. It was discovered in 1905 by the British observer A.S.Williams, and
his preliminary light curve was published the same year.
The variations frequently resemble those of the famous SS
Cygni, with small-scale nova-like outbursts occurring repeatedly at intervals of from 2 to 3 weeks. The total
range is about 3½ magnitudes, and the rise to maximum is
usually accomplished in 2 or 3 days. But at other times
the variations are completely irregular and totally unpredictable. On occasion the star may remain nearly constant
for several months, as shown on the accompanying light
curves.

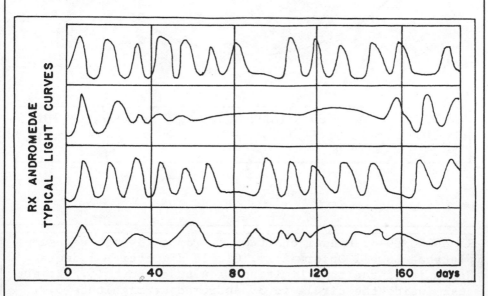

RX ANDROMEDAE
TYPICAL LIGHT CURVES

0   40   80   120   160   days

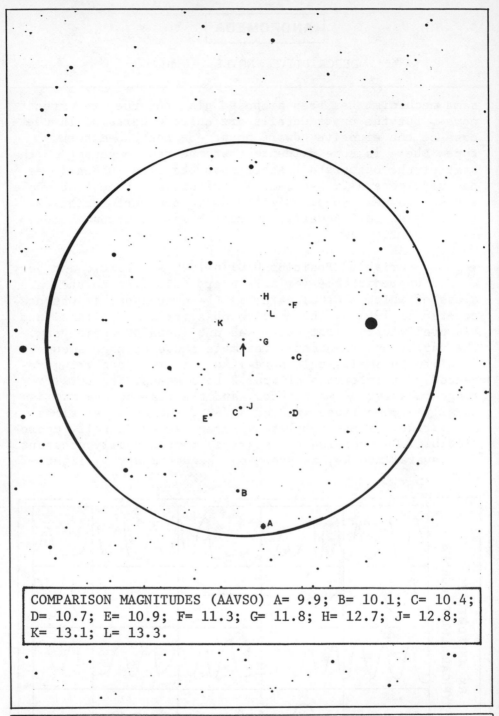

COMPARISON MAGNITUDES (AAVSO) A= 9.9; B= 10.1; C= 10.4;
D= 10.7; E= 10.9; F= 11.3; G= 11.8; H= 12.7; J= 12.8;
K= 13.1; L= 13.3.

RX ANDROMEDAE. Finder chart made from a 13-inch telescope
plate at Lowell Observatory. Circle diameter = 1 degree.
North is at the top. Limiting magnitude about 15. Bright
star inside the circle is 39 Andromedae, magnitude 5.9.

RX Andromedae has a peculiar spectrum, showing bright hydrogen lines against an apparently continuous background, the lines weakening as the star rises to maximum. The color is equivalent to a late A-type star. The spectral features and light curve both show a strong resemblance to Z Camelopardi. RX Andromedae is a very close and rapid binary with a period of 5h 05m, a significant discovery since the similarly acting SS Cygni, AE Aquarii, and U Geminorum are all known to be close dwarf or subdwarf binaries of this same type. A current theory regards the outbursts as a result of an interchange of material between the close components, one of which may be a degenerate star.

R.P.Kraft (1962) classes the blue components of these systems as subdwarfs (sdBe) and finds that their absolute magnitudes lie in the range of +7½ to +9. He also finds some evidence that the intrinsically fainter stars have the shortest orbital periods. The other star in each pair seems to be a red dwarf whose mass is often comparable to that of the Sun, but abnormally underluminous for its mass. Kraft suggests that material ejected from the red star forms a gaseous ring or disc around the blue dwarf. According to one theory, some of this material is eventually brought into contact with the body of the degenerate star, with explosive results. The accuracy of this picture has been questioned, however, by W.Krzeminski (1965) who finds evidence that in the very similar system U Geminorum the outbursts originate in the cooler redder component, rather than in the hot subdwarf. (Refer to U Geminorum)

W.J.Luyten (1965) finds an annual proper motion of about 0.01" for RX Andromedae, suggesting a distance of at least a few hundred light years. At the time of writing, no direct parallax measurement seems to be available. (Refer also to SS Cygni, U Geminorum, AE Aquarii, and Z Camelopardi.)

GRB 34 Groombridge 34 (ADS 246) (BD+43°44) Position 00155n4344. This is a noted red dwarf binary system, and one of the closest double stars to the Solar System. It is located about ¼° north of 26 Andromedae. The star was discovered through proper motion measurements in 1860. The two components are magnitudes 8.1 and 10.9 and are separated by 39". The PA is increasing by about 5° per

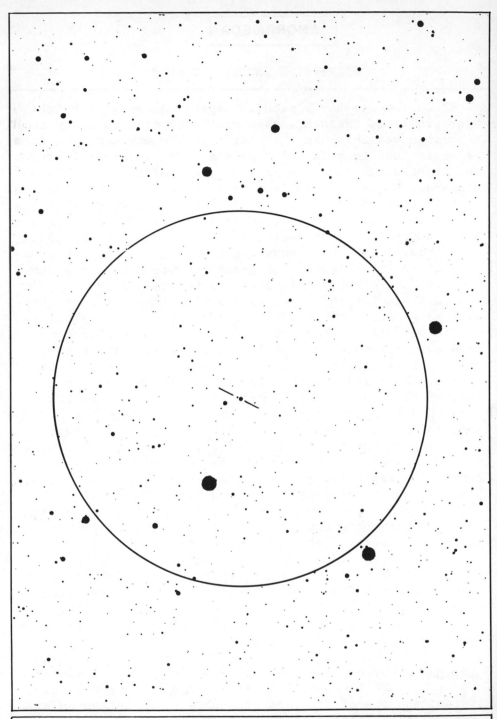

Groombridge 34- Finder chart made from a 13-inch telescope plate at Lowell Observatory. Circle diameter = 1 degree. North is at the top. Limiting magnitude about 15. Bright star inside the circle is 26 Andromedae, magnitude 6.0.

century; the orbital motion is thus so slow that no definite period can yet be derived. According to a preliminary
computation by Hopmann (1957) a period of slightly over
3000 years is suggested, with periastron about 2320 A.D.
Hopmann's orbit has a semi-major axis of 44" and an eccentricity of 0.25.

Groombridge 34 is 11.7 light years distant, and has
the large proper motion of 2.89" annually in PA 82°. The
true separation of the two stars averages about 160 AU.
According to A.H.Joy (1947) the primary is a spectroscopic
binary of uncertain period. The chief facts about the two
stars are given in the following short table:

A	Mag 8.1	Spect dM2	Abs.Mag.	10.3	Lum 0.006 X ⊙
B	10.9	dM4e		13.1	0.00045

A third star of the 11th magnitude, called "C", was
20" distant in 1917, but it is not a physical member of
the system, and does not share the large proper motion of
A & B. The separation in 1961 had increased to about 2.3'
from the gradual drift of the motion pair. The radial velocity of both stars is about 11 miles per second in recession.

M31 (NGC 224) Position 00400n4100. The Great Andromeda Galaxy, the chief object of interest in the
constellation. It is the brightest and nearest of all the
spirals, and the only one which can be considered a definite, obvious naked-eye object. As seen without optical
aid it appears as a small elongated bit of fuzzy light,
about 1° west of the star Nu Andromedae. A pair of good
binoculars will be found very useful in searching for it,
and on a really clear night will enable the full diameter
to be traced out to over 4°. When observing M31 through
the small telescope, a low power wide-field eyepiece is
essential; high powers show only the nuclear condensation.
According to R.H.Allen, the Andromeda Galaxy has
been known at least as far back as 905 A.D. and was mentioned by the Persian astronomer Al Sufi in the 10th Century. It was called the "Little Cloud" and appeared on
star charts long before the development of the telescope

THE GREAT ANDROMEDA GALAXY M31, photographed with the 5"
Cogshall camera at Lowell Observatory. This huge spiral is
probably the largest member of the Local Group of Galaxies.

DESCRIPTIVE NOTES (Cont'd)

in 1609. Simon Marius is usually credited with the first telescopic observation of the object in 1611 or 1612. He compared the soft glow to "the light of a candle shining through horn". For the visual observer, the description is still accurate today, even in the age of the great modern telescopes. The largest instruments reveal little more than an elongated foggy patch which gradually brightens in the center to a nearly star-like nucleus. In a good 8-inch reflector, the prominent dark lane on the northwest edge of the central hub, and the bright star cloud near the south-tip may both be glimpsed, if the sky is dark and clear. But except for these faint details, the soft light of the great galaxy remains a smoothly luminous glow without the slightest hint of resolution. Early observers had thought the "nebula" to be composed of glowing gases; some regarded it as "a solar system in the making" and imagined that our own Sun looked much the same in the days of the primeval dust cloud, when the planets were being formed. Spectroscopic analysis of M31 eventually destroyed all such assumptions, and left no doubt that the light of the enigmatic "Great Nebula" actually came from a multitude of individual stars.

Only long-exposure photographs taken with large telescopes will reveal the true nature of such an object. The "Little Cloud" in Andromeda is actually a vast galaxy, an aggregation of billions of stars like our own Milky Way galaxy. It appears as an elongated oval because it is inclined only 15° from the edge-on position; actually it is round and flat, and has a spiral pattern which classes it as type Sb. This great island universe is the nearest of all the spirals and is probably the largest member of the Local Group of galaxies. The distance is 2.2 million light years. In viewing such an object, we are not only looking out through space across the enormous distance of about 13 thousand quadrillion miles; but we are also looking back through time, to a period about 2 million years ago, when the light of the Andromeda system started on its long journey toward the Earth.

The first hint of the true nature of the Andromeda Galaxy came late in 1923 when several cepheid variable stars were identified in the system. In a study of these objects, made with the 100-inch telescope at Mt.Wilson,

THE ANDROMEDA GALAXY. This striking photograph of M31 was obtained with the 36-inch Crossley Reflector at Lick Observatory.

NORTHEAST SECTION OF THE ANDROMEDA GALAXY M31; from a
photograph obtained with the Crossley Reflector at Lick
Observatory.

SOUTHWEST SECTION OF THE ANDROMEDA GALAXY M31; from a photograph obtained with the Crossley Reflector at Lick Observatory.

DESCRIPTIVE NOTES (Cont'd)

Dr.E.Hubble definitely established the great spiral as an extra-galactic object, and derived a tentative distance of about 900,000 light years. Hubble's discovery was announced at the meeting of the American Astronomical Society in Washington D.C. in December 1924, and dramatically ended the long controversy over the nature of the "spiral nebulae". Further studies, still using the pulsating cepheids as distance indicators, caused a revision of the distance to about 750,000 light years, and until rather recently this remained the most accurate estimate possible. For the method of using cepheid stars as distance indicators, refer to Delta Cephei.

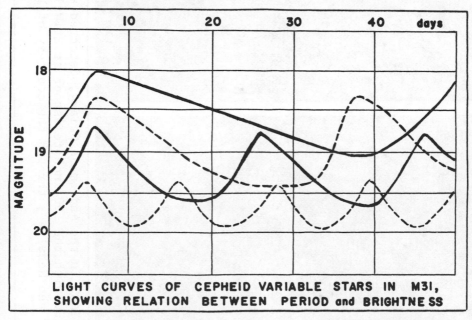

LIGHT CURVES OF CEPHEID VARIABLE STARS IN M31, SHOWING RELATION BETWEEN PERIOD and BRIGHTNESS

In 1953, however, investigation of the Andromeda Galaxy with the newly completed 200-inch telescope proved that the stars of the system are arranged in two different "populations". The spiral arms contained bright blue giant stars and nebulous regions immersed in clouds of dust (Population I) while the nuclear hub is a vast swarm of fainter red and yellow stars (Population II). The discovery was soon made that the cepheid stars are intrinsically different in luminosity in the two populations, a Population I cepheid being the brightest by at least 1.5 magnitudes. As

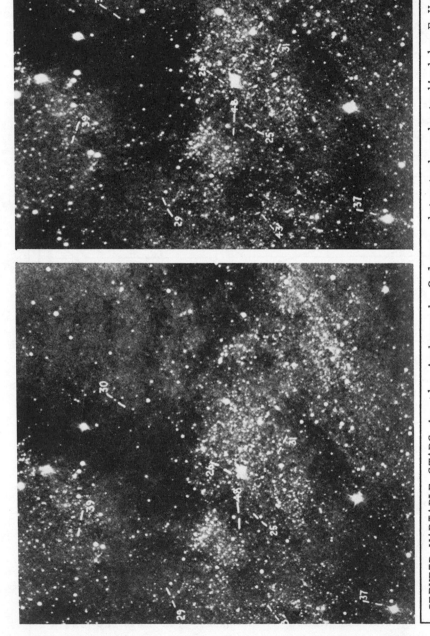

CEPHEID VARIABLE STARS in the Andromeda Galaxy, detected and studied by E.Hubble in 1924, on plates made with the 100-inch telescope at Mt.Wilson. Print reproduced from Hubble's "Realm of the Nebulae", through the courtesy of Yale University Press.

a result, the cepheids used in the distance calibration
were discovered to be more luminous than had been thought,
and the distance scale was seen to be in error by a factor
of 2 or 3. Various lines of evidence now agree in giving
a distance of 2.2 million light years for M31, and placing
the next nearest spiral (M33 in Triangulum) at about 2.4
million light years. It is not expected that any further
great revisions will be required.

The outer parts of M31 were first resolved into stars
on long-exposure photographs with the 100-inch telescope
at Mt.Wilson. Resolution of the nuclear hub proved much
more difficult, and was finally accomplished with special
red-sensitive plates on the 200-inch telescope at Palomar.
This was a triumph of observational astronomy, but it must
be remembered that anything like a complete resolution of
the galaxy is quite impossible. At a distance of over 2
million light years, only the most brilliant stars - the
high luminosity giants - can be seen at all. Our own Sun,
at such a distance, would appear of visual magnitude 29.1,
and would be 200 times too faint to be detected in the
greatest telescopes on Earth.

The Andromeda Galaxy probably contains over 300 billion individual stars. Its computed mass is about 400 billion times that of the Sun. The total luminosity is equal
to 11 billion suns, and the absolute magnitude is given by
A.Sandage as -20.3. This is intrinsically one of the most
luminous galaxies known.

On the best photographs the image of M31 measures a
full 160' X 40', nearly 2.7° across the longer dimension.
This corresponds to an actual diameter of 110,000 light
years. Measurements with the sensitive instrument known as
the densitometer increase the size to 4.5° or 180,000 light
years. Thus M31 must be classed as one of the largest galaxies known. Our own Galaxy, for comparison, is thought to
measure about 100,000 light years in diameter, and the
majority of the known spirals are less than half this size.
The central mass of M31 is a huge elliptical galaxy in itself, and measures about 12,000 light years in diameter.
This hub is a huge globular aggregation rich in red and
yellow giant stars, and comparatively free of dust and gas.
It is classified as a "Pop.II" system. In contrast, the outer portions. containing the spiral arms, comprise a typical

SOUTHERN SECTION OF THE ANDROMEDA GALAXY, photographed with the 100-inch reflector at Mt.Wilson. The prominent star cloud near the top of the print is NGC 206.

DESCRIPTIVE NOTES (Cont'd)

"Pop. I" system, notable for the presence of extremely luminous blue giant stars, extensive bright and dark nebulosity, dust and gas. A detailed study of the Andromeda Galaxy by W.Baade has identified seven distinct spiral arms:
two dust arms near the nucleus, and five outer arms of
coiled star clouds. An interesting feature of M31 is the
system of dark dust lanes which outline the spiral form of
the galaxy, giving us perhaps an idea of the appearance of
our own Milky Way system from a similar distance.

Among the star clouds of M31, one object is to be especially noted. Sufficiently conspicuous to appear in the
"NGC" Catalog as a separate entry, it bears the number
NGC 206. It is easily located near the south tip of the
galaxy and close to the western rim, where it may be detected with a good 8-inch telescope on a dark night. The
dimensions of this cloud of stars are about 2900 X 1400
light years. A few hundred stars in the cloud have luminosities of over 10,000 times that of the Sun, while the
fainter stars are literally uncountable. The brightest
individual stars of the spiral arms have absolute magnitudes of about -7, comparable to such supergiants as Rigel
and Deneb; while for the brightest of the red giants in the
central hub the figure is about -3.5.

In the heart of the central hub lies the actual nucleus of the Andromeda spiral, a sharp star-like condensation which looks nearly stellar even in the largest telescopes. Observations at Lick show that this nucleus has
an apparent size of about 2.5" X 1.5", indicating an actual diameter of some 50 light years. The nucleus seems to
be something in the nature of a super-globular star cluster, containing possibly over 10 million stars. In such a
mass, the separation of the stars would average only a few
hundred AU, and the density would be about 50 or 60 stars
to the cubic light year. In such dense groupings of stars
the possibility of stellar collisions must be considered.
It has been suggested that some of the peculiar phenomena
displayed by galaxies such as M87 in Virgo, with its nuclear "jet", are due to explosions in the nuclei. Such an
explosion could conceivably begin with a collision of two
stars. Another suggestion is that a supernova explosion in
a dense star region acts as a triggering device, and in

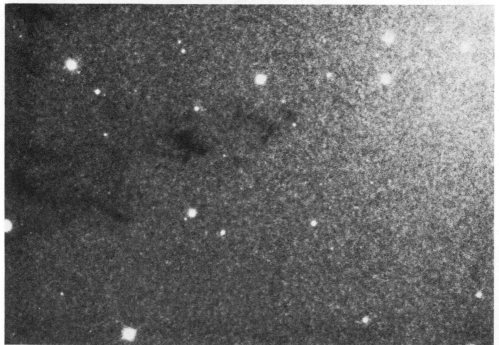

NUCLEUS OF THE ANDROMEDA GALAXY is shown (top) on a plate
made with the 60-inch reflector at Mt.Wilson. Resolution of
the central hub of M31 is evident (below) on a 120-inch
reflector plate made at Lick Observatory.

DESCRIPTIVE NOTES (Cont'd)

some way causes a chain reaction of other explosions. In
some of the strong "radio galaxies" we may be seeing such
phenomena. The intense radio source "Cygnus A" appears to
be a case in point. It is interesting to note that the
center of our own Galaxy has been identified with a very
powerful radio source called "Sagittarius A". This source
is believed to be a very small dense nucleus very similar
to the one observed in M31.

The Andromeda Galaxy has been found to be surrounded
by some 140 objects which have been identified as globular
star clusters from their apparent size, distribution, and
absolute magnitudes. They are apparently comparable with
those in our own Galaxy in size and brightness, and some
have been partially resolved with the 200-inch telescope.
A number of other objects have been identified as planetary nebulae, and many bright emission regions appear along
the spiral arms on red-sensitive photographs. In its content of stars, dust, and gas the Andromeda system strongly
resembles the Milky Way.

The first successful attempt to measure the radial
velocity of the Andromeda system was made by V.M.Slipher
at Lowell Observatory in 1912. At the time the true nature
of the "spiral nebulae" was quite unknown. Using exposures
of up to 7 hours with the 24-inch refractor, Slipher found
a large displacement of the spectral lines toward the blue
end, indicating a high velocity of approach. From measurements of four different spectrograms, the mean velocity
was found to be about 300 kilometers per second.

In announcing the discovery, Slipher stated that "the
magnitude of this velocity, which is the greatest hitherto
observed, raises the question whether the velocity-like
displacement might not be due to some other cause, but I
believe we have at the moment no other interpretation for
it. Hence we may conclude that the Andromeda Nebula is
approaching the Solar System with a velocity of about 300
kilometers per second."

The best of modern measurements give the radial velocity as 266 kilometers per second, but much of this velocity is actually the effect of the motion of our own Sun
in the rotating Milky Way galaxy. Applying this correction,
the true velocity of M31 is reduced to about 35 kilometers
per second in approach. M31 thus does not show a "red—

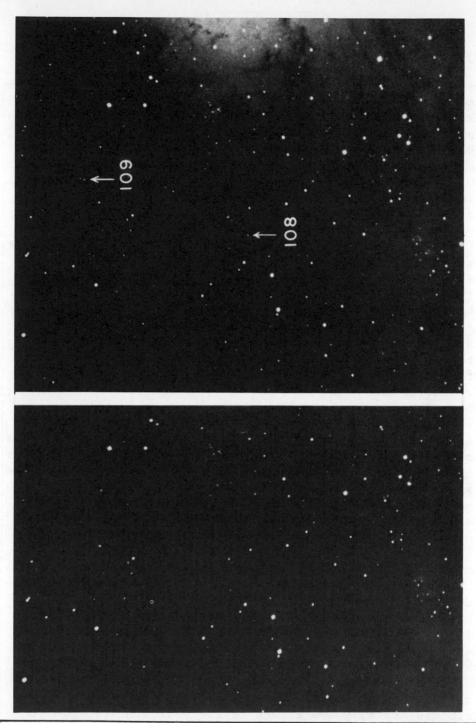

NOVAE IN THE ANDROMEDA GALAXY as identified by E.Hubble in 1932 with the 100-inch telescope at Mt.Wilson. Plate from Hubble's "Realm of the Nebulae", Yale University Press.

shift" as do all the more remote galaxies; it is a member
of the Local Group which contains our own Milky Way, and
the members constitute a gravitationally bound family.

Like our own star system, the Andromeda Galaxy is
known to be in slow rotation about its central mass. It
does not rotate as a solid body, however. The central hub
rotates in only 11 million years, while the outer portions
move more slowly and require from 90 to 200 million years
to make one complete turn. The observed rotation of this
galaxy partly answers the question "Do spiral arms lead or
trail?" In M31 the northwest edge is obviously the nearer
side, and since the radial velocity measurements show that
the southern tip is approaching and the northern tip receding, it is evident that in this particular spiral the
arms trail as the galaxy turns. This is also true for all
the other spirals which are so oriented that a measurement
can be made, and it seems safe to conclude that the question is settled.

NOVAE IN THE ANDROMEDA GALAXY. Since resolution of
M31 was first achieved in 1923, over 100 novae have been
detected in it, and it is estimated that if a constant
watch were kept the total might run as high as 30 per year.
The great distance of M31 makes these stars appear very
faint, from the 15th to the 19th magnitudes. When corrected for distance, however, their actual luminosity is found
to be comparable to the normal novae of our own galaxy.
The faintest of them has a luminosity of about 10,000 suns
and the most brilliant are equal to about 400,000 suns.
The nova of 1925 was, according to Hubble, the brightest
normal nova recorded in M31. Its absolute magnitude was
about -9.3, approximately equal to Nova Aquilae 1918 in
our own galaxy.

Harlow Shapley, in his book "Galaxies", makes the
interesting comment that at the present rate of "novation"
some 50 million novae have probably appeared in M31 in the
last 2 million years. The light waves of all these outbursts are now on their way to the Earth, events of the
far future for us, but of the distant past for the hypothetical inhabitants of the Andromeda Galaxy.

At the distance of M31, even the brightest of the
normal novae is beyond the range of amateur telescopes. In
the year 1885, however, a star appeared near the nucleus

DESCRIPTIVE NOTES (Cont'd)

that exceeded the light of any normal nova by a factor of at least 10,000. Probably near peak brightness when discovered by E.Hartwig on August 20 of that year, the star was near naked-eye visibility with an estimated magnitude of about 6. An observation of the new star was also made by Professor L.Gully in Rouen, France, on August 17, but he failed to realize the significance of the object, and attributed it to a defect in a new telescope! Gully's description leads to a magnitude estimate of 5½ or 6. An analysis of the observations has been made by S.Gaposchkin at Harvard; he finds that the maximum probably occurred on August 17, 1885, and the peak brightness may have been about magnitude 5.4.

The light of the nova decreased for about 5 months, and the star faded from sight in February 1886. According to R.H.Allen it was last seen by A.Hall with the 26-inch refractor at Washington as a 16th magnitude object, on February 1, 1886. This unique phenomenon was of exceptional interest to astronomers; since the nature and distance of the "spiral nebulae" were then unknown, the actual brightness of the new star was a matter of conjecture.

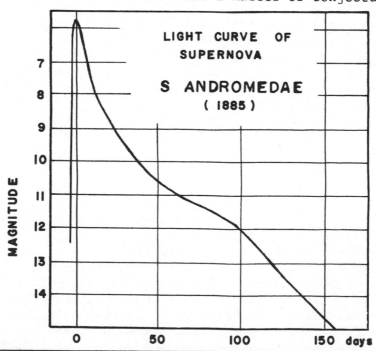

LIGHT CURVE OF SUPERNOVA

S ANDROMEDAE
(1885)

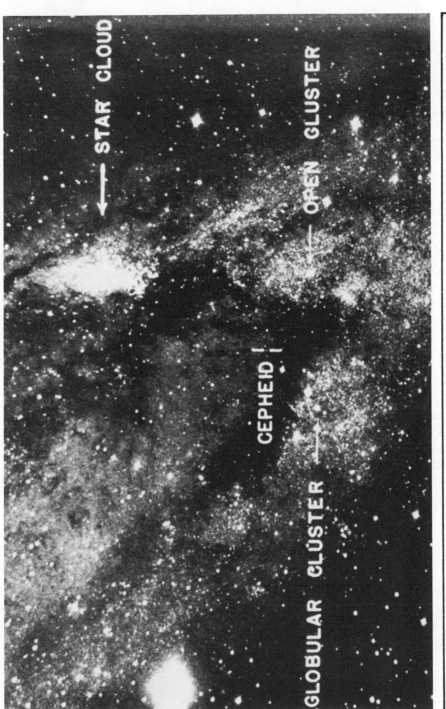

SOUTHERN TIP OF THE ANDROMEDA GALAXY, showing some of the objects identified on 100-inch telescope plates at Mt. Wilson Observatory. This plate originally appeared in E. Hubble's "Realm of the Nebulae" and is reproduced by permission of Yale University Press. The object marked "star cloud" is NGC 206.

Today it is known that the nova of 1885 was in reality one of the most brilliant stars which man has ever viewed; a representative of that wonderful but very rare class of exploding stars known as "supernovae". At a distance of 2.2 million light years, a star which appears as a 6th magnitude object must have an actual luminosity of about 1.6 billion times that of the Sun. The corresponding absolute magnitude would be -18.2. This star, often referred to by its letter designation "S Andromedae", has the distinction of being the first extra-galactic supernova ever observed, though of course its actual significance was not realized at the time. The supernova phenomenon appears to be due to the catastrophic collapse and explosive disintegration of a massive star, an event which may occur, on the average, every 2 or 3 centuries in any one galaxy. In the last thousand years there have been at least four such super-explosions observed in our own Galaxy: the first in Lupus in 1006 AD, the second in Taurus in 1054 AD, now represented by the expanding "Crab Nebula", the third in Cassiopeia in 1572 (Tycho's Star), and the fourth in Ophiuchus in 1604 (Kepler's Star).

M31 AS A RADIO SOURCE. The Andromeda Galaxy has been identified as a source of radio radiation by H.Brown and C.Hazard at Jodrell Bank in England. This radiation, at a frequency of 158.5 megacycles, was detected in 1950 with a paraboloidal antenna 218 feet in diameter, the largest in the world at the time. This was the first detection of radio energy from an external galaxy. A number of other cases are now known, and it is thought probable that every normal galaxy is at least a weak transmitter of radio radiation. The strongest radio sources are definitely abnormal objects such as the peculiar galaxies Cygnus A, NGC 5128 in Centaurus, and M87 in Virgo. In the Andromeda Galaxy, a normal spiral, the radio energy appears to originate in the tenuous gas clouds which occupy the spaces between the stars.

THE COMPANIONS OF THE ANDROMEDA GALAXY. M31 has four small satellite companions, dwarf systems of the elliptical type. All are apparently at the distance of the main system and are gravitationally connected with it. Each is composed of millions of faint stars; resolution of all four systems has been accomplished with red-sensitive exposures on the

NGC 205. The largest of the companions of the Andromeda
Galaxy; resolved into millions of stars on this 200-inch
telescope plate. Mt.Wilson and Palomar Observatories.

NGC 185. One of the dwarf elliptical companions to the Great Andromeda Galaxy. Resolution into stars is shown on this 200-inch telescope plate.   Palomar Observatory

100-inch and 200-inch telescopes. The main facts of interest are given in the following brief table. Luminosities
are in millions of suns, diameters are in light years. M31
itself is listed first for comparison.

NGC	Other	Angular size	Diameter	Mag.	Lum.
224	M31	160' x 40'	110,000	5.0	11,000
221	M32	3.6' x 3.1'	2,400	9.5	70
205	HV 18	8.0' x 3.0'	5,400	10.8	21
185	HII 707	3.5' x 2.8'	2,300	11.8	8
147		6.5' x 3.8'	4,400	12.1	6

For the amateur telescopist, all four of these small
systems are available for observations. M32 may be seen in
field glasses as a fuzzy 9th magnitude "star" just 24' to
the south of the central mass of M31. According to Messier
it was first seen by Le Gentil in 1749. The corrected radial velocity of M32 is about 10 miles per second in recession, probably a result of the orbital motion around the
massive M31. In a 3-inch telescope, NGC 205 is visible as
a larger but dimmer oval blob of light, 35' northwest of
the M31 nucleus. Messier may have been the first to see
NGC 205, in the year 1773, though it was never entered in
his famous catalog. M32 and NGC 205 are conspicuous on
photographs of the Andromeda Galaxy and are well known to
most observers. Both are considerably larger than visual
observations would indicate; densitometer studies of M32
show that the true size is at least 8.5'. These are the
closest galaxies of the elliptical type, and are typical
systems of the Population II class. Only in a small dust
cloud in NGC 205 are any Pop. I stars found.
The other two objects, NGC 185 and NGC 147, are some
distance from the main group, about 7° to the north. They
are much fainter and considerably more difficult to view,
although a good 6-inch telescope is capable of showing
both of them when sky conditions permit. They are 58'
apart and can be seen together in a low power wide-field
ocular. With absolute magnitudes of about -12.8 and -12.5
these are among the intrinsically faintest galaxies known.
A feature of interest is the small irregularly-shaped dust
patch which can be seen on 200-inch telescope photographs

of NGC 185. The remainder of the galaxy appears to be pure
Population II.   NGC 185 and 147 are actually located in
the constellation of Cassiopeia, and are described here
only because of their physical connection with the Great
Andromeda Galaxy.

THE LOCAL GROUP. The Milky Way System and M31 are the
two brightest members of a small cluster of galaxies known
as the Local Group. At least 20 members are now recognized
and additional faint systems may yet be found. For information on other members refer to: The Magellanic Clouds in
Dorado and Tucana, M33 in Triangulum, NGC 6822 in Sagittarius, IC 1613 in Cetus, and the peculiar dwarf galaxies
"Fornax System" and "Sculptor System".

NGC 752   A large scattered cluster of fairly bright
stars, located about 5° south of Gamma Andromeda and slightly west, at 01548n3726. The group is actually more conspicuous in good binoculars than in the average telescope, due to its large area and low density; it
makes its best impression in relatively small rich-field
instruments with wide-angle eyepieces. The apparent diameter is about 45', the members ranging in brightness from
9th to 12th magnitude.

In an early study of the cluster, E.G.Ebbighausen
(1939) obtained proper motions for 125 stars in the group,
identifying 39 stars as almost certain members and 24
others as very probable. The annual proper motion of the
cluster was found to be about 0.012" in PA 160°; the radial velocity is about 2.5 miles per second in approach.
According to a summary by H.Arp (1962) the distance is
close to 1300 light years; the actual diameter must then
be about 17 light years. The brightest stars are listed
in the short table below, according to photoelectric

1.	Mag 8.94;	Spect G7	9.	Mag 9.47;	Spect K0
2.	" 8.94	" K1	10.	" 9.58	" F3
3.	" 8.95	" K0	11.	" 9.70	" A0
4.	" 9.01	" K0	12.	" 9.80	" F5
5.	" 9.04	" K0	13.	" 9.88	" F4
6.	" 9.14	" F5	14.	" 9.91	" F5
7.	" 9.29	" G9	15.	" 9.95	" F4
8.	" 9.35	" K0	16.	" 10.01	" F5

DESCRIPTIVE NOTES (Cont'd)

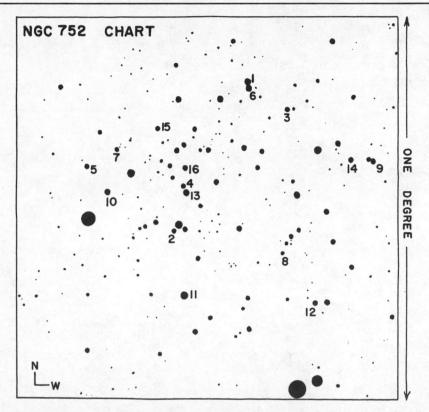

NGC 752 CHART

ONE DEGREE

measurements by H.L.Johnson at McDonald Observatory in
1952. Not included is the apparently brightest star in
the cluster (magnitude 7.1) which has been found to be
a non-member.

Star #1 is a G7 III giant about 40 times brighter
than the Sun; the absolute magnitude is about +0.7. The
membership of Star #2 is somewhat uncertain since the radial velocity seems to be much higher than the cluster
stars; it may very well be a foreground object. Omitting
this star, there are 7 other orange giants in the cluster
but virtually all the other members are F-type subgiants
with the single exception of #11. This A0 star seems to
be confirmed as a cluster member by both proper motion
and radial velocity measurements. With this one exception
the cluster is characterized by a complete absence of
early-type stars.

Although not spectacular visually, NGC 752 is a most

NGC 752. The field of the widely-scattered star cluster
is shown on this plate made with the 13-inch telescope
at Lowell Observatory.

DESCRIPTIVE NOTES (Cont'd)

unusual and interesting cluster. Its stellar population, as shown by the familar H-R diagram or color-magnitude graph, seems to place it somewhere between the normal galactic clusters and the globulars in the matter of age and evolutionary development. Its structure and space motion definitely class it as a galactic cluster, but the stars seem to be evolving toward the typical pattern displayed by the H-R diagram of a globular cluster. The majority of members are F-type subgiants which lie well above the main sequence, and are presumably evolving toward the giant stage. The cluster population appears to end rather abruptly just above absolute magnitude +4; no fainter members are known, and probably do not exist.

In an analysis of the H-R diagram, N.G.Roman (1955) finds that "the differences between the normal and the NGC 752 main sequences is very similar to the change of the theoretical main sequence with time". Current ideas of stellar evolution suggest a minimum age of about 1.5 billion years for the cluster, older than most of the well known galactic clusters, but younger than M67 in Cancer and NGC 188 in Cepheus. Two other clusters which seem to be of comparable age and type are NGC 7789 in Cassiopeia and NGC 2158 in Gemini. H.Arp (1962) refers to these as "intermediate-age star clusters" and suggests that they were formed in the outer regions of the Galaxy where star

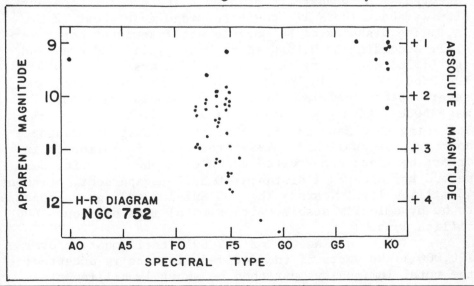

DESCRIPTIVE   NOTES   (Cont'd)

formation is slower and the interstellar gas is less rich
in the atoms of the metals. The fact that the subgiants of
NGC 752 are metal-poor by a factor of 2 has been confirmed
by spectroscopic analysis. It is also possible that the
lack of smaller low-mass stars is due to their gradual escape from the cluster during the long period since its
formation.

In view of the relatively great age of NGC 752, it
would be interesting to identify possible white dwarf members of the cluster. At the distance of the group the most
luminous white dwarfs would appear about 18th magnitude,
and would be difficult to detect. In a preliminary study,
W.J.Luyten (1961) found a number of faint white and bluish
stars in the cluster area; nine of these have apparent
magnitudes of 19 and 20 and may be cluster members, though
no positive proof is yet available.

NGC 891   One of the most striking examples of a spiral
              galaxy seen exactly edge-on; equalling in
interest the more famous NGC 4565 in Coma Berenices. It is
located about midway between Gamma Andromedae and the open
cluster M34 in Perseus, at 02193n4207. The galaxy is not
an easy object in the small telescope since the surface
brightness is quite low, but on a clear night it may be
detected with an aperture of 5 or 6 inches. The apparent
size is about 12' X 1'; the total magnitude about 12.

The distance of NGC 891 is still somewhat controversial. According to M.L.Humason, N.U.Mayall, and A.R.Sand-
age (1956) the galaxy is a member of a small group which
includes NGC 1023 in Perseus, NGC 925 in Triangulum, and
several other members. The computed modulus is about 29
magnitudes, giving a distance of 20 million light years.
According to J.Materne (1974) of the Hamburg Observatory,
however, the published Humason red-shift is seriously in
error; he finds a corrected value of about 435 miles per
second and accepts a distance of 13.1 megaparsecs, or about
43 million light years. The NGC 891-1023-925 group appears
to be dynamically stable, with a total mass of about 800
billion suns.

The true diameter of NGC 891 itself must be over
120,000 light years if the derived distance is accepted;
the total luminosity must then be about 1½ billion times

NGC 891 in ANDROMEDA. A prime example of an edge-on spiral galaxy. 60-inch telescope photograph, Mt.Wilson and Palomar Observatories.

DESCRIPTIVE NOTES (Cont'd)

that of the Sun. The absolute magnitude may be about -18.6. This is several magnitudes fainter than M31.

Photographs taken with great telescopes show a complex system of dark clouds extending in the form of an equatorial band across the entire length of the galaxy, well defined in silhouette as it crosses the nucleus, and breaking up into irregular masses farther out in the region of the spiral arms. Much of these details are revealed only by long-exposure photography, though the dark band itself was known in the time of Lord Rosse, and is clearly shown on his drawings made in 1850 with the 6-foot reflector at Parsonstown, Ireland. According to John Herschel, the visual appearance of the object gave the impression of "a thin flat ring of enormous dimensions, seen very obliquely". One of the very early plates of this galaxy was a fine photograph made by Isaac Roberts with his 20-inch reflector in 1891, and showing the dark lane much as it appears on modern photographs. Measurements by C.K.Seyfert (1940) show that the dark lane is not truly "dark" but is 0.6 to 0.9 magnitude fainter than the surrounding brightness of the galaxy.

In our own Milky Way, similar clouds of dust and dark nebulosity are responsible for the irregularities and dark lanes in the Milky Way. The famous "Great Rift" which runs from Cygnus through Sagittarius is a prime example. When we observe NGC 891, the explanation for such phenomena is quite plain, and it is interesting to note that wide-angle camera photographs of the Milky Way strikingly resemble photographs of NGC 891. (To make the comparison, refer to the section on the Sagittarius Milky Way)

NGC 7662   A bright, slightly elliptical planetary nebula, measuring 32" X 28", bluish-green in color. Position 23234n4212. It can be detected with very small telescopes as a nearly stellar object of magnitude 8½, half a degree southwest of the 5th magnitude star 13 Andromedae. With a 6-inch glass and a magnification of at least 50X it begins to show a softly glowing disc. In a 10-inch glass the darker center gives it an annular appearance; the central star is a difficult object visually, but appears clearly on photographs. Using the 40-inch Yerkes

DESCRIPTIVE NOTES (Cont'd)

refractor, E.Barnard described this nebula as "a beautiful object- a slightly elliptical disc with quite sharply defined outlines. Unsymmetrically placed on this is a roughly elliptical broken ring of greater brightness. The interior of this ring is dark but not black, and in this, approximately central, is ordinarily a faint stellar nucleus."

Like many of the planetaries, NGC 7662 shows some remarkable color effects when seen with a large telescope. The main body of the nebula appears to glow with a bright bluish green color, strongest in the concentric "shells" which enclose the darker center. These details are enclosed in a larger fainter shell which often seems to be of a rosy or pinkish tint, according to some observers, very possibly due at least in part to the effect of contrast. The central star, a hot bluish dwarf, often appears yellow by contrast with the bluish tint of the nebulosity.

The distances of the planetary nebulae are not known with any real accuracy. According to the Skalnate Pleso Catalogue (1951) the distance of NGC 7662 is about 1800 light years, the actual diameter about 20,000 AU. In a survey of the brighter planetaries, C.R.O'Dell (1963) derived a distance of 1740 parsecs or about 5600 light years for this nebula, increasing the actual size to 0.8 light year, or nearly 50,000 AU. The central star is a bluish dwarf with a continuous spectrum and a computed temperature of about 75,000°K. The nuclei of the planetary nebulae are among the hottest stars known.

In the years between 1897 and 1908, E.Barnard found evidence of variability in this central star. His observations, made on nearly 80 different dates, showed a magnitude range of 12th to 16th; the periods of greatest brightness were not long lasting and occurred at irregular intervals. The reality of these changes has been questioned, however, by modern observers. C.R.O'Dell points out that the apparent brightness of a star surrounded by strong nebulosity is critically correlated with the seeing. "As the seeing varies, the ability to discern the star will change because of the superposition of the nebula, while nearby comparison stars will not be affected." As mentioned elsewhere in this book, the central star of the Ring Nebula M57 in Lyra has also been suspected of variability

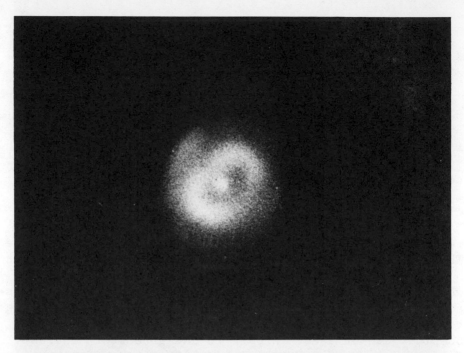

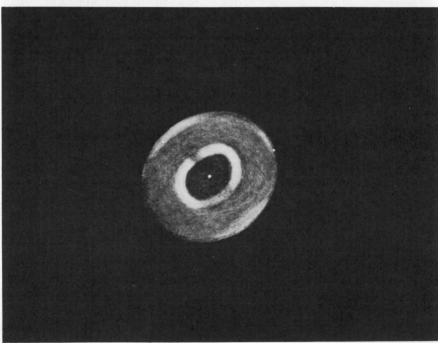

NGC 7662. Top: Direct photograph with the 42-inch reflector at Lowell Observatory. Below: Drawing by Barnard, with the Yerkes 40-inch refractor, Courtesy of the Royal Astronomical Society, from "Monthly Notices" Volume 68 (1908)

DESCRIPTIVE NOTES (Cont'd)

but the physical reality of such changes remains unproved.
Amateurs with fairly large telescopes have an opportunity
to contribute information of value toward a solution of
this controversy. The visibility, or otherwise, of the
central star should be recorded on different nights, with
estimates of the apparent magnitude; then the seeing conditions should be recorded by making critical observations
on various close double stars.

As in all the planetary nebulae, much of the light is
fluorescence, induced by strong ultraviolet from the hot
central star. The characteristic bluish-green glow, however, once attributed to a hypothetical new element "nebu-
lium", is now known to be chiefly due to the so-called
"forbidden lines" of doubly ionized oxygen at 5007 and 4959
angstroms. The great strength of this radiation, excited
by electron collision in the nebula, is possible only because of the extremely low density of the gas, found by
computation to average something like 10,000 atoms per
cubic centimeter. (For a more detailed account of fluores-
cense and collision excitation, refer to the "Ring Nebula"
M57 in Lyra.)

To the astrophysicist the planetary nebulae present
many interesting problems. There is no doubt that the
nebula - a huge globe of rarified gas surrounding a small
super-hot star - has been produced in some way by material
ejected from the star. But it seems clear that one of the
oldest theories, which regarded the planetaries simply as
ancient novae, is quite wrong. The gaseous shells which
appear around a former nova bear a superficial likeness to
planetary nebulae, but expand at enormous rates and seem
to vanish after a relatively short time. In contrast, the
planetaries seem fairly permanent structures, expanding
quite slowly, and in some cases seeming to be maintained
by steady outflow of material from the star. Thus the central stars may be regarded as some variety of eruptive or
"emission" star, possibly related to the Wolf-Rayet stars
or to some of the rapidly evolving red giants which are
known to be ejecting material into space. Another view
regards a planetary nebula as the result of an exceptionally "lazy" nova. (A survey of facts and theories is presented in the section on M57 in Lyra. See Also NGC 7009 and
NGC 7293 in Aquarius, and M27 in Vulpecula.)

LIST OF DOUBLE AND MULTIPLE STARS

NAME	DIST	PA	YR	MAGS	NOTES	RA & DEC
λ113	11.6	135	28	6 - 14	Optical, PA dec; dist inc; spect gK3	09277s2622
ζ^1	8.0	212	52	6 -6½	($\triangle$78) relfix; spect both A0	09286s3140
h4218	5.9	30	54	7½-10½	relfix; spect A0	09311s3611
B185	3.6	205	55	7½-10½	relfix; spect A0	09338s2718
I 202	0.6	148	60	6½ -9½	binary, PA & dist dec; spect F5	09366s3923
I 1519	0.2	6	56	8 - 8	spect F0; no certain change	09447s3906
I 172	1.0	321	59	7½-10½	PA inc, spect F8	09454s3730
H1d 99	1.9	220	54	8 - 9	relfix , spect A0	09456s2722
I 205	1.7	337	59	7 - 10	PA dec, spect F0	09465s2611
h4249	4.3	123	52	8 - 8	neat pair; spect A3, no certain change	09466s3447
Rst 5341	1.2	5	59	6½-10½	spect F8	09507s2704
β215	1.7	343	43	7 - 9½	relfix, spect B9	09518s2746
I 842	3.9	31	43	7 -10½	relfix, spect A2	09524s3440
Arg 23	57.6	82	11	7½-10½	Spect A3	09547s2818
η	31.0	318	20	5 - 12	(h4271) spect F0, cpm pair	09567s3539
I 292	0.3	142	60	7½ - 8	binary, PA & dist dec, spect F8	10020s2808
I 293	0.3	324	60	7 - 8	dist dec, spect A0	10029s2757
Ho 371	6.2	43	38	6½ -12	relfix, spect K0	10035s3039
B194	0.2	45	59	7 - 7	PA inc, spect A0	10098s2821
h4300	9.1	108	41	8½- 9	relfix, spect F8	10135s3302
I 851	0.2	286	60	8½ -8½	PA inc, dist dec, spect F5	10140s2844
h4304	9.5	286	34	7½ -10	relfix, spect A2	10180s3253
I 209	1.2	131	54	8 - 8	slight PA dec, spect F0	10222s3819
I 210	0.9	237	36	7½ -10	relfix, spect F8	10256s3827
δ	11.0	226	32	6 - 9½	relfix, spect B9	10273s3021
I 1202	3.7	131	34	7 - 9½	spect B9	10320s3708
B2001	0.7	63	60	6½ -8½	relfix, spect G5	10386s3529

LIST OF DOUBLE AND MULTIPLE STARS (Cont'd)

NAME	DIST	PA	YR	MAGS	NOTES	RA & DEC
h4342	25.3	54	19	8 - 11	spect A0	10394s3029
h4381	25.8	42	33	7 - 8	spect B9 & A	10523s3829
I 864	1.0	356	47	8½ -8½	relfix, spect G0	10581s3959
	20.9	39	11	-12		

LIST OF VARIABLE STARS

NAME	MagVar	PER	NOTES	RA & DEC
R	7.2--7.8?	---	Uncertain; possibly not variable; spect A0	10076s3729
S	6.3-- 6.8	.6483	Ecl.bin.; W Ursa Major type; spect F0	09301s2824
T	8.8--10.4	5.898	Cepheid; spect G2	09318s3624
U	5.7---6.8	170	Semi-reg; spect Nb	10330s3918
V	7.7--12...	303	LPV. Spect M7e	10189s3433
W	9.0--10...	158	Semi-reg; spect M3	09489s2940
X	8.4--11...	161	LPV. Spect M2	10045s2950
Y	9.4---9.9	3.052	Ecl.bin.; spect F2	10063s3457
Z	9.0--11...	105	Semi-reg	10437s3459
RR	8.0--9...	Irr	Spect M8	09336s3941
WY	9.8--10.8	.5743	Cl.Var.	10138s2928

LIST OF STAR CLUSTERS, NEBULAE, AND GALAXIES

NGC	OTH	TYPE	SUMMARY DESCRIPTION	RA & DEC
2997	50[5]	⊖	Sc; 11.0; 6.0' x 5.0' ! vF,vL,vgvsbMN	09435s3058
3001		⊖	Sb; 13.2; 1.2' x 1.0' F,S,R	09441s3013
3038		⊖	Sa; 12.9; 1.0' x 0.7' pB,pS,R	09492s3232

LIST OF STAR CLUSTERS, NEBULAE, AND GALAXIES (Cont'd)

NGC	OTH	TYPE	SUMMARY DESCRIPTION	RA & DEC
3056		⊖	E0; 12.8; 0.5' x 0.5' pB,S,R,vgbM	09523s2804
----	I.2522	⊖	Sc; 12.9; 1.5' x 1.5' vF,cL,R	09531s3254
3087		⊖	13.0 pB,S,R	09570s3359
3089		⊖	Sp; 13.0; 0.7' x 0.6' pF,pS,R	09573s2804
3095		⊖	SB; 12.7; 2.0' x 1.2' F,L,E,vg1bM	09579s3118
----	I.2537	⊖	S ; 12.8; 2.0' x 1.4' eF,L,cE	10017s2719
3125		⊖	E2; 13.0; 0.5' x 0.4' cF,S,R,vgbM	10042s2941
3175		⊖	Sb; 12.1; 2.0' x 1.5' cB,L,mE,vg1bM	10124s2838
3223		⊖	S ; 12.1; 3.5' x 1.4' pB,vL,1E,ps1bMN	10194s3400
3241		⊖	Sa; 13.0; 1.0' x 0.7' F,pmE,g1bM	10221s3213
3250		⊖	E4; 12.4; 1.2' x 0.7' pB,pL,R,vgbM	10243s3941
3258		⊖	E1; 13.0; 0.9' x 0.8' cF,S,R,ps1bM	10266s3520
3268		⊖	E2; 13.0; 1.0' x 0.8' F,S,1E	10276s3506
3271		⊖	Sa; 12.9; 1.0' x 0.6' pF,S,E, bM	10282s3506
3275		⊖	SB; 12.8; 1.3' x 1.0' F,L,v1E,1bM	10286s3628
3281		⊖	Sb; 12.9; 2.0' x 1.0' F,pL,E, 1bM	10297s3436
3347		⊖	SBb; 12.8; 4.0' x 2.0' pF,L,mE,vmbM	10405s3606
3358		⊖	Sa; 13.0; 0.7' x 0.5' cF,vS,1E	10413s3607
3449		⊖	Sb; 13.2; 2.0' x 0.8' F,S,R	10506s3240

APUS

LIST OF DOUBLE AND MULTIPLE STARS

NAME	DIST	PA	YR	MAGS	NOTES	RA & DEC
h4667	2.4	141	46	8 - 8½	Relfix, spect A0	14180s7320
h4671	5.1	127	40	7½ - 8	relfix, spect F8	14232s7953
I 326	2.3	118	33	7½ -10	slight PA dec;	14264s7604
	14.8	10	33	-13	spect K0	
h4695	17.7	290	18	7 - 12	spect B9	14462s7444
I 236	2.0	115	47	5½ -8½	PA inc, spect G5	14482s7259
Hd 241	30.0	35	00	6½ -13	spect B9	14544s7450
Cp 15	1.6	43	47	7 - 8½	relfix, spect A0	15013s7159
I 969	2.2	319	30	8 -10½	slight dist inc,	15337s7208
h4787	9.9	304	40	7½ -8½	relfix, spect G5	15418s7928
h4790	12.5	348	19	7½-10½	spect A0	15448s7834
I 333	0.3	106	32	7 - 7½	PA & dist dec;	15528s7753
					spect F5	
δ^1 - δ^2	103	12	18	5 - 5	wide cpm pair;	16128s7834
					spect M4, K5	
Hd 255	2.2	246	31	6½ -12	spect K0	16350s7212
I 100	0.7	178	32	7 - 9	spect B9	16537s7321
h4884	34.9	8	40	7 - 9	relfix, spect A0	16593s8215
h4904	7.1	187	40	7½ - 9	relfix, spect F2	17035s7519
I 104	1.5	132	34	6½ -10	slight dist dec;	17146s6959
					spect G2	
h4947	9.7	72	40	8 - 8½	slight dist dec;	17381s8152
					spect K0	
Hd 275	0.6	336	59	7 -7½	PA dec; spect F5	17383s7212
h4974	25.5	120	18	6 - 14	spect K2	17506s7610
h4999	13.1	173	40	7½- 8½	relfix, spect A0	18038s7512
L7507	2.5	243	35	6 - 9½	(Hd 284) PA inc,	18063s7341
					spect F6	

APUS

LIST OF VARIABLE STARS

NAME	MagVar	PER	NOTES	RA & DEC
θ	5.1---6.7	119	Semi-reg; spect M4p	14004s7633
R	5.0---6.3	Irr	Variations uncommon; Spect K--gM0?	14520s7628
S	9.5--15..	---	R Corona Borealis type; Spect R3	15043s7153
T	8.4--15.0	261	LPV. Spect M3e	13509s7734
V	9.5--10.5	---	Irr; spect Mb	14598s7125
W	9.5--11..	---	Irr; spect M6	17079s7406
Z	9.0--11..	---	Irr; spect M	14026s7108
RY	9.0--15..	383	LPV.	14283s7143
SV	9.0--15..	303	LPV.	14339s7218
VY	9.5--10.5	152	Semi-reg	15548s7421
VZ	7.0--15..	377	LPV. Spect Me	16102s7355
WW	7.8--15..	267	LPV. Spect Me	16253s7453

LIST OF STAR CLUSTERS, NEBULAE, AND GALAXIES

NGC	OTH	TYPE	SUMMARY DESCRIPTION	RA & DEC
5612		⊘	S0; 13.0; 1.1' x 0.4' vF,E,gbM	14282s7811
----	I.4499	⊕	Mag 11½; diam 5', class XI; F,L,1C	14527s8202
5967		⊘	SB; 12.9; 2.5' x 1.5' F,pL,1E,gbM	15419s7531
6101	△68	⊕	Mag 10; diam 4'; class X; pF,L,R, stars mags 14.....	16200s7206

AQUARIUS

LIST OF DOUBLE AND MULTIPLE STARS

NAME	DIST	PA	YR	MAGS	NOTES	RA & DEC
1	59.7	221	24	$5\frac{1}{2}$- 11	(h2984) spect K1;	20368n0019
	69.8	36	24	$-11\frac{1}{2}$	both optical	
Ho 135	2.6	224	44	$7\frac{1}{2}$-$12\frac{1}{2}$	relfix, spect F0	20382s1442
Rst 5471	0.6	39	50	$7\frac{1}{2}$-$10\frac{1}{2}$	spect K0	20459n0132
4	1.1	3	61	$6\frac{1}{2}$ -$7\frac{1}{2}$	binary., about 150	20488s0549
					yrs, spect dF3	
7	2.1	166	40	6- $11\frac{1}{2}$	(β1034) relfix,	20542s0953
					spect gK5	
Hwe 55	26.2	72	28	7- $10\frac{1}{2}$	cpm, spect K2	20546n0016
	35.4	116	59	- 13		
Σ2744	1.5	137	61	$6\frac{1}{2}$- 7	PA dec, spect F5	21005n0120
12	2.8	192	50	$5\frac{1}{2}$ -$7\frac{1}{2}$	(Σ2745) relfix,	21014s0601
					cpm; spect gG4, A3	
Σ2752	4.9	167	59	$6\frac{1}{2}$-$10\frac{1}{2}$	PA inc, AB cpm;	21044s1407
	11.7	301	59	-12	(β157) spect K0;	
					AC optical	
β368	0.2	270	61	$7\frac{1}{2}$ -$7\frac{1}{2}$	dist & PA inc;	21048s0826
	12.1	29	00	$-13\frac{1}{2}$	spect A0	
Σ2755	24.3	83	31	$6\frac{1}{2}$- 10	relfix, spect M	21050s0022
Σ2770	7.7	245	37	7- $10\frac{1}{2}$	relfix, spect K0	21090s0320
Σ2775	0.2	123	61	8 - 8	(A883) binary, 78	21121s0102
	21.4	178	61	- 10	yrs; PA inc; ABC	
					cpm; spect A0	
Σ2781	2.8	172	62	8 - 8	relfix, spect F2	21141s0752
Σ2787	22.7	20	31	7 - $8\frac{1}{2}$	relfix, spect A2	21192n0149
	70.5	94	08	- 10		
β	35.5	318	62	3 - 11	(h936) spect G0;	21289s0548
	58.6	187	62	$-11\frac{1}{2}$	both optical (*)	
Ho 288	18.9	278	06	$6\frac{1}{2}$ -13	relfix, spect A2	21315s0436
Σ2809	31.1	163	23	6 - $8\frac{1}{2}$	cpm, relfix;	21350s0037
					spect A2	
24	0.4	296	66	$7\frac{1}{2}$ - 8	binary, 49 yrs;	21369s0017
					(β1212) PA inc,	
					spect F7	
A180	0.8	229	62	9 - 9	no certain change	21403s0238
					Spect F8	
β1263	0.5	212	49	$8\frac{1}{2}$ -10	relfix, spect K0	21422n0236
Σ2825	0.6	130	67	8 - 8	PA inc, spect F2	21443n0037
Σ2838	17.6	184	59	6 - 9	optical, spect F8	21520s0333
β693	0.9	48	59	8 - 10	PA dec, spect B9	21536s0713

LIST OF DOUBLE AND MULTIPLE STARS (Cont'd)

NAME	DIST	PA	YR	MAGS	NOTES	RA & DEC
Σ 2847	1.1	307	62	7½- 8	relfix, spect F0	21555s0343
Hu 282	0.4	38	59	7½- 8½	slow PA inc; spect F0	21579s1359
h5524	9.5	271	13	7 -10½	relfix, spect G5	21588s1551
	103	292	18	-10		
Σ 2851	19.1	121	35	8 - 8½	relfix, cpm; both spect G5	21590s1214
29	3.9	244	50	7 - 7	(S802) relfix, spect A2	21597s1713
Σ 2862	2.5	99	62	7½ -8	PA slight dec; spect G0	22045n0019
β170	1.1	47	59	9 - 9½	PA dec, spect F8	22064s1844
β475	1.3	203	57	7½-10½	PA dec, spect F2	22099s0815
41	5.0	114	59	6 - 7½	(Hh753) slight PA dec, spect K0, F8	22115s2119
Σ 2887	8.6	28	22	9 - 9	relfix, spect K0	22148s0057
51	0.5	320	61	7 - 7	(β172) PA dec, spect A0	22215s0506
	54.4	342	17	-10		
S808	6.8	152	51	8 - 11	relfix, spect F8	22231s2029
53	4.0	325	61	6½ - 7	(Hh762) dist dec,	22238s1700
53 b	46.7	339	01	- 13	PA inc, spect G1, G2; C= 1.8" pair	
ζ	1.7	249	67	4½- 4½	(Σ 2909) spect F2 fine binary (*)	22262s0017
β76	1.7	350	60	8 - 10	PA slow inc, spect F5	22270s0028
β1264	4.1	22	44	8 - 13	relfix, spect A0	22276s0007
Σ 2913	8.2	329	30	7 - 8	relfix, spect F0	22279s0823
β77	2.8	213	61	9 - 10	relfix, spect F0	22314s0202
β770	1.4	343	59	8 -11½	slow PA dec, spect G0	22316s2252
H96	50.9	247	10	7½ -9½	spect K2	22332s2111
A2695	0.3	116	59	7 -8½	spect G0	22358s0809
Σ 2928	3.4	300	61	8 - 8	PA dec, spect G5; dist slight dec.	22369s1252
h3128	10.9	227	34	7 - 12	relfix, spect F8	22373s1927
Kui 114	0.2	124	58	7 - 7½	spect G0	22382s0348
Σ 2936	4.7	49	46	7 - 10	relfix, spect A3	22403n0057
Σ 2935	2.4	310	62	7 - 8	relfix, spect A2	22404s0834
	70.2	357	24	-10½		

LIST OF DOUBLE AND MULTIPLE STARS (Cont'd)

NAME	DIST	PA	YR	MAGS	NOTES	RA & DEC
Σ 2939	11.1	62	09	7½-10½	relfix, spect A5	22427s0954
69	25.3	119	30	6 - 9	(Σ2943) (τ')	22450s1419
					optical, spect B9	
Σ 2944	2.5	276	61	7 - 7½	dist dec, PA inc,	22453s0429
	49.7	106	55	- 8	spect G0, G0	
β 178	0.3	316	61	6 - 8	PA dec, spect gG7	22526s0516
Σ 2970	8.5	38	45	8½- 9	relfix, spect F5	22598s1135
β 384	1.1	65	59	7 - 9	PA dec, spect A0	23000s1849
83	0.1	342	61	6 - 6	(A417) binary,	23026s0758
					22 yrs; spect F2	
86	2.9	83	25	5 - 14	(B588) spect gG9	23040s2401
Σ 2981	3.5	114	62	9 - 9	relfix, spect G5	23069s0906
89	0.4	7	59	5½ - 6	(Rst 3320) PA dec	23072s2244
					Spect sgG2, A2	
Σ 2988	3.5	101	61	7 - 7	relfix, spect K0	23094s1213
β 181	1.4	309	59	7- 10½	spect K2	23112s1341
	18.9	243	59	- 12		
Σ 2993	25.5	177	50	7 - 8	relfix, cpm;	23114s0912
					spect G0, G	
β 715	3.5	254	55	7 -11½	no certain change,	23121s1058
					spect K5	
91	49.6	312	38	4½ -8½	(ψ') (ΣII 12)	23133s0922
91b	0.3	105	58	9 - 9	ABC all cpm;	
	19.7	341	24	-12½	Primary spect K0	
94	13.0	350	58	5 - 7	relfix, cpm;	23164s1344
					spect G5, K2	
95	1.4	167	62	5 - 10	(ψ³) PA dec;	23164s0954
					spect A0	
96	10.6	20	58	6½ -11	(h5394) spect F2,	23168s0524
					slight dist inc.	
h3184	5.4	283	53	7 - 9	relfix, spect G5	23183s1849
97	0.2	79	61	5½ - 7	(Hu295) binary,	23200s1519
					63 yrs, PA inc;	
					spect A3	
Σ 3008	4.0	175	61	7 - 8	optical, PA & dist	23212s0844
					dec, spect K0	
I 1058	1.9	233	13	6½- 10	spect F5	23215s2203
λ 485	5.2	131	33	6 - 12	spect K0	23240s2201
Hu 297	0.5	292	51	7 - 9	PA dec, spect A3	23243s1531

LIST OF DOUBLE AND MULTIPLE STARS (Cont'd)

NAME	DIST	PA	YR	MAGS	NOTES	RA & DEC
101	0.9	126	59	5 - 7	slight dist inc; spect A1	23307s2112
h3202	3.7	239	19	8 - 9½	spect G5	23326s1851
h316	33.1	93	09	6 - 9½	spect gG6	23351s1320
ω²	5.7	86	58	5 - 11	(105) (β3279) cpm relfix, spect A0	23401s1448
β725	4.3	240	33	7 - 11	relfix, spect K0	23402s1136
107	6.6	137	62	5½ -6½	(Hh807) fine pair, dist inc, spect A5, F2	23434s1857
β729	11.5	344	16	8 - 12	relfix, spect F5	23528s1806

LIST OF VARIABLE STARS

NAME	MagVar	PER	NOTES	RA & DEC
χ	5.0---5.3	Irr	Spect gM5	23143s0800
R	5.9--11.4	386	"Symbiotic star"; spect gM7ep (*)	23412s1534
S	7.8--14.2	279	LPV. Spect M4e	22544s2037
T	6.7--13.5	202	LPV. Spect M3e	20473s0520
U	9.5--14..	Irr	period at times ± 250 d.	22006s1652
V	7.8--10.1	245	LPV. Spect M6e	20443n0215
W	7.0--15.0	381	LPV. Spect M7e	20438s0416
X	8.0--14.5	311	LPV. Spect M4e	22159s2109
Y	8.4--14.0	382	LPV. Spect M8	20418s0501
Z	8.0--10.2	137	Semi-reg; spect M1e-M3e	23497s1608
RR	8.3--13.9	183	LPV. Spect M2e	21124s0306
RS	9.0--14.5	215	LPV. Spect Me	21084s0414
RT	8.7--13..	245	LPV. Spect M6e	22205s2218
RU	8.6-- 9.9	69	Semi-reg; spect M5	23218s1736

LIST OF VARIABLE STARS (Cont'd)

NAME	MagVar	PER	NOTES	RA & DEC
RV	8.7--12..	453	LPV. Spect Ne	21033s0025
RW	8.7--13.6	140	LPV. Spect M2e	21205n0037
RX	8.4---9.5	Irr	Spect M5	21100s1436
RY	8.8--10.1	1.967	Ecl.bin.; spect A3	21175s1101
SS	8.6--13.2	208	LPV. Spect M2e	22172s1439
SV	9.0--10.5	Irr	Spect Mb	23202s1105
SZ	9.1--11.3	108	RV Tauri type; spect K5	22401s2127
TT	8.4---9.4	Irr	Spect M5	22523s0939
VX	9.0--12..	369	LPV.	20599s0648
VY	8.4-- 16	---	Recurrent nova, maxima in 1907, 1962	21095s0902
VZ	11.5-- 15		SS Cygni type	21278s0313
WY	9.0--13..	250	LPV.	22127n0125
AE	10.7--11½	Irr	Peculiar erratic star; spect dK0 + sdB? (*)	20376s0103
BK	9.0---9.5	70	Semi-reg	21064s0517
BS	8.8---9.2	.1978	Cl.Var.; Spect F2	23462s0826
BW	9.7--10.2	6.720	Ecl.bin.; spect F7	22206s1535
CY	10½--11½	.061	Dwarf Cepheid (resembles SX Phoenicis) Spect B8---A3 (*)	22352n0117
DV	5.9---6.5	1.576	Ecl.bin; spect A3	20559s1441
DX	6.2---6.8	0.945	Ecl.bin; spect A2	21597s1712
DZ	8.6--10..		Semi-reg; spect M5	22191s0752
EE	8.0---8.7	.5090	Ecl.bin; spect F0	22587s2007

LIST OF STAR CLUSTERS, NEBULAE, AND GALAXIES

NGC	OTH	TYPE	SUMMARY DESCRIPTION	RA & DEC
6981	M72	⊕	Mag 8.6; diam 3', class IX, pB,pL,R,gmCM, rrr (*)	20508s1244
6994	M73	⁂	small triangle of 3 stars; not a true cluster (*)	20562s1251
7009	1⁴	◎	↓ vB,S, Mag 8, diam 25"; bright bluish-green disc; "Saturn Nebula" (*)	21014s1134

LIST OF STAR CLUSTERS, NEBULAE, AND GALAXIES (Cont'd)

NGC	OTH	type	SUMMARY DESCRIPTION	RA & DEC
7089	M2	⊕	Mag 6.0; diam 7'; class II, !! B,vL,mbM,rrr; stars mags 13..... (*)	21309s0104
7171	692[3]	⊘	Sb; 12.8; 2.5' x 1.5' vF,cL,E, vgbM	21583s1331
7184	1[2]	⊘	Sb; 12.0; 5.0' x 1.0' pB,pL,mE,bMN; with Saturn-like inner ring.	21599s2104
7218	897[2]	⊘	Sc; 12.7; 1.8' x 0.7' pB,lE; central mass appears triple	22075s1654
7252	458[3]	⊘	Ep; 13.0; 0.8' x 0.7' F,S,R, faint extension on west side	22180s2456
7293		◎	! pF,vL,vlE; annular; diam 12' with 13th mag central star. "Helical Nebula" (*)	22270s2106
7300		⊘	Sb; 13.2; 1.9' x 0.8' vF,cS,E, vglbM	22283s1417
7302	31[4]	⊘	E3; 13.1; 0.9' x 0.6' F,pS,R,sN	22297s1423
7309	476[2]	⊘	Sc; 13.2; 1.3' x 1.0' vF,pL,glbM; 3-arm spiral	22316s1037
7371	477[2]	⊘	Sb; 12.9; 1.0' x 1.0' vF,pL,R,lbM; faint outer ring.	22434s1116
7377	598[2]	⊘	E1; 12.5; 1.1' x 1.0' pB,S,vlE, gmbM	22451s2235
7392	702[2]	⊘	Sb; 12.6; 1.5' x 0.9' pB,pS,lE,bM (*)	22492s2053
7492	558[3]	⊕	Mag 12.3; diam 3'; class XII; eF,L, 1C	23057s1554
7585	236[2]	⊘	Sp; 12.8; 1.0' x 0.7' pB,pS,R, gbM	23154s0456
7600	431[2]	⊘	E6; 13.1; 1.1' x 0.4' cF,S,E, mbM	23163s0752
7606	104[1]	⊘	Sb; 11.9; 6.0' x 1.6' pF,cL,pmE	23165s0846
7721	432[2]	⊘	Sc; 12.4; 2.6' x 1.0' pF,cL,E,vgbM	23362s0648

LIST OF STAR CLUSTERS, NEBULAE, AND GALAXIES (Cont'd)

NGC	OTH	TYPE	SUMMARY DESCRIPTION	RA & DEC
7723	110[1]	⊖	SBb; 12.1; 3.0' x 2.0' cB,cL,E, gmbM	23364s1314
7727	111[1]	⊖	Sa; 12.0; 2.7' x 2.7' pB,pL,R; mbM; with faint outer arms extending to 5'	23373s1234

DESCRIPTIVE NOTES

ALPHA   Name- SADAL MELIK, sometimes called "Rucbah".
Mag 2.93; spectrum G2 Ib. Position 22032s0034.
The distance of Alpha Aquarii is about 1100 light years,
the actual luminosity about 6000 times that of the Sun.
The star has the same spectral class and surface temperature as the Sun, but is a giant star, whereas the Sun is
a main sequence object. The diameter may be about 80 times
that of the Sun, and the absolute magnitude is about -4.6.
The annual proper motion is very slight, about 0.015"; the
radial velocity is about 4½ miles per second in recession.

BETA   Name- SADAL SUND. Mag 2.86; spectrum G0 Ib.
Position 21289s0548. The computed distance is
1030 light years; the actual luminosity is about equal to
that of Alpha Aquarii, at 5800 times the luminosity of the
Sun. The annual proper motion is about 0.017"; the radial
velocity is about 4 miles per second in recession.
The star has two 11th magnitude companions at 35.5"
and 58.6", both optical attendants only. The closer star
was discovered by John Herschel in 1828, the other by S.W.
Burnham with a 6-inch refractor in 1879. Both separations
appear to have increased slightly since discovery.

GAMMA   Name- SADACHBIA. Mag 3.84; spectrum B9 III.
Position 22191s0138. Parallaxes obtained at
Allegheny and Yale agree in giving the distance as about

DESCRIPTIVE NOTES (Cont'd)

95 light years, leading to an actual luminosity of about
20 suns (absolute magnitude about +1.5.) The star is a
spectroscopic binary with a period of 58.1 days. The yearly proper motion is 0.12"; the radial velocity is about 8
miles per second in approach.

A 12th magnitude star at 49" was noted by John Herschel in 1838; it has no real connection with the bright star
and the separation is decreasing due to the proper motion
of Gamma itself. In the year 2000 the separation will be
about 28".

DELTA Name- SKAT. Mag 3.28; spectrum A3 V. Position
22520s1605. The distance is about 85 light
years (Mt.Wilson and Yale parallaxes). The actual luminosity is about 28 times that of the Sun, the absolute magnitude about +1.2. The annual proper motion is 0.05" and the
radial velocity is 11 miles per second in recession.

EPSILON Name- AL BALI. Mag 3.77; spectrum A1 V. Position 20450s0941. Parallax measurements are rather discordant, but suggest a distance of about 170 light
years, giving an actual luminosity of about 70 suns. The
expected luminosity of an A1 main sequence star is about
absolute magnitude +0.7 (luminosity = 40 suns) which would
decrease the distance to about 135 light years.

The annual proper motion is 0.04"; the radial velocity, suspected to be somewhat variable, is about 10
miles per second in approach.

ZETA Mag 3.66; spectrum F2 IV. Position 22262s0017.
This is the central star of the Y-shaped asterism formed by Gamma, Zeta, Eta, and Pi Aquarii, a figure
often called the "Water Jar" of Aquarius. Zeta is a very
fine close binary star, probably discovered by C.Mayer in
1777, refound by W.Herschel about 1779, and well observed
up to the present day. The two stars are magnitudes 4.42
and 4.59 and are 1.7" apart (1967). The separation appears
to have been decreasing steadily during the past 180 years
since a measurement of 4.56" was made by Herschel in 1781.
The direction of revolution is retrograde, or clockwise.

Although the binary character of this star was recognized by Herschel as early as 1804, the exact period is

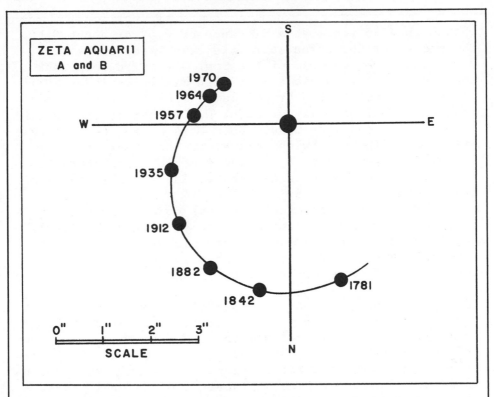

ZETA AQUARII
A and B

still uncertain, and values ranging from 400 years up to over 1600 years have appeared in various texts. According to a computation by Dr.O.Franz (1958) the period is about 600 years, with closest separation of the components in 1972. The computed orbit has a semi-major axis of 4.0" and an eccentricity of 0.45.

Both stars are subgiants of very similar type; the spectra are classed as F2 by some authorities, as F2 & F1 by others. The actual luminosities are 8 and 7 times that of the Sun. The distance of the system is about 75 light years. The true separation of the two stars averages close to 100 AU. Zeta Aquarii shows an annual proper motion of 0.19" in PA 78°; the radial velocity is about 15 miles per second in recession.

According to an analysis by Dr.K.Strand, a third unseen component exists in the system. From the gravitational effect upon the visible pair, the invisible body appears to be a satellite of Zeta B, about which it re-

volves in 25.5 years at a distance of 0.4" or about 9 AU.
The masses of the three stars are calculated to be 1.13,
0.85, and 0.28 the mass of the Sun. The third component
would probably be visible as a 12th magnitude object were
it not so close to the bright pair. From its rather small
mass and low luminosity of about 0.008 that of the Sun, it
seems likely that the unseen Zeta C is a red dwarf of type
dM1 or dM2. The two bright stars, on the other hand, are
definitely over-luminous for their masses, and appear to
be evolving toward the giant stage. The absolute magnitudes
are about +2.6 and +2.8.

The star is an excellent test object for the small
telescope, since the magnitudes are so nearly equal. It
may be resolved with a good 3-inch glass, seeing conditions permitting, and is always an easy object with the 7-
inch Lowell Observatory refractor.

R   Variable. Position 23412s1534. An interesting
          and peculiar variable star, discovered by Harding
in 1811. It sometimes reaches the 6th magnitude at maximum
and has an average period of 386 days, but individual periods may be very erratic. On occasion the star has remained nearly constant during an interval of several years. A
stable period of this sort lasted from 1931 to 1934, when
the magnitude remained close to 9. In the years following,
it returned to its usual pattern of semi-regular variation.
The accompanying light curve shows typical fluctuations
over an 8 year period, and was compiled from the observations of the AAVSO.

R Aquarii is a red pulsating giant star of spectral
type M7e, resembling the long period variables, but with
certain peculiarities which make it a virtually unique
object. From spectroscopic observations the existence of
a companion star seems definitely established. The spectrum of the companion is that of a high-temperature 0 or B
star, and is normally very faint, but becomes strong at
times of unusual fluctuations in brightness. A peculiar
feature of the system is that the amplitude of the "red"
variation seems to decrease when the "blue" component is
most active. There is little doubt that both stars are
intrinsically variable. R Aquarii is thus the typical
example of a rather rare class of variables characterized

THE WATER JAR and COMET KOHOUTEK. The Y-shape asterism called the "Water Jar" appears at the top of this print, made January 13, 1974, with the 13-inch wide-angle camera at Lowell Observatory.

VARIATIONS IN R AQUARII. The top plate was made September
15, 1931; the other on August 25, 1974. Lowell Observatory
photographs made with the 13-inch camera.

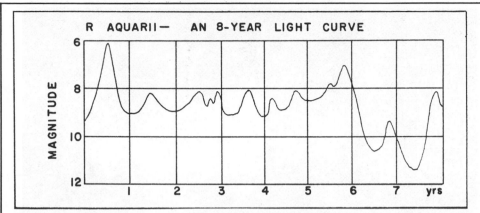

R AQUARII— AN 8-YEAR LIGHT CURVE

by composite spectra, in which the features of a low temperature red giant and a hot subdwarf are combined. P.W.
Merrill has applied the term "symbiotic stars" to objects
of this type.

A revised distance determination of about 800 light
years for R Aquarii allows us the deduce the following absolute magnitudes and luminosities for the system:

M7e Red Star (maximum) = -1.1; luminosity = 230 X Sun
B2? Blue Star (average) = +4.2; luminosity = 1.7 X Sun

The red star appears to be a normal giant of the long
period or semi-regular variable class, similar to Omicron
Ceti (Mira) but the companion is a true subdwarf under-
luminous star which exceeds our Sun only by a magnitude or
so, even at maximum. Theoretical formulae suggest that its
diameter may be about 1/6 to 1/10 that of the Sun, while
the red star must be at least 100 times the size of the
Sun. The separation of the two stars may be on the order
of 1 AU. Radio emission from the R Aquarii system has
been detected at the Algonquin Radio Observatory in Ontario
in April 1973.

THE R AQUARII NEBULA. As early as 1919 it was found
that the spectrum of R Aquarii showed several bright lines
characteristic of the gaseous nebulae. Although these lines
were found to vary greatly in intensity, they showed no
evident correlation with the magnitude changes of the star.
This spectroscopic discovery was confirmed in 1921 by observations made with the 42-inch reflector at the Lowell
Observatory. Photographs obtained by C.O.Lampland revealed
a peculiar lens-shaped nebulosity about 2' in extent, com-

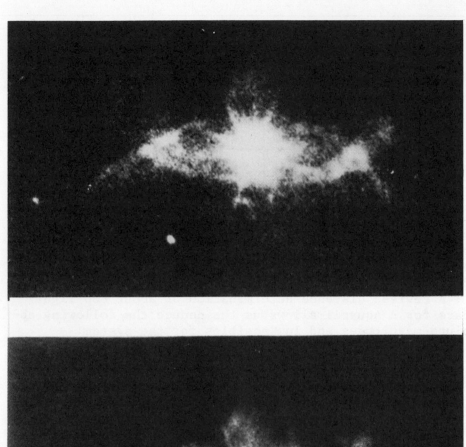

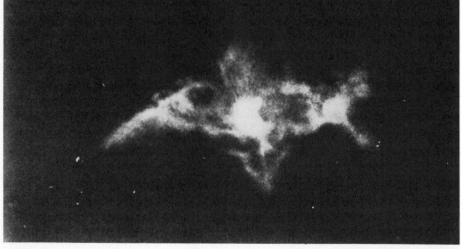

NEBULOSITY SURROUNDING R AQUARII. Top: direct photograph
with the 100-inch reflector. Below: drawing made from
several photographs.   Mt.Wilson Observatory

posed of curved filaments symmetrically placed about the
star. In the center of this cloud, the star itself appeared embedded in a small nebulous disc resembling a miniature planetary nebula, but which seemed to be variable in
brightness.

The expansion of the outer nebula, first suspected
by Lampland, has been confirmed by E.Hubble and W.Baade
at Mt.Wilson, and may indicate that a nova-like outburst
occurred in the system some 6 or 7 centuries ago. This
possibility casts new light on the significance of the
symbiotic stars, and opens up a new and fascinating line
of thought. A number of eruptive stars are now recognized
as members of close binary systems, and there is growing
evidence that the "symbiotic nature" of these stars is in
some way responsible for the nova-like activity.

Z Andromedae is perhaps the most typical example, and
its violent outbursts, such as that of 1914, have been
thoroughly studied. Between explosions the star appears
to act like a normal semi-regular red giant. The bright
outbursts, however, appear to originate in a hot bluish
companion. Two of the actual recurrent novae - T Coronae
and RS Ophiuchi - are close binary systems of the same
type. In each case there is a hot blue subdwarf, mated
with a cooler star of later spectral type. Even the explosive SS Cygni stars seem physically related to the above-
mentioned objects, differing chiefly in the fact that both
components are dwarfs or subdwarfs, and the separation is
much less.

Mention should be made also of two long-period variable
stars, Mira Ceti and X Ophiuchi, which have binary companions actually visible through the telescope. The companion
to X Ophiuchi appears to be a normal K-star, but that of
Mira is a bluish dwarf or subdwarf which is itself variable
by at least two magnitudes, and thus closely resembles the
unseen companions of the other symbiotic stars.

The relationship of the symbiotic stars, erratic variables, and novae, has been the subject of much speculation.
A current theory suggests that the two members of a symbiotic pair may have evolved to the point where an expanding
giant star has begun to engulf its smaller neighbor; the
outbursts of the smaller and partially degenerate star
thus being attributed to accretion of material from the

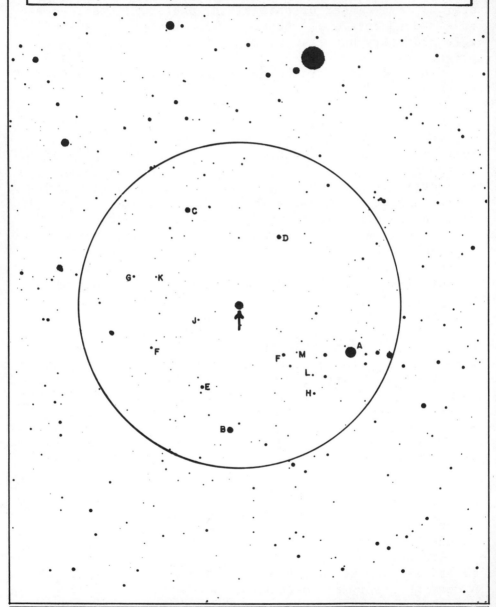

COMPARISON MAGNITUDES (AAVSO) A= 5.6; B= 8.9; C= 9.3; D= 9.9; E= 10.1; F= 10.7; G= 11.0; H= 11.4; J= 11.8; K= 12.1; L= 12.6; M= 13.1.

R AQUARII. Finder chart made from a 13-inch telescope plate at Lowell Observatory. Circle diameter = 1 degree. North is at the top. Limiting magnitude about 15. Bright star at top is Omega-2 Aquarii, magnitude 4.6.

giant. It is difficult to evaluate this picture, and some
recent studies of the SS Cygni stars seem to cast doubt on
the traditional assumption that the outbursts always originate in the bluer and hotter star. However, although the
details are still extremely hazy, it seems most probable
that some process of interaction between the components is
chiefly responsible for the erratic fluctuations of the
symbiotic stars. A similar mechanism may eventually be
identified as a cause of the SS Cygni outbursts, and very
possibly of the recurrent novae. The idea was first suggested for the star AE Aquarii, a peculiar erratic dwarf
which is now known to have a close K-type companion.

(Refer also to: Z Andromedae, AE Aquarii, Omicron
Ceti, BF Cygni, and AG Pegasi. For typical SS Cygni stars
refer to: U Geminorum, SS Aurigae, SU Ursa Majoris, and
SS Cygni. Recurrent novae are chiefly described under:
T Corona Borealis, WZ Sagittae, U Scorpii, T Pyxidis, and
RS Ophiuchi. For novae in general refer to: V603 Aquilae,
GK Persei, DQ Herculis, and CP Puppis.)

AE   Variable. Position 20376s0103. A peculiar faint
         erratic variable star, related probably to the
SS Cygni stars, and possibly to the recurrent novae. The
star is characterized by almost constant activity and the
variations are unusually complex. At times there are very
sudden explosive maxima when the light nearly doubles for
an hour or so; this phenomenon may occur repeatedly at
intervals of about a day. The rise to maximum is frequently very rapid, though on occasion the star will brighten
by only a small amount and then fade back to normal in a
short time. There are also more violent eruptions which
occur at intervals of about a year, when the brightness
may increase by two magnitudes. And finally, recent studies
reveal extremely rapid changes of small amplitude with
average periods of less than an hour.

AE Aquarii is thus an exciting object to observe,
although the greatest visual brightness is rarely as high
as 10th magnitude. The rapid changes make it a difficult
object for adequate spectroscopic study even with large
instruments. At discovery by A.Wachmann in 1931, the star
was thought to be a long period variable of the Mira type.
The sudden outbursts were first detected in a photographic

study by E.Zinner in 1937, and the star was then classed
among the SS Cygni stars, sometimes called "cataclysmic
variables". Although AE Aquarii is much more erratic and
unpredictable than any of the classic SS Cygni stars, it
still appears to be a member of the same general physical
group. Like SS Cygni itself, the star is known to be a
very close binary system in very rapid revolution. A.H.Joy
found the radial velocity to be variable in 1954, with a
range of about 180 miles per second. The masses of the two
stars appear to be very nearly equal. The period is given
by Joy as 0.701 days, but according to M.F.Walker (1965)
the most recent observations indicate a period closer to
10 hours.

The spectrum of the brighter star was classed as dG8 by
Joy, but is now considered to be about dK0. This star has
about a third the luminosity of the Sun, and the absolute
magnitude is about +6. The other component, regarded as
the source of the erratic outbursts, is a dwarfish hot
star with a bright-line spectrum of uncertain class, and
an absolute magnitude of about +7. The actual separation
of the two stars cannot be more than a few hundred thous-
and miles, which implies that they are nearly in contact.
The system is remarkably similar to SS Cygni, and other
such related stars as U Geminorum and Z Camelopardi. In
each case a dwarfish hot star is accompanied by a larger
G or K type companion. These systems are of unusual interest and significance, since they seem to show us a pair

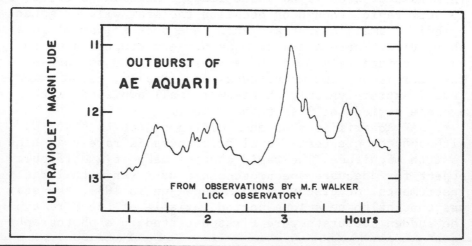

OUTBURST OF
AE AQUARII

ULTRAVIOLET MAGNITUDE

FROM OBSERVATIONS BY M.F. WALKER
LICK OBSERVATORY

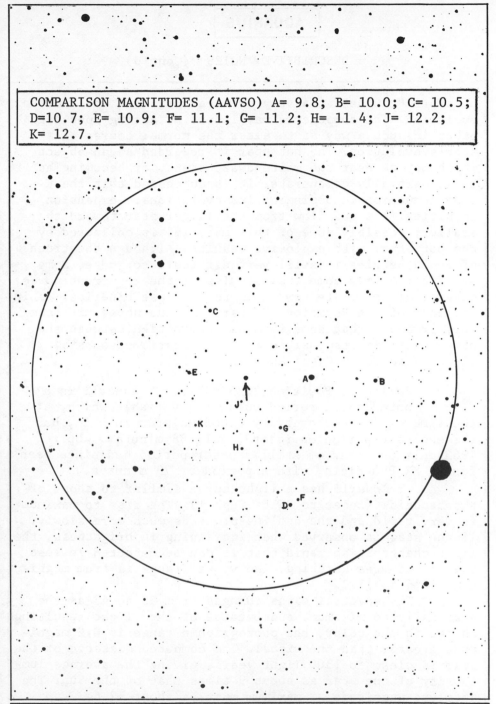

COMPARISON MAGNITUDES (AAVSO) A= 9.8; B= 10.0; C= 10.5; D=10.7; E= 10.9; F= 11.1; G= 11.2; H= 11.4; J= 12.2; K= 12.7.

AE AQUARII. Finder chart made from a 13-inch telescope plate at Lowell Observatory. Circle diameter = 1 degree. North is at the top. Limiting magnitude about 15. Bright star at right is 71 Aquilae, magnitude 4.3.

DESCRIPTIVE NOTES (Cont'd)

of stars "caught" at a very critical phase in their mutual evolution. The components may indeed be acting upon each other in such a way as to alter the normal course of a star's evolution. The hot star is regarded as an object which may be near the white dwarf state, or becoming at least partially degenerate. It is suggested that the K-type companion is beginning its evolutionary expansion, resulting in a gas flow from the larger star toward the smaller; possibly some of this material is collected by the hot star, with explosive results. Although the truth of such theories is still very difficult to judge, some verification may come from studies of the recurrent novae, several of which are now known to be close binaries. As in the case of the "symbiotic" stars and the novae, it seems very probable that some form of interaction between the components is responsible for the erratic outbursts. (Refer also to SS Cygni and U Geminorum)

CY   Variable. Position 22352n0117. CY Aquarii is a noted short-period pulsating variable which, at the time of its discovery by C.Hoffmeister in 1934, had the shortest period on record, only 88 minutes. Up to 1965 only one star of still shorter period had since been found, SX Phoenicis, with a period of 79 minutes.

CY Aquarii has a light curve similar to those of the cepheids (refer to Delta Cephei). The rise to maximum is very rapid and the decline proceeds much more slowly. If the star is observed when increasing in brightness, the light change is so rapid that it can be detected in less than 10 minutes watching. The visual range is from magnitude 10.6 to 11.3.

The spectral class changes from B8 to A3 as the star falls to minimum, and because of the slight resulting change in the color, the photographic range is 0.2 magnitude greater than the visual. The computed distance of the star is close to 1300 light years, giving the average luminosity of the star as about 9 times that of the Sun. The absolute magnitude at maximum must be about +2.6.

DWARF CEPHEIDS. CY Aquarii was formerly classified as a "cluster variable" of the RR Lyrae type. From its small size and lower luminosity, and its position on the H-R diagram, it is now evident that the star cannot be a

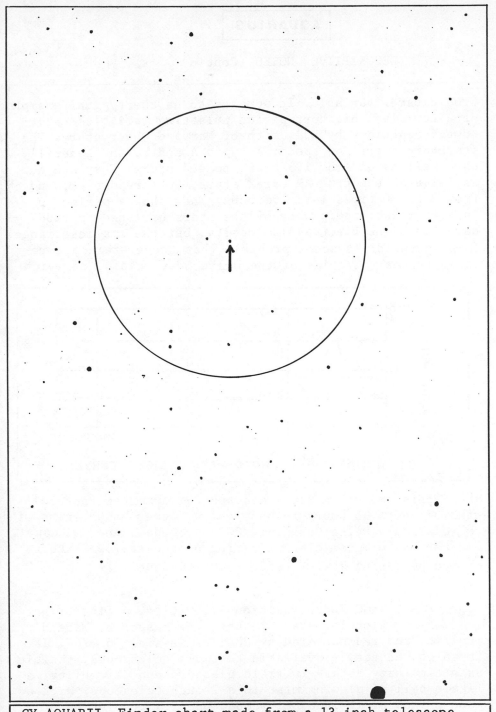

CY AQUARII. Finder chart made from a 13-inch telescope
plate at Lowell Observatory. Circle diameter = 1 degree.
North is at the top. Limiting magnitude about 14. Bright
star at lower right is Eta Aquarii, magnitude 4.1.

true cluster variable. It appears to be the typical example of a new class of short-period pulsating variables called "dwarf cepheids" by H.J.Smith of Harvard Observatory. The stars are dwarfs of spectral types A and F, and generally have periods of from 1.3 hours to 4.7 hours. They are not as large as the true RR Lyrae stars, but are denser, and from 2 to 30 times less luminous. They show a period luminosity relationship whereby the stars of longer period have the highest actual luminosity, but the smallest range in brightness. It seems probable that these stars are related to the variables of the Delta Scuti class, of which

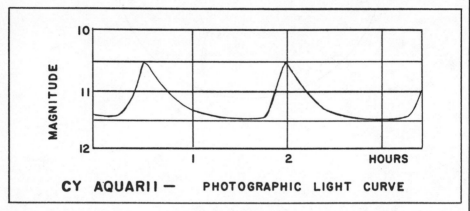

CY AQUARII — PHOTOGRAPHIC LIGHT CURVE

Rho Puppis and Delta Scuti are typical examples (periods of 3.38 and 4.65 hours respectively). Other known stars of the class are Delta Delphini, CC Andromedae, and DQ Cephei. For the H-R diagram classification of the various short-period pulsating stars, refer to Delta Scuti.

M2 (NGC 7089) Position 21309s0104. A fine bright globular star cluster, first seen by Maraldi in 1746, and rediscovered by Charles Messier in 1760. It is an object easily available to small telescopes, visible as a tiny hazy "star" in field glasses, and resembling a little ball of glowing mist in a 2-inch telescope. With a good 8-inch or 10-inch telescope, partial resolution of the cluster may be achieved. The visual diameter of about 7' increases to 11' on the best photographic plates, and the total integrated magnitude is about 6.0.

GLOBULAR STAR CLUSTER M2. A fine object for larger amateur telescopes, partially resolvable in an 8-inch glass. Lowell Observatory 42-inch reflector.

M2 lies at a distance of about 50,000 light years,
considerably farther than the great M13 in Hercules or M5
in Serpens. The actual diameter is about 150 light years.
One of the richer and more compact globular clusters, it
gains in impressiveness through its position in a rather
blank portion of the sky ordinarily devoid of faint stars.
In large telescopes the cluster is a wonderful sight; John
Herschel compared the distribution of the stars to a heap
of fine sand, and considered it to be composed of many
thousands of 14th and 15th magnitude stars. Today it seems
certain that the total population of the cluster is not
less than 100,000 stars, the brightest of which are red
and yellow giants with absolute magnitudes of about -3.
As a standard of comparison it should be remembered that
our Sun at such a distance would appear as a star of magnitude about 20.7, detectable only with the greatest telescopes. The total absolute magnitude of M2 is close to
-10, or about half a million times the luminosity of the
Sun.

According to Sawyer's "Bibliography of Individual
Globular Clusters"(1947) M2 contains 17 known variable
stars, a small total compared to the nearly 200 recognized
in the cluster M3 in Canes Venatici. The majority of these
stars are short-period pulsating variables of the RR Lyrae
class, often called "cluster variables" from their abundance in the globulars. Three classical cepheid variables
have been studied in M2 by H.Arp and G.Wallerstein (1961);
these reach 13th magnitude at maximum, and the periods are
15.57, 17.55, and 19.30 days. A fourth object appears to
be an RV Tauri type star with a 67.09 day period and a
range of slightly over 2 magnitudes.

The integrated spectral class of the cluster is F0;
the radial velocity is very slight, amounting to less than
2 miles per second in approach. (For an account of the
significance of the globulars in the study of stellar evolution, refer to M13 in Hercules).

M72   (NGC 6981) Position 20508s1244. Globular
             star cluster, located in the extreme western
portion of the constellation, 3° WSW of the Saturn Nebula
NGC 7009. This is not one of the more brilliant globulars
and generally may be described as unimpressive except in

large telescopes. It was discovered in August 1780 by M.
Mechain, and confirmed by Messier in October of the same
year. Messier thought the apparent diameter to be about
2' and the observations of Sir William Herschel (1810) gave
about the same apparent size. Modern photography increases
the apparent size to about 5'. Visually, the total magnitude is about 9.

Herschel, with a power of 280X on his great reflector
described M72 as "a very bright object....a cluster of
stars of a round figure but the very faint stars on the
outside of globular clusters are generally a little dis-
persed so as to deviate from a perfectly circular form....
it is very gradually extremely condensed in the centre, but
with much attention even there the stars may be distinguished." M72 is one of the more "open" globulars, and according to H.Shapley has a degree of concentration comparable
to M12 in Ophiuchus and M4 in Scorpius.

In an extensive catalog published by H.S.Hogg (1963)
the apparent diameter is given as 5'.1, the total integrated
magnitude as 10.25 (photographic), and the integrated
spectral type as G2. From current studies the distance
modulus appears to be about 16.3 magnitudes, giving the
distance as about 60,000 light years and the extreme diameter as 85 light years. This cluster is one of the more
difficult globulars to resolve, as the brightest stars do
not quite attain 15th magnitude. K.G.Jones (1968) speaks
of it as "surprisingly difficult to resolve for so large
and unconcentrated a cluster." With the author's 10-inch
reflector, a definite mottling around the edges becomes
noticeable with moderate powers, though averted vision is
necessary to confirm the suspicion of partial resolution.
Walter S.Houston reported much the same impression with a
13-inch aperture telescope.

Forty-two variable stars have been discovered in M72
up to 1973, the majority of which appear to be short-
period pulsating stars of the RR Lyrae class. The cluster
shows an approach radial velocity of about 160 miles per
second.

M73   (NGC 6994) Position 20562s1251. This object,
               not a true cluster, is merely a knot of four
small stars, located about 1.5° E and slightly south from

DESCRIPTIVE NOTES (Cont'd)

M72. It was noted by Messier in October 1780 and described
as a cluster of "three or four small stars which look like
a nebula at first sight; it contains a little nebulosity.."
On this last point Messier was definitely in error, as the
best of modern photographs show no signs of nebulosity in
the group, though of course it is a common experience that
faint double or triple stars often appear fuzzy in small
telescopes or with poor seeing conditions. The over-all
diameter of the asterism is about 1.2, the individual magnitudes about 10.5, 10.5, 11.0, and 12.0. The object is
aptly described by Admiral Smyth's brief note: "A trio of
10th magnitude stars in a poor field - that is M73. I give
it out of respect to Messier's memory."

NGC 7009   Position 21014s1134. A small bright nebula of
the "planetary" class, located about 1 degree
west of Nu Aquarii, and first observed by Sir William
Herschel in 1782. It was called by Lord Rosse the "Saturn"
Nebula from the extending rays or ansae which project from
the main disc on either side. The nebula has a rather high

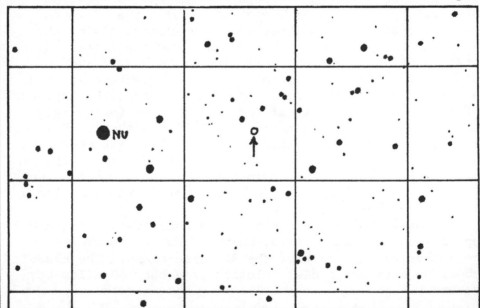

NGC 7009. A finder chart for the Saturn Nebula, showing
stars to about magnitude 9½. Grid squares are 1° on a side
with North at the top.

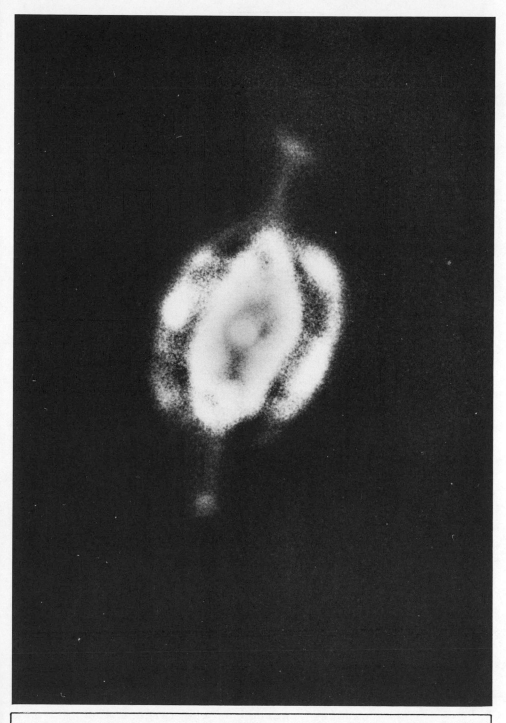

NGC 7009. The "Saturn Nebula" in Aquarius, one of the brighter planetaries. Mt.Wilson Observatory 60-inch telescope photograph.

surface brightness and appears nearly stellar in small
low power telescopes. The total magnitude is about 8; the
central star is about 12th visually, and somewhat brighter
photographically.

   This was one of the first planetary nebulae to be
observed by Rosse with his 6-foot reflector, and is described in his paper "Observations on the Nebulae" (1850).
He saw the nebula as a fairly uniform luminous disc, but
was apparently unable to detect the darker center or the
central star. "It has ansae which probably indicate a
surrounding nebulous ring seen edgeways." Rosse's drawing
portrays the nebula with a somewhat greater degree of
symmetry and perfection than is actually the case; it is
shown as a perfect nebulous miniature of Saturn.

   The nebula is a strikingly beautiful object in large
telescopes, shining with a vivid green fluorescent glow.
The flattened central disc measures about 25" X 17" and is
enclosed in a larger shell about 30" X 26". There is considerable intricate detail in both rings, and the two pro-
jecting rays end in bright condensations about 44" apart.
The ansae may be glimpsed in a good 10-inch telescope. The
central star is an extremely hot bluish dwarf with a continuous spectrum, and a computed temperature of about
55,000°K. Strong ultraviolet radiation from the star is
the cause of the bright fluorescent glow of the nebulosity
and the green tint is due to the radiation of doubly ionized oxygen.

   Distances of planetary nebulae are, in nearly all
cases, only roughly known. According to a study by C.R.
O'Dell (1963) the distance of NGC 7009 is about 3900 light
years, leading to an actual diameter of about 0.5 light
year. The central star has a luminosity of about 20 suns;
the absolute magnitude may be about +1.5. The nebula has
a radial velocity of 28 miles per second in approach. (For
a summary of facts and theories about the planetary nebulae, refer to M57 in Lyra)

NGC 7293   Position 22270s2106. The "Helical Nebula",
                usually regarded as the largest and nearest
of the planetary nebulae. It has a diameter of 12' x 16'
or about half the apparent width of the Moon. Despite its
large size the nebula is faint and has a low surface

DEEP-SKY OBJECTS IN AQUARIUS. Top: The Helical Nebula NGC 7293, photographed with the 42-inch reflector at Lowell Observatory. Below: The galaxy NGC 7392 photographed with the 100-inch reflector at Mt.Wilson.

brightness. The total magnitude is about 6½. Binoculars
will show the object as a large circular hazy spot, and it
is not a difficult object for a small telescope if a low
power ocular is used. Yet it is said that this nebula was
never observed by either of the Herschels with their giant
telescopes! A rich-field instrument with a wide-angle eyepiece is the ideal telescope for objects of this type.

The annular appearance, similar to the Ring Nebula in
Lyra (M57), is not clearly seen visually, but is well shown
on long exposure photographs. The general structure resembles a helix or coil with two turns; a much smaller planetary nebula in Draco (NGC 6543) has a very similar pattern.

The nebula is a spherical shell of very tenuous gas,
illuminated by a tiny but exceedingly hot central star of
the 13th magnitude. As in all the planetaries, the star is
a dense bluish dwarf or subdwarf. The estimated diameter
is about 2% that of our Sun, but the temperature is over
100,000°K. The gases of the nebula are excited to shine by
strong ultraviolet radiation from the star, the typical
bluish-green color of such nebulae being caused by doubly
ionized oxygen. The color is very striking in some of the
smaller brighter planetaries, such as NGC 7009 (also in
Aquarius) but is not noticeable visually in NGC 7293 which
has such a low surface brightness.

A photograph taken in red light with the 200-inch
telescope shows much complex detail in the nebula, and
reveals a peculiar system of spoke-like features on the
inner edge of the ring, all pointing toward the central
star. Each of these small features resembles a tiny comet
with a star-like head and a faint nebulous "tail" stream-
ing outward, away from the central star. The interpretation
of these features is uncertain, but it seems likely that
they are formed by a process analogous to the production of
comet tails, possibly by corpuscular radiation from the hot
central star as the nebula slowly expands.

Photographs made at Mt.Palomar also show a second
faint outer shell of gas, visible most clearly on the north
and east side of the nebula, 11.3' out from the central
star. Multiple shells are known in some of the other planetaries, probably indicating two or more periods of gas
ejection, separated by long intervals of non-activity.

Although NGC 7293 is usually considered the nearest

NGC 7293. The Giant Planetary Nebula in Aquarius, photo-graphed in red light with the 200-inch reflector at Mt. Palomar.

of the planetary nebulae, there is still no general agreement on the exact distance. In his text "Elementary Astronomy", Otto Struve states that a trigonometrical parallax
for the Helical Nebula was obtained by A.van Maanen, giving
a distance of 26 parsecs or about 85 light years. The true
diameter would then be about 0.3 light year, which seems
unexpectedly small. The value reported in the Skalnate
Pleso Catalogue (1951) is 180 parsecs or about 590 light
years. From the angular diameter and surface brightness,
L.Kohoutek (1962) has derived a formula which indicates a
distance of about 86 parsecs, while I.S.Shklovsky (1956)
obtained 50 parsecs from a very similar method. In a table
of the brighter planetaries published by C.R.O'Dell (1963)
the distance is given as 137 parsecs or about 450 light
years. This is close to the value quoted by C.W.Allen in
his "Astrophysical Quantities" (1963). Accepting this figure, the true diameter of the nebula is found to be about
1.75 light year, and the central star has a luminosity of
about 1/15 that of the Sun (absolute magnitude about $+7\frac{1}{2}$).
The total mass of the nebula is estimated to be about 1/10
the solar mass. The radial velocity is about 9 miles per
second in approach.

The place of the planetaries in the picture of stellar
evolution is still uncertain. Superficially, they resemble
the gaseous shells ejected by the novae, but such shells
expand at a violent rate and vanish in a short time. In
contrast, the planetaries seem relatively permanent, with
measured expansion rates of a mere 10 or 20 miles per
second. A typical planetary nebula thus seems to require
about 20,000 years to reach its average size of about 0.5
light year. According to some theorists, the peculiar hot
central stars are former Wolf-Rayet stars which are now
changing to the white dwarf state, the nebula being produced by emission activity during the transition period.
According to another view, a planetary nebula may simply
be an exceptionally "lazy" type of nova, where material is
being ejected quietly, rather than explosively. Nearly 500
planetaries are known, but the total number in our Galaxy
may be 10,000 or more. (For a summary of facts concerning
the planetary nebulae, refer to M57 in Lyra)

AQUILA

LIST OF DOUBLE AND MULTIPLE STARS

NAME	DIST	PA	YR	MAGS	NOTES	RA & DEC
5	13.0	121	42	$5\frac{1}{2}$- $7\frac{1}{2}$	(Σ2379) relfix;	18439s0101
	26.3	146	12	- 11	spect A2, A7; AC optical	
β265	1.1	229	66	7 - 9	PA dec, spect A0	18479n1128
Σ2404	3.5	183	62	6 - 7	relfix; spect K5, K3	18484n1055
Σ2408	2.1	92	62	$7\frac{1}{2}$- $8\frac{1}{2}$	relfix, spect A0	18496n1043
FF	6.8	141	59	6 - 11	slow PA inc, spect F6, primary is cepheid variable	18560n1718
11	17.5	286	57	6 - 9	(Σ2424) optical; spect sgF8	18568n1333
Σ2426	17.1	259	59	7 - 8	relfix, spect K5,	18577n1248
Σ2426b	3.7	167	05	8- $12\frac{1}{2}$	F	
Σ2428	6.9	284	37	8 - 10	slight PA dec, spect F2	18577n1450
Σ2425	31.2	181	25	7 - 8	relfix, spectra both G5	18578s0811
h874	9.7	2	10	7 - 13	spect G5	18582s0032
	22.5	304	01	-$11\frac{1}{2}$		
Σ2432	14.8	93	32	7 - $9\frac{1}{2}$	PA inc, spect B9	18595n1228
Ho 93	1.3	325	24	$7\frac{1}{2}$- 12	PA dec, spect A2	18598n1421
	39.2	210	00	-$12\frac{1}{2}$		
Σ2436	31.9	313	25	$7\frac{1}{2}$ - 8	slow dist dec; spect K0, F8	18598n0841
Σ2434	24.3	106	55	8 - $8\frac{1}{2}$	optical, PA dec; spect G5	19002s0047
Σ2434b	0.9	17	57	$8\frac{1}{2}$-$10\frac{1}{2}$		19002s0047
Σ2443	6.6	312	56	8 - $8\frac{1}{2}$	relfix, spect F5	19018n1443
15	38.4	209	59	$5\frac{1}{2}$ - 7	(Hh598) optical; spect gK1, gK4	19023s0406
Σ2439	22.0	199	15	8 - 9	relfix, spect B8	19023s0713
ζ	6.5	53	34	3 - 12	(17) (β287) spect A0 (*)	19031n1347
Σ2446	9.6	153	51	$6\frac{1}{2}$- $8\frac{1}{2}$	Spect F5, relfix	19034n0628
	34.5	341	05	- $9\frac{1}{2}$		
Σ2447	13.9	344	30	7 - 9	relfix, spect B8	19040s0125
Σ2449	8.0	291	51	7 - 8	relfix, spect F2	19040n0705
Rst 5460	4.6	202	46	8 - 14	spect F5	19065n0117

LIST OF DOUBLE AND MULTIPLE STARS (Cont'd)

NAME	DIST	PA	YR	MAGS	NOTES	RA & DEC
A95	0.2	96	62	$7\frac{1}{2}$ - 8	binary, about 95 years; spect G0	19083s0731
Σ2471	8.3	124	09	8- $10\frac{1}{2}$	PA & dist slight inc; spect A5	19086n0802
β1204	0.2	190	67	$7\frac{1}{2}$- $8\frac{1}{2}$	(Σ2476) spect B9	19095n0232
	31.1	214	06	- 11		
A98	26.0	128	37	$6\frac{1}{2}$- 11	spect K0	19115s0848
A98b	1.2	54	25	11-$11\frac{1}{2}$		
h1376	9.8	123	25	8 - 11	spect A0	19123n1517
0ΣΣ178	89.7	268	25	$5\frac{1}{2}$ -$7\frac{1}{2}$	cpm; spect G5,A1; color contrast	19130n1500
0Σ368	1.0	216	55	$7\frac{1}{2}$- $8\frac{1}{2}$	relfix, spect F0	19138n1604
	16.8	100	07	-		
β140	39.4	321	19	7 -$10\frac{1}{2}$	dist inc, spect	19141s1104
β140b	7.5	209	16	$10\frac{1}{2}$-$10\frac{1}{2}$	G0	
Σ2489	8.2	347	28	$6\frac{1}{2}$- $9\frac{1}{2}$	relfix, spect B9	19142n1427
0Σ370	19.6	15	51	$7\frac{1}{2}$ - 8	relfix, spect K0, A	19147n0915
h881	33.2	341	37	$7\frac{1}{2}$ -$9\frac{1}{2}$	spect K2	19154s0531
	15.7	62	23	-$11\frac{1}{2}$	(Ho574)	
h881b	7.0	307	32	$9\frac{1}{2}$- $9\frac{1}{2}$	relfix	
h5508	19.5	96	03	$7\frac{1}{2}$- 12	spect A5	19154s0102
23	3.1	5	58	$5\frac{1}{2}$- $9\frac{1}{2}$	(Σ2492) PA slow dec, optical; spect K2	19160n0100
	11.3	66	58	-13		
28	59.3	175	00	6 - 9	(0ΣΣ179) spect F0	19173n1217
Σ2494	25.9	85	32	7 - 10	PA slow inc; spect K0	19174s0644
Hu 72	1.1	62	22	$7\frac{1}{2}$-$12\frac{1}{2}$	spect M3	19176s1039
Σ2497	30.0	357	35	7 - 8	relfix, spect G5	19176n0530
Σ2498	12.2	66	49	7 - 8	relfix, spect K0	19177n0357
Hwe 47	0.5	318	58	8 - 8	PA dec, spect A2	19181n0251
Σ2501	19.8	22	34	$7\frac{1}{2}$- 9	relfix, spect F5	19194s0451
	9.6	108	01	- 14		
A1178	4.2	331	26	7 - 13	Spect K2	19195n1049
Σ2510	8.8	181	49	$8\frac{1}{2}$- $8\frac{1}{2}$	relfix, spect A0,	19209n0925
Σ2510b	0.2	191	61	$8\frac{1}{2}$- $9\frac{1}{2}$	A0	
Σ2513	2.3	325	62	8 - $8\frac{1}{2}$	relfix, spect F8	19227n0221
A1181	0.5	199	58	7 - 9	spect A0	19246n1158

AQUILA

LIST OF DOUBLE AND MULTIPLE STARS (Cont'd)

NAME	DIST	PA	YR	MAGS	NOTES	RA & DEC
Σ2519	11.4	124	37	8 - 8	relfix, spect F0	19255s0938
U	35.3	347	62	6½- 12	primary cepheid; spect G0	19267s0709
Σ2531	2.5	242	57	8 - 13	(A2197) relfix,	19270n0259
	31.7	30	27	- 9½	spect B5	
A1653	0.2	224	60	8 - 9	PA dec, spect A3	19273n1218
Σ2533	22.6	214	24	7 - 9	relfix, spect A3	19275s0033
Σ2532	33.7	5	29	6 - 10	relfix, spect K5, orange & gray	19277n0248
Ho 578	29.6	90	59	7 - 12	Dist inc, PA dec, spect G0	19282s0607
Σ2535	1.2	67	59	7 - 11	(D20) relfix	19286s0214
	25.2	298	59	- 10		
β3976	2.2	105	53	7 - 11	relfix, spect A0	19298n0914
β650	6.5	146	27	7½- 11	spect A0	19298n0624
	11.5	330	13	-12½		
	26.7	254	13	- 10		
Σ2537	19.1	133	35	8½- 9	relfix	19310s0417
μ	31.2	280	14	4½- 13	optical; spect K3,	19316n0716
	31.0	292	14	- 13	BC = 6" pair	
β1257	3.8	176	33	7 - 13	spect A2	19338n1102
Σ2543	12.6	155	33	7 - 10	relfix, spect K0, M0	19338n0554
ι	47.0	161	10	4½- 13	(J118) spect B5	19341s0124
Σ2541	5.0	328	59	8½- 10	PA dec, dist inc, spect K0	19341s1033
J133	16.4	61	34	6 - 14	PA inc, spect G5	19345n1110
	20.1	315	34	- 12½		
Lv---	82.1	286	12	7½- 10	Spect A2	19350s1006
Lv--b	4.2	284	56	10- 11		
β3249	0.9	124	52	7 - 9	PA dec, spect A2	19358n0014
Σ2545	3.7	324	59	6 - 8	PA slow inc,	19360s1016
	26.1	165	59	- 11	spect A5	
Σ2547	20.8	331	32	7½- 9	relfix, cpm;	19362s1027
	50.8	143	32	- 10½	spect A0	
QS	0.1	150	59	6½- 6½	(Kui 93) spect B3, ecl.bin.	19388n1342
χ	0.5	77	58	6 - 7	(47) relfix, spect dF3, A3	19402n1142

LIST OF DOUBLE AND MULTIPLE STARS (Cont'd)

NAME	DIST	PA	YR	MAGS	NOTES	RA & DEC
Σ2562	27.1	252	31	6½ - 8	relfix, spect F5	19404n0816
Σ2567	18.4	314	21	7½- 9½	relfix, spect A2	19418n1215
Σ2570	0.3	137	57	7½- 7½	PA dec, spect B3	19426n1039
	4.1	277	55	- 9½		
β1301	56.9	66	33	8½ - 9	spect A2	19439n0411
β1301b	0.6	340	57	9½- 9½		
A1658	0.2	310	62	8 - 8	binary, 90 yrs; spect F5	19464n1456
π	1.4	111	62	6 - 7	(Σ2583) slight PA dec, spect dF2 & A2	19464n1141
51	21.1	117	03	5 - 13	(Ho 275) optical; dist inc, spect F0	19480s1054
α	165	301	25	1 - 10	ALTAIR; optical, dist inc (*)	19483n0844
Σ2587	4.1	100	39	6½- 9	relfix, spect K0	19490n0358
β148	0.6	272	59	8 - 8½	PA dec, spect F2	19493s1029
	27.0	64	25	-13½		
Σ2590	13.5	309	59	7 - 10	(J126) relfix, spect B5	19499n1013
Σ2596	1.9	306	66	7 - 8½	PA dec, spect F8	19517n1510
57	36.0	170	55	5 - 6	relfix, spect B6, B8; cpm (Σ2594)	19519s0821
Σ2597	0.3	81	67	7 - 8	PA & dist dec; spect F2	19526s0652
β	12.8	7	58	4 -11½	(0Σ532) cpm; PA slow dec; spect G8, dM3; (*)	19528n0617
Hd 155	10.1	112	37	7½-10½	spect K2	19552s0912
	13.6	119	19	- 8½		
Hd 155c	5.4	158	06	8½ -11		
β266	15.8	167	09	7 - 11	spect A3	19556n1117
AC 12	1.4	300	59	7 - 8	PA dec, spect F5	19558s0222
Σ2613	4.0	352	56	7 - 7½	relfix, spect F2	19590n1036
HI 93	2.0	295	55	7½- 8	relfix, spect A0	19591s0020
Σ2616	3.3	265	55	7 - 9½	relfix, spect K0	20004n1426
Σ2621	5.7	223	54	8 - 8	relfix, spect B9	20022n0906
β56	1.5	176	52	8 - 9	PA inc, spect F5	20025s0428
h2927	24.3	125	01	7½ -12	spect A0	20028n0019
h2927b	4.8	185	01	12- 13		

LIST OF DOUBLE AND MULTIPLE STARS (Cont'd)

NAME	DIST	PA	YR	MAGS	NOTES	RA & DEC
β357	2.3	120	35	6- $10\frac{1}{2}$	relfix, cpm; spect M2	20031n1522
h1477	20.0	271	24	8- $10\frac{1}{2}$	spect K5	20042n1232
β428	0.8	356	65	7 - $8\frac{1}{2}$	PA inc, spect F2	20044n1248
Σ2629	9.2	188	32	7 - 10	(β58) relfix; spect B9	20050n1556
Σ2628	3.5	341	58	6 - 8	PA slow dec, spect F5	20054n0915
Σ2635	7.4	79	24	7- $10\frac{1}{2}$	AB cpm; relfix,	20077n0818
	73.7	42	24	- $12\frac{1}{2}$	spect F8	
Hh671	55.3	206	23	7 - 8	optical, spect F0, A0	20087s0017
β1205	0.2	95	52	8 - $9\frac{1}{2}$	PA inc, spect F8	20096s0814
Σ2644	3.1	206	58	7 - $7\frac{1}{2}$	relfix, spect A0	20100n0043
Σ2643	3.0	76	37	7 - $9\frac{1}{2}$	PA slow inc; spect A0	20102s0309
Σ2651	1.0	279	66	$8\frac{1}{2}$-$8\frac{1}{2}$	Dist dec, spect F8	20114n1600
OΣΣ202	43.4	193	23	8 - 8	cpm, relfix, spect G5, G	20117n0626
Σ2646	20.0	44	59	7 - $8\frac{1}{2}$	PA & dist dec,	20118s0612
Σ2646b	26.5	106	09	$8\frac{1}{2}$- 12	spect F0	
Σ2654	14.2	233	51	6 - $7\frac{1}{2}$	relfix, spect F0	20126s0340
h910	15.0	323	25	8 - 13	both dist inc;	20130n0242
	29.8	246	25	- $12\frac{1}{2}$	spect F5	
Σ2656	9.5	234	10	7 - 12	relfix, spect A2	20132n0739
Σ2677	32.7	29	27	6 -$10\frac{1}{2}$	relfix, spect A0	20221n0054
S749	59.9	189	20	$6\frac{1}{2}$- 7	cpm; both spectra	20249s0216
	43.6	316	08	- $10\frac{1}{2}$	F8	
A170	1.6	217	41	$6\frac{1}{2}$- 11	relfix	20294s0525

LIST OF VARIABLE STARS

NAME	MagVar	PER	NOTES	RA & DEC
σ	5.0---5.2	1.95	Ecl.bin.; Spect B3 (*)	19367n0517
η	3.7---4.5	7.177	Cepheid, spect F6--G4 (*)	19499n0052
R	5.3--12.0	300	LPV. Spect M7e (*)	19040n0809
S	8.4--12.1	147	LPV. Spect M3e	20093n1528
T	8.8--10.0	Irr	Spect M5; very red star	18433n0841
U	6.1-- 6.9	7.024	Cepheid; spect G0---G6	19267s0709
V	6.6---8.1	350	Semi-reg; spect N6, deep red color	19017s0546
W	7.5--13.9	487	LPV. Spect Se	19127s0708
X	8.3--14.5	348	LPV. Spect M6e	19490n0420
Y	4.9--4.95	1.302	(18 Aquilae) ellipsoidal variable, spect B7	19046n1100
Z	8.6--13.8	129	LPV. Spect M3e	20125s0618
RR	8.4--13.8	394	LPV. Spect M6e	19550s0201
RS	8.7--15..	412	LPV. Spect M7e	19564s0801
RT	7.1--14.2	326	LPV. Spect M7e	19357n1136
RU	8.1--14.3	274	LPV. Spect M5e	20104n1250
RV	8.6--14.6	219	LPV. Spect M3e	19383n0949
RW	8.4--		variability uncertain, spect F3	20096n1554
SY	9.0--14.5	357	LPV. Spect M5e	20048n1248
SZ	8.2-- 9.6	17.138	Cepheid; spect G0---K5	19021n0114
TT	7.2---8.0	13.75	Cepheid; spect F8---K0	19057n0113
TV	9.2--12.8	240	LPV. Spect M4e	20104n0608
TZ	8.5---9.5	Irr	Spect M6	20277s0455
UV	8.6---9.6	340	LPV. Spect N	18563n1417
UW	9.0---9.6	Irr	Spect M0	18550n0023
UU	11--16.8	75:	SS Cygni type, spect Gep	19546s0927
WZ	9.3--12.8	319	LPV. Spect M5e	20122n0437
XY	9.5 13..	398	LPV. Spect M8	19124n0409
AK	9.3--15..	297	LPV. Spect M6e	18578s0700
DO	8.6--16.5	---	Nova 1925	19287s0632
DY	9.5--11.8	131	RV Tauri type; spect G5 ---K0	19438s1104
EL	5.5-- 19	---	Nova 1927	18534s0323
FF	5.8---6.2	4.471	Cepheid; spect F6; 11th mag companion at 6.8"	18560n1718
FM	8.1---8.8	6.114	Cepheid; spect F5--G0	19069n1028
FN	8.5---9.8	9.482	Cepheid; spect F8--G2	19103n0328
KN	8.4---9.9	139	Semi-reg; spect M5e	20277n0142

AQUILA

LIST OF VARIABLE STARS (Cont'd)

NAME	MagVar	PER	NOTES	RA & DEC
KO	8.2---9.0	2.864	Ecl.bin.; spect A0	18448n1042
KP	9.7--10.5	1.684	Ecl.bin.	19003n1543
LU	9.5--11..	106	Semi-reg; spect M4	19366n1537
MS	9.0--10.8	Irr	Spect M	19388n1141
OO	9.2--10.1	.5068	Ecl.bin.; spect G5	19458n0911
QS	5.8---6.0	2.513	Ecl.bin.; spect B3; also close visual binary	19388n1342
V337	8.8---9.5	2.734	Ecl.bin.; spect B0	19016s0206
V339	8.9--11..	217	LPV.	19358s1004
V342	9.0--12.5	3.391	Ecl.bin.; spect A	19147n0915
V346	9.0--10.4	1.106	Ecl.bin.; spect A0	20076n1012
V356	7.0--16.5	---	Nova 1936	19147n0138
V368	5.0--16.5	---	Nova 1936	19241n0730
V450	6.3---6.9	64	Semi-reg; spect M7	19313n0521
V496	8.2-- 9.1	6.807	Cepheid; spect G5	19057s0731
V500	6.1---17	---	Nova 1943	19500n0821
V528	7.4--18.5	---	Nova 1945	19168n0032
V599	6.5---6.6	1.849	Ecl.Bin.; spect B5,B8	18598s1048
V603	-1.4---12	---	Nova Aquilae 1918 (*)	18464n0031
V604	8.0---17	---	Nova 1905	18595s0431
V606	6.0--15.5	---	Nova 1899	19178s0014
V805	7.8---8.5	2.408	Ecl.bin.; spect A2	19035s1144
V822	6.8-- 7.2	2.641	Ecl.bin.; spect B8	19287s0213
V830	9.9--10.7	Irr	Spect M5	19444n1554
V842	8.4--10..	Irr	Spect M6	19114n0232
V843	9.8--10.2	1.498	Ecl.bin.; spect B9	19027s0653
V844	9.0--10..	369	Semi-reg; spect M6ep	19045n0704
V865	8.5--12..	360	LPV. Spect S	20213n0047
V889	8.7-- 9.3	11.12	Ecl.bin. spect B9	19166n1609
V913	8.0--9.2-	50:	Semi-reg; spect M5	18521n1034
V915	8.8--9.5..		LPV? Spect M5	19008n1211
V923	6.3--6.42	.8518	Shell star, spect B5 p	19290n0320

LIST OF STAR CLUSTERS, NEBULAE, AND GALAXIES

NGC	OTH	TYPE	SUMMARY DESCRIPTION	RA & DEC
6709		⊙	Mag 8, diam 12'; pRi,1C; about 40 stars; class D (*)	18491n1017
6741		◎	Mag 12, diam 8" x 7", nearly stellar; 17^m central star	19001s0031
6751		◎	Mag 12, diam 20"; S,B; annular; 13^m central star	19032s0605
----	B133	▣	Starless dark spot 10' x 5'; in Aquila Milky Way (*)	19045s0655
6755	19^7	⊙	vL,vRi,1C; diam 10'; about 50 stars mags 12... class D	19054n0409
6756		⊙	S,Ri,1C; mag $10\frac{1}{2}$, diam 3'; about 25 stars mags 11.... class F	19062n0435
6760		⊕	Mag 11, diam 2'; class II; pB,pL,vg1bM	19086n0057
6772	14^4	◎	vF,L,R; mag 14, diam 75" x 55"; 18^m central star	19120s0248
----	I.4846	◎	stellar; mag 12; diam 2"	19137s0909
6778		◎	S,E, mag 13, diam 20"; hazy disc; 15^m central star	19157s0143
6781	743^3	◎	F,L,R, mag 12.5; diam 105"; $15\frac{1}{2}^m$ central star (*)	19160n0626
6790		◎	eS,B, nearly stellar; mag $11\frac{1}{2}$; diam 2"; 18^m central star	19204n0124
6803		◎	eS, stellar, mag 11, diam 4" 14^m central star	19289n0958
6804	38^6	◎	cB,S, mag 13, diam 60"; with central star about 13^m	19292n0907
6807		◎	eS, mag 14, diam 2", stellar	19321n0535
----	B143	▣	! L,Irr; diam 30' dark neby with 2 extensions to west (*)	19380n1100
6814	744^3	⊘	Sb; 12.2; 2.0' x 2.0' pF,pL,R,bM (*)	19399s1025

DESCRIPTIVE NOTES

ALPHA   Name - ALTAIR. Mag 0.77; spectrum A7 V. The
position is 19483n0844. Altair is the 12th
brightest star in the sky. Opposition date (midnight culmination) is July 18.

Altair is located at a distance of 16 light years,
and is thus one of our nearer neighbors among the brighter
stars. It is white main sequence star rather similar to
Sirius in type, about 1½ times the size of our Sun and 9
times brighter. The absolute magnitude is +2.2. The star
shows an annual proper motion of 0.66" in PA 54°; the
radial velocity is 16 miles per second in approach.

The 10th magnitude optical companion at 165" was
first measured by F.G.W.Struve in 1836 when the separation
was 152" in PA 322°. The star has no actual connection
with Altair itself, and the separation is increasing from
the proper motion of the primary.

The most remarkable fact about Altair is its very
rapid rotation, one of the fastest known. The rotational
speed at the equator, measured by the widening of the spectral lines, amounts to 160 miles per second, and the star
completes one turn in about 6½ hours. In comparison, the
rotation period of our Sun is 25.4 days. As a result of
its rapid spinning the star probably has the shape of a
flattened ellipsoid, the equatorial diameter being nearly
twice the polar diameter.   (Photograph on page 230)

BETA   Name - ALSCHAIN. Mag 3.71; spectrum G8 IV.
Position 19528n0617. The distance is some 40
light years, the actual luminosity about 4 times that of
the Sun (absolute magnitude +3.5.) The measured radial velocity is 24 miles per second in approach. Beta Aquilae is
a convenient comparison star for the nearby cepheid variable Eta Aquilae.

Beta is a visual double star, but a rather difficult
one due to the faintness of the companion which is 12.8"
distant. The two stars form a physical pair with a fairly
large annual proper motion of 0.48" in PA 175°. First
detected by O.Struve in 1852, the faint star is a red dwarf
of spectral type dM3, about 1/300 the luminosity of the
Sun. The absolute magnitude is about +11. The PA of the
pair has decreased by 10° in the last century, and nothing
can be known of the period of revolution except that it

must require many centuries. On the assumption that the total mass of the system is about one solar mass, the period of revolution may be in the range of 1000 to 2000 years. The projected separation of the pair is 160 AU.

GAMMA   Name - TARAZED. Mag 2.67; spectrum K3 II.
Position 19439n1029. Rather discordant parallax measurements suggest a distance in excess of 300 light years. From the spectral characteristics the absolute magnitude appears to be about -2.4 (luminosity = 750 suns) and the resulting distance is close to 340 light years. The annual proper motion is 0.01"; the radial velocity is about 1 mile per second in approach.

Gamma Aquilae is situated in an interesting section of the summer-time Milky Way, near the Great Rift. About $1\frac{1}{2}°$ to the west is the curious dark nebula B143. (*)

DELTA   Mag. 3.36; spectrum F0 IV. The position is
19230n0301. Parallaxes obtained at McCormick, Allegheny and Yale agree in giving a distance of about 53 light years. The actual luminosity is then about 10 times that of the Sun, and the absolute magnitude about +2.3. The star has an annual proper motion of 0.27" in PA 73°; the radial velocity is 18 miles per second in approach.

Spectral variations with a period of about 4 hours have been detected in this star, and are probably due to some sort of atmospheric pulsation, rather than to binary motion as was first suggested. In addition, there is an unseen companion with a period of 3.42 years, detected by systematic variations in the proper motion. According to H.L.Alden (1944) the maximum expected separation of the pair is about 0.3", but since the expected difference in brightness is several magnitudes, it is unlikely that the companion will be detected visually.

ZETA   Mag 2.99; spectrum A0 V. Position 19031n1347.
The distance, from Allegheny and McCormick parallaxes, is about 90 light years. The star has an absolute magnitude of +0.8 (luminosity = 40 suns). The annual proper motion is 0.10"; the radial velocity is $15\frac{1}{2}$ miles per second in approach. The spectral lines are unusually wide and diffuse, probably indicating a very high rate of

rotation. Zeta Aquilae is also a visual double star, a
close and difficult pair discovered by S.W.Burnham with
the 26-inch refractor at Washington in 1878. The separation has increased slightly since discovery (4.9" in 1878)
and the PA is decreasing at a rate of about 13° per century. The period is unknown, but must be many centuries.
The faint star is a dwarf of uncertain type, with a luminosity about 1/100 that of the Sun. The projected separation of the pair is about 175 AU.

ETA   Variable. Position 19499n0052. A bright variable star of the cepheid class, discovered by
Pigott in 1784, shortly after Goodricke's discovery of the
variations of Delta Cephei itself. It is one of the most
easily observed of the cepheids, being exceeded in apparent brightness only by Polaris and Delta Cephei. The period, precise as fine clockwork, is 7.17644 days. During
this time the magnitude changes slowly and smoothly from
a minimum of 4.5 to a maximum of 3.7, the rise requiring
slightly over 2 days, and the fall about 5. The light
changes can be easily seen without a telescope, and may be
followed by comparing the star with the nearby Beta Aquilae
which is magnitude 3.71.
    The light variations are attributed to an actual
pulsation of the star, though the exact details are uncertain and the cause is controversial. When the star is
growing in brightness, the spectroscope shows that the
surface is approaching the Earth, while at minimum it
appears to be receding; the star thus seems to be alternately expanding and contracting. These pulsations are
accompanied by a cyclic change in temperature, color, and
spectral class, the range being from type G4 at minimum
to F6 at maximum. The spectroscopic features are those of
a supergiant of luminosity class Ib. The absolute magnitude of the star at mid-range is about -3.2 photographic
or -3.8 visual, equivalent to about 2800 suns. At peak
brightness the luminosity reaches about 4000 times that
of the Sun. The diameter, which varies somewhat during
the cycle, lies in the range of 70 to 80 times that of
the Sun. The luminosities of the cepheid stars are proportional to their periods, allowing these stars to be
used as distance indicators for very remote objects such

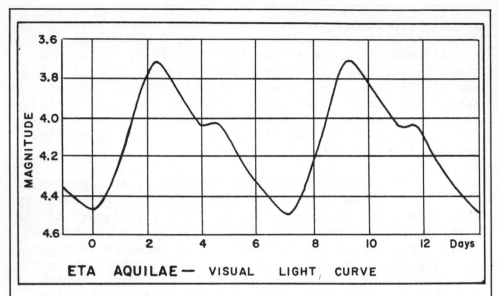

ETA AQUILAE— VISUAL LIGHT CURVE

as star clusters and even the nearer galaxies. From the period-luminosity relation, the absolute magnitude may be found, and the comparison with the apparent magnitude then gives the distance. By this principle, Eta Aquilae itself is estimated to be about 1300 light years away.

A peculiar feature of the light curve is the noticeable "hump" on the descending branch, indicating a 9-hour interruption in the fading of the star toward minimum. In comparing this light curve with those of other cepheids, it is found that this feature is present in many other stars whose periods lie in the 7 to 10 day range. It is fully as apparent in W Geminorum, S Sagittae, and S Muscae. While not fully understood, it seems likely that the explanation for this "standstill" in the light is connected with the phenomenon of multiple pulsations. If various layers of the star's atmosphere are rising and falling at different speeds there may be times, always occuring at the same phase, when two interfering pulsation waves temporarily cancel out.

The measured parallax of Eta Aquilae is about 0.005" which is too small to be a reliable distance indicator; the annual proper motion is about 0.01"; the radial velocity averages 9 miles per second in approach. (For a more detailed discussion of the cepheid variable stars, refer to Delta Cephei)

DESCRIPTIVE NOTES (Cont'd)

THETA   Mag 3.25; spectrum B9.5 III. The position is
            20087s0058. The distance derived from a par-
allax obtained at McCormick is about 325 light years, in
fair agreement with that computed from spectroscopic features. The absolute magnitude is about -1.7 (luminosity =
400 suns). The annual proper motion is only 0.03"; the radial velocity is about 16 miles per second in approach.
            Theta Aquilae is a spectroscopic binary with a
period of 17.124 days. Both spectra are visible, and are
both close to type B9 or A0. According to C.U.Cesco and
O.Struve (1946) the spectroscopic orbit has the rather
high eccentricity of 0.60. The separation of the two stars
may average about 15 million miles.

LAMBDA   Mag 3.44; spectrum B9 V. Position 19036s0458.
            Allegheny and Yale parallaxes give the dis-
tance as about 160 light years, in good agreement with the
distance derived from the spectroscopic parallax method.
The spectral type is somewhat uncertain since the lines are
abnormally wide and diffuse, presumably indicating a very
rapid rotation. The absolute magnitude is about -0.1 (luminosity = 90 suns). The annual proper motion is 0.09"; the
radial velocity is 8½ miles per second in approach.
            Lambda Aquilae is located in an interesting area
of the Milky Way, just to the northeast of the great Star
Cloud in Scutum. The wide optical double 15 Aquilae is less
than 1° to the north and slightly west, and the very red
N-type variable V Aquilae lies a degree to the southwest,
near the faint annular nebula NGC 6751.

SIGMA   Mag 5.17 (slightly variable); spectrum B3 V.
            Position 19367n0517. A rapidly revolving
eclipsing binary star, discovered at Mt.Wilson in 1912. In
1916 the first spectroscopic orbit was computed by F.C.
Jordan, and the light variations were detected photometrically by E.Dershem at the University of Illinois in 1918.
In a study by C.Wylie and J.Stebbins in 1920, the light
curve was identified as that of an eclipsing binary of the
lyrid class. Both components are B-type giants of very
similar type, classed as B3 by some authorities, and as B3
and B4 by others. The orbit appears to be very nearly circular and the period is 1.95027 days. The actual distance

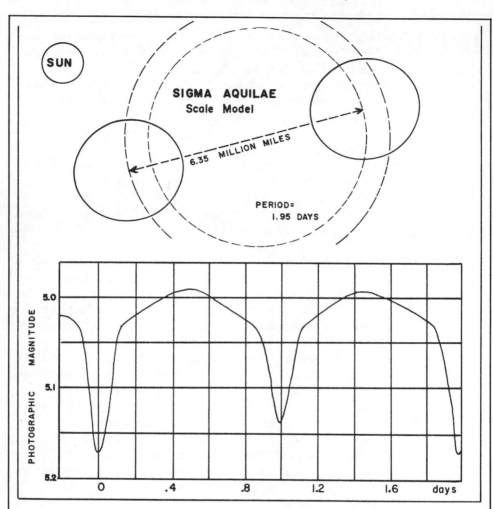

between the two stars is calculated to be 6.35 million miles, and the orbital velocities are 49 and 62 miles per second. The mutual eclipses of the system are small partial obscurations; primary maximum has a depth of about 0.2 magnitude. The light curve is of the Beta Lyra type, in which the magnitude varies continually, and it is evident that both stars are somewhat ellipsoidal in shape by reason of rapid rotation and tidal effects. The equatorial velocity of rotation of the brighter star has been measured at about 75 miles per second. The chief facts about the two components are given in the following table:

	Spect.	Diam.	Lum X☉	Abs.Mag.	Dens	Mass
A	B3	3.70	440	-1.8	0.14	6.9
B	B4	3.65	275	-1.3	0.10	5.5

Diameters, masses, etc., are given in terms of the Sun. From the computed luminosities, the distance appears to be approximately 950 light years. Sigma Aquilae shows no evident proper motion; the radial velocity is about 3 miles per second in approach.

An optical companion of magnitude 12½ was recorded by John Herschel in 1830, at 48" in PA 328°. There has been no change since discovery, and the faint star is probably not physically related to Sigma.

R   Variable. Position 19040n0809. The brightest of
the long-period variable stars in Aquila, discovered at Bonn, Germany, in 1856. It is a pulsating red giant of the Mira class, often attaining naked-eye visibility at maximum. The star is located in the Great Rift of the Aquila Milky Way, about 5½° south of Zeta Aquilae, and is easily recognized in the telescope by its fine red color which grows in intensity as the star fades. The spectral type changes from M5e at maximum to M8e at minimum, with a corresponding drop in the temperature from about 2350°K to 1890°K. This is one of the coolest stars known.

The visual range is about 6½ magnitudes (400 times) in brightness, but radiometric measurements show that the total energy emitted in all wavelengths changes by only

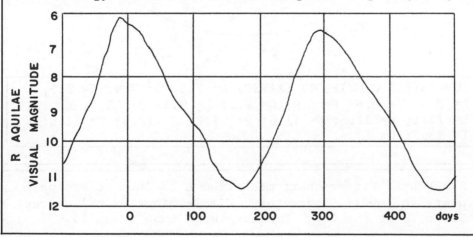

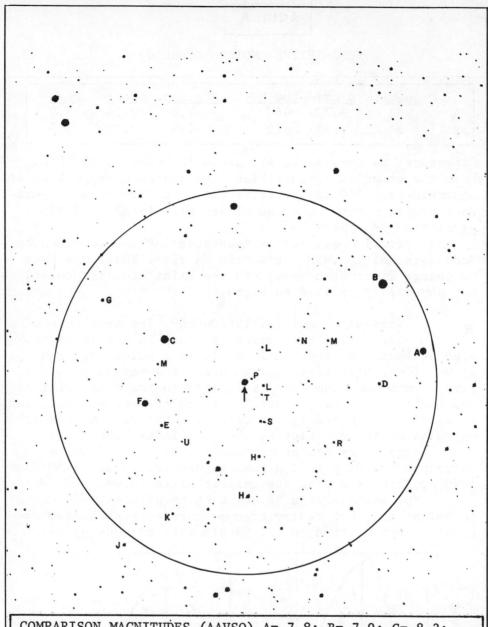

COMPARISON MAGNITUDES (AAVSO) A= 7.8; B= 7.9; C= 8.3; D= 8.5; E= 8.8; F= 8.9; G= 9.1; H= 9.5; J= 10.4; K= 10.9; L= 11.1; M= 11.5; N= 11.6; P= 11.8; R= 12.3; S= 12.6; T= 13.2; U= 13.6.

R AQUILAE. Finder chart made from a 13-inch telescope plate at Lowell Observatory. Circle diameter = 1 degree. North is at the top. Limiting magnitude about 15.

DESCRIPTIVE   NOTES   (Cont'd)

0.9 magnitude. The apparent fading at minimum results from
the fact that a large part of the energy output has shifted
over to the invisible infrared portion of the spectrum. The
total "radiometric magnitude" of R Aquilae at the maximum
is about +1.8; the star would thus appear brighter than
Polaris if the human eye was sensitive to radiation at all
wavelengths.

R Aquilae shows the unusual feature of a slowly changing period; in the last 80 years it has decreased from 350
days to about 300 days. Only a few such cases are known;
the most definite other examples are R Hydrae and R Centauri. The explanation awaits more complete knowledge of the
mechanics of the pulsations and the physical processes at
work; it seems safe to say only that a change in period
implies some sort of readjustment in the star's internal
structure. Possibly R Aquilae is entering a stage where the
evolution proceeds with great rapidity. A radio "Flare" of
this star was detected on October 8, 1973.

The distance of no long-period variable is known very
accurately. In a statistical study of many stars of the
type, R.Wilson and P.W.Merrill (1942) derived a probable
absolute magnitude in the range of -0.2 to -1.0 for stars
of spectral type M5e--M8e at maximum. The peak luminosity
of R Aquilae may thus be about 200 times that of the Sun;
the resulting distance is close to 600 light years. The
star shows a small annual proper motion of 0.07"; the
radial velocity is about 19 miles per second in recession.

NOVA AQUILAE 1918   (V603 Aquilae) The most brilliant
nova recorded during the last 300
years. Position 18464n0031. The nova was first noticed on
the night of June 8, 1918, as an object of the 1st magnitude, some 6° north of the Scutum star cloud in the Milky
Way. Among the early discoverers was E.Barnard, who was
then in the state of Wyoming for the purpose of observing
a total solar eclipse which had occurred only a few hours
previously! At the same hour the new star was independently discovered by a youth of 17, who later became America's
champion comet-discoverer and variable star observer,
Leslie C. Peltier of Delphos, Ohio.

At discovery the nova was already brighter than Alpha
Aquilae (Altair) but within a matter of hours it had taken
its place as the leading star of the northern sky, and

outshone every star in the heavens with the single exception of Sirius. Examination of plates taken previously of
the region showed that the star had been an 11th magnitude
object up to June 3. On June 7 it had risen to 6th magnitude, and on June 9 attained peak brilliance of magnitude
-1.4. From that brightness it slowly faded to 4th magnitude by the end of June. In March 1919 it was about 6th
magnitude and at the limit of naked-eye visibility.

Spectroscopically, the nova was a remarkable object.
During the period of greatest brilliancy, spectroscopic
analysis showed successive shells of gas being blown into
space with velocities of from 1000 to 1400 miles per second. A few months after maximum, a gaseous nebulosity was
detected about the star; its diameter increased for some
years at a rate of about 2" per year, so that the former
nova began to resemble a planetary nebula. Then this gaseous shell faded and eventually vanished into space. Nova
Aquila today is a bluish star of magnitude 11.95 (1968),
apparently much smaller and denser than our Sun.

The distance of Nova Aquilae is calculated to be in
the neighborhood of 1200 light years, implying that the
explosion witnessed in 1918 had actually occurred about
700 A.D. The actual luminosity at maximum was some 440,000
times that of the Sun, probably among the brightest normal
novae on record. The absolute magnitude was about -9.3. In
its present 12th magnitude state the star is still about 2
times brighter than our Sun. The light increase during the
outburst was thus about 100,000 times, and was accomplished in only 6 days!

THE PHENOMENA OF NOVAE. These stellar outbursts are
popularly referred to as "exploding stars", but the term
should not be taken to imply that the star is destroyed in
the blast. This may indeed be true in the case of the more
violent "supernovae" to be described later. But for the
ordinary or "classical" novae at least, the phenomenon
appears to be restricted to the outer layers of the star,
and leaves the main stellar body essentially unchanged.
In a typical normal nova the following phenomena are observed:

First, a many thousand-fold increase in brightness
in the course of a few days, reaching at maximum a luminosity of roughly 10,000 to 450,000 times that of the Sun.

DESCRIPTIVE NOTES (Cont'd)

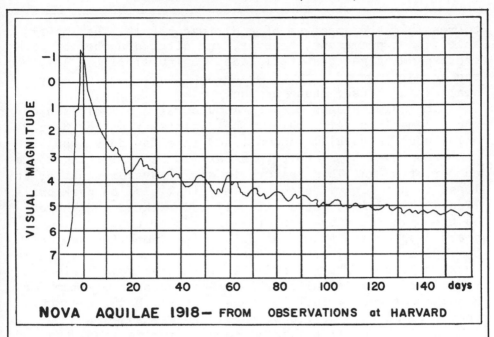

NOVA AQUILAE 1918— FROM OBSERVATIONS at HARVARD

The absolute magnitude at maximum may range from -5.0 to
about -9.5. This state of maximum brilliance rarely lasts
for more than a few days, however, and after passing its
peak brilliance the nova begins to decline, and returns to
its original faintness after a few years. The fading may
not proceed uniformly and the star is often subject to
minor fluctuations and pulsations. Even after reaching a
final minimum it may still show measurable variations for
a number of years.

Remarkable changes in the spectrum are observed simul-
taneously with the light changes. As the nova rises to its
maximum the spectrum shows absorption lines of hydrogen
and ionized atoms of iron, calcium, and other metals; these
soon show a large displacement toward the violet end of the
spectrum, indicating a very violent expansion of the outer
layers of the star. At maximum, measurements of the lines
reveal that successive shells of gaseous matter are being
blown into space with enormous expansion velocities which
range from a few hundred up to over 2000 miles per second.
After a year or so this expanding cloud may actually become
visible in the telescope. Also, as the nova starts to fade,
bright lines appear in the spectrum, which eventually comes

to resemble the spectrum of a diffuse nebula. The appearance is not permanent, however, and after a number of years
the expanding gaseous shell fades to invisibility, leaving
the former nova as a faint white-hot star, considerably
smaller and denser than the Sun.

The chief stages of a nova outburst are indicated on
the schematic diagram below. The "initial rise" of 8 or 9
magnitudes is usually accomplished in a few days, and is
often followed by a short "pre-maximum halt" lasting for a
number of hours. This feature is not always present, or at
least is not always detected. The final rise to peak brilliancy then follows at a slightly lower rate, and increases the brightness by another magnitude or two. Various
stars display individual peculiarities; Nova Aquilae and
Nova Puppis (1942) were typical "fast novae" with a single
sharp maximum and a rapid decline, whereas Nova Aurigae
(1891) and Nova Herculis (1934) remained near maximum for
a number of weeks. In 1925, Nova Pictoris had three maxima
spaced out over a period of nearly 10 weeks.

During the first 1 to 3 months after maximum, the
typical nova fades by about $3\frac{1}{2}$ magnitudes. This period is
called the "early decline" and is followed by a 2 to 3
month "transitional phase" where stars again show their
individual behavior. Some merely continue to fade rather

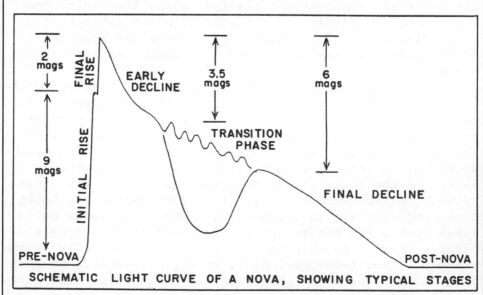

SCHEMATIC LIGHT CURVE OF A NOVA, SHOWING TYPICAL STAGES

NOVA AQUILAE 1918. The star is shown before the outburst, and shortly after maximum in June 1918. Photographed with the 42-inch reflector at Lowell Observatory.

steadily as did Nova Puppis, others begin a series of oscillations with a period of several days (Nova Persei 1901)
and still others drop suddenly by 8 or 9 magnitudes, only
to re-brighten again by some 5 magnitudes before beginning
the final decline. Examples of this "dip and recovery" type
were Nova Aurigae (1891) and Nova Herculis (1934).   Nova
Aquilae itself displayed definite oscillations during the
transition period, though those of Nova Persei were   much
more striking.

The fading of a nova to the final "post-nova" stage
usually requires several years, and in the case of some of
the very slow novae the star may not reach its normal minimum for several decades. Nova Aquilae, a fairly   typical
case, required 7 years to fade to its normal 11th magnitude
state. Nova Aurigae (1891) required 15 years, Nova Cygni
(1876) about 8 years, and Nova Persei (1901) about 15 years.

After reaching the post-nova stage, some novae may
continue to show slight variations, while others remain
quite steady. Nova Aquilae and Nova Aurigae have remained
nearly constant for a number of years, whereas Nova Cygni
(1876) and Nova Persei (1901) still show rapid variations.
Those of the latter occasionally exceed two magnitudes.
Post-nova variations are known for both fast and slow novae
and there appears to be no evident correlation with the
type of outburst or amplitude of the light curve.

Virtually all known post-nova stars are objects of
the same peculiar type, hot bluish subdwarfs of small radius and high density, apparently intermediate between the
main sequence stars and the true white dwarfs. M.Humason
(1938) obtained the spectra of 16 old novae and classified
them all as type O or early B. The majority had   strong
continuous spectra; some showed emission lines.   On   the
assumption that the surface brightness of a post-nova is
the same as that of a normal O-type star, the densities of
some of the old novae have been computed. For Nova Aquilae
the result is 70 times the density of the Sun, for   Nova
Persei about 200 times.

SOME NOVA HISTORY. Approximately 100 novae have been
recorded throughout history, the majority within the last
century. Ancient records are in many cases vague and   uncertain, and frequently contain too little information to
permit a definite identification of the reported object.

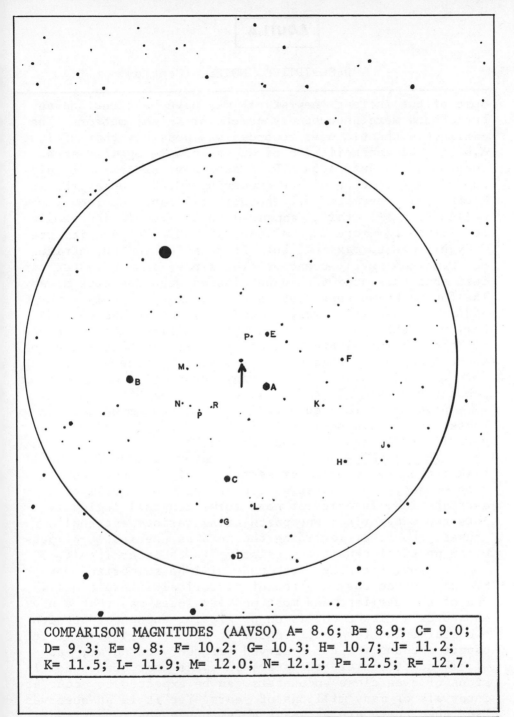

COMPARISON MAGNITUDES (AAVSO) A= 8.6; B= 8.9; C= 9.0;
D= 9.3; E= 9.8; F= 10.2; G= 10.3; H= 10.7; J= 11.2;
K= 11.5; L= 11.9; M= 12.0; N= 12.1; P= 12.5; R= 12.7.

NOVA AQUILAE 1918. Finder chart made from a 13-inch telescope plate at Lowell Observatory. Circle diameter = 1°;
limiting magnitude about 14. North is at the top. Bright
star in the field is GC 25805, magnitude 6.3.

DESCRIPTIVE NOTES (Cont'd)

Some of the ancient "new stars" may have been real novae,
but others were undoubtedly comets or bright meteors. The
earliest authentic nova recorded was possibly that of 1006
A.D. in the constellation of Lupus, now recognized as a
supernova. In July 1054 A.D. a supernova explosion resulted in the formation of the expanding "Crab Nebula" (M1 in
Taurus.) In November 1572 another brilliant supernova, now
called "Tycho's Star", appeared in Cassiopeia and remained visible for more than a year. In 1600 the peculiar star
P Cygni rose to maximum, but after fading to 5th magnitude
has remained nearly constant ever since, and it is not certain that this star should be classed with the true novae.
The fourth known supernova in our Galaxy in the last thous-
and years (Kepler's Star) blazed up in Ophiuchus in 1604.
Following this, no other nova attained first magnitude until the maximum of Eta Carinae in 1827; but here again the
identification of this peculiar star as a true nova is uncertain. It is classified by many authorities as an erratic
nebular variable. In the first half of the 20th Century,
five novae have reached a brilliance of 1st magnitude: Nova
Persei 1901, Nova Aquilae 1918, Nova Pictoris 1925, Nova
Herculis 1934, and Nova Puppis 1942.

THE CAUSE OF NOVAE. To the amateur and professional
alike, the nova problem presents one of the most fascinating subjects for investigation in the whole realm of
astrophysics. Theories of nova formation fall logically
into two categories; one postulating various external
causes, the other regarding the nova as the result of peculiar physical conditions existing in the star itself.

Theories of the first type will be summarized very
briefly, since they are now of historical interest only.
One of the earliest and most popular ideas was that a nova
showed us a direct collision or grazing encounter between
two stars. The fatal difficulty in this theory is that it
cannot account for the observed frequency of novae; so
widely scattered are the stars in space that direct collisions or even close encounters can be expected to occur at
intervals of many millions of years. Yet it is an observed
fact that there are at least a few novae seen every year
in our Galaxy, and the true total probably amounts to 30
or 40 per year.

Another theory attributed nova outbursts to the rapid

DESCRIPTIVE   NOTES   (Cont'd)

passing of a star through thick nebulosity; according to
this view the sudden blazing of the star would be the effect of friction, much as a meteor is heated to the point
of incandescence by its plunge through the atmosphere of
the Earth. A serious objection to this theory is the observed similarity of all nova outbursts, the fact that their
light curves are nearly identical in form, and the fact
that post-nova stars are all objects of the same rare type.
It is difficult to see how chance encounters between various types of stars with nebulae of widely varying density
and size could lead in every case to such strikingly similar results.

   As a result of the recent rapid advance in astrophysical knowledge, there is now a general agreement that nova
outbursts are due to no external cause, but to some instability in the star itself. Perhaps it would be well to
qualify this statement by admitting that a number of post-
nova stars are known to be extremely close binaries, and
it appears likely that the presence of a close companion
is connected in some way with the outbursts. In this sense
only, some nova outbursts may have an "external cause".
This question will be discussed in more detail after a
brief review of the probable reasons for instability in a
single star.

   First, let us consider the general picture of stellar
evolution as it presents itself to us today. The stars are
"atomic furnaces", globes of intensely heated gas in which
energy is produced by various nuclear reactions, chiefly
the transmutation of hydrogen into helium. The resulting
internal radiation pressure prevents the star from unlimited contraction under the action of gravitation. Thus the
normal stars are maintained in a state of equilibrium, and
their energy is released at a fixed rate. It can be calculated that the hydrogen supply of the Sun is sufficient for
more than 10 billion years of energy production in this
balanced state. The more massive stars will radiate their
substance away at a correspondingly higher rate, and the
heaviest known stars will last less than a million years.
Thus it is among the high-luminosity giants that we find
the stars of shortest life expectancy.

   In the article on Betelgeuse (Alpha Orionis) we have
traced the evolution of a star from the main sequence

TABLE OF BRIGHT NOVAE

DESIGNATION	Year	Max.	Min.	Remarks
CM Tauri	1054	-4	16:	Supernova; produced "Crab Nebula" NGC 1952 (M1)
B Cassiopeiae	1572	-4½	?	Supernova; "Tycho's Star"; no visible remnant
P Cygni	1600	3	?	Nova-like variable = permanent nova
V843 Ophiuchi	1604	-2½	?	Supernova; "Kepler's Star"
CK Vulpeculae	1670	2.7	?	Uncertain, probably slow nova
V841 Ophiuchi	1848	4.3	12.6	Slow nova; slight oscillations in early decline
T Corona Borealis	1866	2.3	10.6	Recurrent nova; second outburst in 1946
Q Cygni	1876	3.0	14.8	Very fast nova; smooth decline
T Aurigae	1891	4.4	14.8	Slow nova; deep dip during transition phase
V1059 Sagittarii	1898	4.5	16.5	Fast nova; smooth decline
GK Persei	1901	0.2	13.5	Very fast; large oscillations during fading
DM Geminorum	1903	4.8	16.5	Very fast, smooth decline
DI Lacertae	1910	4.3	14.4	Fast, smooth decline
DN Geminorum	1912	3.3	14.7	Fast, oscillations during decline
V603 Aquilae	1918	-1.4	11.9	Very fast, oscillations during decline
V476 Cygni	1920	2.0	17.0	Very fast, smooth early decline
RR Pictoris	1925	1.2	12.7	Slow nova, large oscillations at maximum
DQ Herculis	1934	1.3	15.0	Slow nova; light curve resembles T Aurigae
CP Lacertae	1936	2.2	15.3	Very fast, smooth decline
V630 Sagittarii	1936	4.3	15.5	Very fast, smooth decline
CP Puppis	1942	0.4	18..	Very fast, unusually large range= 18 magnitudes
V446 Herculis	1960	3.0	15..	Fast Nova, smooth decline
V533 Herculis	1963	3.0	15..	Moderately fast nova
V1500 Cygni	1975	1.8	21..	Very fast, unusually large range= 19 magnitudes

through the red giant stage. This material will not be repeated here, as we are now interested in the events which
follow the red giant stage. Having no internal energy supply, the star is now contracting, and growing denser and
hotter. The final result will be a star of planetary size
and of incredibly high density - a white dwarf, as such an
object is called.

White dwarfs are no figment of the scientific imagination. Over 200 stars of the type are now known, and they
are truly remarkable objects. Densities of several tons to
the cubic inch are typical for such objects. The connection
between these degenerate dwarfs and the spectacular novae
lies in the fact that a massive star cannot contract into
a white dwarf without becoming unstable in the process. It
is not difficult to see why this is believed to be so. A
star of small mass will contract to the point where the
increasing pressure of the highly compressed interior bal-
ances any further tendency toward contraction. But when a
stellar mass exceeds about 1.25 solar masses, there is no
balancing point, and such a star will have an unlimited
tendency toward contraction.

Unlimited contraction implies that the rotation velocity of the star must increase steadily in accordance with
the law of the conservation of angular momentum, and also
that the internal pressure and temperature will increase
without limit. It is possible that in both of these factors
we have conditions which lead to catastrophe. As the speed
of rotation increases, the centrifugal force increases
also, and if it becomes stronger than the star's gravitational force the star will be disrupted. This theory of
rotational instability has been suggested as a possible
cause of nova formation. Another suggestion is that certain nuclear reactions come into play once a certain limit
of density and temperature is passed. Detailed develop—
ments of these ideas have been presented by Schatzman,
Hoyle, Gamow, Gaposchkin, and other students of the nova
problem. Some theorists favor the idea of a shock-wave being propagated from the interior of the star to the surface; others assume a series of nuclear reactions which
lead to a star's increasing instability and final explosion. These theories are often reviewed in current texts;
the student is referred also to the authoritative book

DESCRIPTIVE NOTES (Cont'd)

"The Galactic Novae", by C.P.Gaposchkin. Fred Hoyle also presents many informative and entertaining speculations concerning the novae in his "Frontiers of Astronomy".

NOVAE AS DOUBLE STARS. An important addition to our knowledge of novae was provided recently by the discovery that a few novae are close binaries. Duplicity was first found for Nova Herculis 1934, an extremely close binary in which the components eclipse each other in a cycle of 4.65 hours, one of the shortest periods known. Nova Aurigae 1891 is a very similar system with a period of 4.9 hours. Nova Persei 1901 is now recognized as a double with a period of 1.9 days, and Nova Aquilae was found in 1962 to be a rapid binary with a period of 3h 20m. Many of the other old novae remain to be studied, but there is a growing suspicion that all of them may prove to be unusually close binaries, and that the outbursts are connected in some way with the duplicity. If so, many of the older theories concerning novae may be drastically revised. Additional evidence comes from two other types of objects: the "recurrent novae" and the eruptive variables of the SS Cygni and U Geminorum type. There is growing evidence that these stars also, are very close binary systems.

RECURRENT NOVAE are defined as stars which have shown more than one outburst. As of 1965 there are 7 known examples, of which the most famous are T Corona Borealis, RS Ophiuchi, and WZ Sagittae. Aside from the feature of repeating outbursts, these stars are notably different from the classical novae; the amplitudes of the outbursts are much less and the duration of the maxima are short. The brightest example, T Corona, rises from 10th magnitude to 2nd in less than a day, but fades below naked-eye range in a week, and reaches the post-nova stage in a few months. The star is a 228-day binary in which the nova component is much the fainter of the two except during an outburst. The other star is type M, and is presumably a red giant. At maximum, the actual luminosity of T Corona is comparable to that of a classical nova; it is thus rather puzzling to find that another example, WZ Sagittae, is a white dwarf which reaches a luminosity of a mere 30 suns at maximum. Obviously the recurrent novae are not all objects of the same type, and probably do not form a true physical group. Yet, the expected duplicity has been detected for WZ

Sagittae; the orbital period of 81.6 minutes is at present the shortest known. The two stars must be of abnormally small size and revolve nearly in contact.

SS CYGNI STARS are bluish dwarfs which show repeated outbursts on a small scale; a typical example (the star U Geminorum) flares up several times a year, brightening by a factor of about 100, and remaining at maximum for a few days. It might appear that these feeble objects bear only a superficial resemblance to the violent classical novae; yet the same type of mechanism may be operating, as once again the expected duplicity seems to be an essential feature. For SS Cygni the revolution period is about $6\frac{1}{2}$ hours, for U Geminorum about $4\frac{1}{2}$ hours.

DISCUSSION. There are thus three types of objects to consider: the full-scale "classical" novae, the recurrent novae, and the SS Cygni stars. In all three cases, typical examples are recognized as close and very rapid binaries. Although it may be that at least one member of such a pair is intrinsically unstable, it seems likely that novae and nova-like outbursts are connected in some way with the presence of the close companion. The simplest picture is to assume that the unstable star is triggered by accretion of material from the companion, an idea first proposed for the erratic star AE Aquarii. But even if this theory is basically correct, the exact details of the nova-process will undoubtedly be the subject of much speculation for years to come.

An interesting point must be made in connection with the recurrent novae and the SS Cygni stars. These objects appear to show a period-amplitude relation, whereby those stars of longer period show the more violent outbursts. If the relation is assumed to hold true for the classical novae as well, we might guess that stars like Nova Aquilae will explode again, but only at intervals of several thousand years. The suggestion that all novae are recurrent leads on to further speculation: Is there an evolutionary connection between the three types of stars, and if so, in which direction does it proceed? Will SS Cygni, for example, eventually develop into a full-scale nova? Or will Nova Aquilae ultimately settle down with periodic small-scale eruptions of the SS Cygni class? And, if the classical novae actually repeat their violent outbursts, how

many explosions are required to reduce the star's mass to
the stable point? The material ejected during each maximum
appears to be only a millionth of the solar mass; thus the
star may undergo thousands of such explosions before its
career is ended. In any case, the final result, it is believed, is a super-dense white dwarf star, or perhaps a
pair of such stars, if the duplicity discovered for some
novae is an essential feature of all. Assuming, however,
that a single star can become violently unstable and erupt
as a nova, it is interesting to speculate on the possibility that many of the known white dwarfs were once novae.
The famous companion to Sirius, in a nova outburst, would
appear about as bright as a full moon to observers on the
Earth!

SUPERNOVAE. A typical nova, as we have noted, may
reach a luminosity of several hundred thousand suns. At
rare intervals, an exploding star appears which exceeds
the brightness of a normal nova by a factor of 10,000 or
more. These objects, the "supernovae" seem to average one
in about 3 centuries in any one galaxy. Only four have
been definitely recorded in our own Galaxy, the last in
Ophiuchus in 1604. The supernova phenomenon is probably
restricted to the rather massive stars, and it seems that
the star may be almost totally destroyed in the outburst.
Although no supernova has appeared in our Galaxy since the
invention of the telescope, they are detected from time to
time in the other galaxies. The best known example was the
star of 1885 which appeared in the nucleus of the Great
Andromeda Galaxy M31. (For a discussion of supernovae,
refer to "Tycho's Star" - also known as B Cassiopeiae)

VAN BIESBROECK'S STAR (LFT 1467) (Ross 652b) (Wolf 1055b)
A famous red dwarf star, which has
the lowest visual luminosity known
for any star. It was discovered photographically with the
82-inch reflector of the McDonald Observatory in Texas in
1943. It is a distant companion of the 9th magnitude star
BD + 4°4048, whose position is 19145n0506. The separation
of the pair is 74" in PA 150°, corresponding to a true separation of about 400 AU. The distance from the Earth is 19
light years. Both stars show the large annual proper motion
of 1.47" toward PA 203°. The primary is a dwarf M3 V star

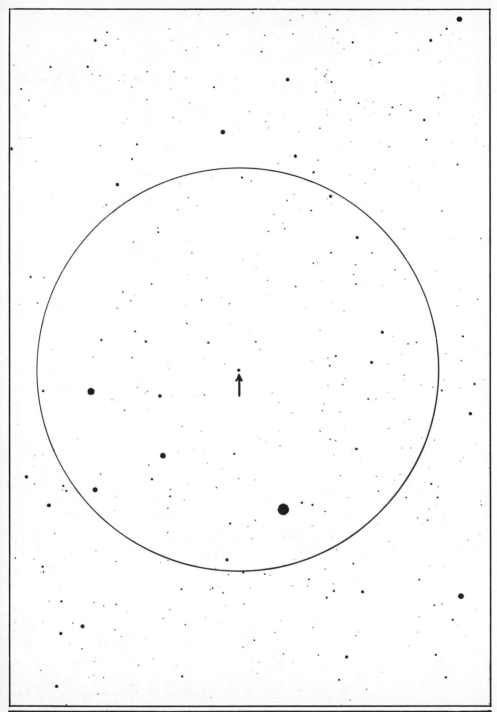

FIELD OF VAN BIESBROECK'S STAR, centered on BD +4°4048.
Circle diameter = 1°; north at the top, limiting magnitude
about 14. Bright star in the field is 22 Aquilae, mag 5.4.
Chart made from a Lowell Observatory 13-inch plate.

LIST OF LOWEST-LUMINOSITY STARS

STAR	Gliese	Con.	Mag & Spect		Abs.Mag.	Mu & PA		RA & DEC
Van Biesbroeck's	752b	Aqil	18.0	dM6	+19.3	1.47"	203°	19145n0504
LFT 544	283b	Mono	17.6	m	18.7	1.26	117	07380s1717
LFT 912	467b	Musc	17.7	k - m	18.6	1.17	339	12256s7113
LFT 911	467a	Musc	15.7	k - m	16.6	1.17	339	12256s7113
Ross 614b	234b	Mono	14.8	m	16.8	1.00	131	06268s0246
Wolf 359	406	Leo	13.7	dM6	16.8	4.71	235	10541n0720
LFT 507	261	Mono	16.0	DA	16.8	0.82	185	06596s0623
CD -37°10765b	618b	Scor	16.0	dM7	16.5	1.22	325	16168s3725
LFT 555	293	Voln	15.0	DF	16.2	2.05	135	07528s6738
LFT 1514	777	Cygn	16.5	dM6	16.1	0.86	128	20014n2944
WX Ursa Majoris	412b	UMaj	14.8	dM8	16.0	4.53	282	11030n4347
L726-8a	65a	Ceti	12.4	dM6	15.2	3.37	80	01364s1813
L726-8b (UV Ceti)	65b	Ceti	12.9	dM6	15.9	3.37	80	01364s1813
LFT 1583	810b	Aqar	16.2		15.6	1.48	107	20528s1415
Ross 193b	812b	Aqar	16.5		15.5	0.82	105	20541s0502
LFT 215	102	Arie	14.8	dM6	15.4	0.68	176	02308n2443
Ross 92	359	Leo	15.7	dM6	15.4	0.65	130	09382n2216
Wolf 457	492	Virg	15.9	DC	15.4	1.05	210	12576n0346
Wolf 489	518	Virg	14.7	DK	15.3	3.87	253	13343n0358
Proxima Centauri	551	Cent	10.7	dM5	15.1	3.85	282	14263s6228
LFT 1211	589b	Serp	15.2	dM6	15.0	1.20	264	15331n1753
HL4 (LP 658-2)	---	Orio	15.1	DK	16.4	2.38	167	05527s0409
LP 357-186	---	Taur	18.0	D	16:	0.45	144	04094n2347
LP 768-500	---	Ceti	18.5	DC	17:	1.18	188	01458s1726
LP 321-98	---	Coma	19:	D	18:	0.55	234	12398n3014
LP 464-53	---	Pisc	20:	dM	19:	1.02	141	00170n1238
LP 9-231	---	Drac	15:	D	18:	3.59	337	17570n8244

whose visual luminosity is about 1/250 that of our Sun.
(Apparent magnitude = 9.1) The companion is the remarkable
member of the pair, having a luminosity of about 1/570,000
that of the Sun. The apparent magnitude is 18.0 visual, and
+19.3 absolute. The star's apparent luminosity is about 700
times the light of Jupiter. If put in place of the Sun, it
would appear slightly brighter than the full moon. The
actual size of Van Biesbroeck's Star is not definitely
known, but the spectral type (dM6e) and the luminosity indicate that it is only a fraction of the mass and diameter
of our Sun. According to current ideas of stellar structure
such a star cannot maintain a sufficiently high internal
temperature to operate the hydrogen-to-helium reaction
which powers most stars. It is suggested that gravitational contraction may be a source of at least a part of the
energy output, and that such a star may be slowly cooling
and approaching the "black dwarf" state.

Although only a few dozen stars are known with absolute
magnitudes fainter than 15, it must not be inferred that
such stars are actually rare in space. Obviously, we can
detect only the nearby specimens. The table on Page 228
gives the chief information concerning the 27 stars which
presently rank as the faintest known. The information was
compiled from the lists of W.J.Luyten and W.Gliese (1957).
Spectra or estimated spectral classes are given when known.
Absolute magnitudes of the last five entries are somewhat
uncertain, since the distances are not yet accurately determined. It is possible that LP 464-53 may replace Van
Biesbroeck's Star as the faintest known, when the distance
is eventually measured.

It is also interesting to note that the extreme faintness of Van Biesbroeck's star is due in part to its color.
Much of the radiation is in the infrared, and the total
or "bolometric magnitude" is thus about 3 magnitudes brighter. Considered from this standpoint, there are probably
several stars known at present whose luminosities are lower. The record-holder, however, may not be the white dwarf
star LP9-231, whose bolometric absolute magnitude was once
thought to be about +17.4, based on a preliminary trigonometric parallax by W.J.Luyten. From recent measurements
this star now appears to be more remote (about 60 light
years) than was originally thought, and certainly has a
higher bolometric luminosity than Van Biesbroeck's Star.

DEEP-SKY OBJECTS IN AQUILA. Top: The 1st magnitude star
Altair. Below: The open star cluster NGC 6709. Lowell
Observatory photographs with the 13-inch camera.

DEEP-SKY OBJECTS IN AQUILA. Top: The planetary nebula NGC 6781. Below: The many-armed spiral galaxy NGC 6814. Both photographs were made with the 200-inch Palomar telescope.

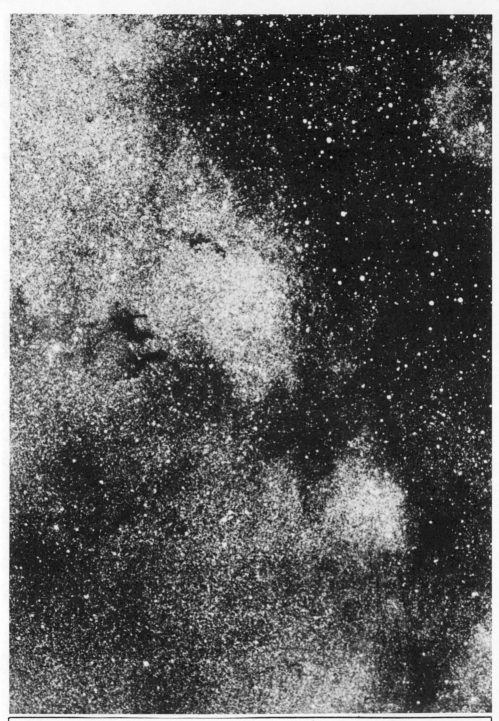

STAR CLOUDS IN THE MILKY WAY near Gamma Aquilae. The dark
nebula B143 is the curious two-pronged marking at left
center.   Lowell Observatory Photograph

DESCRIPTIVE NOTES (Cont'd)

B143   Dark Nebula. Position 19380n1100. Observers of the summer skies will find one of the richest and most spectacular fields for exploration in the Milky Way. from Cygnus to Sagittarius. The star clouds of Aquila are remarkable for the great profusion and complexity of dark nebulous matter; the long belt of interstellar dust clouds known as the "Great Rift" is most conspicuous in this constellation, visually dividing the Milky Way into two parallel streams. The eastern branch continues on to the rich regions of Scutum and Sagittarius toward the south and the other branch diverges to the west where it loses itself in the vast spaces of Ophiuchus.

Some of the dark nebulae of Aquila call for special comment. About $1\frac{1}{2}°$ to the west of Gamma Aquilae and slightly north is the curious dark marking B143, a strangely shaped dust cloud some 30' in diameter with two sharply-outlined "prongs" pointing westward. Another 30' to the south is another similar prong, which bears a separate number, B142. This double dark nebula is one of the few in the sky which can be appreciated in amateur telescopes. Appearing not merely as a starless area, it actually gives a strong impression of an obscuring mass suspended between the observer and the star-strewn background. A rich-field telescope is essential for such objects; an 8" or 10" will provide a fine view when skies are dark and clear.

Another prominent dark nebula, B133, is shown on Page 234, on a photograph made with the Mt.Wilson 100-inch telescope. An object of this sort would probably be completely undiscoverable were it situated in some blank portion of the sky, far from the Galactic plane. In the densest part, measuring 9' x 5', not a single star image can be detected on 100-inch telescope plates with a 4-hour exposure!

Distances of such dark nebulae are usually quite uncertain, and the only safe statement is the obvious one, that the dark cloud is evidently nearer than the bright region against which it is seen projected. The star clouds of this region lie at a distance of possibly 5000 light years, and the dark nebulosities may lie anywhere between, but are probably at least a thousand light years distant. Guessing at a probable distance of about 2500 light years, the actual diameters are found to be something like seven light years for B133 , and over 20 light years for B143.

DARK NEBULA B133. A prominent dark cloud in the Aquila
Milky Way, about 2° south of Lambda Aquilae. Mt.Wilson
Observatory 100-inch telescope photograph.

LIST OF DOUBLE AND MULTIPLE STARS

NAME	DIST	PA	YR	MAGS	NOTES	RA & DEC
I 374	2.7	299	36	7½ -11	relfix, spect B8	16322s5108
R	3.6	123	33	6 - 8½	(h4866) primary	16356s5654
					ecl.bin; spect B9	
h4871	1.3	165	59	8 - 8	(Slr 12) PA dec;	16362s4741
	30.0	45	14	- 10	Spect F5	
B1818	1.4	35	52	6½- 9	Spect F8	16364s6021
Slr 21	1.6	320	42	7½- 9½	relfix, spect B8	16374s4739
h4876	1.6	14	38	6 - 8½	chief star of the	16376s4840
	9.6	266	38	- 7	cluster NGC 6193;	
	13.4	160	38	- 10	AC spect = 05, 06	
Cor201	2.9	43	59	7 - 7½	relfix, spect A8	16468s4957
h4890	30.6	323	59	8 - 8½	wide cpm pair;	16503s4651
					spect B8, A0	
Hd 261	7.0	315	10	6 - 13	spect A0	16507s6311
Rst 5070	0.2	335	42	7 - 8	spect B8	16530s6031
λ 316	0.7	177	43	6½- 7½	slight PA dec;	16567s4835
					spect G5	
h4901	2.8	130	52	8 - 8	relfix, spect B8	16568s5847
Hn 131	2.2	133	43	6½- 9½	relfix, spect B9	16570s5629
I 1306	0.1	215	59	7 - 7½	PA inc, spect A3	16579s5104
I 997	0.6	161	43	8 - 8	relfix, spect F5	16588s5829
Cor 206	8.0	234	33	7 - 8	relfix, spect A0	16591s5006
Hd 264	6.2	247	34	6½- 12	relfix, spect A2	17000s4706
h4916	9.8	277	31	8½- 8½	relfix, spect B5	17047s4924
△213	8.1	167	34	7 - 8½	relfix, spect B2	17066s4641
△214	30.1	356	18	6 - 9½	optical; dist &	17083s6708
					PA inc; spect K0	
h4920	3.0	325	43	7 - 9	slight PA dec;	17086s5832
					spect F2	
L7194	4.3	208	36	5½- 8½	(Brs 13) binary,	17153s4636
					about 550 yrs, PA	
					inc; spect G8, M0	
h4931	1.1	256	42	8 - 8	relfix, spect A0	17162s5924
h4936	7.4	77	59	8½- 9½	relfix, spect K0	17164s4608
Co 213	9.4	283	17	7 - 10	spect A5	17185s5825
Hd 270	2.0	187	38	6 - 11	relfix, spect K0	17186s5758
I 385	0.4	149	42	7½- 8	PA dec, spect A0	17204s5910
	17.8	211	01	- 13		
γ	17.9	328	34	3½ -10	(h4942) relfix;	17212s5620
	41.6	66	13	-12	spect B1 (*)	

LIST OF DOUBLE AND MULTIPLE STARS (Cont'd)

NAME	DIST	PA	YR	MAGS	NOTES	RA & DEC
h4949	2.2	257	53	6 - 7	PA slow dec; spect B9	17232s4548
I 598	1.2	135	44	7 - 9	relfix, spect F0	17249s6144
I 1323	0.2	157	59	8½- 8½	PA inc, spect F8	17258s4945
I 40	18.0	210	33	6½- 10	relfix, spect F8	17281s4600
I 106	0.9	37	59	7½- 8½	PA dec; spect B9	17334s4913
h4970	7.9	69	38	7½- 8½	relfix, spect F2	17384s4837
	18.2	233	33	- 10		
Hd 276	1.2	108	44	7 - 9	relfix, spect B9	17392s5700
R 303	3.4	109	54	7½- 8½	relfix, spect A0	17410s5408
h4978	12.3	269	33	6 - 9½	relfix, spect B3	17464s5336
h4982	41.9	59	13	7 - 9½	spect K0	17467s4816
Rmk 22	2.5	93	52	7 - 8	PA slow inc; spect F8	17531s5522
h5015	3.9	258	35	6½- 11	relfix, spect B8	18048s4547

LIST OF VARIABLE STARS

NAME	MagVar	PER	NOTES	RA & DEC
R	6.0-- 7.0	4.425	Ecl.bin; spect B9; also visual double h4866	16356s5654
T	9.5--11..	Irr	Spect N	16585s5500
U	7.7--14.1	225	LPV. Spect M3e--M5e	17496s5140
V	9.5--13..	379	LPV. Spect M6e--M7e	17511s4818
W	9.0--11..	125	Semi-reg; spect M7	17531s4948
X	8.0--13.5	176	LPV. Spect M5e--M7e	16324s5518
Y	8.5--12..	241	LPV. Spect M2e--M4e	16347s5943
Z	8.6--12..	289	LPV. Spect M3e--M5e	16501s5611
RR	9.5--12..	206	LPV. Spect M3e	17239s4952
RS	9.0--13..	200	LPV.	16542s6422
RT	9.5--10.8	38	Semi-reg	17222s5512
RU	9.0--11..	252	LPV. Spect M5e	17246s6051

ARA

LIST OF VARIABLE STARS (Cont'd)

NAME	MagVar	PER	NOTES	RA & DEC
RV	9.5--15..	292	LPV.	17281s6415
RW	8.7--12.1	4.367	Ecl.bin; spect A0	17306s5707
RX	9.5--10..	Irr	Spect M5	16480s6100
RY	8.6--11.0	144	RV Tauri type; spect G5---K0	17171s5104
SS	9.1--10..	Irr	Spect K5	17544s5009
SZ	9.0--13..	221	LPV. Spect N	17065s6153
UW	9.7--10.6	3.297	Ecl.bin; spect A0	17438s4844
OY	6.0--17..	---	Nova 1910	16369s5220
V340	9.8--11.1	20.81	Cepheid	16413s5115

LIST OF STAR CLUSTERS, NEBULAE, AND GALAXIES

NGC	OTH	TYPE	SUMMARY DESCRIPTION	RA & DEC
----	H11	⊙	Diam 7'; 60 faint stars; class D	16316s4930
6188		▢	F, vL, Irr neby; 20' x 12' with cluster NGC 6193 & triple star h4876 (*)	16359s4855
6193	△413	⊙	Diam 20', vL,1Ri,1C; about 30 stars & neby NGC 6188. Class E; stars mags 6.....	16376s4840
6208	△364	⊙	L,Ri, diam 20'; about 50 stars mags 9...12; class E	16455s5344
6204	△442	⊙	pRi, diam 5'; about 25 stars mags 11.... CM; class F	16457s4656
6215		⊘	SB; 11.2; 1.7' x 1.3' pF,1E, vglbM; Eta Arae 80" to west	16468s5855
6221		⊘	SB; 11.4; 2.7' x 2.0' pB,v1E,cL,glbM	16485s5908

LIST OF STAR CLUSTERS, NEBULAE, AND GALAXIES (Cont'd)

NGC	OTH	TYPE	SUMMARY DESCRIPTION	RA & DEC
6250		⊙	L,1Ri; diam 11'; about 15 stars mags 8....12	16543s4543
6253	△374	⊙	Diam 6'; pL,Ri,mC; about 70 stars mags 13.... Class F	16551s5238
----	H13	⊙	Diam 15'; about 70 faint stars; class D	17017s4806
----	I.4642	◎	vS,F, mag 12.4; diam 15"; appearance nearly stellar	17076s5520
6300		⊘	SB; 11.4; 3.9' x 3.5' F,L, 1E	17123s6246
6326		◎	Mag 12; diam 15" x 10" pB,vS	17168s5142
----	I.4651	⊙	L,pC, diam 15'; about 70 stars mags 10...14; class E	17207s4954
6352	△417	⊕	pF,L; mag 9; diam 8'; Class X1; stars mags 14.....	17216s4826
6362	△225	⊕	cB,L, Mag 8, diam 9'; Class X, stars mags 14......	17266s6701
6397	△366	⊕	B,vL,Ri; mag 7, diam 19'; class IX; stars mags 10.... Possibly nearest globular cluster (*)	17368s5339

DESCRIPTIVE NOTES

ALPHA   Mag 2.95; Spectrum B2 V. Position 17280s4950.
          The computed distance is about 390 light years,
leading to an actual luminosity of 760 times that of the
Sun (absolute magnitude = -2.4). The annual proper motion
is 0.08"; the radial velocity is about 1 mile per second
in approach. Alpha Arae is a spectroscopic binary star of
uncertain period.

BETA   Mag 2.87; spectrum K3 Ib. Position 17211s5529.
The distance is estimated to be about 1030
light years, and the star is a supergiant with about 5700
times the luminosity of the Sun. The computed absolute magnitude is about -4.6. The annual proper motion is 0.03";
the radial velocity is very slight, less than ½ mile per
second in approach.

GAMMA   Mag 3.32; spectrum B1 III. Position 17212s5620.
The computed distance is about 680 light years
which leads to an actual luminosity of about 1700 suns; the
corresponding absolute magnitude is -3.3. The star shows
an annual proper motion of 0.02"; the radial velocity is
2½ miles per second in approach.
The 10th magnitude companion at 17.9" was discovered
by John Herschel in 1835. There has been no certain change
in separation or angle in over a century, and the faint
star is probably not a true physical companion to Gamma.
The projected separation of the pair would be about 3760
AU, an unusually wide pair perhaps, but no greater than
many others whose physical connection is undoubted. The
companion, if at the same distance as the bright star, has
a luminosity of about 3½ suns.

ZETA   Mag 3.16; spectrum K5 III. Position 16545s5555.
The distance is about 90 light years; the true
luminosity about 35 times that of the Sun. The absolute
magnitude is about +1.0. Zeta Arae shows an annual proper
motion of 0.04"; the radial velocity is 3½ miles per second in approach.

NGC 6188   Position 16359s4855. A wonderful field of
bright and dark nebulosity, located near the
central line of the Milky Way, some 7° south and west of
Zeta Scorpii. The brightest portion of the nebulosity was
discovered by John Herschel in 1836, and has the form of a
very irregular triangle, measuring about 20' x 12'. On the
northeast side, near the apex, is located the galactic star
cluster NGC 6193, whose giant stars supply the illumination
for the entire cloud. Long-exposure photography reveals a
wealth of spectacular details in the nebulosity. The dark

NGC 6188. An interesting region of mixed bright and dark nebulosity, photographed with the 60-inch reflector of Harvard Observatory's southern station in South Africa.

DESCRIPTIVE NOTES (Cont'd)

obscuring masses are bordered by bright rims which seem to be reflecting the glare of the involved stars, and the whole unearthly picture is strongly reminiscent of the famous "Horse-head" in Orion. Some astronomers have proposed that the bright-rim nebulae mark the fronts of advancing shock waves as a dark cloud expands into space, sweeping up the interstellar dust and gas. If we are actually seeing a collision zone of this sort in NGC 6188, the region affected must be at least 15 light years in extent.

In a survey of bright emission nebulosities at the Commonwealth Observatory in Australia, C.S.Gum (1955) found that the group is merely the center of a vast nebulosity which has a full diameter of over 3°, or about 160 light years. From a study of the involved stars, a distance modulus of 9.6 magnitudes is derived, giving the distance as about 2700 light years.

The involved cluster, NGC 6193, is a remarkable and brilliant aggregation, about 15 light years in diameter. The brightest star is the visual double h4876 (HD 150136), an O-type giant whose apparent magnitude of 5.9 implies an actual luminosity of about 3000 suns, before making any correction for interstellar absorption. The spectral type is given by various authorities as O5, O6, or O7; the absolute magnitude may be about -3.7. The 7th magnitude companion at 9.6" was discovered in 1836, and has shown no definite change in separation or PA since that time; it has a spectral class of O6. The closer companion, at 1.6", was first measured in 1878. Neither star has shown any relative motion since discovery; the projected separations are: AB = 1300 AU; AC = 8000 AU. There is also a 10th magnitude companion at 13.4" in PA 160°, and a fourth star at 13.9" in PA 15°, magnitude 11. In addition, the primary star appears to be a spectroscopic binary of uncertain period; the mean radial velocity is about 14 miles per second in recession. (Refer also to the "Horse-head" nebula B33 in Orion, and the nebula M16 in Serpens)

NGC 6397   Position 17368s5339. A bright globular star cluster, located on the left edge of the Milky Way, some 10½° south of Theta Scorpii. It was first observed by Lacaille in 1755. It is an object of special interest from recent studies which indicate that it may be the nearest of all globulars to the Solar System. Unfortunately the far southern declination places it beyond the reach of observers in the United States. The most thorough studies of the cluster have been made with the 74- inch reflectors at Mt.Stromlo in Australia, and at Radcliffe Observatory in South Africa.

NGC 6397 is not one of the richer globulars, but has a rather loose, scattered structure which permits easy resolution in relatively small telescopes. The extreme diameter is close to 20', and the total integrated magnitude is 7.3. The two dozen brightest stars (mags 10--12) show no evident concentration toward the cluster center, but seem to be distributed in random groups and curving rows across the background of fainter members. The cluster closely resembles the better-known M4 in Scorpius in apparent size, brightness, and structure. According to H.B. Sawyer's "Bibliography of Individual Globular Clusters" (1947) the integrated spectral type is F5; the radial velocity about $6\frac{1}{2}$ miles per second in recession.

A peculiar feature of NGC 6397 is the apparent absense of short-period pulsating variables which are often so common in the globulars. One such star, with a period of 0.331 day, was detected in a study of the cluster by H.Swope and I.Greenbaum (1952) but it appears to be two magnitudes too faint to be a cluster member, and is thus probably a background star. Two other variables lie within the apparent borders of the cluster - a long-period variable and a semi-regular type - but it is not certain that either is a true cluster member. In contrast, some globulars contain dozens of variable stars; Omega Centauri possesses over 160, and M3 has nearly 200.

The brightest members of NGC 6397 are red giants of absolute magnitude about -2, or some 500 times the luminosity of the Sun. The total luminosity of the cluster is about 8000 times the light of the `Sun; much fainter than many of the well known globulars. The true diameter may be about 50 light years. These figures are based on a study of NGC 6397 by L.Searle and A.W.Rodgers at Mt.Stromlo in 1965. They derive a true distance modulus of about 12 magnitudes for the cluster, with an estimated uncertainty of about 0.3 magnitude. The resulting distance is about 8200 light years, which makes this the closest globular cluster

NGC 6397. This globular star cluster in Ara is believed to
be the closest cluster of this type to the Solar System.
Radcliffe Observatory

known, significantly nearer than the two great clusters
Omega Centauri (NGC 5139) and 47 Tucanae (NGC 104) which
are usually considered the nearest. There is some reason
to believe that the cluster M4 in Scorpius is also nearer
than either of the two giants, but it lies in a region so
heavily obscured by interstellar dust that reliable measurements are almost impossible to make.

In a study of NGC 6397 at Radcliffe by R.Woolley, J.B.
Alexander, L.Mather, and E.Epps (1961) accurate magnitudes
and colors for nearly 1000 stars in the cluster were obtained, down to magnitude 15. The resulting H-R diagram
is shown below. As explained elsewhere in this book (see
M13 in Hercules) the appearance of the diagram is very
different from that of a typical galactic cluster, and indicates that the globulars are very ancient star groups.
This view is supported also by spectroscopic studies made
at Mt.Stromlo (1965) which show that the stars of the
cluster are deficient in the atoms of the metals by a
factor of about 100, when compared to normal stars of
Population I. This appears to be another effect of age;
the material available for star formation today is growing
increasingly metal-rich through the activity of massive
stars which build up the heavier atoms by nuclear reactions
in their interiors, later releasing these atoms into space
in supernova explosions. "First-generation stars" were thus
presumably formed from material which was metal-poor, and
this indicates that NGC 6397 may be one of the most ancient
star clusters known. (Refer also to M13 in Hercules, Omega
Centauri, 47 Tucanae, M4 in Scorpius, and M5 in Serpens)

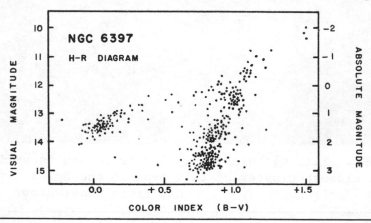

ARIES

LIST OF DOUBLE AND MULTIPLE STARS

NAME	DIST	PA	YR	MAGS	NOTES	RA & DEC
1	2.7	166	62	6 - 7½	gold & blue; cpm, spect K0, A6, no certain change	01474n2202
Σ175	22.8	355	61	8 - 8½	optical; PA & dist inc; spect F0	01483n2052
Ho 311	0.4	202	61	7½- 7½	PA inc, spect F5	01484n2424
Σ178	2.9	201	62	8 - 8	PA slight inc; spect F0	01494n1034
γ	7.8	360	59	4½- 4½	(Σ180) fine pair, both white (*)	01508n1903
λ	37.4	46	33	5 - 7½	easy pair; relfix cpm; spect F0, dG1	01551n2320
Σ196	2.2	48	58	8½- 11	PA dec, spect G	01568n2046
	30.9	163	18	- 9	AC dist dec	
β515	1.4	267	34	7½-12½	spect F2	01584n1619
Σ200	8.2	124	19	8½ - 9	relfix, spect G0	01588n2351
10	0.5	281	61	6 - 8½	(Σ208) binary, about 390 yrs; dist dec; spect dF4; PA inc	02008n2542
Σ212	1.9	163	61	8 - 8½	relfix, spect A0	02034n2452
11	1.6	342	43	6½- 12	(Ho 312) PA inc, spect B8	02040n2528
Σ221	8.4	146	35	8 - 9	relfix, spect A3	02069n2007
	63.4	229	04	- 12		
Σ224	6.1	246	61	7½ - 8	slight dist inc; spect G5	02082n1327
Σ226	1.8	240	62	8 - 10	Dist & PA dec, spect G5	02094n2344
Σ237	14.5	238	24	8½ - 9	relfix, spect G0	02130n1032
Σ240	4.8	51	44	7½ - 8	relfix, spect F0	02144n2339
Σ244	4.5	289	51	9 - 9	relfix, spect F5	02148n2200
Σ261	2.9	71	62	8½ -8½	relfix, spect F0	02217n1117
Σ271	12.4	182	35	6½- 11	relfix, cpm; spect dF4	02276n2501
Σ273	6.9	358	56	8 - 9	relfix, spect F5	02293n1809
30	38.6	274	37	6 - 7	(ΣI 5) wide cpm pair (*)	02341n2426
Σ287	6.4	74	51	7½- 10	relfix, spect K0	02362n1438

LIST OF DOUBLE AND MULTIPLE STARS (Cont'd)

NAME	DIST	PA	YR	MAGS	NOTES	RA & DEC
O Σ43	1.1	20	61	7 - 8½	binary, about 470 yrs; PA slow dec; spect F5	02378n2625
33	28.6	360	21	6 - 8½	(Σ289) relfix; spect A3	02378n2651
Σ291	3.3	120	61	7½ - 8	relfix, spect B9	02383n1835
	65.7	241	23	- 9½		
μ	19.1	264	34	6 -12½	(β522) cpm pair; spect A0	02396n1948
β306	3.0	18	38	6½ -11	relfix, cpm; spect A3	02410n2526
	50.3	90	25	-12		
Σ300	3.1	311	63	8 - 8	slow PA inc; cpm; spect F0	02416n2915
Σ305	3.7	314	61	7 - 8	binary, about 720 yrs; PA dec spect F9	02446n1910
β307	15.4	315	14	7- 11½	spect B5	02447n2928
π	3.2	120	53	5 - 8½	(Σ311) AB cpm; relfix, spect B6	02465n1715
	25.2	110	38	-10½		
41	24.6	277	22	4 - 11	(OΣ47) optical; AB dist inc, AC dist dec; spect B8	02470n2703
	31.3	213	22	- 11		
O Σ46	4.9	74	35	7 - 10	(h656) relfix; spect F0	02470n3019
	20.1	168	00	- 13		
AG 56	6.8	109	26	9 - 9	relfix, spect G5	02502n1028
Σ326	6.3	219	63	7½ -9½	cpm; slight dist dec; spect dK2 & dM1 (*)	02527n2640
β1173	0.2	251	62	7½- 7½	PA inc, spect K0	02557n2356
	4.6	282	62	- 13		
β525	0.4	254	68	7½ -7½	binary, about 430 yrs; PA inc; spect A3	02560n2125
ε	1.4	203	66	6 - 6	(Σ333) slow PA inc; spect A2, A2	02564n2108
O Σ49	2.1	54	58	7 - 10	PA dec, spect A0	02577n1749
Σ338	19.4	201	26	8 - 8	slight dist dec; spect F8	02591n1040
Σ342	3.1	305	39	8½- 9	relfix, spect F5	03021n2743

LIST OF DOUBLE AND MULTIPLE STARS (Cont'd)

NAME	DIST	PA	YR	MAGS	NOTES	RA & DEC
52	0.1	191	61	6 - 6	(Σ346) PA dec;	03025n2504
	5.1	356	38	- 11	spect B7, B7	
β1030	0.6	120	66	8½- 8½	PA dec, spect F0	03072n2133
Σ366	47.2	40	15	7 - 10	spect K0	03114n2246
Σ366b	1.7	191	43	10-10½	(β530) relfix	
A2224	0.7	334	67	8 - 9½	PA inc, spect K0	03135n1909
Σ375	2.4	315	61	7½- 9½	relfix, spect A5	03174n2331
	64.9	290	11	- 13		
Σ376	7.0	252	49	8 - 8	relfix, spect A2	03175n1933
Σ381	0.9	105	66	7 - 9	PA inc, spect G5	03204n2048
Σ394	6.8	162	49	7 - 8	relfix, spect A3	03250n2017
66	1.1	68	63	6 - 12	(β878) PA dec; cpm, spect sgG6	03255n2238

LIST OF VARIABLE STARS

NAME	MagVar	PER	NOTES	RA & DEC
R	7.3--13.8	187	LPV. Spect M3e	02133n2450
S	9.2--15.9	295	LPV. Spect M4e	02019n1217
T	7.4--11.0	314	LPV. Spect M6e--M8	02455n1718
U	6.6--15.2	370	LPV. Spect M5e	03083n1437
V	8.0-- 8.6	75:	Semi-reg; spect R4	02123n1200
W	9.5--14..	---	Nova 1855	03177n2847
X	9.0--10.2	.651	Cl.Var.; Spect A0--A7	03058n1015
Y	9.2--10.4	109:	Semi-reg; spect M5e--M7	02380n3059
RR	6.0--	?	Variability uncertain; Spect gG8	01530n2320
RX	9.4---9.7	1.029	Ecl.Bin.	02125n2221
RY	9.5--10.5	Irr	Spect M6	01597n1602
RZ	5.9---6.3	Irr	(45 Arietis) Spect gM6	02530n1808
ST	9.0--10.6	100:	Semi-reg; spect M4	03072n1314
SU	9.5--15..	---	Nova 1854	02457n1709
SX	5.75--5.82	.728	(56 Arietis) Spect A0p; Alpha Canum Ven. type	03093n2704
UV	5.16--5.19	.037	(38 Arietis)(HR 812) Delta Scuti type, Spect A7	02422n1214
53	6.1--6.23	.1527	(UW Arietis) Beta Canis Maj type, spect B2 (*)	03046n1741

LIST OF STAR CLUSTERS, NEBULAE, AND GALAXIES

NGC	Oth	TYPE	SUMMARY DESCRIPTION	RA & DEC
697	179[3]	⊘	Sb/Sc; 12.5; 2.0' x 1.0' F,cL,E,bM; flattened spiral (incorrect position on Skalnate Pleso Atlas)	01486n2206
772	112[1]	⊘	Sb; 12.0; 5.0' x 3.0' B,cL,E,gbM; asymmetric in form, strong spiral arm on north-west side	01566n1846
821	152[1]	⊘	E2; 12.7; 1.0' x 0.9' pB,S,vlE,vmbM; 10^m star 1'np	02056n1046
877	246[2]	⊘	Sc; 12.5; 1.8' x 1.2' pF,pL,lE,pgbM; 9^m star $4\frac{1}{2}$'sf	02153n1419
972	211[2]	⊘	Sc/pec; 12.3; 2.7' x 1.0' pB,cL,lE,gmbM; coarse spiral pattern; irr. dust lanes (*)	02313n2906
976		⊘	Sb; 12.7; 0.7' x 0.7' F,vS,R	02312n2044
1156	619[2]	⊘	I; 12.5; 2.0' x 1.5' pB,pL,cE (*)	02567n2503

DESCRIPTIVE NOTES

ALPHA   Name- HAMAL, "The Head of the Sheep". Mag 2.00; spectrum K2 III. Position 02043n2314. Hamal is about 75 light years distant (Mt.Wilson parallax) and has an actual luminosity of about 70 Suns. The absolute magnitude is +0.2. The annual proper motion is 0.24"; the radial velocity is $8\frac{1}{2}$ miles per second in approach.

BETA   Name- SHERATAN, "The Sign". Mag 2.65; spectrum A5 V. The computed distance is about 52 light years, giving an actual luminosity of 17 Suns, and an absolute magnitude of about +1.7. The annual proper motion is 0.15"; the radial velocity is about 1 mile per second in approach. Position 01519n2034.

Beta Arietis was discovered to be a spectroscopic binary by H.C.Vogel in 1903, and the first orbit
was computed by H.Ludendorff in 1907. The period is 106.997
days and the orbit has the high eccentricity of 0.89. In
his analysis of the system, R.M.Petrie (1938) stated that
"the results of this study show that the orbit of Beta
Arietis is by far the most eccentric of any known spectroscopic binary, and is exceeded by very few of the visual
systems. The orbital elements show no definite variation
over a period of 30 years". The mean separation of the two
stars appears to be in the range of 15 - 20 million miles.

GAMMA   Name- MESARTHIM. Mag 3.90; spectra B9 V and
               AOp. Position 01508n1903. This is one of the
best known double stars, and one of the earliest to be
discovered, found accidentally by Robert Hooke in 1664
while he was following a comet. The components have shown
no change in angle in 3 centuries, but the separation may
have decreased slightly since the time of F.G.W.Struve,
whose measurement of 8.6" was made in 1830. The present
separation is about 7.8"; the individual magnitudes are
4.75 and 4.83. The two stars share a common proper motion
of 0.14" annually in PA 141°. Some indication of orbital
motion may be found in the difference in the radial velocities of the two stars: A = less than 1 mile per second
in approach; B = about $2\frac{1}{2}$ miles per second in recession.

According to a parallax obtained at Allegheny, the
distance of Gamma Arietis is about 160 light years, giving
the system a total luminosity of about 50 suns. The projected separation of the pair is about 385 AU. According
to the Moscow General Catalogue (Supplement 1974) the
southern component of Gamma Arietis is a magnetic variable
of the Alpha Canum type; period 2.607 days and amplitude
of 0.02 magnitude. The star has a somewhat unusual spectrum, containing very prominent lines of silicon.

In addition to the physical pair, a third star of
the 9th magnitude (Gamma C or β512) lies 221" distant in
PA 84°. In 1878 S.W.Burnham found it to be a close double;
magnitudes 9 and 13, separation 1.7". This star does not
share the motion of the bright pair, and the separation
from Gamma A&B is slowly decreasing from an early measurement of 228" in 1823.

DESCRIPTIVE NOTES (Cont'd)

30　　　Mag 6.57; spectrum dF5. Position 02341n2426. A
wide and easy double star, the components sharing
a proper motion of about 0.15" per year in PA 90°. There
has been no change in separation or angle since the first
measurements were made in 1835. The magnitudes are 6.57
and 7.37, spectra dF5 and dF6. Both stars are yellow, but
many observing lists refer to the smaller star as bluish
or lilac. The brighter star is a spectroscopic binary with
a period of 9.851 days. The mean radial velocity of the
system is about 9½ miles per second in recession.

　　　Trigonometric parallaxes obtained at Allegheny,
McCormick, and Mt.Wilson give the distance as about 190
light years; the resulting absolute magnitudes are +2.8
and +3.6 (luminosities = 6 and 3 X Sun).

53　　　Mag 6.09; spectrum B2 V. Position 03046n1741.
　　　This is one of three so-called "Runaway stars"
which are characterized by abnormally high space velocities
and appear to be moving outward from the region of the
Orion Nebula association. A study of the three stars has
suggested that such objects are escaped members of the
group of young stars connected with the Great Nebula. If
so, the stars were ejected only a few million years ago,
possibly by the explosions of supernovae. The space velocity of 53 Arietis is approximately 35 miles per second,
somewhat lower than the other two known stars of the type.
The expulsion from the Orion region is estimated to have
occurred about 5 million years ago. The annual proper
motion of the star is about 0.025"; the radial velocity is
17 miles per second in recession.

　　　The two other Runaway stars are Mu Columbae and
AE Aurigae, with space velocities of over 70 miles per
second, and computed separation ages of about 2.0 and 2.7
million years, respectively. The most unusual star of the
three is AE Aurigae; it is an erratic variable with an
amplitude of about 0.7 magnitude, and is presently located
in the midst of a large diffuse nebulosity, IC 405. The
studies of these objects may supply valuable information
concerning the birth of stars and their formation in groups
and expanding associations.

　　　The most interesting question concerning such stars
is of course the problem of their "escape" from the groups

GALAXIES IN ARIES. Top: The coarse spiral NGC 972. Below:
The irregular system NGC 1156.
Palomar Observatory 200-inch telescope.

in which they were formed. Various mechanisms have been
proposed, but none has been entirely satisfactory in the
attempt to explain the acceleration of a star to a high
velocity. The explosion of a supernova could not in itself
produce such an effect, but still might be the answer to
the problem in another sense. If we suppose that the pre-
supernova star was a member of a close binary pair, the
orbital velocities would have been very high; the sudden
explosion of one star would then "release" the other star
which would continue out into space at the same high velocity. In the case of 53 Arietis, this explanation is
rendered somewhat questionable by the fact that the star
may be a close binary at the present time. The radial
velocity has been reported to be variable, and the star
is listed as a spectroscopic binary in R.E.Wilson's Catalog (1953). However, the star has recently been identified
as a Beta Canis type variable (period= 3^h40^m) and it now
appears that the variable velocity may be due to that
cause, rather than binary motion. (For a diagram of all
three Runaway stars, refer to AE Aurigae, page 288.)

Σ326   Position 02527n2640. Double star, discovered
           by F.G.W.Struve in 1831. The components are
moving together through space at the rate of 0.31" per
year in PA 124°. The separation of the pair has decreased
somewhat since discovery, when the first measurement of
about 9" was made. The projected separation of the pair
is about 100 AU. Both stars are dwarfs, smaller and fainter than the Sun. The computed absolute magnitudes are +6.3
and +8.5; the spectral classes are dK2 and dM1; and the
distance of the system is about 55 light years, according
to a Yale trigonometric parallax.
       A third faint component is not mentioned in Aitken's
ADS Catalog, but was found in the course of proper motion
studies at Lowell Observatory in 1959. It is a red dwarf
of the 15th magnitude, about 43" from the main pair, at
PA 260°. This star has a calculated luminosity of about
1/4000 that of the Sun, and the actual separation from
the double primary must be at least 790 AU. The radial
velocity of the whole system is about 20 miles per second
in recession.

AURIGA

LIST OF DOUBLE AND MULTIPLE STARS

NAME	DIST	PA	YR	MAGS	NOTES	RA & DEC
h348	31.0	284	21	$7\frac{1}{2}$-$10\frac{1}{2}$	relfix, spect K0	04398n3350
Σ603	8.3	239	38	8 - 8	relfix, spect G0	04504n4930
Σ613	14.2	103	58	8 - 9	optical group; all	04552n4404
	21.9	57	58	-$10\frac{1}{2}$	dist & PA dec; spect F0	
ω	5.4	359	50	5 - 8	(4 Aurigae) (Σ616) PA slow inc; spect A0	04558n3749
5	3.4	267	51	6 - $9\frac{1}{2}$	(0Σ92) binary; PA inc, spect dF3	04569n3919
Σ619	4.2	140	56	9 - 9	PA inc, spect G	04575n5011
ε	28.6	224	25	3 - 14	(β554) primary	04584n4345
	42.9	275	25	-$11\frac{1}{2}$	ecl.bin. (*)	
	45.3	317	24	- 12		
A1023	0.4	67	51	$6\frac{1}{2}$- $8\frac{1}{2}$	Spect F8; relfix	05018n4651
9	5.2	82	58	$5\frac{1}{2}$- 13	(β1046) PA dec;	05028n5132
	90.1	61	23	- $9\frac{1}{2}$	spect F0; ABC cpm	
0Σ94	17.9	304	21	$7\frac{1}{2}$- 11	dist inc, spect B9	05035n5014
	25.4	64	21	- 11		
0Σ96	21.0	104	12	$6\frac{1}{2}$- 11	spect A3	05053n4904
Σ644	1.5	221	63	$6\frac{1}{2}$ - 7	relfix, spect B2, K3; cpm; color contrast pair	05069n3714
Σ648	4.8	66	55	$7\frac{1}{2}$ - 8	AB cpm; PA dec,	05078n3159
	40.2	114	27	- 12	spect G5; AC dist	
	38.4	66	27	- 13	inc	
0Σ101	5.8	184	34	$7\frac{1}{2}$- 10	relfix, spect B8	05101n4655
A1031	0.3	30	51	7- $10\frac{1}{2}$	PA inc, spect G5	05115n4707
14	11.1	352	09	5 - 11	(Σ653) slow PA	05122n3238
	14.6	226	33	- 7	inc; AC relfix; spect A9	
α	723	141	00	0 - 10	CAPELLA. cpm (*)	05130n4557
R	47.5	339	18	var-$8\frac{1}{2}$	R = LPV, spect M7e & G0	05132n5332
Hu1101	0.4	287	56	7 - 9	relfix, spect K0	05138n3925
Σ666	3.0	74	53	8 - 8	relfix, spect A3	05138n3317
Σ657	1.0	296	59	8 - 9	PA slow inc; F5	05148n5247
16	4.2	56	36	5 - 11	(0Σ103) relfix; cpm; spect K3	05149n3319

LIST OF DOUBLE AND MULTIPLE STARS   (Cont'd)

NAME	DIST	PA	YR	MAGS	NOTES	RA & DEC
Σ669	9.9	276	36	8 - 8½	relfix, spect F0	05153n4512
Es 59	14.0	10	19	8½ - 9	spect A3	05154n3328
λ	29.1	274	00	5 - 13	spect G0; optical	05156n4004
	41.7	268	34	- 12	group; dist inc	
Sei 180	7.2	358	00	7 -10½	spect B8	05159n3451
18	3.9	169	63	7- 12½	spect A5; slight PA inc (Ho 18)	05161n3356
Σ681	23.2	181	28	6½ -8½	relfix, spect F0, A	05169n4655
UV	3.4	4	33	8- 11½	relfix; primary N-type variable	05185n3228
Σ684	1.5	139	58	8 - 10	relfix, spect A0	05185n4502
Σ687	17.3	68	57	8 - 9½	(β886) spect B3;	05190n3345
	48.8	154	57	- 10	C = 0.9" pair	
OΣ104	19.7	190	58	7 - 11	dist inc; spect K5	05195n4659
σ	8.7	167	22	5 - 11	(β888) spect K4	05212n3720
	27.3	330	14	- 13	slight PA dec.	
Σ698	31.2	346	51	6½- 7½	relfix, cpm spect K1, F6	05219n3449
Σ699	8.8	343	52	8 - 8½	relfix, spect A0	05222n3800
Σ706	3.2	44	63	8 - 9	PA slow inc; spect F5	05231n3018
Hu 217	0.6	251	60	7 - 8½	PA dec, spect B5	05264n3520
Σ715	0.9	201	62	8 - 9	relfix, spect A0	05267n4115
	19.5	51	04	- 11½		
Σ 719	1.0	330	55	7 - 9½	slight dist inc;	05269n2931
	15.0	351	35	- 9	spect G5	
Σ 711	8.1	229	62	8 - 9½	cpm; slight PA & dist dec; spect G0, K2; C = 0.6" pair	05274n5438
	191	245	08	- 10		
Σ718	7.7	75	55	7 - 7	relfix, spect F5	05285n4921
Sei 323	8.8	265	00	8 - 11	spect B1	05303n3626
β1267	0.5	202	63	8½- 8½	PA dec, spect F5	05319n3054
Hu 1229	1.8	197	48	7½- 13	spect A5	05320n3752
Σ737	10.7	305	15	8½- 9	In star cluster M36, relfix, spect B	05331n3406
Σ736	2.4	354	53	7½- 8½	PA slow inc, spect F8	05336n4148

LIST OF DOUBLE AND MULTIPLE STARS (Cont'd)

NAME	DIST	PA	YR	MAGS	NOTES	RA & DEC
26	0.2	323	62	$5\frac{1}{2}$ - 6	(β1240) (Σ753)	05354n3028
	12.4	267	33	- 8	AB binary, 53 yrs;	
	33.1	114	15	$-11\frac{1}{2}$	PA dec; ABC cpm;	
					spect G5, A3, F0	
OΣ112	0.8	59	62	$7\frac{1}{2}$- 8	PA dec, spect B9	05365n3756
Hu 825	0.4	343	52	8 - 8	relfix, spect A0	05367n3559
Σ764	26.0	14	32	$6\frac{1}{2}$ - 7	easy pair; relfix,	05382n2928
					cpm; spect B8, A0	
β14	5.7	194	33	$7\frac{1}{2}$-$10\frac{1}{2}$	relfix, spect G0	05393n2950
Ho 509	11.5	204	23	7 - 12	spect F5	05397n3318
Σ768	18.6	221	05	7 - $9\frac{1}{2}$	spect B8	05398n4106
Σ778	3.2	185	56	$7\frac{1}{2}$- 9	relfix, spect B8	05409n3055
β560	1.4	137	62	8 -$8\frac{1}{2}$	PA dec, dist inc;	05442n2939
					spect F8	
Ho 19	7.0	346	58	$6\frac{1}{2}$- 12	relfix, spect G0	05446n3509
OΣ117	11.9	30	14	7 - 10	relfix, spect K5	05450n3031
τ	39.5	352	16	5 - 12	(β192) spect G8	05457n3910
	49.6	35	26	- 12		
Σ791	4.8	91	38	$8\frac{1}{2}$ - 9	relfix, spect A0	05466n3934
	60.6	213	09	- 10		
Σ796	3.8	62	43	7 - 8	relfix, spect A3	05467n3146
	207	324	09	- 10		
ν	54.6	206	11	4 - $9\frac{1}{2}$	(32 Aurigae) spect	05480n3908
					K0	
Es 1321	5.6	351	14	$7\frac{1}{2}$- $9\frac{1}{2}$	spect K0	05486n4648
Σ799	0.8	170	62	$7\frac{1}{2}$- $8\frac{1}{2}$	PA dec, spect B8	05488n3833
Σ802	2.8	107	58	8 - $8\frac{1}{2}$	relfix, spect G0	05490n4009
Σ807	2.2	147	34	8 - 10	slow PA inc;	05496n3426
					spect F8	
Σ808	2.8	166	47	$8\frac{1}{2}$ -11	relfix, spect G0	05496n2946
	16.1	58	37	- 9	(D10)	
β1053	1.4	350	62	$7\frac{1}{2}$- $9\frac{1}{2}$	PA & dist inc;	05501n3720
					spect F5	
Σ811	5.0	231	34	8- $9\frac{1}{2}$	relfix, spect B5	05510n3029
OΣ122	0.1	221	62	$7\frac{1}{2}$ -8	binary; about 320	05524n3656
					yrs; PA inc, spect	
					A5	
A1726	5.0	274	19	7 - 14	spect A0	05557n4537
θ	3.4	318	62	3 - $7\frac{1}{2}$	(OΣ545) AB cpm;	05563n3713
	51.5	298	63	- 11	spect B9, G (*)	

LIST OF DOUBLE AND MULTIPLE STARS   (Cont'd)

NAME	DIST	PA	YR	MAGS	NOTES	RA & DEC
β1055	1.7	324	53	6½- 11	(Hh209) AB cpm;	05567n4435
	34.2	331	39	- 9	slight PA dec;	
					spect K2	
Σ825	8.3	146	59	8 - 9	relfix, spect A0	05582n3631
	37.8	139	06	- 11½		
0Σ127	1.5	333	58	7 -10½	relfix, spect G5	05586n3843
0Σ128	39.0	13	44	6½- 9	(35 Cam) cpm;	06005n5135
					spect A5	
0Σ128b	0.6	321	51	9 - 10	(Hu 559) PA dec	
A1729	0.4	51	60	7 - 9	PA inc, spect A2	06009n4536
Σ834	22.9	308	15	8 - 9	relfix, spect A	06014n3014
β893	17.9	130	25	6 - 12	spect F8	06016n3758
	83.9	140	16	- 11		
J17	2.6	153	58	9 - 9½	relfix	06016n4303
Es1234	9.8	267	13	7 -11½	spect B0	06021n4815
0Σ129	10.0	209	58	6 - 11	relfix, spect M3,	06032n2931
					F7	
0Σ131	1.5	280	34	7 - 10	relfix, spect B9	06040n3617
0Σ130	0.4	204	60	7 - 8	PA inc, spect B5	06043n4240
0Σ132	1.6	328	63	6½- 10	PA slow inc;	06048n3759
					spect A2	
h379	9.4	116	26	7½-11½	spect A2	06059n3116
41	7.7	356	63	5 - 7	(Σ845) relfix,	06078n4844
					cpm; spect A2, A5	
Es---	15.4	215	13	6½-10½	spect G5	06079n4311
Σ861	63.7	18	35	8 - 8	dist dec, spect A	06081n3041
Σ861b	1.6	318	42	9 - 9	relfix	
Σ862	6.5	336	06	7 - 11	relfix, spect G5	06089n2930
Σ872	11.3	217	49	6 - 7	relfix, easy cpm	06123n3610
					pair; spect dF4,F0	
Σ879	7.5	70	03	9- 10½	relfix	06131n3006
Σ883	3.3	264	54	8 - 8½	relfix, spect F0	06156n3948
	28.4	258	24	- 10		
A2116	1.5	33	52	7- 13½	PA inc, spect F0	06163n3827
Σ888	0.2	180	62	7½- 7½	(β895) AB binary,	06168n2827
	3.0	260	55	- 9	54 yrs; PA inc;	
					spect A3; all cpm;	
					AC PA slow inc.	
Σ884	9.1	270	18	8½- 8½	relfix, spect A0	06168n4709

LIST OF DOUBLE AND MULTIPLE STARS   (Cont'd)

NAME	DIST	PA	YR	MAGS	NOTES	RA & DEC
A1319	0.7	139	62	7 -9½	relfix, spect A2	06187n4613
Σ896	18.7	82	16	8 - 8½	spect A5	06219n5154
A1732	3.5	14	18	7½-12½	spect G5	06248n5229
Σ906	6.3	336	57	8½- 9½	relfix, spect F5	06254n3725
	55.0	141	10	-10½		
β896	1.0	184	62	7 - 10	PA dec, spect B9	06284n3212
	18.1	210	20	- 13		
A2519	4.1	70	21	7 - 12	spect A2	06295n3950
Σ918	4.7	332	53	6½- 7½	PA inc, spect A3;	06300n5230
					neat cpm pair	
0Σ147	43.2	73	26	7 - 8½	relfix, spect K0;	06309n3807
	46.3	117	26	- 9½	C = 0.5" pair	
Σ928	3.5	133	52	7½ - 8	relfix, spect F5	06312n3835
0Σ148	2.6	74	58	7 - 11	relfix, spect G5	06317n3706
Σ929	6.0	25	49	7 - 8	relfix, spect G5	06320n3746
Σ933	25.6	75	15	8 -8½	spectra both A2	06333n4111
Σ935	3.4	323	10	8 - 9	relfix, spect G5	06346n5220
Σ941	2.0	81	59	7 - 8	relfix, cpm;	06352n4138
	82.8	134	09	- 10	spect B9	
Σ941c	6.4	316	05	10- 13		
54	0.9	36	60	6 - 8	(0Σ152) relfix;	06364n2819
					spect B6	
Σ945	0.5	296	62	7 - 8	binary, dist dec,	06368n4101
					PA inc, spect F2	
A218	0.1	141	62	8½- 8½	PA dec, spect F5	06386n3044
0Σ154	23.2	103	63	6½- 8½	optical; PA & dist	06408n4041
					dec, spect M4	
56	36.2	31	58	6 - 9	(ψ^5) optical, PA	06431n4338
					inc, dist dec,	
					spect G0	
Es 15	28.8	275	16	7 - 10	spect K0	06464n4615
59	22.3	224	34	6½ -10	(Σ974) relfix	06496n3856
	25.8	219	08	-13	spect A7	
Σ978	17.0	88	18	7 - 9½	optical, dist inc,	06523n3759
					PA dec, spect K0	
Σ979	7.4	209	58	8 - 9	relfix, spect A0	06529n4636
Σ994	26.5	56	49	7 - 7½	slight dist inc;	06561n3710
					spect B	
Σ1022	5.6	129	36	7 - 10	relfix, spect A0	07061n3639

AURIGA

LIST OF DOUBLE AND MULTIPLE STARS   (Cont'd)

NAME	DIST	PA	YR	MAGS	NOTES	RA & DEC
65	11.4 39.7	10 36	34 31	5 - 12 - 12	(β901) AB cpm; slight dist inc, spect G8. C = optical	07187n3651
Σ1086	12.2	102	14	7½- 9	relfix, spect K0	07250n4251

LIST OF VARIABLE STARS

NAME	MagVar	PER	NOTES	RA & DEC
β	1.9---2.0	3.960	Ecl.Bin.; spect A2 (*)	05559n4457
ε	3.0---3.8	9883	Supergiant Ecl.bin.; spect F0. (*)	04584n4345
ζ	3.8---3.9	972	Ecl.Bin.; spect K4,B7; giant system (*)	04590n4100
ψ^1	5.0---6.1	Irr	(46 Aurigae) spect M0	06210n4919
R	6.8--13.7	459	LPV. Spect M7e	05132n5332
S	8.3--12.2	590	Semi-reg; spect N3e	05238n3406
T	4.4--15.5	---	Nova 1891 (*)	05288n3025
U	7.5--15.5	407	LPV. Spect M7e	05389n3201
V	8.5--12.9	353	LPV. Spect N3e	06203n4744
W	8.2--15.3	274	LPV. Spect M3e	05235n3652
X	8.1--13.5	164	LPV. Spect M3e	06083n5015
Y	8.8---9.9	3.859	Cepheid; spect F0--G1	05250n4224
Z	9.3--11.8	135	Semi-reg; spect G0e--G6e	05577n5318
RR	9.0--15..	307	LPV. Spect M3e	06085n4311
RS	8.5--11.0	170	Semi-reg; spect M4	06002n4618
RT	4.9-- 5.9	3.728	(48 Aurigae) cepheid (*)	06254n3032

AURIGA

LIST OF VARIABLE STARS   (Cont'd)

NAME	MagVar	PER	NOTES	RA & DEC
RU	9.0--14.5	468	LPV. Spect M8e	05367n3737
RV	9.5--10.8	229	LPV. or Semi-reg; spect N	06312n4233
RW	9.0--12..	Irr	Erratic, resembles T Tauri; spect dG5e (*)	05046n3020
RX	7.3---8.2	11.62	Cepheid; spect F8---G8	04079n3953
SS	10--14.8	Irr	Dwarf nova; SS Cygni type Peculiar spectrum (*)	06096n4746
SU	9.2--10.5	Irr	RW Aurigae type, spect G2	04528n3029
SV	9.5--10.1	Irr	Spect M1	06000n4626
SX	8.4---9.1	1.210	Ecl.Bin.; spect B3+B3	05082n4206
SY	9.3--10.4	10.14	Cepheid; spect F8--G2	05091n4246
SZ	9.1--13..	453	LPV. Spect M8e	05385n3854
TT	8.5---9.4	1.333	Ecl.Bin.; spect B4+B4 Lyrid type	05063n3931
TU	8.5---9.6	73	Semi-reg; spect M5	06319n4540
TW	8.2---9.7	150:	Semi-reg; spect M5	05534n4530
TX	8.5---9.2	Irr	Spect N3	05057n3856
UU	5.1--7...	235:	Semi-reg; spect N3	06331n3829
UV	7.5--10.1	393	Semi-reg; spect Ne; 11½m companion at 3.4"	05185n3228
UW	9.6--12.0	530	Semi-reg; spect R6	06538n4111
UX	8.4---9.5	90:	Semi-reg; spect M4	05121n4929
UZ	8.7---9.9	65:	Semi-reg or Irr; spect M3	05117n4004
VW	8.3--10.2	220	Semi-reg; spect M6	06139n3314
VX	8.5--11.4	322	LPV. Spect M4e---M6e	07251n4105
WW	5.5---6.2	2.525	Ecl.Bin.; spect A7+A7	06292n3230
AB	7.2---8.4	Irr	Spect A0e; RW Aurigae type	04526n3028
AC	9.0--13..	311	LPV. Spect M5e	05231n5005
AE	5.4---6.1	Irr	Erratic; spect 09.5 (*)	05130n3415
AI	9.2---9.8	64	Semi-reg; spect M5	06306n3018
AL	9.4--14..	385	LPV. Spect M7e	06300n3138
AR	5.8---6.5	4.135	(17 Aurigae) Ecl.Bin.; spect A0+A0	05150n3343
AZ	8.5--13..	420	LPV. Spect N0e	05577n3940
BF	8.7---9.4	1.583	Ecl.Bin.; spect B5	05015n4113
CO	7.4---8.2	39	RV Tauri type? Spect F5	05571n3519
CQ	9.2--10.1	10.62	Ecl.Bin.; spect G0	06006n3120
CZ	9.2--10.5	Irr	Spect M4	07255n3602
EO	7.7---8.0	4.066	Ecl.Bin.; Spect B3+B3	05150n3634

LIST OF STAR CLUSTERS, NEBULAE, AND GALAXIES

NGC	OTH	TYPE	SUMMARY DESCRIPTION	RA & DEC
1664	59^8		pL,1Ri,1C; diam 15'; about 40 stars mags 11....class E	04474n4337
1778	61^8		Irr; 6' x 3'; 10 stars mags 10....	05050n3700
----	I.405		L,F neby; 18' x 30' with variable star AE Aurigae in center (*)	05130n3416
1857	33^7		Diam 9'; 45 stars mags 8... pL,pRi,pC; class D	05166n3918
1883	34^7		Diam 3'; F,pRi,pC; about 20 stars mags 13.... class F	05222n4630
----	I.410		F neby, diam 20', encloses cluster NGC 1893	05193n3328
1893			Diam 12', 1C; about 20 stars mags 9...12; class D; neby I.410 involved	05193n3321
1907	39^7		Diam 5'; pC,pRi; 40 stars mags 10... class F	05247n3517
1912	M38		B,vL,vRi; diam 20', about 100 stars mags 8... (*)	05253n3548
1931	261^1		B,L,R, 3' x 3' with several stars involved	05281n3413
1960	M36		B,vL,1C; 60 stars mags 9... diam 12', class F (*)	05329n3407
1985	865^3		F,S,R,bM; diam 1', with 13^m star	05345n3158
2099	M37		!B,vRi,mC; 150 stars mags 9. ... Diam 20'; class F (*)	05490n3233
----	I.2149		S,vB, diam 10", mag 10; central 07-star 14^m	05526n4607
2126	68^8		Diam 5', scattered group of 20 stars mags 11...14; 6^m star on NE edge; class D	05581n4955
2192	57^7		Diam 5'; 25 stars mags 12.. cL,C; class F	06110n3950
2281	71^8		L,B,vlC, diam 15', about 30 stars mags 7... class E	06458n4107

DESCRIPTIVE NOTES

ALPHA   Name- CAPELLA, "The Goat Star". Mag 0.06; the
6th brightest star in the sky. Spectrum G8 III
+ F (composite), color golden yellow. Position 05130n4557.
Opposition date (midnight culmination) is December 12.

Capella is the nearest to the Pole of all the
first magnitude stars, and from the latitude of the United
States is visible at some hour of the night throughout the
year. The star is 45 light years distant, according to
trigonometrical parallaxes obtained at Allegheny, Yerkes,
Mt.Wilson, and McCormick. The resulting luminosity equals
160 suns (absolute magnitude about -0.6). The annual proper motion is 0.44" in PA 169°; the radial velocity is 18½
miles per second in recession. The motion closely matches
that of the Taurus moving group associated with the Hyades
star cluster, and Capella may be an outlying member of the
group.

Capella has been described as a red star by several ancient and medieval writers including Ptolemy and
Riccioli. It seems unlikely that the color has actually
changed since ancient times, and R.H.Allen in his "Star
Names and Their Meanings" suggests that a yellow or orange
star might seem red to "those whose eyes are specially
sensitive to that tint".

The star is a binary, too close for telescopic
observation. The duplicity was first detected with the
spectroscope at Lick Observatory in 1899, and the separation was first directly measured by J.A.Anderson with the
interferometer on the 100-inch telescope at Mt.Wilson in
December 1919. The two stars are about 70 million miles
apart, and revolve about their common center of gravity in
a retrograde direction in a period of 104.022 days. The
maximum apparent separation is only about 0.05", and the
computed orbit is very nearly circular, with an eccentricity of 0.01. The chief facts about the two stars are
given in the following brief table.

A	Spect G8 III?	Diam. 13 x ⊙	Mass 3.0	Lum. 90
B	F6 ?	7	2.8	70

Capella shows a number of spectral peculiarites which make
it difficult to classify the components accurately. The
majority of the spectral features are produced by the G-

CAPELLA. A "Close-up" of the 6th brightest star. The 10th magnitude companion is indicated by the small circle at lower left. Lowell Observatory 13-inch telescope plate.

DESCRIPTIVE NOTES (Cont'd)

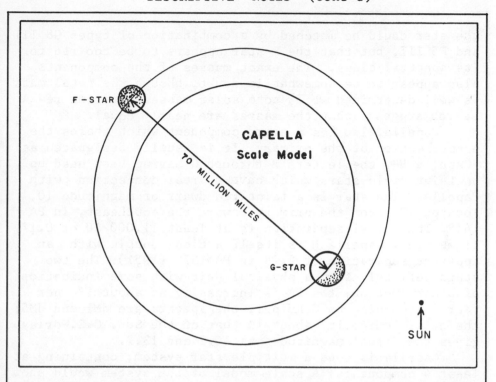

F-STAR

CAPELLA
Scale Model

70 MILLION MILES

G-STAR

SUN

star, classed by various authorities as G0,G5, or G8. Only
a few spectral lines can be attributed to the F-star. This
would ordinarily imply that the G-star is much the brighter of the two, but this explanation is contradicted by the
interferometer observations, which show that the difference
cannot be more than a few tenths of a magnitude. In their
study of the system, O.Struve and S.M.Kung (1951) found
evidence that the lines of the F-star are greatly broaden-
ed - apparently to the point of obliteration - by turbulent motions in the star's atmosphere; the motions are
"perhaps as large as in any star yet observed".

In a study by K.O.Wright at the Dominion Astrophysi-
cal Observatory in 1953, spectral types of G5 III and G0
were derived, and the same classes appear in C.E.Worley's
Catalog of Visual Binary Orbits (1963). W.W.Morgan in the
Yerkes Atlas of Stellar Spectra assigned types of G5 and
F6, but with considerable uncertainty.

In a spectrophotometric study of Capella, K.L.Frank-
lin (1959) found that the observed energy distribution of

DESCRIPTIVE   NOTES   (Cont'd)

the star could be matched by a combination of types G8 III
and F5 III, but that the F-star appears to be too red for
its spectral class. The exact masses of the components
also appear to be somewhat in doubt, though the total mass
is well determined at 5½ to 6 solar masses. Current research suggests that the masses are nearly equal.

Capella also has a third component which shares the
proper motion of the primary. It is usually designated as
"Capella H", the letters B through G having been used up
on faint field stars which have no real connection with
Capella. The star is a faint red dwarf of magnitude 10,
located 12' from the primary toward the southeast, in PA
141°. The actual separation is at least 11,000 AU or 0.17
light year. Capella H is itself a close double with an
apparent separation of 2.7" in PA 137° (1951). The two
stars defintely form a physical pair with some indication
of binary motion; the PA is increasing at about 1° per
year. According to G.Kuiper, the spectra are dM1 and dM5;
the total luminosity about 1% that of the Sun. C.E.Worley
gives the visual magnitudes as 10.2 and 13.7.

Capella is thus a multiple star system, containing at
least 4 components. A scale model of the system would show
Capella A and B as two globes 13 and 7 inches in diameter
and 10 feet apart; the components of Capella H would then
be each 0.7 inch in diameter, 420 feet apart, and 21 miles
from the main pair A & B !

BETA NAME- MENKALINAN. Magnitude 1.90 (variable);
Spectrum A2 IV or A2 V. Position 05559n4457. The
computed distance is about 90 light years; the actual luminosity about 110 times that of the Sun (absolute magnitude -0.3). The annual proper motion is 0.05"; the radial
velocity is 11 miles per second in approach.

Beta Aurigae is a short-period eclipsing binary, in
which two stars of very nearly equal size and brightness
revolve in their orbits in a period of 3.96003 days. The
star was one of the first spectroscopic doubles to be discovered, identified by A.Maury in 1890. The light variations were detected photometrically by J.Stebbins in 1910
and his orbital elements were published the following year.
The orbit is virtually circular, and is oriented about 13°

from the edge-on position. The eclipses are of small amplitude; the photographic range being from magnitude 1.92 to
2.01.   There are two eclipses of almost identical depth in
each revolution of the system, and about 25% of the diameter of each star is occulted at mid-eclipse. Each star is
approximately 2.6 times the diameter of the Sun, and the
computed masses are 2.35 and 2.25. Both spectra are A2. The
separation of the pair is about 7⅓ million miles, or about
1/12 the separation of the Earth and Sun.

Beta Aurigae shows very nearly the same space motion
as Sirius, and appears to be a member of a widely scattered moving stream of at least 70 members, including other
bright stars such as Alpha Ophiuchi and Delta Leonis. The
space motion of this stream is rather similar to that of
the well known Ursa Major cluster, but the connection of
the two groups is not certain. It is generally believed
that the true cluster and the larger stream are associated
only temporarily. (Refer to the Ursa Major cluster)

Beta Aurigae has a distant optical companion of magnitude 10½ at 184" in PA 40°, first recorded by Sir William
Herschel in 1783.   There is also a closer attendant of the

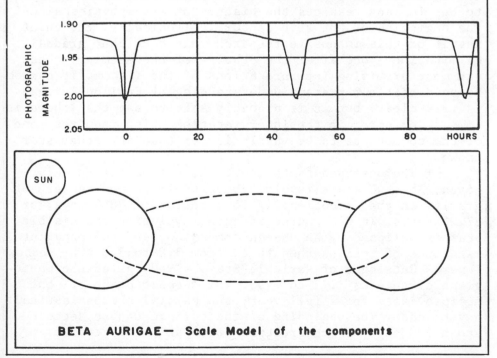

BETA AURIGAE— Scale Model of the components

14th magnitude, discovered by E.Barnard in 1901, when the
separation was 12.6" in PA 181°. The most recent observations of this star, reported in the Lick "Index Catalogue",
show no change in separation, but a slight decrease in the
PA to about 174°. The measures suggest common proper motion
with the primary, and give the luminosity of the faint star
as about 1/630 that of the Sun. The projected separation of
the pair is about 350 AU.

EPSILON   Mag 3.00 (variable); Spectrum about F0, but
given by various authorities as A8, F0, or F2.
Supergiant, luminosity class Ia. Position 04584n4345. This
is one of three stars forming the flattened triangular
group called "the Kids"; the other two are Eta and Zeta.
Epsilon is the northernmost of the three, and the nearest
to Capella. It is located about 3° distant from Capella,
toward the southwest.

Epsilon Aurigae is a noted eclipsing binary star, one
of the most remarkable and puzzling of all known eclipsing
variables. It has been the subject of so many studies and
investigations that O.Struve (1962) declared its history
to be "in many respects the history of astrophysics since
the beginning of the 20th century." Ironically, the chief
result of this intensive research has been the gradual
elimination - one after the other - of the seemingly best
and most promising interpretations of the system. It cannot
be said that our present understanding of Epsilon Aurigae
is very clear, but it is probably safe to say that there is
some major error in the interpretation which requires one
of the components to be vastly larger than any other star
known.

The observed facts about the system are quickly
given. The two stars revolve about their common center of
gravity in the exceptionally long period of 9883 days, or
27.06 years. In the course of each revolution the visible
star is eclipsed by an unseen companion, and the apparent
magnitude of Epsilon then falls from 3.0 to 3.8. The Moscow
General Catalogue of Variable Stars (1958) gives the photographic range as 3.73 to 4.53. The deepest phase of the
eclipse lasts for a full year; the partial phases last half
a year each. The beginning of the eclipse can be detected
about 190 days before greatest obscuration is reached.

THE KIDS. Epsilon Aurigae is at center; Eta and Zeta are
the two stars near the bottom. Capella is the bright star
at upper left. Lowell Observatory 5-inch camera plate.

DESCRIPTIVE NOTES (Cont'd)

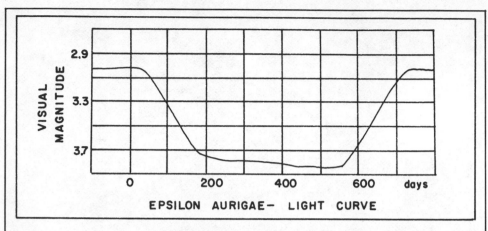

EPSILON AURIGAE— LIGHT CURVE

During minimum, the light is usually said to remain nearly
constant, but observations at the most recent eclipse, in
1955-1957, showed a slight fading of about 1/10 magnitude
between second and third contacts. The cause of this is
not known. There are also slight secondary irregularities
of about the same order of magnitude which become noticeable during an eclipse, and for several years before and
after. At the 1928 eclipse, some of these irregularities
seemed to repeat at intervals of about 11 months, though
other observations have revealed no real evidence of true
periodicity.

As an eclipsing binary, Epsilon Aurigae is an interesting example of a type in which a long "atmospheric eclipse"
precedes and follows the actual hiding of the star by its
companion. Thus it may be that the secondary fluctuations
are due to large-scale irregularities in the outer atmosphere of the eclipsing star. Other well known stars of
this class are VV Cephei, Zeta Aurigae, and 31 Cygni.

The first recorded minimum of Epsilon Aurigae was that
of 1821, observed by K.Fritsch. At the next eclipse, in
1847-1848 the variability was confirmed by Schmidt, Heis,
and Argelander. Schmidt continued his observations, and
recorded a third minimum in 1874-1875. In 1912 an analysis
of the accumulated observations was published by H.Luden-
dorff, it being then evident that the star was an eclipsing
binary of unusually long period and extraordinary interest.
Although the eclipses of 1928-1930 and 1955-1957 were very
widely observed, the main problem remains unsolved: What
is the true nature of the mysterious companion which causes

the eclipses, and which by some calculations may be the largest star known ?

THE BRIGHT COMPONENT, which gives all the visible light of the system, is a supergiant whose type is close to F0; the spectral characteristics suggest a luminosity which probably equals that of Rigel, at about 60,000 times the light of the Sun. The computed absolute magnitude is about -7.1. The diameter may be about 180 times that of the Sun. From these figures, the estimated distance is about 3300 light years, too great for reliable parallax measurements. An attempt made at Allegheny, however, yielded a result of 0.001", equivalent to 3260 light years. Needless to say, such a result cannot be taken at face value, and proves only that the distance is very great. The annual proper motion of the star is less than 0.01"; the radial velocity averages about 1.8 miles per second in approach.

From the radial velocity measurements, the mean orbital speed of the visible star is in the range of 9 to 10 miles per second, and the orbit is found to be considerably non-circular, with an eccentricity of about 0.33. The orbit is about 15 AU in radius, or about 1.4 billion miles. Very similar results were obtained by Dr.K.Strand (1959) by astrometric measurements of Yerkes 40-inch telescope plates; he obtained a semi-major axis of 0.014", corresponding to about 1.25 billion miles or 14 AU. The orbit of the star seems to be oriented about 18° from the edge-on position. The total mass of the system is believed to be about 30 solar masses, with the visible star having somewhat the greater mass. The 27-year period then implies a mean separation of about 30 AU, comparable to Neptune and the Sun.

THE ECLIPSING COMPONENT has never been observed directly or detected spectroscopically, and would have remained entirely unknown except for its periodic transits across the bright primary. According to the usual or "traditional" interpretation, first introduced in 1937 by G.Kuiper, B. Stromgren, and O.Struve, the star may be a low density supergiant of exceptional characteristics, perhaps the largest, coolest, and most rarified star known. It would be 15 times the size of its companion, or about 2800 times the diameter of the Sun. The average density, about one-billionth that of the Sun, would approach what we would call an absolute vacuum. According to this interpretation,

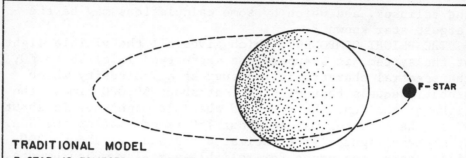

TRADITIONAL MODEL
F-STAR IS ECLIPSED BY SUPERGIANT INFRARED STAR WHOSE DIAMETER MAY BE OVER 2500 TIMES THE SIZE OF THE SUN.

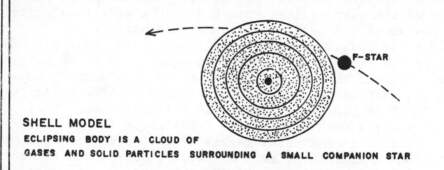

SHELL MODEL
ECLIPSING BODY IS A CLOUD OF GASES AND SOLID PARTICLES SURROUNDING A SMALL COMPANION STAR

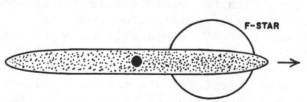

TRANSIT MODEL
ECLIPSING BODY IS A FLATTENED RING WHICH IS SEEN EDGE-ON DURING ITS TRANSIT OF THE F-STAR. THERE ARE NO TOTAL ECLIPSES.

POSSIBLE MODELS FOR THE PECULIAR SYSTEM EPSILON AURIGAE.

DESCRIPTIVE NOTES (Cont'd)

the eclipsing star is normally invisible, partly because
its faint light would be lost in the glare of the highly
luminous primary, but also perhaps because it may be at too
low a temperature to emit much visible radiation. A surface
temperature of less than 1500° K has been obtained through
indirect calculation, and indicates that the star radiates
chiefly in the far infrared, emitting virtually no visible
light.

        Still following this interpretation, we find another
peculiarity: the star seems to be partially transparent,
at least in the outer layers. This is made evident by the
fact that the visible star does not disappear completely
when eclipsed; it merely dims to about half its normal
light. And, although the shape of the light curve seems to
imply a total eclipse, the spectrum of the eclipsed star
remains visible throughout "totality"and is essentially
unchanged except for a definite strengthening of the absorption lines. A doubling of the lines, before and after
eclipse, has been observed, and may be attributed to gas
streams between the components. A more difficult feature
to explain is the fact that the eclipsed star fades without changing color; the eclipsing body apparently acts as
a "neutral filter" and absorbs all wavelengths equally. To
explain this feature, and also the nearly constant light
during maximum eclipse, it has been proposed that the outer layers of the eclipsing star are ionized by radiation
from the F-star, and the actual eclipsing body is this relatively thin ionized layer. This is the model presented in
the first diagram on the opposite page.

        If it is actually a star, this strange object may well
be the largest star known, and would fill up much of the
Solar System out to beyond the orbit of Saturn.   Other interpretations have been suggested, however, and at present
it seems likely that our ideas about this strange system
will soon be drastically revised. All attempts to detect
the infrared radiation of the strange companion have failed, and it now seems more likely that the eclipsing body is
not a star at all, but rather a vast cloud of gases, dust,
or solid particles, surrounding a relatively small star
which cannot itself be detected. M.Hack (1961) has proposed
that the eclipsing body is a shell or ring of ionized gases
surrounding a hot 0-type or B-type star which may be about

2 magnitudes fainter than the primary, and is therefore undetectable spectroscopically. S.Huang (1965) suggests that the secondary star is encircled by a flattened disc of rotating gases which is viewed edge-on, and which passes horizontally across the primary star to produce the eclipses. According to this model, shown in the third of the diagrams on page 270, the nearly flat bottom of the light curve does not imply that the eclipse is total, and there is thus no need to explain why the eclipsed star is still visible all through "totality". With these newer interpretations, we may be near a solution of the mystery of Epsilon Aurigae, though at the cost of demoting this remarkable object from its ranking position among the largest known stars.

Epsilon Aurigae also has a faint visual companion of magnitude 14, discovered by S.W.Burnham with the 18½-inch refractor at Dearborn Observatory in 1891. According to the Yale "Catalogue of Bright Stars" (1964) the two stars probably form a physical pair. The projected separation is about 30,000 AU, or close to 0.5 light year. (Present apparent separation about 28.6")

ECLIPSE TIME-TABLE for EPSILON AURIGAE		
A June 9, 1928	July 1, 1955	July 22, 1982
B Nov 30, 1928	Dec 21, 1955	Jan 11, 1983
C May 30, 1929	June 20,1956	July 12, 1983
D Dec 4, 1929	Dec 25, 1956	Jan 16, 1984
E May 14, 1930	June 4, 1957	June 25, 1984

A = partial phase begins   B = "total" phase begins
C = mid-eclipse   D = total phase ends
E = partial phase ends

ZETA   Name- SADATONI. Mag 3.76 (variable); spectrum K4 II + B7 V. Position 04590n4100. One of the three stars forming the small triangular group called "the Kids", located about 2.75° south of Epsilon Aurigae. It is an eclipsing variable, first recognized as a spectroscopic double by A.Maury in 1897, and confirmed as a binary by W.W.Campbell in 1908. Zeta Aurigae consists of a relatively small blue-hot star and a K-type giant companion,

DESCRIPTIVE   NOTES   (Cont'd)

orbiting about their common center of gravity in a period
of 972.176 days, or about 2.66 years. The computed separation of the components is in the range of about 500 million
miles, and the eccentricity of the orbit is 0.40. The main
facts about the two stars are given in the table below.

	Spect.	Diam.	Mass	Lum.	Abs.Mag.	Temp.
A	K4 II	160	8.3	2100	-3.5	3200° K
B	B7 V	4	6.8	400	-1.6	15000

The diameter and luminosity of the primary are perhaps
the most uncertain figures in the table. A few authorities
have classed the star as a supergiant of class Ib, which
could raise the absolute magnitude to as high as -4.4. The
diameter of 160 X ⊙ should be regarded as a minimum; some
estimates have ranged up to 300 solar diameters. The mass
figures are according to recent studies (1960). From the
derived luminosities the distance of the system appears to
be about 1200 light years.   The annual proper motion is
about 0.03"; the radial velocity is 8 miles per second in
recession.

The eclipse of the smaller B7 star by its giant companion   occurs once every 2 years and 8 months. For a period
of about a month before the actual eclipse begins, the
light of the small star must come through progressively
deeper layers of the giant's atmosphere, and spectroscopic
study at this time has revealed much information about the
structure and composition of the star's atmospheric layers.
There appear to be local condensations and irregularities
in the giant's chromosphere, perhaps comparable to solar

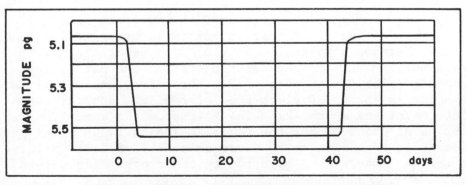

ZETA AURIGAE— PHOTOGRAPHIC LIGHT CURVE

prominences. These studies also reveal the chemical strat-
ification of the star's atmosphere. Lines of the neutral
metals appear to be produced in the lowest levels of the
atmosphere, while the atoms of ionized metals are found in
higher levels. Hydrogen and ionized calcium are abundant
at all levels, out to the detectable limits of the star's
atmosphere, more than 20 million miles above the surface.

The actual eclipse of the B-star is total for 38 days
and is preceded and followed by partial phases lasting 32
hours each. During totality the spectrum of the B-star
vanishes completely. Because of the great difference in the
colors of the components, the amplitude of the light curve
depends critically upon the method of observation. As a
visual variable, the star is of small interest, since the
range is less than 0.15 magnitude. Photographically it is
about ½ magnitude, and rises to nearly 2 magnitudes when
observed in the ultraviolet. The photographic range is
about 5.0 to 5.6.

ECLIPSE TIME-TABLE FOR ZETA AURIGAE	
Beginning Ending	Beginning Ending
Dec 15, 1947--Jan 22, 1948	Aug 1, 1966--Sept 8, 1966
Aug 13, 1950--Sep 20, 1950	Mar 31, 1969--May 7, 1969
Apr 11, 1953--May 19, 1953	Nov 27, 1971--Jan 4 , 1972
Dec 9, 1955--Jan 16, 1956	Jul 26, 1974--Sep 3 , 1974
Aug 7 , 1958--Sep 14, 1958	Mar 26, 1977--May 3 , 1977
Apr 5, 1961--May 13, 1961	Nov 23, 1979--Dec 31, 1979
Dec 2, 1963--Jan 10, 1964	

ETA Mag 3.18; spectrum B3 V. Position 05030n4110.
The third star in the small group called "the
Kids", forming a naked-eye pair of 3/4° separation with
Zeta Aurigae. Eta is the eastern star of the pair. The
computed distance is about 370 light years, the actual luminosity is about 580 times that of the Sun, and the absolute magnitude about -2.1. The star shows an annual proper
motion of about 0.08"; the radial velocity is 4½ miles per
second in recession. The close proximity of Zeta Aurigae
offers an interesting opportunity to compare star colors,
particularly noticeable when Zeta itself is in eclipse.

THETA   Mag 2.65; spectrum B9 V. Position 05563n3713.
The computed distance is about 110 light years, leading to an actual luminosity of about 85 suns, and an absolute magnitude of about +0.1. The annual proper motion is 0.10"; the radial velocity is 17½ miles per second in recession. Theta Aurigae is often called a "silicon star", from the abnormal strength of that element in the lines of its spectrum.
The star has two companions for the small telescope, the closer pair forming a slow retrograde binary of uncertain period. The earliest recorded measurements appear to be those of O.Struve in 1871, and the PA of the pair has turned through about 40° in the last 70 years. Thus the period may be about 7 or 8 centuries. The separation has remained at about 3½" for many years. The small star, of magnitude 7½, is about equal to our Sun in luminosity and spectral type. The projected separation of the pair is about 110 AU.
The second companion, at about 52", was noted by O. Struve in 1852. It is not physically connected with the close pair, and the separation is increasing due to the proper motion of Theta itself.

IOTA   Mag 2.67; spectrum K3 II. Position 04537n3305.
The computed distance is about 330 light years; the actual luminosity about 750 times that of the Sun. The corresponding absolute magnitude is -2.4. The annual proper motion is only 0.02"; the radial velocity is 10½ miles per second in recession.

T   Nova Aurigae 1891. Position 05288n3025, about 2° north and east of Beta Tauri. The nova was discovered on the night of January 23, 1892, by the amateur observer T.D.Anderson of Edinburgh, Scotland. (This same dedicated observer was later the discoverer of Nova Persei in 1901.) Previous photographs of the region of Nova Aurigae show that the star had been bright for some 6 weeks, and had apparently gone unnoticed. Photographs taken up to December 8, 1891 do not show it, but on a plate of December 10 it is magnitude 5.4. Thus the rise to naked-eye visibility must have taken place in a period

DESCRIPTIVE NOTES   (Cont'd)

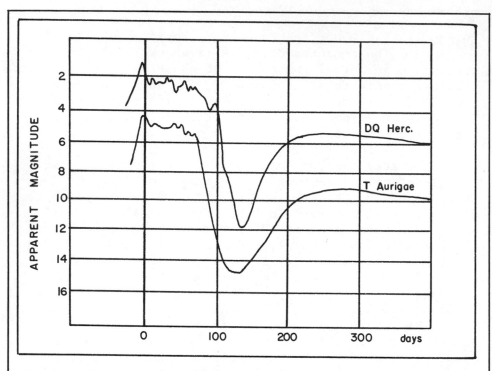

of scarcely more than 24 hours. The further brightening, by a factor of 2½ times, occupied the next 9 days.   The maximum probably occurred about Dec. 20, at magnitude 4.4. The nova had faded to 5.0 at the time of its discovery.

   C.P.Gaposchkin has referred to this star as the "first well observed nova of modern times". At Lick Observatory a very complete series of spectroscopic observations was made by W.W.Campbell with the 36-inch refractor. Much of our modern knowledge of Nova Aurigae has resulted from a very thorough analysis of this material by D.B.McLaughlin.

   The nova faded slowly during January and February of 1892; in March the brightness began to decrease rapidly, and fell to magnitude 15 by late April. In August the star began to brighten again, and reached magnitude 9½, at which it remained for 3 years. In 1897 it had faded to 11½, and in 1903 it was about 14th. Finally, some 33 years after the outburst, it reached a constant minimum of magnitude 15½. There have been no definite changes since 1925.

   The spectrum of the nova at discovery showed numerous bright bands displaying high approach velocities,   some

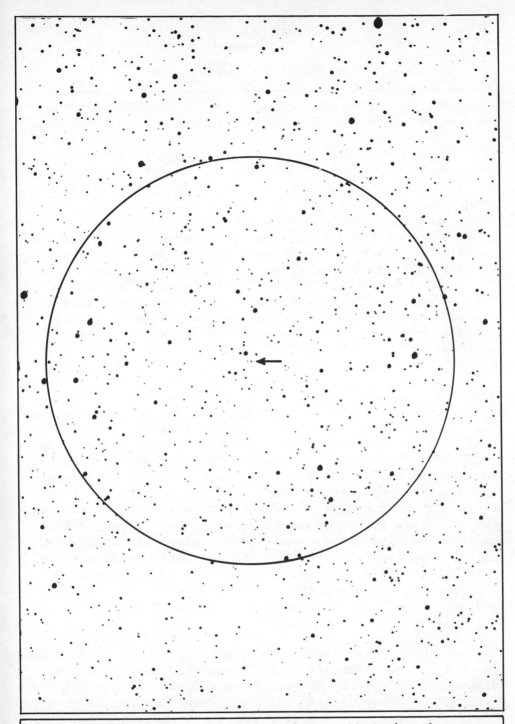

NOVA AURIGAE. Finder chart made from a 13-inch telescope plate at Lowell Observatory. Circle diameter = 1 degree. North is at the top. Limiting magnitude about 15.

DESCRIPTIVE NOTES (Cont'd)

exceeding 600 miles per second. At the reappearance in August, the spectrum had changed to resemble a planetary nebula. Using the 36-inch Lick refractor, E.Barnard found that the image of the nova appeared as a diffuse nebulous disc, measuring about 3" in diameter. In 1943 this shell had increased to a diameter of about 12". The nova is now about 14.8 magnitude (photographic), bluish in color, and shows an 0-type spectrum with some emission lines.

The computed distance of T Aurigae is about 4100 light years; the absolute magnitude at maximum was about -6.2, corresponding to 25,000 times the light of our Sun. The light curve was a rather rare type, characterized by a maximum lasting some three months, a sudden drop at about 100 days, and a subsequent recovery to a lower secondary maximum. Nova Herculis 1934 (DQ Herc) is another example of this class, usually called "slow novae". The two light curves are compared on page 276.

In 1954, M.F.Walker at Mt.Wilson made the surprising discovery that DQ Herculis is a close binary with the very short period of 4.65 hours. The recurrent nova WZ Sagittae was later found to be a very similar system, suggesting that nova activity might be connected in some way with the duplicity. This theory is now strengthened by new findings; Nova Persei (1901) and Nova Aquilae (1918) are both close binaries, and Walker's studies of T Aurigae reveal it as a close and rapid double also. The period is 4.905 hours, and the two stars form an eclipsing pair in which the primary eclipse (partial) lasts about 40 minutes. This is another of those strange systems in which two dwarf stars are revolving almost in contact; to interpret such a system and explain the nova phenomenon is indeed a challenge for the modern astrophysicist. (For a discussion of novae in general, refer to Nova Aquilae 1918)

RT   (48 Aurigae) Variable. Position 06254n3032.
A bright cepheid variable star which was discovered in 1905 by T.H.Astbury, a member of the British Astronomical Association. It is easily located, slightly less than midway along a line drawn from Epsilon Geminorum to Theta Aurigae· As in all the classical cepheids, the cycle of variations is characterized by split-second precision, the exact period being 3.728261 days and the visual

amplitude about 1 magnitude. The rise to maximum requires about 1½ days and the decline about 2½ days. The photographic range is 5.3 to 6.5.

The variations of the star appear to be due to an actual stellar pulsation, the star expanding at maximum and contracting at minimum. Spectroscopic measurements show that the radial velocity varies by about 35 miles per second in the course of the cycle, the largest approach velocity coinciding with maximum brightness. As shown on the graph below, the light curve is virtually a mirror-image of the radial velocity curve. The variations are accompanied by a change in color and spectral class, from F5 to about G0. The star is a supergiant of luminosity class Ib, with a maximum visual luminosity of about 2300 suns. The absolute magnitude at maximum is about -2.6 photographic, or -3.1 visual. The mean radial velocity is 13 miles per second in recession; the very slight proper motion has been measured at 0.017" annually.

Although the causes of the pulsations are still controversial, it is known that there is a relationship between the periods and luminosities of these stars, intrinsically brighter cepheids requiring a longer time to complete the

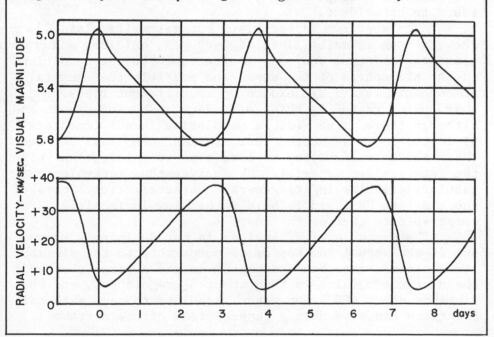

cycle. Thus the true luminosities of distant cepheids may
be determined from the observed periods, and the distances
derived. This principle was used in the first attempt to
determine the distance of the Andromeda Galaxy in 1923. In
the case of RT Aurigae, the derived distance is about 1600
light years, more than 1350 times closer than the Andromeda
system. RT Aurigae would appear as a 21st magnitude object
if it was at the distance of the Andromeda Galaxy! (For a
general account of the cepheid variables, refer to Delta
Cephei).

RW   Variable. Position 05046n3020. A peculiar var-
           iable star, often considered the prototype of
its rare class. It was discovered by Ceraski at the Moscow
Observatory in 1906. The position is about 1° southwest of
the midpoint of a line drawn from Iota Aurigae to Beta
Tauri. The light changes are large and often rapid; the
star may sometimes change by as much as a magnitude in a
few hours. The maximum visual range is about $3\frac{1}{2}$ magnitudes
or some 25 times in brightness. The Moscow General Catalog
(1958) gives the photographic range as 9.6 to 13.6. The
variations are quite irregular, and no definite periodicity
seems to be evident.

The spectrum is peculiar, but resembles class G5 and
shows strong emission lines of hydrogen, calcium, and other
elements. Spectrum analysis indicates turbulent conditions
in the atmosphere of the star, and possibly that several
atmospheric layers are expanding at different rates. The
surrounding region is thick with dark obscuring clouds and
although there is no visible nebulosity in the immediate
vicinity, it is suggested that stars of the class may owe
their sudden variability to some type of interaction with
the interstellar material. RW Aurigae thus resembles the
nebular variables in its general characteristics. Stars of
the type are believed to have rather low luminosities,
comparable to that of the Sun.

The distance of RW Aurigae is not definitely known,
but if the actual luminosity is comparable to the similar
star T Tauri, the absolute magnitude may average about +5;
the distance modulus is then about $5\frac{1}{2}$ magnitudes, and the
distance about 400 light years. Needless to say, such calculations can give only a general idea of the distance.

A small number of stars which fluctuate in a rather similar manner are frequently called "RW Aurigae type" variables, though the resemblance may in many cases be superficial, and the members do not form a real physical group. Spectral types range from B to M, with and without emission lines, the light variations are erratic, the stars are generally main sequence objects rather than giants, and many of them are associated with regions of bright or dark nebulosity. Among stars of the type, the rapidly varying RR Tauri is one of the best known examples, and its light curve closely resembles that of RW Aurigae. The spectral type, however, is A2. Perhaps the most interesting of all is the star T Tauri, associated with the variable nebula NGC 1555. This star has given its name to a fairly well-defined class of nebular variables which may be a sub-class of the RW Aurigae type. Knowledge of many of these stars is still fragmentary, but the T Tauri stars are currently believed to be newly formed from the nebulous clouds where they are found. It may be that the RW Aurigae stars are also in an early stage of development, and have not yet reached a stable state. The finding of many low-luminosity erratic variables in nebulous clusters, such as the Orion complex, NGC 6611 in Serpens, etc., seems to lend support to this view. These strange objects may eventually teach us much about star formation. (See also T Tauri, RR Tauri, NGC 6611, and R Monocerotis, associated with the variable nebula NGC 2261)

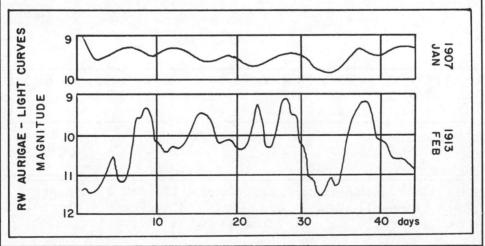

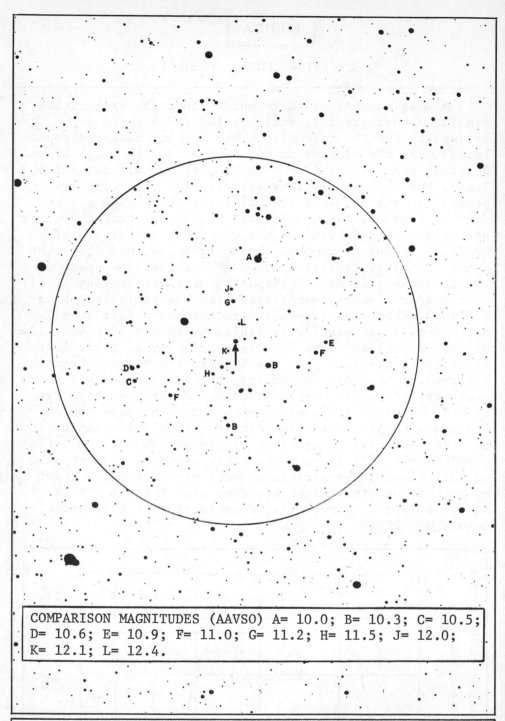

COMPARISON MAGNITUDES (AAVSO) A= 10.0; B= 10.3; C= 10.5;
D= 10.6; E= 10.9; F= 11.0; G= 11.2; H= 11.5; J= 12.0;
K= 12.1; L= 12.4.

RW AURIGAE. Finder chart made from a 13-inch telescope
plate at Lowell Observatory. Circle diameter = 1 degree.
North is at the top. Limiting magnitude about 15.

SS   Variable. Position 06096n4746. An explosive dwarf variable star of the SS Cygni class, discovered in 1907 by E.Silbernagel at Munich, by a comparison of plates made in 1901 and 1903. The star is located about 3½° northeast of Beta Aurigae. It is characterized by a nearly constant minimum, interrupted by violent nova-like outbursts at intervals ranging from 50 to 100 days or more. The average period, however, is about 55 days. At these times the light increases by a factor of about 60 times; in about 24 hours the star may rise to maximum brilliancy. The magnitude at minimum is approximately 15, and is therefore beyond the reach of smaller amateur telescopes, but the star can be located when near peak brightness if the position is accurately known.

The light curve is very similar to those of SS Cygni and U Geminorum, but in addition to the regular outbursts the star occasionally exhibits more rapid and irregular fluctuations. Such an erratic period, shown on the second section of the light curve below, usually lasts about 100 days; the star then returns to its normal cycle.

Like the other well-studied stars of this rare class, SS Aurigae is an extremely close binary star whose period, recently determined to be about 4h 20m, is among the shortest known. The components are tiny sub-dwarf stars; the

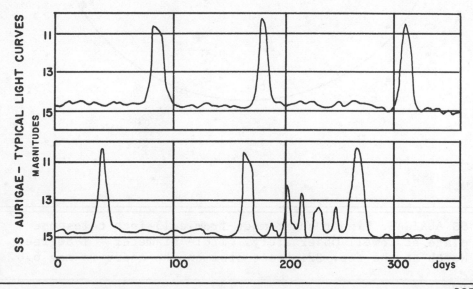

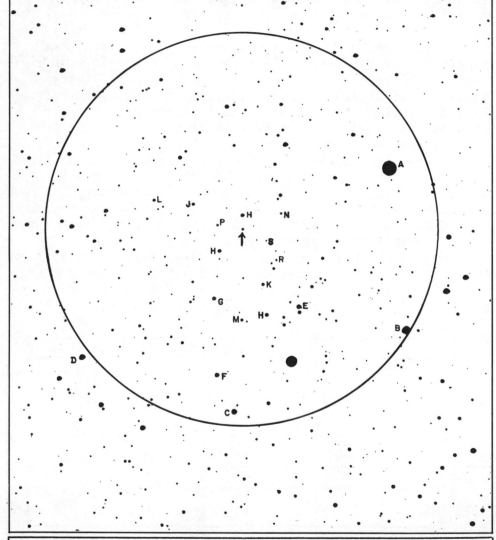

COMPARISON MAGNITUDES (AAVSO) A= 6.8; B= 7.6; C= 8.0; D= 8.9; E= 9.4; F= 10.1; G= 10.5; H= 10.8; J= 11.0; K= 11.2; L= 11.5; M= 11.8; N= 12.4; P= 12.5; R= 13.0; S= 13.2.

SS AURIGAE. Finder chart made from a 13-inch telescope plate at Lowell Observatory. Circle diameter = 1 degree. North at the top. Brightest star on the chart is mag 6.8.

explosive member of the pair is usually thought to be the bluer component, classified as a Be type subdwarf, but even this essential point is in dispute. There is some evidence that in the case of U Geminorum, the outbursts originate in the cooler of the two stars, and not in the hot dwarf. (Refer to U Geminorum). The members of these unusual pairs appear to have absolute magnitudes in the range of +7½ to +9; during an outburst the total light rises to about the luminosity of the Sun. A rough calculation, based upon the "distance modulus" method, then indicates a distance in the range of 350--400 light years. W.J.Luyten (1965) has measured an annual proper motion of 0.03" for SS Aurigae, consistent with the derived distance. The relationship of these stars to the novae and recurrent novae has been the subject of much speculation. (Refer also to SS Cygni, U Geminorum, AE Aquarii, WZ Sagittae, and Nova Aquilae 1918)

AE   Variable. Position 05130n3415. Spectrum 09.5 V.
        An unusual O-type variable star, normally of the 6th magnitude, but subject to irregular variations of small amplitude. The computed distance is about 1600 light years, leading to a luminosity of about 900 suns. The average absolute magnitude may be about -2.5.
        AE Aurigae illuminates the diffuse nebulosity IC 405, often called the "Flaming Star Nebula". This turbulent cloud is some 18' in extent , corresponding to an actual diameter of about 9 light years. The present association of star and nebula, however, appears to be the result of a chance encounter. Radial velocity measurements show that the star is receding at about 36 miles per second, and the nebula at only 13 miles per second. Photographs of the region also support the inference that the star has only recently entered the nebula. About a degree southwest of the star is a faint nebulosity (sometimes identified in catalogs as S126) which shows a sharp eastern boundary, parallel to the motion of AE Aurigae. The appearance seems to suggest that this boundary is the edge of a zone which has been swept clear of nebulosity by the northward motion of the star.
        A comparison of red and blue plates reveals some peculiar features. On blue exposures the most prominent detail is the bright twisted filament running out from the star on

AE AURIGAE and the nebula IC 405; photographed with a 12½-inch reflector by Evered Kreimer of Prescott, Arizona. This print is oriented with northwest at the top.

the southeast side. According to G.H.Herbig (1958) the
spectrum of this feature indicates that the composition is
chiefly dust, associated with very little free gas. Virtually all the details visible on blue photographs show a
continuous spectrum. Red plates show an entirely different
pattern of emission features, where the radiation of ionized gas is predominant. The presence of dust clouds so near
an O-type star again indicates that the star and nebula
have been associated a relatively short time; presumably
the structure and appearance of the nebulosity will eventually be greatly modified by the star's radiation.

The interesting fact about the motion of AE Aurigae is
that the star seems to be moving directly outward from the
region of the Great Nebula M42 in Orion, suggesting the
possibility that the star is an escaped member of the huge
Orion association of O and B-type stars. The annual proper
motion of 0.03" indicates a considerable space velocity
(about 80 miles per second) when allowance is made for the
distance of some 1600 light years. If the speed of recession has remained reasonably constant, it can be estimated
that the separation occurred about 2.7 million years ago.
In addition, there are at least two other stars known which
show high space velocities outward from the Orion region;
these are 53 Arietis and Mu Columbae. The three objects are
often referred to as "Runaway Stars". The plotted paths of
the three are shown on the diagram on page 288. The space
velocity of 53 Arietis is about 35 miles per second, while
that of Mu Columbae is close to 75 miles per second. The
chief difficulty in the "escape theory" is the lack of a
suitable accelerating process to explain the high velocities of the stars. The explosion of a supernova has been
suggested, but such an explosion, by itself, would not
produce such an effect. According to a modification of
this idea, it may be possible that the star was once a
member of a close binary pair, with high orbital velocities; the explosion of the companion would then free the
other star which would continue out into space at the same
high velocity. In favor of this idea is the fact that many
supernova explosions would be expected to occur in such a
region as the Orion complex, where massive rapidly evolving
stars are plentiful. (Refer also to 53 Arietis and Mu
Columbae).

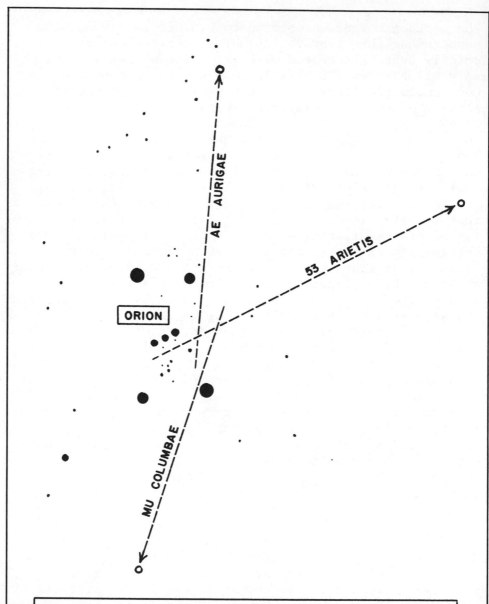

AE AURIGAE

53 ARIETIS

ORION

MU COLUMBAE

DIAGRAM SHOWING THE PLOTTED PATHS OF THE THREE "RUNAWAY STARS" AND THEIR RECESSION FROM THE REGION OF THE ORION ASSOCIATION.

AE AURIGAE and the NEBULA IC 405. Top: Blue photograph.
Below: Photograph in red light.
            Crossley Reflector, Lick Observatory

ANDREWS' STAR   (GC 7066) (BD +31°1048) Suspected flare star. Position 05374n3120. Spectrum B7. This star appears to be an unusual variable of unknown type, discovered photographically by A.D.Andrews of Armagh Observatory in Ireland. The star is located just 1° north and somewhat east of 26 Aurigae, where it may be found as an unlabeled 6th magnitude star on both Norton's Atlas and the Skalnate Pleso Atlas. The variations were first detected on plates of the region made on March 1, 1964, with the Armagh 12-inch Schmidt telescope. The first plate showed the star nearly 3 magnitudes brighter than normal, but on plates obtained only two hours later the magnitude was back to about 6. A spectacular increase in brightness was observed again on March 14 when the star was found to be about 2 magnitudes brighter than normal; it returned to its usual brightness in less than an hour.

The star is not listed as a known variable in any of the standard catalogs, and a check of Armagh plates going back to 1955 revealed no earlier variations. Changes of such rapidity have not previously been discovered in any B-type star. The variations suggest a flare star, but all known examples are M-type red dwarfs. According to the Yale "Catalogue of Bright Stars" (1964) the spectral type of Andrews' star is B7 V. No parallax or proper motion data are recorded.

Andrews suggests the possibility of a faint M-type companion, which might be lost in the glare of the B-star except during a flare. Evidently this is an object well worth further study. Comparison magnitudes of some nearby field stars are: Chi Aurigae= 4.77; 26 Aurigae= 5.40; 136 Tauri= 4.61.

M36   (NGC 1960) Position 05329n3407. The first of 3 bright galactic star clusters in the Auriga Milky Way, discovered by Le Gentil in 1749. It lies about 5° southwest of Theta Aurigae and some 2.3° distant from the cluster M38; the two clusters may be viewed together in the field of a wide-angle low power telescope. M36 is the smaller but brighter of the two, and contains about 60 stars of magnitudes 9 to 14. The central knot of bright stars measures about 10' in diameter, and includes the easy double star Σ737, separation 10.7". The group makes its best impression with a fairly low power (20X to 50X) on a

A FIELD OF STAR CLUSTERS IN AURIGA. M37 is at upper left, M36 below center, and M38 at lower right. Photographed with a 5-inch camera, Lowell Observatory.

STAR CLUSTER M36. A bright galactic cluster in the star
clouds of the Auriga Milky Way, photographed with the 13-
inch telescope at Lowell Observatory.

DESCRIPTIVE NOTES   (Cont'd)

6-inch or 8-inch telescope. M36 is one of the younger galactic star clusters, containing bright B-type stars among
its members, and would be as splendid a group as the famous
Pleiades if it were some 10 times closer. According to the
photoelectric measurements of H.L.Johnson and W.W.Morgan
(1953) the 15 brightest stars have the following magnitudes
and spectra:

8.86	B2V	9.23	B2V	9.92	---
8.96	B3V	9.36	B2V	10.01	B2 III
9.04	B3V	9.52	B3V	10.37	---
9.13	B2V	9.58	B2	10.44	B8
9.14	B2V	9.76	B2V	10.65	B6V

The brighter members are all B-type stars, including
both main sequence stars, subgiants, and several giants of
luminosity class III. No red giants exist in this cluster.
The brightest members have absolute magnitudes of about
-1.6 (luminosity = 360 suns). The majority of these stars
show very broad spectral lines, attributed to rapid rotation, another point of similarity to the Pleiades. From
photometric studies of M36 made at Lowell Observatory, H.L.
Johnson (1957) has derived a distance of about 1260 parsecs
or about 4100 light years. The true diameter of the group
is then about 14 light years, and the total luminosity is
equivalent to about 5000 suns.

M37   (NGC 2099) Position 05490n3233. A superb galactic
            star cluster for telescopes of all sizes, usually
considered the finest of the three Messier open clusters
in Auriga, and apparently first observed by Messier himself
in 1764. It will probably look like a nebula in instruments
smaller than 1½-inch aperture, but in anything larger than
a 2-inch, some of the individual stars will be seen easily.
"A diamond sunburst", as C.E.Barns described it, this
striking cluster is a virtual cloud of glittering stars.
"Even in small instruments", says T.W.Webb, "it is extremely beautiful, one of the finest of its class". The great
observer Smyth called it "a magnificent object, the whole
field being strewed as it were with sparkling gold-dust;
it resolves into infinitely minute points of lucid light".
The Earl of Rosse commented on the "wonderful loops and
curved lines of stars", which seem also to be a feature of
some other galactic clusters, as M35 in Gemini, for example.

STAR CLUSTER M37. One of the finest of the galactic star
clusters; photographed with the 13-inch telescope at
Lowell Observatory.

M37 contains about 150 stars down to magnitude 12½; the total population may be in excess of 500 stars. The stellar population of this cluster is significantly different from that of M36, and suggests an older and more evolved group of stars. An age of somewhat over 200 million years is indicated by current knowledge of stellar evolution. The earliest type star in the cluster is of spectral class B9 V, and the majority of the other bright members are main sequence A stars with absolute magnitudes of about -1. But the cluster also contains at least a dozen red giants. The brightest of these has a visual magnitude of about 9½ and stands out near the cluster center "like a ruby on a field of diamonds".

The distance of the cluster is about 4600 light years according to a study by F.R.West (1964); this agrees well with an earlier determination of 4700 light years, published by Harlow Shapley in 1931. The actual diameter of the cluster is about 25 light years, the total luminosity about 2500 times the light of the Sun.

M38   (NGC 1912) Position 05253n3548. A large star cluster in the Auriga Milky Way, located about 2.3° northwest of M36, discovered by Le Gentil in 1749. It is a scattered group of irregular form, with the brightest stars in a pattern resembling an inverted letter "Pi". To Webb it was "a noble cluster arranged as an oblique cross" with a pair of stars in each arm. "Larger stars dot it prettily with open doubles. Glorious neighborhood".

The full diameter of M38 is about 20' and the total membership must be well over 100 stars. The earliest type members are giants of spectral class B5, with absolute magnitudes of about -1.5. The cluster also contains a number of A-type main sequence stars and several giants of type G. The brightest star of the cluster is a yellow G0 giant with a visual magnitude of about 7.9, and an actual luminosity of about 900 suns. As a useful standard for comparison it may be remembered that our Sun would appear as a star of magnitude 15.3 at the distance of M38, some 4200 light years. The true diameter of the cluster is about 25 light years, comparable to M37. A number of other fainter clusters will be found in this rich region of the sky.

STAR CLUSTER M38, as photographed with the 13-inch telescope at Lowell Observatory. The smaller cluster NGC 1907
is near the lower right edge.

LIST OF DOUBLE AND MULTIPLE STARS

NAME	DIST	PA	YR	MAGS	NOTES	RA & DEC
β612	0.3	211	62	6 - 6	Binary, 22 yrs; PA inc, spect A6	13371n1100
1	4.6	137	63	6 - 9	(Σ1772) PA dec, Spect A1, dF8	13383n2012
τ	5.4	7	58	$4\frac{1}{2}$-$11\frac{1}{2}$	(OΣ270) PA inc, dist dec, cpm, Spect F7, dM2	13449n1742
Σ1785	2.8	145	61	7 - $7\frac{1}{2}$	Binary, 155 yrs, PA inc, spect K6.	13468n2714
O$\Sigma\Sigma$126	85.9	208	00	$6\frac{1}{2}$- 7	Spect both G0	13480n2131
Σ1793	4.6	242	50	7 - 8	Relfix, spect A6	13568n2604
OΣ276	0.5	203	66	$7\frac{1}{2}$- $8\frac{1}{2}$	Slight PA inc, spect G0	14061n3659
	9.7	72	53	- 10		
OΣ275	5.0	353	43	7 - 10	relfix, spect K0	14067n0737
OΣ277	0.4	19	57	8 - 8	(Σ1812) Spect F2	14102n2857
	14.2	108	57	- 9		
	72.6	153	53	- 12		
OΣ278	0.2	338	67	$7\frac{1}{2}$- $7\frac{1}{2}$	Binary, about 210 yrs, PA dec, spect F2	14103n4425
OΣ279	2.2	253	52	7 - 9	Relfix, spect K0	14114n1214
$\varkappa$	13.3	236	58	$4\frac{1}{2}$- $6\frac{1}{2}$	Fine pair, relfix, Spect A7, F2. Primary variable	14117n5201
Σ1816	0.8	87	68	7 - 7	Dist dec, spect F0	14117n2920
15	1.0	120	67	$5\frac{1}{2}$- 8	(Kui 66) spect K0	14124n1020
Σ1829	5.4	150	38	$7\frac{1}{2}$- 8	relfix, spect F5	14136n5040
Σ1825	4.4	163	62	7- $8\frac{1}{2}$	slight dist inc, PA dec, spect F4	14142n2021
ι	38.5	33	42	5 - $7\frac{1}{2}$	cpm, relfix, spect A7, A2	14144n5136
β1271	2.4	350	46	7 - 12	cpm, spect G5	14154n5447
Σ1834	1.1	100	67	7 - 7	Binary, about 320 yrs, spect F2	14185n4844
Ho 262	5.3	272	25	7 - 13	relfix, spect K0	14205n3244
Σ1835	6.2	192	58	$5\frac{1}{2}$- 7	slight PA inc, spect A1, F2	14209n0840
Σ1835b	0.2	239	62	$7\frac{1}{2}$- $7\frac{1}{2}$	(β1111) Binary, 40 yrs, PA inc, spect F2	14209n0840
Σ1838	9.1	334	46	7 - 7	relfix, spect F5	14216n1128

LIST OF DOUBLE AND MULTIPLE STARS   (Cont'd)

NAME	DIST	PA	YR	MAGS	NOTES	RA & DEC
Σ1843	19.9	187	22	7½- 9	relfix, spect F5	14228n4803
A2069	0.2	194	62	8½- 8½	binary, 44 yrs; PA dec, spect F8	14244n1638
Σ1850	25.6	262	58	6 - 7	relfix, spect A0, A0	14264n2831
Σ1854	25.8	256	58	7- 10½	relfix, spect B9	14277n3201
A570	0.2	215	62	6½- 6½	binary, 30 yrs; PA dec, spect A7	14301n2654
γ	33.4	111	25	3 -12½	(β616) optical; spect A7 (*)	14301n3832
Σ1858	2.7	36	55	7 - 8	relfix, spect G5	14316n3548
	33.2	315	25	- 13		
Σ1863	0.5	72	66	7½- 7½	PA dec, spect F2	14364n5148
π	5.6	108	62	5 - 6	(Σ1864) PA slow inc; spect B9, A5	14384n1638
Σ1867	1.0	5	60	7½ - 8	PA dec, spect F5	14386n3130
ζ	0.9	307	66	4½ - 5	(Σ1865) binary,	14388n1357
	99.3	259	11	- 10½	about 125 yrs; spect both A2	
Σ1871	2.0	306	57	7 - 7	PA inc, spect F0	14398n5137
Σ1870	4.4	230	37	8 -10½	relfix, spect F2	14405n0817
ε	2.9	338	62	2½ - 5	(Σ1877) superb colored double; spect K0, A2 (*)	14428n2717
0Σ285	0.3	181	62	7 - 7½	binary, 88 yrs; PA dec, spect F5	14436n4235
Σ1879	1.4	95	66	7½- 8½	binary, about 225 yrs; PA dec, spect G2	14438n0952
	53.0	208	25	- 12		
Ho 263	0.5	212	00	7 - 10	spect K0; faint star not seen in recent years.	14454n2418
Σ1884	1.8	56	62	6½ - 8	cpm; spect F5	14462n2434
A1110	0.5	250	68	7½- 7½	PA dec, spect F5	14472n0811
	19.6	203	16	-12		
	23.0	237	16	-12½		
Σ1889	15.7	88	58	6½- 10	relfix, spect F4, G7	14479n5135
39	2.9	45	62	6- 6½	(Σ1890) relfix; spect dF5, dF6	14480n4855

LIST OF DOUBLE AND MULTIPLE STARS   (Cont'd)

NAME	DIST	PA	YR	MAGS	NOTES	RA & DEC
Σ1886	7.7	227	37	7 - 9	Relfix, spect K0	14486n0956
ξ	6.9	347	62	5 - 7	(Σ1888) Binary, Spect G8, K4 (*)	14491n1918
OΣ287	1.0	161	67	7½- 7½	PA inc, spect G0	14496n4508
Ho 389	1.5	95	57	7 - 9½	Relfix, spect A0	14498n2030
OΣ288	1.5	175	66	6½- 7	Binary, 215 yrs; Spect F9, F9	14510n1554
OΣ289	4.6	112	63	6½- 10	cpm, spect A3	14539n3230
Σ1895	12.5	43	34	8 - 8½	relfix, spect A2	14556n4022
Sh 191	40.5	342	40	7 - 7½	cpm, spect F0, F2	14581n5404
OΣ291	35.6	156	32	6 - 8½	spect A0p	14589n4728
Hu 745	0.4	25	55	7½- 9½	relfix, spect K0	15022n2002
44	0.5	312	68	5½- 6	(Σ1909) binary, about 220 yrs, (*)	15022n4751
47	6.2	254	58	5½- 13	(β1068) relfix, cpm, spect A0, A5	15038n4821
A2385	0.1		58	6½- 6½	Rapid binary, 8 yrs, spect A2	15050n1838
Σ1910	4.2	211	62	7 - 7	relfix, spect G5	15052n0925
Σ1916	9.9	330	25	7 - 9½	relfix, spect F8	15080n3910
Σ1921	30.4	283	25	7 - 7	Spect A2, A2	15101n3851
Es---	25.8	343	08	7 - 10	Spect K0	15111n4846
Σ1926	0.5	247	63	6 - 8½	PA & dist dec, spect F0	15130n3829
δ	105	79	23	3½- 7½	wide cpm pair, spect G8, G0 (*)	15135n3330
μ	108	171	56	4½- 6½	(Σ1938) cpm (*)	15226n3733
μ b	2.0	19	68	7 - 7½	Binary, 260 yrs, PA dec, spect G1	15226n3731
OΣ296	1.9	284	67	7½- 9	cpm, PA dec, spect G5	15247n4411
	67.3	316	11	- 12		
β944	10.8	127	12	6½-12½	Spect A0	15278n4753
	56.2	68	13	- 12½		
ν²	0.1		60	5½- 5½	(53) (A1634) Spect A3, PA uncertain	15300n4104
OΣ298	1.0	191	66	7½- 7½	Binary, 56 yrs, PA inc, spect K4	15343n3958
OΣ301	3.9	31	35	7 - 10½	relfix, spect K0	15445n4237
β621	0.5	34	66	8 - 9½	PA dec, spect A0	15482n4440

BOOTES

LIST OF VARIABLE STARS

NAME	MagVar	PER	NOTES	RA & DEC
$\varkappa$	4.54 ± 0.04	.069	Spect A7, probably Delta Scuti type; also visual double star	14117n5201
γ	3.05--3.10	.29	Spect A7; spectrum variable (*)	14301n3832
44b	6.5--7.1	.2678	(i Bootis) ecl.bin.; spect G2+G2 (*)	15022n4751
R	6.7--12.8	223	LPV. Spect M3e	14350n2657
S	8.0--13.7	271	LPV. Spect M3e	14212n5402
T	9.7-----	---	Nova 1860 (*)	14118n1918
U	8.5--12..	200	Semi-reg; spect M4e	14520n1754
V	7.0--11.3	258	LPV. Spect M6e	14277n3905
W	5.0---5.5	Irr	Spect M3; constant for long periods	14412n2644
X	8.3--- ?		variability uncertain; spect K0	14218n1633
Z	8.3--14.5	281	LPV. Spect M5e	14041n1343
RR	8.2--12.8	195	LPV. Spect M3e	14452n3932
RS	9.5--11.0	.3773	Cl.Var.; spect B8--F0	14314n3158
RT	7.4--12..	274	LPV. Spect M8e	15153n3632
RV	7.5---8.8	137	Semi-reg; spect M6e	14372n3245
RW	7.6---9.5	210	Semi-reg; spect M5	14391n3147
RX	7.0---9.2	78	Semi-reg; spect M8e	14220n2556
RY	7.0--7.4		class uncertain; spect F6	14475n2314
SS	9.5--10.4	7.606	Ecl.Bin.; spect dG5+dG8	15117n3845
UV	7.4---8.7	Irr	Spect F5	14203n2546
XZ	8.8---9.6	Irr	Spect M5	13515n1731
ZZ	6.8---7.5	4.992	Ecl.Bin.; spect F0+F0	13539n2610
AB	4.5---	---	Nova 1877	14047n2059
AC	8.3--8.9	.3524	Ecl.bin; W Ursa Maj type, Spect F9	14547n4634
AD	9.0--9.6	1.034	Ecl.bin; spect G0	14330n2451
BP	5.5 ±.01	1.306	Alpha Canes Ven. type, spect A0p	15425n5233

LIST OF STAR CLUSTERS, NEBULAE, AND GALAXIES

NGC	OTH	TYPE	SUMMARY DESCRIPTION	RA & DEC
5248	34[1]	⊘	Sc; 11.0; 3.2' x 1.4' B,L,E, psbM (*)	13351n0908
5466	9[6]	⊕	Mag 9.0; diam 5', class XII L,vRi,pC, stars mags 11....	14032n2846
5523	134[3]	⊘	Sb; 12.8; 5.0' x 0.8' F,pL,mE, nearly edge-on spiral galaxy	14126n2534
5533	418[2]	⊘	S ; 12.6; 1.8' x 0.8' pB,E,mbM	14141n3535
5548	194 [2]	⊘	Sa; 12.9; 0.5' x 0.5' cF,pS,R, vsvmbM	14157n2522
5557	99[1]	⊘	E1; 12.6; 0.9' x 0.8' smbM, cB,S,1E	14164n3643
5600	177[2]	⊘	Sc; 12.4; 1.0' x 0.9' pB,pS,gbM	14214n1452
5614	420[2]	⊘	Sa; 12.9; 2.1' x 0.8' pB,S,1E, mbM; extending filament on NW side (*)	14220n3505
5633	185[1]	⊘	Sa; 12.8; 0.9' x 0.5' cB,pS,E, pg1bM	14256n4622
5641		⊘	SBb; 13.1; 2.1' x 0.9' pB,pS,1E, mbM	14271n2902
5653	330[2]	⊘	Sa; 12.9; 0.5' x 0.4' pF,pS,R,bM	14280n3125
5660	695[2]	⊘	Sc; 12.3; 2.2' x 2.2' pB,L,iR,vgbM	14281n4950
5665	27[2]	⊘	Sc; 12.7; 1.0' x 0.8' pB,pL,1E,gbM	14299n0818
5669	79[2]	⊘	Sc; 12.5; 2.5' x 2.0' F,L,R,1bM	14303n1008
5676	189[1]	⊘	Sc; 11.9; 3.0' x 1.5' B,L,E, pgbM	14310n4941
5687	808[2]	⊘	S0; 12.7; 0.6' x 0.4' pF,S,E, lenticular	14333n5442
5689	188[1]	⊘	SBa; 12.6; 2.0' x 0.6' cB,pL,E, psbM	14337n4857
5739	171[1]	⊘	Sa; 13.1; 1.1' x 0.9' pB,S,R, smbM	14406n4203

LIST OF STAR CLUSTERS, NEBULAE, AND GALAXIES (Cont'd)

NGC	OTH	TYPE	SUMMARY DESCRIPTION	RA & DEC
5820	756^2	⊘	E5; 12.8; 0.7' x 0.3' B,E,sbM; faint extension on SE side; bright double star Sh 191 lies 8' east.	14572n5405
5899	650^2	⊘	Sb; 12.4; 2.3' x 0.6' cB,pL,pmE, smbMN	15132n4214

DESCRIPTIVE NOTES

ALPHA   Name- ARCTURUS, "The Guardian of the Bear".
    The 4th brightest star in the sky, formerly given 6th place, but shown by modern measurements to out- shine both Vega and Capella. Magnitude -0.06; spectrum K2 III. Position 14134n1927. Opposition date (midnight cul- mination) is April 27.
    Arcturus is located at a distance of about 37 light years, one of the Sun's nearer neighbors in space. The diameter of the star is estimated to be about 20 million miles, roughly 25 times the diameter of the Sun. The lum- inosity is about 115 times that of the Sun, and the abso- lute magnitude is -0.3. The computed mass of the star is about 4 times the solar mass, leading to a density in the range of 0.0003 the solar density. The spectrum is that of a K-type giant, and rather resembles the spectrum of a sun- spot. With modern infrared recording devices, the heat re- ceived from the star can be measured, and is found to equal the heat of a single candle at a distance of about 5 miles. The actual surface temperature of Arcturus is approximately 4200°K. The color of Arcturus is usually described as a golden yellow or "topaz"; Smyth called it reddish yellow.
    A remarkable fact about Arcturus is its great annual proper motion of 2.29" in PA 209°, the largest proper

DESCRIPTIVE NOTES (Cont'd)

motion shown by any of the 1st magnitude stars with the
exception of Alpha Centauri. The motion was first detected
by Halley in 1718. The actual space velocity of Arcturus
is almost 90 miles per second in the direction of the constellation of Virgo. This motion has been bringing the
star closer to the Earth ever since it first became visible
to the naked eye nearly half a million years ago. At the
present time, Arcturus is almost at its minimum distance
from the Solar System, about 37 light years. The star still
shows an approach radial velocity of about 3 miles per second, which will gradually diminish to zero as the star
passes us several thousand years from now. Arcturus will
thereafter continue to recede from us as it continues its
motion toward Virgo, and will have faded below naked-eye
visibility in the course of another 500,000 years.

Arcturus is a "Population II" star, a member of the
great spherical halo which is centered on the hub of our
galaxy. This explains the large apparent motion, and the
rapid passage through our part of the heavens; Arcturus is
moving in a highly inclined orbit around the center of the
galaxy, and is presently cutting through the galactic
plane. The Sun, on the other hand, is moving with the general "stream of traffic" in the plane of the galaxy; thus
the large relative motion between the two objects. From
the viewpoint of an Arcturian, it would be the Sun and the

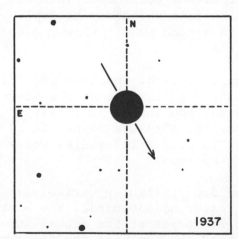
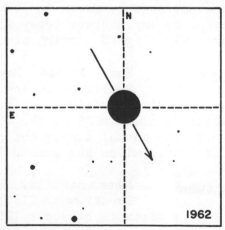

THE PROPER MOTION OF ARCTURUS — From Lowell Observatory 13-inch
Telescope Plates.   Scale= 19" of arc/mm.

rest of the general stream which is moving so rapidly.

As the brightest star of the northern skies, Arcturus has been known and admired since ancient times. R.H.Allen states that the star was one of the first to be given a name, and devotes several pages of his classic work "Star Names and Their Meanings" to a discussion of the various titles and mythological references concerning Arcturus. In ancient times it was known as the "Watcher" or the "Guard-ian"; the Arabs knew it under two names which may be trans-lated "the Lance-Bearer" and "the Keeper of Heaven". It is sometimes called "Job's Star" from the reference to it in the Book of Job, although the reference now appears to be a mis-translation, and probably refers to the Great Bear or Big Dipper instead.

Arcturus became a famous object - in the popular sense of the word - in the spring of 1933 when the "Century of Progress" Exposition opened in Chicago. The light of the star was focused by telescopes on photoelectric cells, and the current generated was used to activate the switch to turn on the flood-lights at the exposition grounds. Arctu-rus was chosen for the purpose because its distance was then estimated to be 40 light years; the light reaching the Earth in 1933 had started on its journey about 1893, when another fair had been in progress in Chicago.

Smyth stated that Arcturus was the first star on rec-ord to be observed in daylight with a telescope. This was accomplished by Morin in 1635, a feat which may be dupli-cated by any amateur today with a good small telescope and properly aligned setting circles.

BETA   Name- NEKKAR. Magnitude 3.48, spectrum G8 III. Position 15001n4035. The distance of the star is approximately 140 light years, the actual luminosity about 70 times that of the Sun. The absolute magnitude is +0.3. The annual proper motion is 0.06"; the radial velo-city is 12 miles per second in approach.

GAMMA   Name- SEGINUS. Mag 3.05 (slightly variable); Spectrum A7 III. Position 14301n3832. The star is at a distance of about 120 light years; the actual lum-inosity is about 75 times that of the Sun, and the absolute magnitude about +0.2. The annual proper motion is 0.19";

ARCTURUS. A "close-up" of the brightest star north of the Celestial Equator. This one-hour exposure, in red light, was made with the 13-inch astrograph at Lowell Observatory.

the radial velocity is 21 miles per second in approach. The star shows a slight variability in a period of about 7 hours, with an amplitude of a few hundredths of a magnitude (photographic range = 3.20 to 3.25). The exact classification is somewhat uncertain, but the star is probably related to the δ Scuti variables or the Alpha Canum Venaticorum type. There is also an optical companion of magnitude $12\frac{1}{2}$ at 33" distance, discovered by S.W.Burnham in 1878 with the $18\frac{1}{2}$-inch Dearborn refractor. The PA and separation are both increasing from the proper motion of the primary, and the apparent separation was at a minimum (19") about 1780.

DELTA   Mag 3.47; spectrum G8 III. Position 15135n3330. The distance of Delta Bootis is approximately 140 light years according to parallaxes obtained at Allegheny and McCormick; the resulting luminosity is about 70 times that of the Sun (absolute magnitude +0.3). The star shows an annual proper motion of 0.15"; the radial velocity is close to 7 miles per second in approach.

At a distance of 105" is the distant 7th magnitude companion, first measured by F.G.W.Struve in 1835. The two stars form a wide common proper motion pair, with a projected separation of about 4560 AU. The companion is a G0 main sequence star, very similar to our sun in type and luminosity. Its computed absolute magnitude is about +4.6.

EPSILON   Name- MIRAK or IZAR. Mag 2.37; spectrum K0 II or K1 II. Some authorites however, suggest a luminosity class of III. Epsilon Bootis is one of the most beautiful of the double stars, though generally a difficult object for a 3-inch glass and not exactly easy for beginners even with a 6-inch. T.W.Webb, however, observed the images clearly separated with a $2\frac{1}{4}$-inch achromat, and also states that Buffham resolved the pair with a 9-inch mirror stopped down to 1 7/8 inches. The star was discovered by F.G.W.Struve in 1829, who honored it with the poetic title "Pulcherrima" in appreciation of the fine color contrast. The primary, magnitude 2.47, is yellow-orange in color, and the smaller star, magnitude 5.04, is bluish but often seems slightly greenish. The spectral class is about A2.

Parallax measurements have been somewhat discordant, but suggest a distance in the range of 200- 300 light

DESCRIPTIVE NOTES (Cont'd)

years. Some individual published measurements are given
here:
  Allegheny (trigonometric) 0.009" = 360 light years.
  Yale   "   0.016 = 200   "   "
  Mt.Wilson (spectroscopic) 0.019 = 170   "   "

The Yale "Catalogue of Bright Stars" (1964) gives the parallax as 0.013", corresponding to a distance of about 250
light years. This gives the K-star an absolute magnitude
of about -1.9 (luminosity = 500 suns), very close to the
accepted value for a K0 II giant. The companion, an A-type
main sequence star, then has an absolute magnitude of +0.6
(luminosity = 45 suns). The annual proper motion of the
pair is only 0.05"; the radial velocity is 10 miles per
second in approach.
     The two stars definitely form a physical pair, but
the relative motion is extremely slow. There has been no
definite change in separation since discovery, but the PA
appears to be very gradually increasing, from 321° in 1829
to 338° in 1962. The projected separation of the pair is
about 230 AU. If one star is actually being seen far beyond
the other, the true separation may be much greater. The
lack of definite orbital motion in more than 130 years
suggests that such is the case. The period may be at least
several thousand years.

ETA   Name- MUPHRID. Mag 2.69; spectrum G0 IV. The
             position is 13523n1839. The distance of this
star is about 32 light years, the standard distance for
calculating absolute magnitudes; thus the apparent and the
absolute magnitudes are the same - +2.7. The star is a G-
type subgiant with a luminosity of about 7 suns. The annual
proper motion is 0.37" in PA 190°; the radial velocity is
very slight, less than 0.1 mile per second in approach.
The star is a spectroscopic binary with a period of 495
days.

MU   Name- ALKALUROPS. Mag 4.30 and 6.50; spectra
           F0 IV and dG1. Position 15226n3733. Mu Bootis
is a wide common proper motion pair with a separation of
108", discovered by F.G.W.Struve in 1826. Trigonometric
and spectroscopic parallaxes agree in giving the distance

as about 95 light years. The projected separation is then about 3170 AU, and the absolute magnitudes are +2.0 and +4.2. The annual proper motion is 0.17" in PA 300°; the radial velocity is about 6 miles per second in approach.

The fainter star is a close binary with a period of about 260 years. According to a computation by Baize (1952) the orbit has a semi-major axis of 1.46" and an eccentricity of 0.59. The motion is retrograde, with periastron in 1865. Both stars resemble the Sun in type and luminosity. The mean separation is about 43 AU.

XI   Mag 4.54; spectrum G8 V. Position 14491n1918.

A well known and attractive binary star, discovered by Sir William Herschel in 1780. It is among the nearer double stars with a distance of 22 light years. The orbital period is computed to be 149.9 years, with periastron occurring in 1909. The semi-major axis of the orbit is 4.9", and the eccentricity is 0.50. The true separation of the two stars averages about 33 AU. The apparent separation varies from 1.8" (1912) to 7.3" (1984). The two

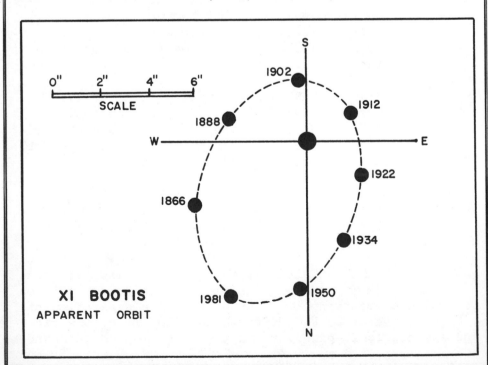

XI BOOTIS
APPARENT ORBIT

stars show a fine color contrast, usually described as yellow and reddish-violet. Information of interest about the components is given in the brief table below:

	Mag.	Spect.	Mass.	Lum.	Abs.Mag.	Diam.
A	4.7	G8 V	0.87	0.50	+5.5	0.90
B	6.8	K4 V	0.76	0.07	+7.6	0.80

The annual proper motion of the system is 0.17" in PA 129°; the radial velocities of the two components are 2½ and 4 miles per second in recession.

Astrometric studies by K.Strand (1943) also indicate a third unseen component in the system, revolving about one of the visible stars in a period of 2.2 years. The unseen star has a computed mass of about 0.1 solar mass, an unusually small value. Even at the relatively small distance of Xi Bootis, a star of such a low mass would probably be too faint to be detected visually in the glare of the two bright components. The expected apparent magnitude would be 14th or fainter.

44   (i Bootis) Mag 4.76; spectrum dG1 + dG2. The
              two stars of this pair form an interesting
binary system, discovered by F.G.W.Struve in 1832. The
position is 15022n4751. The apparent orbit is a very elongated and narrow ellipse which allows the apparent separation to vary from 4.7" (1880) to less than 0.4" (1969).
The period is still uncertain by a number of years. Orbits
by K.Strand (1937) and A.Gennaro (1940) give the following results:

	Period	Semi-major axis	Eccentr.	Periastron
Strand	219 yr	3.6"	0.42	1790
Gennaro	254 yr	4.0"	0.30	1783

The orbit diagram on page 310 is plotted from the results of computations by Strand. The distance of the pair is about 40 light years; the annual proper motion is 0.40" in PA 274°; the radial velocity is 15 miles per second in approach.

The primary star is very similar to our Sun in size, luminosity, and type; the absolute magnitude may be about +4.4. The mean separation of the two stars is about 45 AU. The fainter star is an object of special interest; it is a

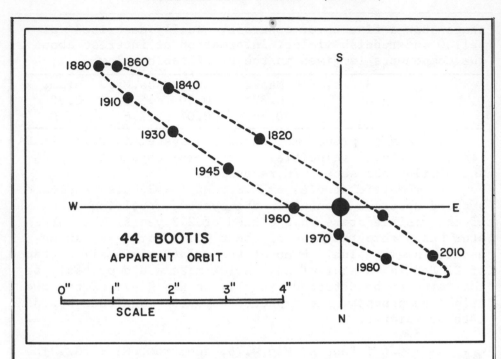

44 BOOTIS
APPARENT ORBIT

very close binary in which the two components form a dwarf
eclipsing system resembling such pairs as W Ursae Majoris
and U Pegasi. The two stars are revolving in a virtually
circular orbit, in the exceptionally short period of 6.427
hours, and the components eclipse each other at every revolution. The drop in light during each eclipse is about half
a magnitude. The computed separation of the two stars is in
the neighborhood of 3/4 of a million miles, or about three
times the separation of the Earth and Moon. The total mass
of the pair is 1.5 solar masses, and both stars are dwarfs
of spectral type dG2, comparable to our sun in size and
brightness.

      Systems of this type are of considerable interest, as
some recent studies make it seem probable that they are
the ancestors of the eruptive "dwarf novae" of the U Geminorum and SS Cygni variety. Evidence for an exchange of
material between the components is already present, in the
form of gaseous streams between the stars. One example,
U Pegasi, has already shown flares or eruptions of small
amplitude, perhaps a preview of its future violent activity. (See also W Ursae Majoris, U Pegasi, and SS Cygni)

T   Nova Bootis 1860. Position 14118n1918, about 25'
from Arcturus in PA 250°. The only definite
observations of this enigmatic object were those of Joseph
Baxendell in April 1860. While searching for new variable
stars he found this object at about magnitude 9.75 on April
9; the brightness was recorded as approximately the same on
April 10 and 11. On April 22 the magnitude had fallen to
12.8, and the following night the star could not be found
in a 13-inch reflector. Despite thorough searches by
Schoenfeld, Winnecke, Pickering, Daniel, Hartwig, Zinner,
and others, the star has never been seen again. Modern
plates of the field show no star as bright as 17th magnitude near the position.

Granting the reality of this object, the nova
appears to have had an amplitude of at least 7 magnitudes,
an unusually rapid decline of about a magnitude in four
days, and a position unusually far from the central plane
of the Galaxy (Galactic latitude about +68°). From these
facts J.Ashbrook (1953) suggests that the star was probably
a recurrent nova, which implies that several additional
outbursts may have been missed in the last century. One
alternate explanation, that the star was a distant variable of the U Geminorum type (dwarf nova) does not seem
likely in view of the many searches made by experienced
observers. Interested amateurs should make periodic checks
of the field and any reappearance of the star should be
reported to a major observatory without delay.

T BOOTIS FIELD, showing stars to about magnitude 9½. Grid
squares are 1° on a side with north at the top.

GALAXIES IN BOOTES. Top: The normal spiral NGC 5248.
Below: The unusual spiral NGC 5614 which has a somewhat
distorted structure.   Palomar Observatory 200-inch plates.

LIST OF DOUBLE AND MULTIPLE STARS

NAME	DIST	PA	YR	MAGS	NOTES	RA & DEC
I 271	2.3	145	48	7½- 10	relfix, spect G5	04201s4255
I 1146	6.3	42	39	7 - 11	spect B9	04225s4010
h3650	3.0	183	37	7 - 8½	relfix, spect A0	04250s4039
β747	2.9	220	42	7½- 9½	relfix, spect A2	04312s3823
h3674	39.8	208	19	7½- 12	spect K2	04372s3726
h3678	41.3	328	13	8 - 9½	relfix, spect F8	04382s4508
α	6.6	121	33	4½- 13	cpm; dist inc; spect F2	04389s4157
Rst 5208	0.2	352	47	8 - 8	spect F5	04394s3941
h3697	13.6	268	55	6 - 10	(ν) spect F0; PA & dist dec	04486s4125
h3699	17.9	145	33	7½- 11	slight PA inc; spect F5	04492s4544
λ41	9.4	122	27	7½- 12	relfix, spect A2	04507s3044
γ	2.9	308	42	4½- 8	(Jc 9) spect gK3, slight PA dec.	05026s3533

LIST OF VARIABLE STARS

NAME	MagVar	PER	NOTES	RA & DEC
R	6.7--13.7	391	LPV. Spect M6e	04388s3820
S	9.7--10.5	100	Semi-reg; spect K	04551s3314
T	7.0-- 9.8	156	Semi-reg, spect N4	04455s3618

LIST OF STAR CLUSTERS, NEBULAE, AND GALAXIES

NGC	OTH	TYPE	SUMMARY DESCRIPTION	RA & DEC
1679		⊘	Sc?; 13.5; 1.2' x 0.8' B,L,R, 1bM	04479s3204

CAMELOPARDALIS

LIST OF DOUBLE AND MULTIPLE STARS

NAME	DIST	PA	YR	MAGS	NOTES	RA & DEC
Σ362	7.1	142	55	$7\frac{1}{2}$- 8	Relfix, spect A0	03123n5951
	26.1	43	15	$-10\frac{1}{2}$		
0Σ52	0.5	84	59	$6\frac{1}{2}$- 7	PA dec, spect A2	03131n6529
Hu 1056	1.0	269	65	8 - 8	PA dec, spect F8	03150n6703
Σ368	2.2	340	53	$8\frac{1}{2}$- $8\frac{1}{2}$	relfix, spect F8	03169n6819
Σ374	10.9	295	23	7 - $8\frac{1}{2}$	relfix, spect F8	03195n6717
A978	2.1	243	25	8 -$13\frac{1}{2}$	spect B8	03212n6018
A980	0.2	81	62	7 - 8	PA dec, spect B8	03243n6005
Σ385	2.4	162	37	$4\frac{1}{2}$- 9	relfix, spect B9	03250n5946
Σ389	2.8	67	53	7 - 8	relfix, spect A0	03261n5912
Σ390	14.8	159	20	5 - 9	relfix, cpm pair;	03262n5517
	110	172	10	- 10	spect A1	
Σ392	25.8	347	21	$7\frac{1}{2}$- $9\frac{1}{2}$	spect K0	03266n5244
0 Σ54	23.6	358	47	7 - $8\frac{1}{2}$	slight dist dec;	03273n6725
					spect F0	
Σ396	20.4	243	27	$6\frac{1}{2}$ - 8	relfix, cpm pair;	03295n5836
					spect A2, F	
Σ400	0.8	254	66	7 - 8	binary, about 290	03309n5952
	92.3	236	62	- 10	yrs; spect F4	
Σ402	12.6	161	05	8- $10\frac{1}{2}$	spect K0	03316n6308
0$\Sigma\Sigma$36	46.1	69	23	7 - 8	spect F5; B=	03355n6343
					Hu1062	
Hu1062	0.1	170	53	9 - 9	binary, PA dec	03355n6343
Σ419	3.0	75	55	7 - 7	relfix, spect A3	03376n6941
Σ419b	0.5	154	18	8- $10\frac{1}{2}$	(A984) PA dec	
Pi 97	54.8	35	25	6 - $8\frac{1}{2}$	spect gK5, B8	03386n5949
	21.4	95	13	- $13\frac{1}{2}$		
	34.8	300	15	- 13		
Σ421	12.4	234	11	7 - 11	spect G5	03408n7128
0$\Sigma\Sigma$39	58.3	75	24	6 - 7	cpm pair; spect	03453n5658
					B9, A0	
Σ445	3.0	257	10	8 - 9	relfix, spect B5	03465n5958
Σ455	11.9	166	22	8 - $8\frac{1}{2}$	relfix, spect G5	03519n6922
0 Σ67	1.7	45	62	5 - $8\frac{1}{2}$	relfix, cpm pair;	03529n6058
					spect gK4, A0	
Σ480	3.2	326	37	$8\frac{1}{2}$- $8\frac{1}{2}$	relfix, spect K2	04005n5536
Ho 221	4.8	94	21	7- $11\frac{1}{2}$	spect F5	04007n5456
Σ472	6.8	14	10	9 -$9\frac{1}{2}$	relfix, spect A5,	04010n7154
					G0	
Σ484	5.3	132	25	9 -$9\frac{1}{2}$	relfix, in cluster	04034n6212
					NGC 1502	

CAMELOPARDALIS

LIST OF DOUBLE AND MULTIPLE STARS (Cont'd)

NAME	DIST	PA	YR	MAGS	NOTES	RA & DEC
Σ485	18.1	304	57	6 - 6	relfix, spect B0;	04034n6212
	6.0	257	25	- 12½	B0; in cluster NGC	
	11.6	359	10	- 13	1502; A = variable	
					star SZ Cam.	
Σ474	23.3	146	15	8½- 8½	relfix, spect G5	04060n7606
Σ490	4.6	58	62	8½- 9	relfix, spect F0	04062n6002
Σ503	4.4	226	20	8½- 8½	relfix	04125n6403
β1233	5.1	38	26	7½-12½	spect B8	04129n6658
Σ511	0.4	133	62	7½ - 8	binary, about 220	04137n5840
					yrs; PA dec,	
					spect A0	
Σ509	11.8	20	04	7½- 11	relfix, spect B9	04137n6147
	38.0	248	19	-8½		
0Σ75	0.6	170	62	7½ - 8	PA slow inc;	04143n6022
					spect B9	
Kui 16	1.3	210	62	6- 10½	spect A0	04187n5930
	32.3	59	25	- 8½		
Σ526	5.7	54	21	8 - 8½	relfix, spect F0	04215n6009
Σ531	1.0	312	59	7½ - 8	PA inc, spect F0	04227n5532
1	10.3	308	55	5 - 6	(Σ550) relfix;	04281n5348
					spect B0	
Σ557	23.4	126	15	8 - 8½	relfix, spect F2	04330n6253
Σ557b	0.3	153	24	9 - 11	(Hu1083) PA inc	
3	3.8	297	19	5 - 12	(β1043) relfix;	04360n5259
					spect K0	
2	0.1	13	62	5½ - 7	binary, 26 yrs;	04360n5323
	1.0	254	62	- 7½	(β1295) (Σ566)	
	22.6	215	11	- 13	AC binary, about	
					425 yrs; spect A5	
A1013	0.3	356	62	7 - 7	PA inc, spect A3	04390n5926
Hu612	0.4	330	62	7 - 8½	binary, 165 yrs;	04438n5313
					PA inc, spect F2	
Σ587	21.0	185	25	7 - 8½	relfix, spect A3	04441n5302
Σ584	11.8	122	07	7½-10½	relfix, spect K0	04451n6627
Hu1087	1.3	110	23	7½- 12	spect F2	04488n6724
5	12.9	245	33	5½- 13	(β1187) cpm;	04510n5511
					spect A0	
0Σ88	0.9	311	43	6½- 8	PA inc, spect G0	04527n6141
7	0.9	284	20	4½- 8	all cpm; PA dec;	04533n5340
	25.8	240	33	- 11	(Σ610) spect A1	

LIST OF DOUBLE AND MULTIPLE STARS   (Cont'd)

NAME	DIST	PA	YR	MAGS	NOTES	RA & DEC
Σ604	2.2	39	34	8 - 9	relfix, spect G0	04544n6959
Hu1093	5.4	5	20	7 - 12	spect F5	04571n6101
OΣ89	0.6	296	57	6 - 7½	PA slow dec; spect A2	04583n7400
Σ617	12.6	121	27	8½- 9	relfix, spect F8	04588n6256
Σ618	32.6	211	25	7½- 7½	relfix, spect both G0; Σ617 in field	04588n6300
β	80.8	208	23	5 - 9	(10 Cam) relfix, spect G0, A5	04590n6022
β b	14.8	168	19	9 - 11		
11 - 12	180	8	00	5½ - 6	wide optical pair (ΣI 13) spect B2, gG5	05018n5854
Es888	6.7	190	25	7 - 11	spect G5	05023n5420
	26.5	246	25	- 12		
A841	47.7	340	19	7½- 9	spect A0	05032n7537
A841b	0.5	216	29	9 - 10		
β749	1.2	234	56	8 - 10	PA inc, spect F8	05034n5528
Σ633	12.3	342	22	6½-10½	relfix, spect F0	05060n6332
Hu1097	1.5	116	24	6½- 11	relfix, spect B9	05075n7625
Σ638	5.2	222	34	7½- 8½	relfix, spect K0	05088n6946
Σ634	10.4	91	43	5 - 9	(19H) optical, dist dec from 34" in 1833; spect F6, F5	05143n7912
Σ676	1.2	268	62	8 - 9	PA dec, spect F8	05199n6441
Σ677	0.7	172	67	7½- 8	binary, about 390 yrs; PA dec, spect G0	05200n6321
A846	1.1	347	37	7- 10½	spect A0	05211n7431
19	1.3	47	41	6½- 10	(Hu1107) relfix, cpm; spect B9	05324n6407
Σ695	10.3	157	37	8½- 9	relfix, spect G5	05330n7918
Σ695b	2.0	172	37	10 - 10		
A1037	0.8	357	30	7- 11½	spect F0	05366n7358
	27.1	345	22	- 14		
Σ3115	0.9	359	63	6½- 7½	PA dec, spect A2	05444n6248
Σ780	3.8	104	54	6½- 8	relfix, spect F8	05460n6544
	12.3	150	54	- 10		
29	25.1	131	24	6½- 9½	spect A2	05463n5654

LIST OF DOUBLE AND MULTIPLE STARS   (Cont'd)

NAME	DIST	PA	YR	MAGS	NOTES	RA & DEC
Hu1112	0.2	316	46	7½ - 8	spect F8	05516n8245
0Σ121	0.2	138	59	7½- 8½	PA dec, spect F8	05588n7400
Σ831	11.8	76	37	8½- 8½	relfix, spect F8	06058n6759
Σ784	1.2	205	58	8½- 8½	PA inc, spect F8	06073n8412
Σ868	3.6	43	38	8½- 9	relfix, spect F8	06178n7357
0Σ136	5.7	80	37	6½- 10	cpm; spect A2	06226n7034
M1b 343	2.6	158	24	8- 12½	spect B8	06344n6612
Σ973	12.5	31	56	6½- 7½	cpm; slight dist & PA inc, spect G0	06575n7519
Σ1006	29.6	73	22	7 - 8	relfix, spect G5	07023n6237
47	2.1	317	62	6- 10½	(Σ1055) PA dec, spect A	07179n6000
Σ1051	1.1	284	53	6½- 8½	PA inc, spect F0	07206n7310
	31.5	82	35	- 6½		
Σ1122	15.4	5	28	7 - 7	relfix, spect F2	07412n6517
Σ1127	5.3	340	56	6 - 8	relfix, spect A2	07424n6411
	11.3	175	56	- 9		
Hu1247	0.2	127	62	8 - 8	binary, 18.8 yrs; PA inc, spect F5	07437n6025
Σ1136	6.2	227	35	7½-11½	PA & dist dec; spect K5	07483n6502
0ΣΣ90	48.6	82	24	6 - 6½	both spect G0	07580n6314
Σ1169	20.7	15	55	7½ - 8	slight PA inc; spect G0	08089n7939
0Σ188	10.5	193	14	6½- 10	spect K0	08162n7459
0Σ192	1.8	234	37	6½- 10	cpm; relfix, spect A	08310n7454
Hu883	3.6	74	20	7½- 12	spect F8	10552n7957
Σ1479	4.6	22	55	8 - 9	relfix, spect F8	10580n8330
Σ1539	19.0	313	19	8 - 9	relfix, spect F5	11269n8119
0ΣΣ117	66.7	76	24	6 - 8	spect K0	12088n8159
Σ1625	14.4	219	53	6½- 7	relfix, spect both F0	12141n8025
Σ1694	21.6	326	58	5 - 5½	easy cpm pair; spect A2, A0	12486n8341
Σ1720	1.8	334	58	8½- 9	relfix, spect A0	12586n8312
Kui 61	1.0	178	58	6½- 10	spect G5	13119n8044

CAMELOPARDALIS

LIST OF VARIABLE STARS

NAME	MagVar	PER	NOTES	RA & DEC
R	7.9--14.4	270	LPV. Spect Se	14211n8404
S	8.1--11.0	326	LPV or semi-reg; spect R8e (*)	05356n6846
T	7.2--14.1	374	LPV. Spect S3e	04352n6603
U	7.7-- 9.5	412	Semi-reg; spect N5	05375n6229
V	8.5--15.5	522	LPV. Spect M7e	05559n7430
W	9.8--14.2	284	LPV.	06191n7529
X	7.4--13.6	143	LPV. Spect M3e	04392n7501
Y	9.9--11.7	3.305	Ecl.bin.; spect A7	07345n7612
Z	9.9--14.3	Irr	Eruptive variable (*)	08197n7317
RR	9.6--11.3	124	Semi-reg; spect M6	05294n7226
RS	8.0--9.6	88	Semi-reg; spect M4	08441n7909
RT	8.8--13..	365	LPV. Spect M6e	06305n6408
RU	8.3--9.2	22.13	Cepheid; W Virginis type spect K0---R2 (*)	07163n6946
RV	7.1---8.2	182	Semi-reg; spect M4	04265n5718
RW	8.8--10.0	16.42	Cepheid; spect F5--G0	03503n5831
RX	7.5--8.2	7.912	Cepheid; spect G2--K2	04008n5831
RY	7.8---9.6	136	Semi-reg; spect M3	04261n6420
ST	7.0---8.4	195	Semi-reg; spect N5	04460n6805
SU	9.5--13..	285	LPV. Spect M	06318n7358
SV	9.1---9.7	.5931	Dwarf ecl.bin.; W Ursae Majoris type; spect dG5 & dG3	06306n8219
SY	9.5--10.8	400:	Semi-reg; spect M4	06024n8005
SZ	7.3--7.6	2.698	(Σ485) in NGC 1502; Ecl.bin.; lyrid type; spect B0	04034n6212
TU	5.0---5.2	2.933	(31 Cam) Ecl.bin; spect A0; lyrid type	05505n5953
UV	8.2---8.9	294	Semi-reg; spect R8	04015n6140
UX	8.1-- 9.3	Irr	Spect M6	05054n6837
VZ	4.7---5.2	237	Semi-reg; spect M4	07208n8231
XX	8.0--10..	---	R Coronae Bor type; spect gG1e	04048n5314
ZZ	7.4---8.1	Irr	Spect M5	04133n6213
AA	7.6---8.3	Irr	Spect M5	07096n6854
AS	8.2--8.6	1.715	Ecl.Bin; spect A0	05243n6928
AW	8.0--8.6	.7713	Ecl.Bin; spect A0	06419n6941
AX	5.95-6.08	8.015	Alpha Canum type; spec A2	07575n6028

CAMELOPARDALIS

LIST OF STAR CLUSTERS, NEBULAE, AND GALAXIES

NGC	OTH	TYPE	SUMMARY DESCRIPTION	RA & DEC
----	I.342	⊖	Sc; 12.0; 15' x 15' vL,F,R,vSBN (*)	03419n6757
1501	53[4]	◎	Mag 12, diam 55" x 48"; pB,S,vlE, bluish disc with central star mag 13½ (*)	04026n6047
1502	47[7]	∴	Irregular cluster, diam 8'; pRi,cC, about 25 stars mags 8.... class E; includes double stars Σ484 & Σ485	04030n6211
1530		⊖	SB; 13.0; 2.5' x 1.5' pB,L,E	04169n7511
1569	768[2]	⊖	I; 11.8; 2.3' x 0.7' pB,S,lE,bMN	04260n6445
1961	747[3]	⊖	Sb; 11.6; 3.7' x 2.5' cF,pL,mbMN	05368n6924
2146		⊖	Sb/pec; 11.6; 5.0' x 2.5' pB,lE, distorted spiral arms	06107n7823
2268		⊖	Sb; 12.2; 2.2' x 1.5' pF,pL,E	07013n8430
2314		⊖	E1/pec; 12.9; 0.7' x 0.6' vF,S,lE	07038n7519
2347	746[3]	⊖	Sb; 12.7; 1.0' x 0.8' vF,S,lE	07116n6454
2336		⊖	Sb; 12.4; 5.0' x 3.0' pB,pL,E; fine multiple-arm spiral	07162n8020
2366	748[3]	⊖	I; 12.6; 6.0' x 3.0' vF,pL,mbM,mE; probably distant member of M81 group in Ursa Major (*)	07236n6908
2403	44[5]	⊖	Sc; 8.8; 16' x 10' ! cB,eL,vmE,vgmbM; large well-resolved spiral (*)	07320n6543
2441		⊖	Sc; 12.7; 1.6' x 1.6' vF,pS,bN	07471n7306
2460		⊖	Sb; 12.7; 1.0' x 0.7' F,S,lE, faint outer arms	07527n6031
2523		⊖	SBb; 12.7; 1.8' x 1.4' pB,pL,lE; θ structure (*)	08092n7345

LIST OF STAR CLUSTERS, NEBULAE, AND GALAXIES   (Cont'd)

NGC	OTH	TYPE	SUMMARY DESCRIPTION	RA & DEC
2551		⊘	Sb; 13.1; 1.2' x 0.8' vF,S,E	08188n7335
2633		⊘	SBb; 12.6; 2.2' x 1.1' F,S,1E,bM	08427n7418
2646		⊘	E/pec or S0; 12.8; 0.5' x 0.3'; vF,S,1E	08446n7340
2655	288[1]	⊘	Sa; 11.6; 4.0' x 4.0' vB,cL,1E,bN; spiral arms diffuse & amorphous	08494n7825
2715		⊘	Sc; 12.1; 4.5' x 1.2' pB,L,mE	09020n7816
2732		⊘	Sa?; 12.9; 2.0' x 0.8' pB,S,mE, lenticular	09073n7924
2748		⊘	Sc; 12.4; 2.1' x 0.7' pB,pL,mE,vg1bM	09082n7641
----	I.3568	◎	Mag 11.6; diam 18", with faint central star.	12324n8251

DESCRIPTIVE NOTES

S   Variable. Position 05356n6846. A long-period pulsating variable star, discovered by T.E. Espin in 1891. The star has an average period of about 326 days, and the amplitude occasionally reaches 3 magnitudes. The greatest visual brightness is about magnitude 8.0. The star is a noted object because of its spectral class of R8. It was one of the first R-type variables to be recognized; another typical example is RU Virginis. These stars are similar to the N-type "carbon stars" such as R Leporis, except that they are at a somewhat higher temperature and the carbon bands are considerably weaker. They do not show

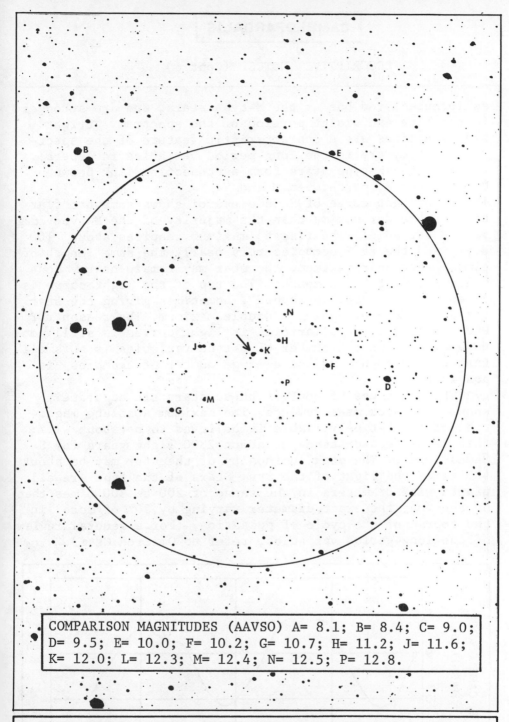

COMPARISON MAGNITUDES (AAVSO) A= 8.1; B= 8.4; C= 9.0; D= 9.5; E= 10.0; F= 10.2; G= 10.7; H= 11.2; J= 11.6; K= 12.0; L= 12.3; M= 12.4; N= 12.5; P= 12.8.

S CAMELOPARDI. Finder chart made from a 13-inch telescope plate at Lowell Observatory. Circle diameter = 1 degree. North is at the top. Limiting magnitude about 15.

as intense a red hue as the N-type stars, and are stronger in the blue and violet portion of the spectrum. Bright hydrogen lines are a characteristic feature of the spectrum, as they are in the long-period variables in general. Possibly the R-type stars form a connecting link between types K and N, or between M and N.

The light curve of S Camelopardi shows somewhat flatter and broader maxima than the majority of the M-type long period variables. The rise to maximum requires about 100 days, and the fall occupies very nearly the same length of time. When near maximum, the star may remain nearly constant for some three months. The top of the light curve is often slightly "saddle-shaped", sometimes giving the star the appearance of having a double maximum. The reason for such individual peculiarities in light curves is still not understood, and the number of R-type variables is so small that no thorough study of a large number of examples is possible.

The distances of none of these stars are accurately known by direct measurements. The maximum absolute magnitude of an R8-type variable is believed to be about -1.7; this leads to a distance of about 2800 light years for S Camelopardi. The peak luminosity of the star may be about 400 times the light of the Sun. Stars of the type are all giants with diameters in the range of 200 to 300 times that of the Sun, the exact diameter varying by 30% or more in the course of the cycle of pulsation. (For a general review of the long-period variables, refer to Omicron Ceti).

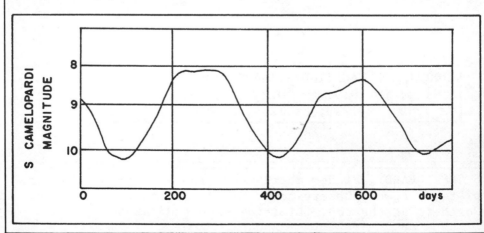

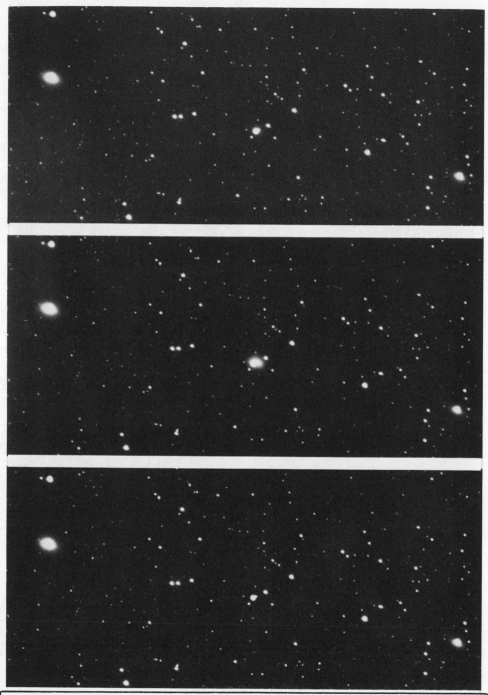

S CAMELOPARDALIS. The long-period variable star is shown near minimum (top) in December 1940, and brightening to maximum (center) in April 1941. These photographs were made with the 13-inch telescope at Lowell Observatory.

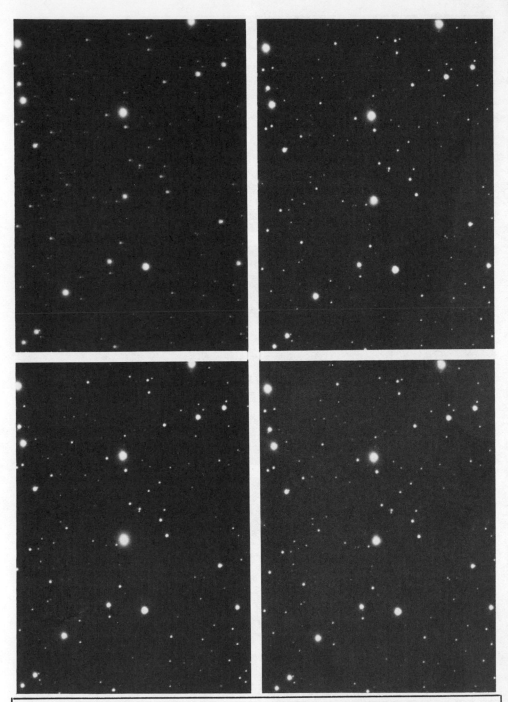

Z CAMELOPARDALIS. The strange eruptive variable is shown here during its outburst of January 1942. From Lowell Observatory plates made with the 13-inch telescope.

Z   Variable. Position 08197n7317. One of the
strange eruptive variable stars or "dwarf nova"
types, usually considered the typical example of its class.
It was discovered at Greenwich Observatory in 1904, during
the course of work for the Astrographic Catalogue. The star
shows constant eruptions which resemble a nova outburst on
a small scale; these explosions repeat persistently at intervals of 2 to 3 weeks. The rise to maximum is normally
completed in less than 2 days, and the normal light range
is about 3 magnitudes. At other times, however, the star
may remain constant or nearly so for several months, neither at maximum or minimum, but at some intermediate brightness. In addition, there are occasional periods of very
erratic changes, which follow no predictable pattern.

The Z Camelopardi stars closely resemble the better
known SS Cygni stars, and are distinguished from them only
by shorter average periods, smaller amplitudes, and the
occasional periods of constant brightness at an intermediate magnitude. The two classes gradually merge into each
other, however, and spectroscopic studies reveal no actual
difference between them. Z Camelopardi itself has a peculiar spectrum resembling class G5 at minimum, but with several bright lines which weaken as the star rises to maximum. The very similar star RX Andromedae has the same type
of spectrum. Like SS Cygni itself, these stars appear to
be close sub-dwarf binaries; the orbital period of Z Cam
has recently been determined to be about 6^h58^m. Outbursts
of such systems are attributed to some type of interaction
between the components, but the details are still uncertain
and the connection of these stars with the more violent
classical novae is speculative. (Refer also to SS Cygni)

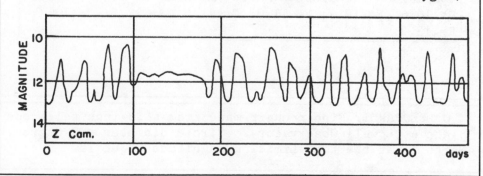

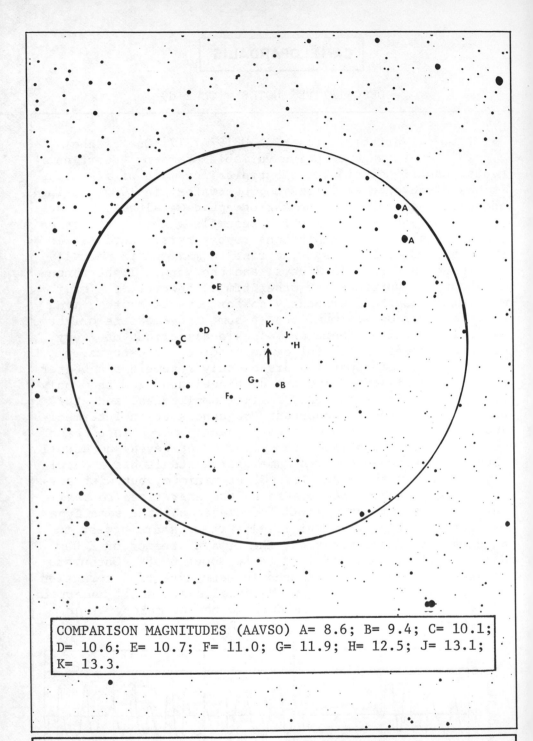

COMPARISON MAGNITUDES (AAVSO) A= 8.6; B= 9.4; C= 10.1; D= 10.6; E= 10.7; F= 11.0; G= 11.9; H= 12.5; J= 13.1; K= 13.3.

Z CAMELOPARDI. Finder chart made from a 13-inch telescope plate at Lowell Observatory. Circle diameter = 1 degree. North is at the top. Limiting magnitude about 15.

DESCRIPTIVE NOTES (Cont'd)

RU   Variable or ex-variable. Position 07163n6946.
        A former cepheid star which has apparently now
ceased to pulsate, a case which appears to be unique in the
annals of astrophysics. The star was originally discovered
by L.Ceraski at Moscow in 1907; it had a period of 22.13
days and a visual light range of magnitude 8.3 to 9.2. One
of the Population II cepheids, the star resembled W Virginis, and had a somewhat unusual light curve with a broad
dome-shaped maximum; the rise and fall being nearly equal
in duration. The spectrum at maximum was near K0 Ib, but at
minimum it was usually classed as type R, another unique
feature. Strong bands due to carbon compounds are prominent
in the spectrum. And in addition, RU Camelopardi was one of
the few cepheids in which definite changes in period had
been detected; from 22.216 days to 22.097 days. In recent
years the period had been given in standard catalogs as
22.134 days.
        The last normal pulsations of RU Camelopardi appear
to have been recorded in 1962. Early in 1965, observations
of the star were obtained by S.Demers and J.D.Fernie at the
David Dunlap Observatory in Canada; they found only slight
irregular fluctuations. In 1966 no variations greater than
0.04 magnitude have been detected. At Sonneberg Observatory
in Germany a series of patrol plates has shown that the
amplitude of the variations was still normal in 1961 and
1962, decreasing in 1963 and 1964, and becoming virtually
constant in 1965. The present magnitude is about 8.5.
        Since no other case of this type is known, it is not
possible to offer a ready explanation, or to answer the

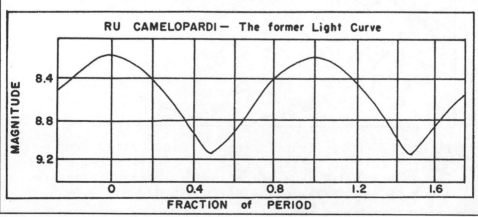

RU CAMELOPARDI— The former Light Curve

question: Has RU Camelopardi stopped pulsating for good, or will the variations eventually reappear? From theoretical studies of cepheids, it has always seemed that the pulsations must eventually die away, but over a time interval of at least 1000 years! Is the case of RU Camelopardi explainable by some random accident to the star, or has it perhaps entered a phase of its evolution where its physical characteristics change with abnormal rapidity? This is one star which should be carefully watched for any possible renewal of activity. (For a discussion of cepheids, refer to Delta Cephei; for an account of the Population II type cepheids, refer to W Virginis).

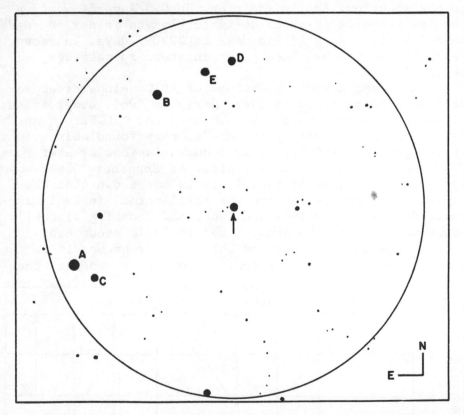

RU CAMELOPARDI; Identification field, traced from a 13-inch telescope plate at Lowell Observatory. Circle diameter = 1 degree. Comparison star magnitudes are: A = 8.05; B = 8.73; C = 8.94; D = 9.07; E = 9.09.

NGC 1502 in CAMELOPARDALIS. This compact cluster contains
the two easy double stars Σ484 and Σ485.   Lowell
Observatory photograph with the 13-inch telescope.

NGC 2523 in CAMELOPARDALIS. A barred spiral galaxy of unusual structure. Palomar Observatory photograph with the 200-inch telescope.

NGC 2403. A large Sc-type spiral galaxy, one of the nearest systems beyond the Local Group. This photograph was
made with the 61-inch astrometric reflector of the U.S.
Naval Observatory at Flagstaff. OFFICIAL U.S. NAVY PHOTOGRAPH

CENTRAL REGION OF NGC 2403. The heart of the great spiral is shown on this plate made with the 200-inch reflector.
Mt.Wilson and Palomar Observatories.

NGC 2403   Position 07320n6543. A large, loose-structured spiral galaxy lying in a rather blank region of the northern heavens, and now recognized as one of the nearest of the spirals beyond the Local Group. It may be seen easily as a large hazy spot in binoculars, and is frequently so detected by comet hunters. The magnitude of 10.2 in the Shapley-Ames Catalogue seems definitely too faint, and should be corrected to about 8.8. The apparent size is approximately 16' x 10'. A definite degree of mottling becomes apparent with larger amateur telescopes, hinting at details which are fully revealed only on the photographic plate.

The structure of NGC 2403 greatly resembles that of the "Pinwheel" galaxy M33 in Triangulum. The central nucleus is small, and the spiral arms are coarse and irregular with many bright condensations, star clouds, and nebulous regions. The galaxy appears to be at approximately the same distance as the M81 - M82 group in Ursa Major, and is very probably an outlying member of that group, which is centered some 14° away. The distance is thus about 8 million light years, and the total luminosity about 4 billion times that of the Sun. The apparent diameter of 16' corresponds to about 37,000 light years. The absolute magnitude of the system is close to -18. NGC 2403 is too close to show a large red-shift; the corrected radial velocity is about 112 miles per second in recession.

According to A.R.Sandage, this galaxy was the first system beyond the Local Group in which cepheid variable stars were identified. In 1960, 27 variables had been detected in the system, and periods had been determined for 10 of them. No novae have been recorded in the galaxy. On 200-inch telescope plates made at Palomar, at least 100 hydrogen emission regions have been identified, the largest having a diameter of some 880 light years. A similar nebulosity exists in M33, but nothing quite as gigantic has been identified in our own galaxy. In addition, a large number of blue giant stars populate the spiral arms of NGC 2403, revealing to the modern astronomer the fact that star formation is still in progress in that distant island universe. This is evidently a rather "young" galaxy, perhaps comparable in age to M33 in Triangulum.

IC 342   Position 03419n6757. A large round spiral galaxy of type Sc; difficult for small telescopes but of great interest, since it may be a member of the Local Group of galaxies which includes our own Milky Way System. It was discovered by W.F.Denning about 1890 and reported to J.L.Dreyer who included it in the first "Index Catalogue" (a supplement to the NGC) in 1895. E.Hubble and M.Humason (1934) detected the spiral pattern and called attention to the large apparent size, revealed by densitometer measurements to be about 40' E-W and 33' N-S. In apparent size, IC 342 is thus one of the largest spirals in the sky, and is probably among the half dozen nearest galaxies. M31 and M33 are the only spirals likely to be closer to the Milky Way system.

IC 342 may be observed in an 8-inch glass under good conditions, and appears as a small fuzzy 12th magnitude nucleus surrounded by a very large and faint hazy glow. On photographs this outer glow reveals itself as a beautiful pattern of spiral arms curving about the nucleus. Oriented almost face-on, the object is as perfect in form as the great M101 in Ursa Major, and evidently much nearer to our own galaxy. An interesting fact about IC 342 is its location only 10° above the galactic plane, well within the star clouds of the Milky Way. This undoubtedly results in a heavy degree of obscuration, and the distance is therefore indeterminate. The observed red-shift is very nearly zero, but after correcting for the solar motion the true value is found to be about 106 miles per second in recession. This is comparable to the velocities measured for some other members of the Local Group, as NGC 6822 in Sagittarius. If accepted as a member, IC 342 is the fourth known spiral in the Local Group; the other dozen or so members are all dwarf elliptical systems and irregulars. Tentatively accepting a distance comparable to M31 (the Andromeda Galaxy) the actual size of IC 342 is found to be about 25,000 light years, and the total luminosity possibly about 10 million times the light of the Sun. These figures can be regarded as little more than intelligent guesses, since the distance and exact degree of obscuration are both unknown. Another suspected member of the Local Group is NGC 6946 in Cepheus, which, however, shows about twice as large a red-shift.

IC 342. Spiral Galaxy in Camelopardus, identified as a member of the Local Group of galaxies.
60-inch telescope photograph, Mt.Wilson Observatory.

DEEP-SKY OBJECTS IN CAMELOPARDUS. Top: The planetary
nebula NGC 1501. Below: The irregular galaxy NGC 2366.
Mt.Wilson and Palomar Observatories.

CANCER

LIST OF DOUBLE AND MULTIPLE STARS

NAME	DIST	PA	YR	MAGS	NOTES	RA & DEC
A2882	1.9	215	23	$7\frac{1}{2}$- 14	spect A0	07536n1100
Σ1162	9.1	328	07	8 - 10	relfix, spect G5	07546n1321
Σ1167	12.0	228	04	9 - 11	relfix, spect G	07557n1636
A2884	2.5	113	22	8 -$13\frac{1}{2}$	spect F8	07564n1658
Σ1170	2.3	104	62	8 - 8	relfix, spect F5	07570n1350
Σ1171	2.3	326	36	6 - 11	PA dec, spect K1	07580n2343
h2423	5.6	262	25	8 - 12	spect F5	07584n1944
	61.4	230	00	- 11		
Σ1173	10.0	51	32	8 - $9\frac{1}{2}$	relfix, spect G0	07585n1705
A2540	1.4	164	23	$7\frac{1}{2}$- 13	spect K0	07585n2403
J732	2.6	192	54	9 - 9		07595n1017
Ho 349	9.8	226	11	8 - 12	spect K0	07598n1236
	63.2	290	00	- 12		
OΣ186	0.9	74	63	$7\frac{1}{2}$- 8	relfix, spect A2	08003n2625
A2956	4.5	172	21	8 - 14	spect K2	08005n1005
Hu 848	1.9	154	21	8 - 13	spect G0	08006n1349
Ho 350	4.8	198	29	8 - 12	PA inc, spect F8	08007n1219
β581	0.4	139	61	8 - 8	binary, 45 yrs;	08016n1226
	4.9	210	61	- $10\frac{1}{2}$	spect K2	
Σ1179	21.4	202	39	$8\frac{1}{2}$- $8\frac{1}{2}$	dist inc;	08020n1213
Σ1179b	4.0	56	37	9 -13	(β582) relfix	
Σ1177	3.4	351	62	$6\frac{1}{2}$- $7\frac{1}{2}$	relfix, cpm pair;	08026n2740
					spect A0	
Σ1181	5.2	140	27	7 - 9	relfix, cpm pair;	08027n0821
					spect G5	
11	3.2	218	34	7 -10	(Σ1186) relfix;	08058n2738
					spect K0	
J375	7.2	146	62	9 - 9	dist & PA inc	08061n1221
	14.4	347	62	- 14		
Σ1188	16.1	201	24	8 - 9	relfix, spect G0	08063n3030
Σ1187	2.5	29	62	7 - 8	dist inc, PA dec,	08064n3222
					spect F2	
Σ1191	3.2	72	44	$8\frac{1}{2}$- 9	relfix, spect G5	08079n1911
h777	11.0	351	26	$8\frac{1}{2}$ -10	spect A0	08088n1050
Hu851	2.2	232	22	$7\frac{1}{2}$- 14	spect G5	08090n1336
ζ	0.9	337	68	$5\frac{1}{2}$- 6	(Σ1196) famous	08093n1748
	5.8	83	62	- $5\frac{1}{2}$	triple system (*)	
Σ1197	1.8	102	39	8 - 9	relfix, spect A2	08096n2942
Σ1201	6.6	183	27	8 - $9\frac{1}{2}$	relfix, spect A5	08102n0944
Σ1202	2.4	311	60	$7\frac{1}{2}$- $9\frac{1}{2}$	PA dec, spect F8	08108n1100

LIST OF DOUBLE AND MULTIPLE STARS (Cont'd)

NAME	DIST	PA	YR	MAGS	NOTES	RA & DEC
Ho 38	7.4	82	05	$7\frac{1}{2}$- 12	spect F2	08108n2756
	30.0	140	04	- 12		
ADS 6668	1.4	343	48	$8\frac{1}{2}$- 9	relfix, spect G0	08110n1547
β1243	1.5	332	57	7 - 12	PA dec, spect F0	08113n1750
	63.0	301	24	- 9		
Σ1206	12.6	200	32	9 - $9\frac{1}{2}$	relfix, spect A2	08120n0720
Ho 524	4.0	340	33	8 - 11	spect K0	08131n1851
Es 293	4.6	214	37	9 - $9\frac{1}{2}$		08136n3225
β	29.2	294	10	$3\frac{1}{2}$- 14	(β1065) cpm; spect K4	08138n0921
Σ1212	5.4	238	31	$8\frac{1}{2}$- $9\frac{1}{2}$	relfix, spect F2	08150n3059
Σ1218	4.3	269	34	9 - 11	relfix, spect F2	08205n2321
Σ1219	11.6	260	18	$8\frac{1}{2}$- $8\frac{1}{2}$	relfix, spect G5	08203n0748
21	1.0	259	34	$6\frac{1}{2}$- 11	(A2961) spect gM2	08212n1048
Σ1220	29.0	214	18	$8\frac{1}{2}$- 10	spect K0, relfix	08223n2431
β1066	2.4	188	44	7 - 13	relfix, spect K0	08223n0935
HVI 109	30.4	343	62	$5\frac{1}{2}$-$9\frac{1}{2}$	spect G8	08232n0744
J73	3.8	212	50	$8\frac{1}{2}$- 12	spect F0	08233n0759
24	5.8	47	57	7 - $7\frac{1}{2}$	(Σ1224) slow PA inc; spect F1, F6	08237n2442
24b	0.2	332	62	8 - 8	(A1746) binary, 22 yrs; spect F6	
φ²	5.1	217	62	$5\frac{1}{2}$- 6	(Σ1223) cpm; PA slow inc; spect A3, A4	08238n2706
Σ1227	24.6	163	21	8 - $9\frac{1}{2}$	spect A0, relfix	08245n2319
Σ1228	9.0	350	38	8 - $8\frac{1}{2}$	spect F2, relfix	08246n2744
Σ1231	24.8	211	18	8 - $8\frac{1}{2}$	spect F5, relfix	08268n3133
Σ1236	36.9	113	25	8 - $8\frac{1}{2}$	spect F5, A2	08283n3206
A2896	1.9	343	21	$7\frac{1}{2}$- 13	spect F0	08312n1046
Σ1245	10.3	25	55	6 - 7	relfix, spect dF6	08332n0648
	93.2	120	10	- $10\frac{1}{2}$	& dG5	
Σ1246	10.4	115	34	$8\frac{1}{2}$- $9\frac{1}{2}$	relfix, spect G0	08332n1005
β584	1.2	291	62	7 - 12	relfix, spect A0;	08371n1944
	45.2	156	52	- 7	in cluster M44	
	92.9	241	52	- $6\frac{1}{2}$		
Σ1254	20.5	54	56	$6\frac{1}{2}$- 9	group in M44;	08375n1951
	63.2	342	56	- 8	spect K0, G0, A1,	
	82.6	43	56	- 9	& F5	

LIST OF DOUBLE AND MULTIPLE STARS (Cont'd)

NAME	DIST	PA	YR	MAGS	NOTES	RA & DEC
β585	0.5	94	66	7½ - 9	PA dec, spect A2	08384n2039
Σ1262	6.7	201	31	8 - 10	relfix, spect K0	08390n2359
Σ1266	23.4	64	16	8½- 9½	cpm; spect G0	08414n2838
δ	38.4	90	58	4 - 12	optical, dist & PA dec, spect K0	08418n1821
ι	30.5	307	58	4½- 6½	(Σ1268) relfix; cpm; yellow-blue; spect G8, A3	08437n2857
Σ1276	12.5	354	38	8 - 8	relfix, spect A0	08445n1121
β1068	0.2	122	54	7½- 8½	PA dec, spect A5	08468n0905
	17.7	312	03	- 13		
Σ1283	16.4	123	33	7 - 8	relfix, cpm pair; spect F0	08472n1501
A2473	0.3	15	60	7½- 7½	PA inc, spect G5	08478n1811
51	4.4	272	34	6 - 13	(Hu 1125) (σ¹)	08495n3240
	79.0	23	24	- 9½	spect A	
Σ1288	7.4	259	10	9 - 9	relfix, spect G0	08498n2839
57	1.5	318	62	6 - 6½	(Σ1291) PA slow	08512n3046
	55.6	199	53	- 9	dec, cpm; spect G7, K0	
0Σ195	9.6	139	55	7½- 8	relfix, cpm pair; spect F8	08513n0837
A2131	0.3	342	24	7 - 8	binary, 44 yrs;	08520n2624
	44.0	3	25	- 13	AC optical, dist inc. Spect G1	
61	0.1	225	45	7 - 7	(Ho 252) PA inc, spect dF3	08549n3026
α	10.9	322	62	4½ -11	(h110) relfix, cpm; spect A	08558n1203
64	89.6	295	14	5½ - 9	(σ³) spect G9	08565n3237
A1975	3.5	82	20	6½- 14	Spect A2	08570n2638
Σ1297	4.7	160	34	8½- 9½	relfix; cpm	08576n2255
66	4.6	137	55	6 - 8	(Σ1298) (σ⁴) relfix, spect A3	08583n3227
Σ1300	5.0	187	63	9 - 9	cpm; PA slow dec; spect K5	08585n1528
Σ1301	10.0	359	07	8½- 9	relfix, spect F8	08590n2624
Ag 162	3.9	107	42	9 - 9	spect K0	09042n3051
Σ1311	7.5	200	56	6½ - 7	relfix, cpm;	09046n2311
	27.9	118	06	- 12½	spect dF4, dF3	

CANCER

LIST OF DOUBLE AND MULTIPLE STARS (Cont'd)

NAME	DIST	PA	YR	MAGS	NOTES	RA & DEC
Σ1322	1.8	54	57	7½- 8	relfix, spect A0	09099n1644
Σ1327	7.8	61	62	8 - 9	dist dec, PA dec;	09126n2807
	27.3	19	62	- 9	spect F8	
Σ1327c	2.2	351	13	9 - 14		
0Σ198	13.9	136	58	8 -12½	PA & dist dec;	09133n2337
					spect K0	
Σ1332	5.8	26	62	7 - 7½	PA inc, spect F5	09144n2352
Σ3121	0.5	5	62	7½- 7½	binary, 34 yrs;	09149n2847
					PA inc; spect dK4	
h128	21.8	278	58	6½- 13	optical dist dec;	09152n1143
					spect A3	

LIST OF VARIABLE STARS

NAME	MagVar	PER	NOTES	RA & DEC
R	6.1--11.5	362	LPV. Spect M6e--M8e	08138n1153
S	8.3--10.5	9.485	Ecl.bin.; spect A0, G5	08411n1913
T	7.8--10.6	482	Semi-reg; spect N3	08538n2002
U	8.7--15..	305	LPV. Spect M2e	08329n1904
V	7.5--13..	272	LPV. Spect S2	08189n1727
W	7.5--14.4	393	LPV. Spect M7e	09070n2527
X	6.2---7.5	170	Semi-reg; spect N3	08526n1725
Z	8.1---9.4	104:	Semi-reg; spect M6	08196n1509
RR	8.3--13..	298	LPV. Spect M3e	08081n2318
RS	5.3---6.4	120	Semi-reg; spect M6e	09076n3110
RT	7.1---8.3	90:	Semi-reg; spect M5	08555n1102
RU	9.9--11.5	10.173	Ecl.bin.; spect dF9 & dG9	08346n2344
RX	8.1--10.2	120	Semi-reg; spect M8	08117n2453
RZ	8.2---9.4	21.64	Ecl.bin.; spect gK2 & gK5	08360n3158

CANCER

LIST OF VARIABLE STARS   (Cont'd)

NAME	MagVar	PER	NOTES	RA & DEC
SY	9.5--12..	Irr	Eruptive, resembles Z Camelopardi	08582n1805
SZ	8.9--14..	315	LPV. Spect M2	08186n1410
TU	9.9--12.2	5.562	Ecl.bin.; spect A0	08496n0917
TW	8.5---9.3	70.76	Ecl.bin.; spect G9, F5	08269n1237
UU	9.2-- 9.8	97.52	Ecl.bin.; spect K4	07597n1518
UV	8.0---9.5	Irr	Spect M4	08359n2120
UY	9.9--16..	229	LPV.	08333n1323
VV	9.2--10..	Irr	Spect M3	08084n1918
VW	9.4--14..	366	LPV.	08109n1021
VZ	7.2---7.9	.1784	Cl.Var.; spect A7--F2	08382n1000
WY	9.4--10.2	.8294	Ecl.bin.	08590n2653
WZ	9.2--10..	120:	Semi-reg; spect M2	08060n1654
XX	8.9--10..	Irr	Spect M4	08404n2047
ZZ	9.3--10.8	25.595	Ecl.Bin; spect A3	07544n1109

LIST OF STAR CLUSTERS, NEBULAE, AND GALAXIES

NGC	OTH	TYPE	SUMMARY DESCRIPTION	RA & DEC
2545	627[2]	⊖	Sb; 13.0; 1.1' x 0.9' F,S,1E, F outer ring; 8^m star 4' west	08113n2130
2608	318[2]	⊖	Sc; 12.8; 1.7' x 0.8' F,vL,E,mbM	08322n2838
2623		⊖	S pec; 14.0; 2.0' x 0.4 Extending filaments (*)	08354n2556
2632	M44	⠶	!! Praesepe Cluster; mag 4.5; diam 80' (*)	08375n1952
2672		⊖	E1; 12.6; 0.4' x 0.3' pB,pL,1E,mbM; S ellip gal NGC 2673 on east edge	08466n1916
2682	M67	⠶	! vB,vL,vRi,1C; mag 7, diam 15'; stars mags 10... (*)	08483n1200
2749		⊖	E2; 13.2; 0.8' x 0.6' pF,S,1E	09025n1831
2764	236[3]	⊖	E3; 13.3; 0.9' x 0.5' eF,S,1E	09054n2139
2775	2[1]	⊖	Sa; 11.5; 2.2' x 1.5' cB,cL,1E,vgvsmbM	09077n0715

DESCRIPTIVE NOTES

ZETA   Name- TEGMENI. Mag 5.10; spectra F8 V and
GO V. Position 08093n1748. This is one of the
most remarkable of all known multiple star systems, discovered by T.Mayer in 1756. It was listed as a double star
until 1781, when Sir William Herschel discovered a third
component. The closer pair, A and B, form a binary system
with a period of 59.6 years. The orbit is retrograde, and
the apparent separation varies from 0.6" to about 1.2",
with widest separation occurring in 1960. The semi-major
axis of the computed orbit is 0.88" and the eccentricity
is 0.32. The actual separation of the pair averages about
19 AU, comparable to Uranus and the Sun. Both components
are yellowish main sequence stars; the apparent magnitudes
are 5.6 and 5.9.

The third component, Zeta C, revolves about the pair
at a distance of 5.8", in a computed period of about 1150
years. The magnitude is 6.02, the spectral class is dG2.
Orbital elements of the wide pair are uncertain; the separation has remained nearly constant since the early measurements of O.Struve in 1826, but the PA has changed by

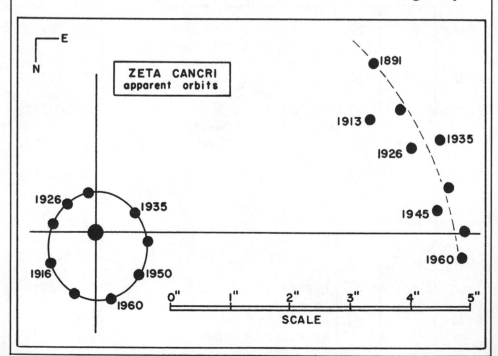

ZETA CANCRI
apparent orbits

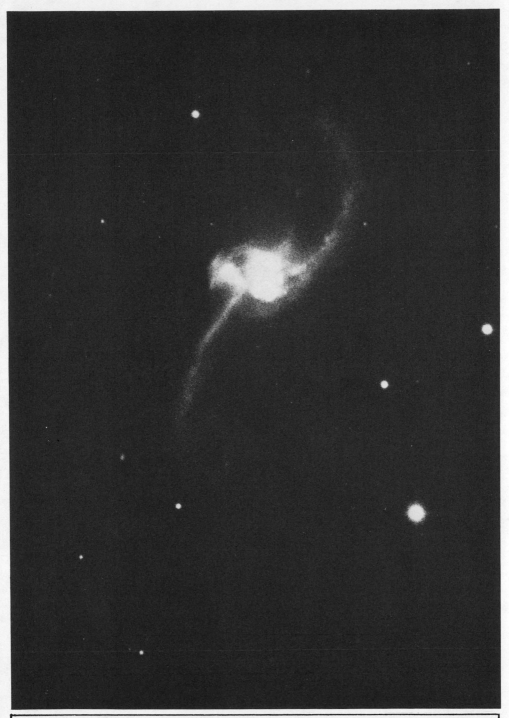

GALAXY NGC 2623 in CANCER. This odd object with extending
filaments resembles the "Ring-tail" galaxy NGC 4038 in
Corvus. Palomar Observatory 200-inch telescope photograph.

PRAESEPE. Star Cluster M44 in Cancer. This historic photograph is a portion of the first plate made with the 13-inch wide-angle telescope at Lowell Observatory, on April 6, 1929.

about 70° in the last 135 years. The computed semi-major
axis of the A-C system is about 8", equivalent to a mean
separation of about 175 AU. The eccentricity is 0.26.

From irregularities in the motion of star C, it has
been found that this star has an unseen companion with a
period of 17.64 years; the average separation being about
0.25" or close to 5 AU. The companion has not been detected visually in any telescope, and must be a dwarf star of
low luminosity, probably a white dwarf. Some astronomers
have suspected the existence of a fifth star in the system.

In his analysis of the Zeta Cancri system, C.Gasteyer
(1954) derived masses of 0.99, 0.88, 0.90, and 0.90 suns
for the four stars. The adopted parallax of 0.047" leads to
a distance of 70 light years. The primary star is then seen
to be about twice the brightness of our Sun, with an absolute magnitude of about +3.9. The B and C stars are each
slightly brighter than the Sun. The annual proper motion of
the system is 0.16"; the radial velocity is about $3\frac{1}{2}$ miles
per second in approach.

M44 (NGC 2632) Position 08375n1952 . "Praesepe" or
the "Beehive" star cluster, sometimes called
the Manger, one of the largest, nearest, and brightest of
the galactic star clusters. It is clearly visible to the
naked eye, and appears as a nebula, but even an opera-glass
will reveal its stellar nature. The group is over a degree
in apparent size, and needs a low-power telescope and a
wide-field eyepiece. A rich-field telescope is excellent
for such an object. Good binoculars will also give a very
pleasing view.

According to legend, Praesepe was used in ancient
times as a weather indicator. Aratus and Pliny have both
stated that the invisibility of the object in an otherwise
clear sky was considered to forecast the approach of a violent storm. Praesepe was one of the few clusters mentioned
in antiquity, though of course its true nature was not recognized. Hipparchus (130 B.C.) called it a "Little Cloud"
and Aratus (about 260 B.C.) refers to it as a "Little Mist".
According to R.H.Allen, it appeared on Bayer's charts of
about 1600 under the designation "Nubilum" or "Cloudy One".

The actual nature of Praesepe remained a mystery until
the year 1610, when the newly invented telescope revealed

PRAESEPE. The Beehive star cluster, photographed with the 13-inch telescope at Lowell Observatory.

DESCRIPTIVE   NOTES   (Cont'd)

that the object consisted of myriads of small stars.
Galileo was the first to view the Beehive through the telescope, and was astonished and delighted by the first sight
of the glittering cluster. T.W.Webb states that Galileo
counted 36 bright stars in the group; later observers have
recorded over 350, down to fainter than 17th magnitude.

The cluster is too distant for parallax measurements
to be reliable, but modern studies have established the
distance, through more indirect methods, as about 525 light
years. The bright central portion of the cluster is about
13 light years in diameter, but some of the more distant
members increase the total size to something like 40 light
years. About 200 stars are recognized as physical members
of the group, the magnitudes ranging from 6.3 to 14. For
the 15 brightest stars, the following magnitudes and spectra have been obtained:

1. 6.30; gA6	6. 6.67; sgA7	11. 6.90; K0 III
2. 6.39; K0 III	7. 6.75; sgA7	12. 7.32
3. 6.44; K0 III	8. 6.78; sgA5	13. 7.45; A6 V
4. 6.59; K0 III	9. 6.78; sgA5	14. 7.54; sgA6
5. 6.61; A2	10. 6.85; gA8	15. 7.54

Star #1 is Epsilon Cancri, the brightest member of the
cluster; it has a luminosity of 70 suns, and an absolute
magnitude of about +0.2. Eighty stars in the cluster are
brighter than 10th magnitude, and about 100 are brighter
than the Sun. As a standard of comparison, it may be remembered that our Sun would appear as a star of magnitude
10.9 at the distance of Praesepe. The great majority of
the stars are normal main sequence objects, ranging in type
from spectral class A2 to K6. There are four orange giants
of type K0 III in the group, and 5 known white dwarfs with
several more suspected. An interesting member is the faint
variable TX Cancri, a dwarf eclipsing binary of the W Ursae
Majoris type. The period is 0.38 day, the spectral type is
dF8, and the photographic range is 10.5 to 10.8.

The annual proper motion of M44 is 0.037" in PA 249°,
the actual space velocity being about 25 miles per second.
The motion appears to be very nearly equal and parallel to
that of the Hyades Cluster in Taurus, and it has been proposed that the two groups had a common origin. In favor of

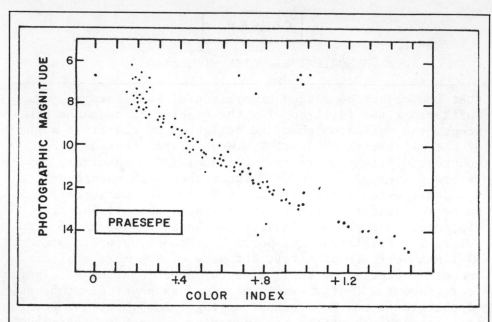

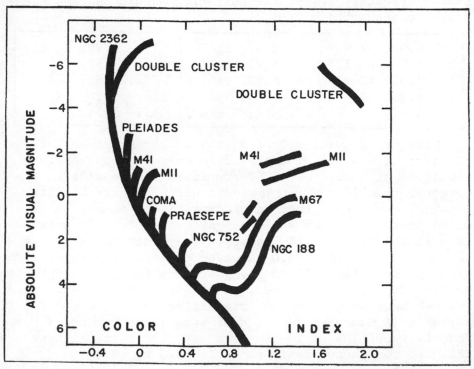

ABOVE- The Color-Magnitude diagram for M44, from observations by H.L.Johnson. BELOW- Comparison of diagrams for
various clusters, based on studies by Allan Sandage.

DESCRIPTIVE NOTES (Cont'd)

this idea is the fact that the two clusters seem identical
in age; against it is the fact that the present separation
is over 450 light years. Thus the question is not definitely settled. (Refer also to the Hyades cluster in Taurus.
For a discussion of cluster age-dating, see M13 in Hercules).

M67   (NGC 2682) Position 08483n1200. A rich gal-
              actic star cluster, located 1.8° west of the
star Alpha Cancri, and about 9° south of Praesepe. It is a
compact group, some 15' in diameter, and containing 500
or more members, from the 10th to the 16th magnitudes. The
known spectral types range from B9 to K4. The cluster is
some 2500 light years distant, according to recent studies
by O.J.Eggen and A.Sandage (1964). The true diameter is
some 12 light years. A peculiar feature of M67 is the great
distance above the plane of the galaxy, nearly 1500 light
years. The majority of the open clusters are distributed
generally along the central plane of the Milky Way.
      M67 has a stellar population quite unlike that of a
typical galactic star cluster, and was the subject of a
detailed study by H.L.Johnson and A.Sandage in 1954. With
the 82-inch telescope at McDonald and the 60-inch at Mt.
Wilson, accurate colors and magnitudes were obtained for
all the brighter members. Additional studies by Eggen and
Sandage (1964) with the 100-inch and 200-inch reflectors
extended this survey to the 16th magnitude; and accurate
colors and magnitudes are now known for 500 stars in the
cluster. When plotted on the familiar H-R diagram, these
stars reveal a color-magnitude array unlike that of any
other galactic cluster known up to 1954. The faint cluster
NGC 188 in Cepheus has since been found to show a similar
pattern, which resembles that of a typical globular star
cluster, rather than a galactic type. In both clusters, the
evolution of the brighter stars has carried them away from
the main sequence; in the case of M67 this "turn-off point"
is near absolute magnitude +3.5, implying an age of about
10 billion years. Probably only NGC 188 is known to have a
greater age among galactic clusters. (For an explanation
of the use of H-R diagrams in cluster age-dating, refer to
M13 in Hercules).
The brightest member of M67 is a 10th magnitude B9 star,

STAR CLUSTER M67. One of the most ancient known galactic
clusters; photographed with the 13-inch telescope at
Lowell Observatory.

whose true luminosity is equivalent to about 50 suns. The
eleven K-type giants in the cluster are nearly of comparable luminosity, with absolute magnitudes ranging from +0.5
to about +1.5. An interesting feature of the H-R diagram is
the scattering of brighter stars (mags 11- 11½) forming a
horizontal branch across the diagram. These are stars which
have evidently passed through the red giant stage, and are
now evolving back toward the left on the diagram, growing
bluer and hotter. A similar "horizontal branch" is a well
known feature of the H-R diagram of a globular cluster;
but in M67 this feature lies about a magnitude lower on the
diagram, near absolute magnitude +1.5 rather than +0.5. The
evolved stars of M67 thus appear to have only half the luminosity of similar stars in a globular. The explanation is
not entirely clear, but is usually attributed to a difference in chemical composition. Typical globulars seem to be
deficient in the atoms of the metals, but the composition
of M67 is fairly comparable to that of the Sun.

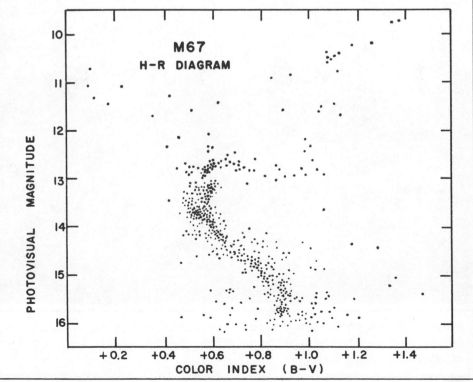

STAR CLUSTER M67. The unusual galactic cluster, as it appears on a photograph made with the 200-inch reflector at Palomar Observatory.

LIST OF DOUBLE AND MULTIPLE STARS

NAME	DIST	PA	YR	MAGS	NOTES	RA & DEC
Σ1606	0.4	288	68	6½- 7	PA & dist dec; spect A3	12083n4010
Σ1607	27.6	14	58	8½- 9	PA inc, dist dec	12090n3622
Σ1609	11.0	206	54	8 - 9½	relfix, spect F2	12092n5107
	68.6	146	11	- 13		
Σ1613	1.3	10	60	8½- 9	relfix, spect F5	12100n3602
A1999	1.0	351	60	8½- 10	cpm; PA inc; spect K2	12130n4024
2	11.4	260	58	5½ - 8	(Σ1622) relfix; spect M1, F7	12136n4056
A1781	2.7	302	55	9 - 9	spect G5	12138n4500
Σ1624	6.1	150	16	7 - 10	relfix, spect A2	12142n3952
Σ1632	10.2	193	30	6½- 9½	relfix, spect K0	12178n3811
Ho 356	3.5	97	38	8½- 9½		12182n3517
0Σ250	0.4	344	62	7½- 8	slow PA inc, spect F0	12220n4322
Σ1642	2.5	180	63	8 - 9	relfix, spect F5	12233n4501
Σ1646	5.2	257	36	8½- 11	relfix	12257n3658
Σ1645	10.2	158	63	7 - 7½	neat pair, relfix; cpm; spect F5 NGC 4460 in field	12257n4504
0Σ251	0.2	22	21	8 - 9	binary, about 600 yrs; spect F8	12266n3140
J1023	4.8	172	54	9½- 9½	relfix	12283n3656
Σ1653	7.8	342	28	8½- 8½	relfix, spect F2	12309n3219
Ho 256	0.8	113	59	7 - 9	PA inc, spect A3	12417n3602
Σ1679	5.7	207	37	8½- 9	relfix, spect F8	12437n5006
Σ1688	14.4	345	30	8½- 10	relfix, spect G0	12512n3815
α	19.6	228	58	3 - 5½	(12) (Σ1692) cpm; fine object, spect A0, F0; relfix (*)	12537n3835
β925	6.9	211	26	7 - 12	spect A0	12543n4349
0Σ257	13.0	353	26	8 - 8½	relfix, spect F2	12545n4553
Σ1702	35.6	83	25	8½- 10	relfix, spect K0	12562n3833
Σ1718	0.4	223	58	9- 10½	(Hu643) PA inc;	13033n5115
	13.2	272	22	- 9	spect F8	
β930	2.7	119	58	6 - 12	slow PA inc, spect K1	13036n4532

CANES VENATICI

LIST OF DOUBLE AND MULTIPLE STARS (Cont'd)

NAME	DIST	PA	YR	MAGS	NOTES	RA & DEC
Σ1723	6.6	8	34	8 - 9½	relfix, spect K0	13060n3901
15 + 17	284	297	22	6 - 6	wide optical pair; spect B7, F0	13074n3848
17	1.2	276	58	6 - 10	(β608) slight PA dec, spect F0	13078n3846
0Σ261	2.3	340	66	7 - 7½	PA dec, dist inc; spect F8	13097n3221
0Σ263	2.1	134	55	7½- 8½	relfix, spect G5	13145n5050
Hu 644	1.2	90	62	8½- 9	binary, 49 yrs; spect dM2	13176n4802
Σ1755	4.1	132	62	7 - 8	relfix, spect G5	13301n3705
0Σ269	0.3	229	66	6½ - 7	binary, 55 yrs; spect A3	13306n3510
Σ1758	3.7	299	55	8 - 8	slow PA dec, cpm; spect G0	13308n4924
h1234	30.6	14	60	6½- 11	optical, slight PA dec, spect A3	13322n3903
25	1.8	107	66	5½- 7½	(Σ1768) PA dec; binary, 240 yrs; spect A7	13352n3633
Σ1776	7.4	200	37	8 - 8	relfix, spect G0	13397n4628
0ΣΣ125	71.3	238	25	6 - 8½	spect gG5	13448n3848
h1244	6.8	131	58	6½- 11	cpm; slight PA dec, spect A0	13513n4226
Σ1789	6.5	328	63	8 - 8	relfix, spect F8	13518n3304
0Σ272	1.6	7	54	7 - 10	PA dec, spect F2	13522n3010
0Σ274	13.1	59	58	7 - 10	PA dec, spect K0	14045n3501

CANES VENATICI

LIST OF VARIABLE STARS

NAME	MagVar	PER	NOTES	RA & DEC
α	2.9--2.95	5.469	Cor Caroli; magnetic spectrum variable (*)	12537n3835
R	7.2--12.8	328	LPV. Spect M6e--M8e	13468n3947
S	9.6-- ?		Uncertain, possibly not variable; spect gK4	13108n3738
T	8.6--12.6	290	LPV. Spect M6e	12277n3147
U	7.2--11..	345	LPV. Spect M7e	12449n3839
V	6.8---8.8	192	Semi-reg; spect M4e--M6e	13173n4547
W	9.8--10.4	.5518	Cl.Var.; Spect A6--F6	14044n3804
Y	5.0---6.4	158	Semi-reg; spect N3; very red star (*)	12428n4543
RS	8.1--9.4	4.798	Ecl.bin.; spect F4 + dG8	13083n3612
RW	9.0--10..	100:	Semi-reg; spect M7	13574n3726
TT	8.5--9.2	Irr	Spect R6p	12570n3805
TU	5.8---6.7	Irr	Spect M4---M6	12527n4728
TV	9.4--10.5	Irr	Spect M5	13130n4232
TX	8.3--10.7	Irr	Z Andr.type? Spect K2+B5	12423n3702
VX	8.6---9.2	360:	Semi-reg; spect gM4	13275n4121
VZ	8.5-- 9.0	.8425	Ecl.Bin; spect G5	13297n2850
AI	5.97--6.15	.1709	Delta Scuti type, spect F0 (4 Canum)	12213n4249
20	4.72--4.75	.1217	Delta Scuti type? F0	13153n4050

LIST OF STAR CLUSTERS, NEBULAE, AND GALAXIES

NGC	OTH	TYPE	SUMMARY DESCRIPTION	RA & DEC
4111	195[1]	⊖	E7; 11.6; 3.4' x 0.8' lenticular or edge-on S0; vB,pS,mE; Ursa Major border	12045n4321
4138	196[1]	⊖	E4; 12.3; 1.4' x 0.8' B,pL,lE, gbM	12070n4358
4143	54[4]	⊖	E4; 12.1; 1.3' x 0.8' cB,lE, vgvsmbM	12071n4249
4145	169[1]	⊖	Sc; 12.2; 5.2' x 3.2' B,vL,E, vglbM, SN; fine multiple arm spiral	12075n4010
4160		⊖	S ; 12.6; 3.5' x 0.8' pB,mE,sBN; thin narrow ray; edge on galaxy	12091n4401

LIST OF STAR CLUSTERS, NEBULAE, AND GALAXIES   (Cont'd)

NGC	OTH	TYPE	SUMMARY DESCRIPTION	RA & DEC
4151	165[1]	⊖	SB or S/pec; 11.2; 2.5' x 1.6'; vB,S,E,vsmbM, BN	12080n3941
4190	409[2]	⊖	I; 13.2; 1.1' x 0.8' cF,pS,1E,vg1bM	12111n3654
4214	95[1]	⊖	I; 10.5; 7.0' x 4.5' cB,cL,E, possibly an early barred spiral	12131n3636
4217	748[2]	⊖	Sb; 11.9; 4.0' x 1.0' pF,L,mE, edge-on spiral with equatorial dust lane	12133n4722
4220	209[1]	⊖	Sa or S0; 12.3; 2.5' x 0.6' cB,pL,pmE,psbM, lens-shaped	12137n4810
4242	725[3]	⊖	S ; 11.5; 4.0' x 3.0' vF,cL,R,vgbM; dim, spiral pattern ill-defined	12149n4554
4244	41[5]	⊖	Sb; 10.7; 13.0' x 1.0' pB,vL,eE; narrow streak; edge-on spiral (*)	12150n3805
4258	43[5]	⊖	Sb; 9.0; 19.5' x 6.5' ! vB,vL,vmE,sbM,BN (*)	12165n4735
4346	210[1]	⊖	E6; 12.4; 1.7' x 0.7' F,S,mE,mbMN	12210n4716
4369	166[1]	⊖	Sa or S0; 12.4; 1.3' x 1.2' cB,S,R,mbMN	12221n3939
4389	749[2]	⊖	SB/pec; 12.8; 1.8' x 0.8' pB,pL,E,vg1bM	12231n4558
4395	29[5]	⊖	S ; 11.0; 10.0' x 8.0' vF,vL; 3-branch spiral of loose structure, triskelion shape.	12234n3349
4449	213[1]	⊖	I; 10.5; 4.2' x 3.0' (*) vB,cL,mE, rectangular shape	12258n4422
4460	212[1]	⊖	E7; 12.6; 2.0' x 0.5' B,pL,E, psbM	12264n4508
4485	197[1]	⊖	E/pec or I; 12.5; 1.3' x 0.7'; B,pS,E	12282n4158
4490	198[1]	⊖	Sc; 10.1 5.0' x 2.0' vB,vL,mE, pear-shaped; "Cocoon galaxy"; NGC 4485 is 3' to north.	12283n4155

LIST OF STAR CLUSTERS, NEBULAE, AND GALAXIES   (Cont'd)

NGC	OTH	TYPE	SUMMARY DESCRIPTION	RA & DEC
4618	178[1]	⦵	Sc/pec; 11.2; 3.0' x 2.5' B,L,E,mbM; ringtail pattern	12392n4125
4631	42[5]	⦵	Sc?; 9.7; 12.5' x 1.2' ! vB,vL,eE,bMN; edge-on spiral (*)	12398n3249
4656	176[1]	⦵	I/pec; 11.0; 19.5' x 2.0' pB,L,vmE, irregular bar with curved ends (*)	12416n3226
4736	M94	⦵	Sb; 8.9; 5.0' x 3.5' vB,L,lE,vsmbM,vBN (*)	12486n4123
4800	211[1]	⦵	SB; 12.2; 1.2' x 0.9' pB,cS,lE,psbM	12524n4648
4861	30[4]	⦵	I; 12.8; 3.0' x 1.0' vF,pL,vmE	12567n3508
4868	644[2]	⦵	Sb; 13.1; 1.1' x 1.0' pB,S,R,mbM	12568n3735
4914	645[2]	⦵	E2; 13.0; 1.0' x 0.8' pB,smbM,lE	12584n3735
5005	96[1]	⦵	Sb; 10.8; 4.7' x 1.6' vB,vL,vmE,vsmbM (*)	13085n3719
5033	97[1]	⦵	Sb; 11.0; 8.0' x 4.0' vB,pL, E; smbM	13112n3651
5055	M63	⦵	Sb; 9.8; 9.0' x 4.0' vB,pL,pmE, multiple-arm spiral, BN (*)	13135n4217
5112	646[2]	⦵	Sc; 12.6; 3.2' x 2.3' F,L,E,vglbM; 3-branch spiral	13196n3900
5194	M51	⦵	Sc; 8.7; 10.0 ' x 5.5' !! Whirlpool Galaxy; fine spiral (*)	13278n4727
5195	186[1]	⦵	I?; 11.0; 2.0' x 1.5' L,B,lE; connected to north arm of M51	13279n4731
5198	689[2]	⦵	E2; 12.9; 0.7' x 0.5' pF,pS,lE,mbM	13282n4656
5272	M3	⊕	!!! Mag 6; diam 18'; eB,vL, eRi,eC, rrr; class VI; stars mags 11... superb object (*)	13399n2838
5273	98[1]	⦵	E1; 12.7; 0.9' x 0.8' cB,pL,R,psmbM	13399n3555

LIST OF STAR CLUSTERS, NEBULAE, AND GALAXIES (Cont'd)

NGC	OTH	TYPE	SUMMARY DESCRIPTION	RA & DEC
5297	180[1]		Sb; 13.0; 5.2' x 0.8'	13443n4405
			cB,L,pmE,gbM, nearly edge-on	
5301	688[2]		Sb; 13.0; 3.4' x 0.5'	13450n4624
			cF,L,vmE; nearly edge-on	
5313	711[2]		Sb; 13.0; 1.3' x 0.7'	13477n4013
			pB,pS,vlE,glbM	
5326	712[2]		Sa or Sb; 13.1; 1.5' x 0.4'	13487n3949
			cF,S,vlE, sbM	
5347	424[2]		SBb; 13.2; 1.1' x 0.9'	13511n3343
			pF,cL,R,lbM	
5350	713[2]		Sb; 12.9; 2.0' x 1.2'	13512n4037
			cF,pL,bM	
5351	697[2]		Sb; 13.0; 2.4' x 1.1'	13512n3809
			cF,L,lE,vgbM	
5353	714[2]		E5; 12.3; 1.1' x 0.4'	13513n4031
			pB,S,E; NGC 5354 is 1' n	
5354	715[2]		E3; 13.0; 0.9' x 0.7'	13513n4032
			F,S,lE; 1' pair with 5353	
5362	671[2]		Sb; 13.2; 1.8' x 0.7'	13528n4130
			pB,pL,E	
5371	716[2]		Sb; 11.5; 3.7' x 3.0'	13536n4043
			pB,L,R,bM; fine spiral	
5377	187[1]		Sa; 12.0; 3.0' x 0.6'	13543n4727
			B,L,mE,smbMN; outer ring	
5380	698[2]		SB; 13.2; 0.5' x 0.5'	13548n3751
			F,cS,R,mbM	
5383	181[1]		SBb; 12.7; 2.2' x 2.0'	13550n4205
			cB,cL,R,gbM. fine barred	
			spiral (*)	
5395	190[1]		Sb; 12.7; 2.1' x 1.0'	13565n3739
			cF,cL,E,lbM; distorted	
			structure; interacting pair	
			with 5394?	
5394			Sb; 13.5; 0.5' x 0.5'	13565n3740
			S,F; on n tip of 5395	
5406	699[2]		Sb; 13.0; 1.4' x 1.0'	13582n3909
			F,pS,vlE; lbM	
5444	417[2]		E1; 13.1; 0.7' x 0.6'	14012n3522
			pB,pL,vlE, vsmbM	

DESCRIPTIVE NOTES

ALPHA   (12 Canum) Mag 2.89 (slightly variable);
Spectrum A0p or B9.5p. Position 12537n3835.
Name- COR CAROLI, "the Heart of Charles". The popular story
is that the star was so named by Halley in honor of King
Charles II of England. According to R.H.Allen, "This was
done at the suggestion of the court physician Sir Charles
Scarborough, who said that it had shone with special brilliance on the eve of the King's return to London, May 29,
1660." According to Deborah J.Warner of the National
Museum of History and Technology in Washington, however,
the original name was "Cor Caroli Regis Martyris" honoring
the executed Charles I; the name, however, was probably not
in wide use until the restoration of the British monarchy
under Charles II, following the Cromwellian period. The
attribution of the name to Halley appears in a report published by J.E.Bode at Berlin in 1801, but seems to have no
other verification.
    Alpha Canum also marks the position of "Chara", one of
the two hunting dogs in the mythological outline of the
constellation. The other dog is named "Asterion" and is
marked by Beta.
    Cor Caroli is one of the most attractive double stars
for the small telescope, and is a favorite of observers,
despite the fact that the color contrast - if any- is very
slight. R.H.Allen called them flushed white and pale lilac;
to Miss Agnes Clerke they were "pale yellow and fawn". T.W.
Webb states that John Herschel saw no color contrast in the
pair whereas Dembowski recorded the odd color impression of
"pale olive blue" for the fainter star. Webb himself called
the color of the fainter star a "pale copper" which agrees
better with the known spectral type of F0 V. The two stars
are slightly under 20" apart, and the magnitudes are 2.89
and 5.60. No change in either PA or separation has been
detected since the early measurements of F.G.W.Struve in
1830, but the stars definitely form a physical pair, as
they share a common proper motion of 0.24" per year in PA
282°. The projected separation is about 770 AU.
    Yale parallax measurements indicate a distance of about
120 light years, giving actual luminosities of about 80 and
7 suns. A slightly greater distance of about 135 light
years is quoted by Donald H.Menzel in his "Field Guide to
the Stars and Planets". The smaller distance appears to

agree better with the assumed luminosity calculated from
the spectral features, about 0.0 to +0.6. Alpha Canum shows
a radial velocity of about 1.8 miles per second in approach
and the space motion suggests that the star may be a member
of the Taurus stream associated with the Hyades cluster.

The primary star of this classic pair is of special
interest to the astrophysicist; it is the standard example
of a magnetic spectrum variable. H.Ludendorff in 1906
reported the variability of certain metallic lines in the
spectrum; in 1913 it was found by A.Belopolsky that there
are periodic changes in the intensity of various spectral
lines- notably those of chromium and europium. One group
of lines grows strong and the other faint, in an alternate
rhythm with a period of 5.46939 days. A slight variation
in light, of about 0.05 magnitude, was detected in the
course of this cycle by P.Guthnick and R.Prager in 1914.
Maximum intensity of the europium features coincides with
maximum light. There is also a slight change in color; the
star is bluest at minimum. Cor Caroli is also noted for the
overabundance of the atoms of the metals in general, and of
the "rare earths" elements in particular. The star has a
remarkably intense magnetic field which varies periodically
with the changes in the spectral lines. H.Babcock and S.
Burd (1952) found a range of +5000 to -4000 gauss. These
changes are all very complex, and while the processes
responsible are still largely unknown, it now appears that
it may be possible to analyse the star in terms of a "magnetic oscillator model". This view regards the spectral
changes as the result of the motions of stratified layers
of the star in response to the varying magnetic field.
These objects form a very rare class of variable stars;
the Moscow General Catalogue (1958) listed no more than 9
known examples, but in the newer (1971) edition the number
had grown to 28. Among the brightest other examples are
Epsilon Ursa Majoris, Iota Cassiopeiae, Chi Serpentis,
Kappa Piscium, 56 Arietis, and Beta Corona Borealis. The
peculiar star Gamma Bootis, sometimes regarded as a member
of this class, is still something of a puzzle, but is not
a typical member. J.S.Glasby (1969) lists it among the Beta
Canis Majoris stars, while the 1971 Moscow Catalogue has it
placed among the Delta Scuti variables! The exact classification of many of these odd stars is still uncertain.

DESCRIPTIVE NOTES (Cont'd)

Y   Variable. Position 12428n4543. Mag 4.8 (max).
      Spectrum N3. A bright semi-regular variable
star, often honored with the somewhat poetic title given by
Father Secchi- "La Superba", in reference to the splendid
appearance of its spectrum. Miss Agnes Clerke (1905) spoke
of the "extraordinary vivacity of its prismatic rays, separated into dazzling zones of red, yellow, and green, by
broad spaces of profound obscurity". The star is #152 in a
list of unusually red stars compiled by Schjellerup in
1866. It lies in a rather blank region of the constellation
between Cor Caroli and the stars of the Great Dipper, but
can be located and identified without difficulty from its
unusual color. The position is about 35% of the distance
along a line drawn from Cor Caroli to Delta Ursa Majoris.
This is one of the reddest of all the naked-eye stars, and
shows a truly odd and vivid tint in large telescopes.
According to the Arizona-Tonantzintla Catalogue (1965) the
star has a color index (B-V) of 2.55 magnitudes, and the
difference between the visual and ultraviolet magnitudes
is 9.16 mags! The extreme faintness of the blue and ultraviolet portion of the spectrum is due chiefly to very
strong molecular absorption, apparently by the tri-atomic
molecule C_3 , according to studies by A.McKellar and E.H.
Richardson (1954). La Superba is thus one of the "carbon
stars" of spectral type N, the reddest of all known stars.
The famous "Crimson Star" R Leporis belongs to this class,
in which the bands of carbon compounds appear in the spectrum. Water vapor has also been recently detected in the
atmospheres of some of these stars, one of the results of

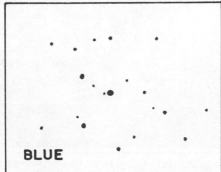
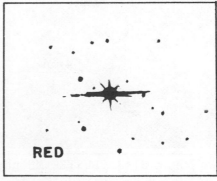

BLUE   RED

Y CANUM. The unusual color of the star is illustrated here
by a comparison of red and blue exposures.

GLOBULAR STAR CLUSTER M3 photographed (top) with an 8-inch reflector by Kent de Groff, and (below) with the 200-inch telescope at Palomar.

an abnormally low surface temperature. Some investigators
have suggested that the stars of types N and R should be
assigned to the same new spectral class to be called "C",
the Carbon Stars. Subdivisions range from C0 to C9, with a
subscript sometimes added to indicate the carbon abundance.
With this system the classification of Y Canum is C5$_4$.

The maximum visual brightness of the star is about 4.8
and the visual range is about 1.5 magnitudes in a semi-
periodicity averaging 160 days. According to the new Moscow
Catalogue (1971) a longer superimposed period of about 2100
days may be involved in the cycle. The photographic range
is 8.2 to about 10.0.

An attempt at a direct trigonometrical parallax has
yielded no result, proving that the distance must be in the
range of 400 light years or more. J.H.Moore at Mt.Wilson
found evidence in 1923 that the absolute magnitudes of the
N stars range from -1.5 to about -2.4; Y Canum is thus a
giant star, perhaps comparable to Mira in size. It is also
among the coolest stars known, with a surface temperature
of about 2600°K. The annual proper motion has been measured
at 0.01"; the radial velocity is about 7.5 miles per second
in recession. (Refer also to R Leporis, TX Piscium, and
S Cephei)

M3   (NGC 5272). Position 13399n2838. A beautiful
            bright globular star cluster, one of the most
splendid in the sky. It was discovered by Messier in 1764
and can be seen as a hazy 6th magnitude "star" in field
glasses. The small telescope shows it as a round nebulous
object about 10' in diameter, but the apparent size is
nearly doubled on the best photographic plates. At least
a 4-inch telescope is needed to partially resolve the outer edges, and a good 6-inch glass with a fairly high power
will reveal hundreds of stellar points. Large telescopes
show an incredible swarm of countless star images, massing
to a wonderful central blaze, with glittering streams of
stars running out on all sides. The seeming arrangement of
the outer members into radiating streams and branches was
noticed by both the Herschels and Lord Rosse; a similar
pattern is evident in the great Hercules cluster M13 and
in other bright globulars. Rosse found several small dark

GLOBULAR STAR CLUSTER M3. One of the finest clusters of this type. This photograph was made with the 120-inch reflecting telescope at Lick Observatory.

obscuring patches in the central mass, more or less verified on modern photographs, but more definitely present in
the Hercules system M13. Their presence is of much interest
since globulars are usually thought to contain virtually no
dust or gas. True association with the cluster, however, is
obviously very difficult to prove. Regarded as foreground
objects, it may be argued that such small dark nebulae are
present elsewhere in the sky but cannot be detected; such
star swarms as M3 and M13 obviously make perfect backdrops
to reveal their presence.

M3 contains many thousands of stars, from magnitude
11 or so to the limit of detectability. The number of variable stars in M3 is greater than in any other globular,
189 of these stars having been detected up to 1963. Periods
of the majority of these stars average about half a day,
and the light may change so rapidly that in one case it
doubles in less than 10 minutes. Stars of this type, called
"cluster variables", appear to be a sub-class of the well
known cepheids. The typical example is the star RR Lyrae.
In M3, these stars have apparent magnitudes of $15\frac{1}{2}$. Since
their absolute magnitudes are known from other studies to
lie in the range 0.5 to +1.0, the distance modulus is seen
to be about 15 magnitudes. From this method, the distance
of M3 is in the range of 35,000 to 40,000 light years. The
actual diameter is about 220 light years, the total luminosity about 160,000 times the light of the Sun, and the
absolute magnitude close to -8.2. According to H.B.Sawyer's
"Bibliography of Individual Globular Clusters" (supplement
1963) the integrated spectral type of M3 is F7, the total
photographic magnitude is 7.2, and the radial velocity is
about 90 miles per second in approach.

At Palomar Observatory, more than 45,000 stars have
been counted in this cluster, down to a magnitude of $22\frac{1}{2}$.
The faintest stars reached in this survey were about 1/6
the luminosity of the Sun; if the Sun could be removed to
the same distance it would appear of magnitude 20.4. From
the work of the Palomar astronomers, the total mass of M3
is about 140,000 times the mass of the Sun, and the total
population is probably at least half a million stars. It is
interesting to note, however, that the brightest stars of
the cluster are not the chief contributors to its mass.
Over 90% of the light of the swarm is given by relatively

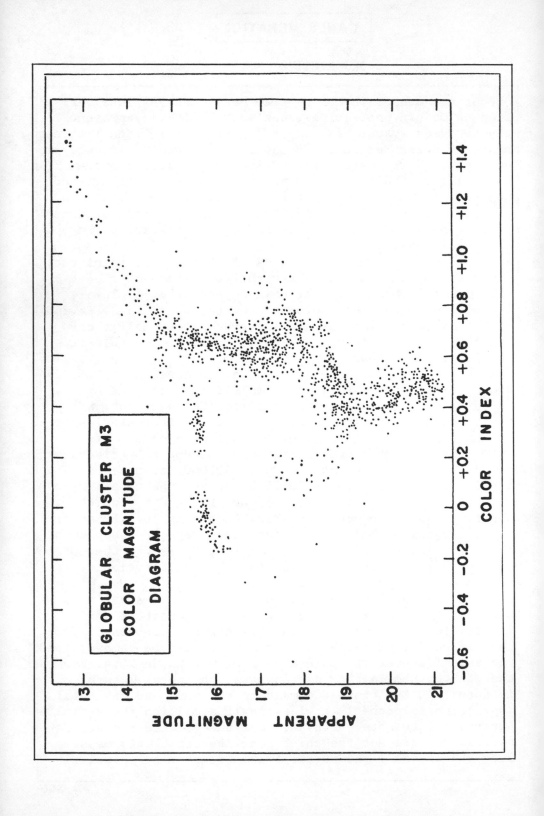

GLOBULAR CLUSTER M3
COLOR MAGNITUDE
DIAGRAM

few stars, those of absolute magnitude $+3\frac{1}{2}$ and brighter.
But the total mass of the cluster is accounted for mainly
by the vast numbers of fainter stars.

M3 shares with M13 and M5 the distinction of being one
of the three brightest globulars in the northern sky, and
has probably been more thoroughly studied than any other
cluster of the type. As a result of the work of A.Sandage,
H.Arp, and W.Baum at Palomar, the color-luminosity relationship of all the members above 21st magnitude is known
with good accuracy, and is shown in the H-R diagram on
page 366. The pattern of plotted points is quite unlike
that of a typical H-R diagram for stars near the Sun, and
indicates that the globulars are extremely ancient star
groups. In the article on M13 in Hercules, the fundamental
facts about cluster age-dating are given in some detail;
here it may be of some interest to call attention to the
characteristic features of the M3 diagram. Stars of 19th
magnitude and fainter are still main sequence objects. In
a normal group of Population I stars, this main sequence
would continue on toward the upper left portion of the diagram, into the region of bright blue stars. In M3, stars
of this type are missing; instead there is a population of
subgiants and giants forming a branch which leads to the
bright red giants in the upper right portion of the graph.
Evidently, all stars brighter than 19th magnitude have
evolved into subgiants and giants, and the form of this
giant branch reasonably duplicates the theoretical evolutionary tracks of older stars. In this way, it can be seen
that the features of the graph imply great age. Finally
there is the so-called "horizontal branch" at apparent
magnitude $15\frac{1}{2}$; these are evidently stars which have formerly passed through the red giant stage, are now in the
next or "helium-burning" stage, and have become hotter and
bluer. In the prominent "gap" in this branch occur the
pulsating cluster type variables, again verifying the theory that this type of variability is a feature of the later
stages of a star's evolution. From the various lines of
evidence, it is believed that a star cluster such as M3
may be something like 10 billion years old. Only a very few
galactic clusters, such as NGC 188 in Cepheus and M67 in
Cancer- appear to be of comparable age. (See also M13 in
Hercules, M5 in Serpens, and NGC 5139 in Centaurus)

SPIRAL GALAXY M51. The field of the "Whirlpool Galaxy", photographed with the 13-inch telescope at Lowell Observatory.

M51   (NGC 5194) Position 13278n4727, about 3½° SW from Eta Ursa Majoris, the end star in the handle of the Great Dipper. This is the famous "Whirlpool Galaxy", the first galaxy found to show a spiral form. It was discovered by Messier in October 1773, and the intriguing spiral pattern was first detected by Lord Rosse with his giant 6-foot reflector at Parsonstown, Ireland, in 1845. Rosse published his drawing of the object in 1850; it seems that he had observed the galaxy previously with a 3-foot telescope and had missed the spiral pattern. Sir John Herschel, with his 18-inch reflector, had described a "very bright round nucleus surrounded at a distance by a luminous ring". The discovery of the spiral pattern aroused much interest, and was regarded by some 19th century students of cosmology as a confirmation of Laplace's Nebular Hypothesis. Thus the "spiral nebulae" were at first thought to be new solar systems in the process of formation, and it was not until 1923 that the question was settled with finality. The spirals were now recognized as external galaxies, and the modern picture of the Universe began to emerge.

M51 is an Sc type spiral, about 35 million light years distant, of the 8th magnitude visually, and about 10' in apparent diameter. As one of the nearest and brightest of the galaxies, and the one which shows the best-defined spiral structure, the Whirlpool is of great interest to all observers, though very little detail may be seen except in fairly large telescopes. A good pair of binoculars will show the object on a clear dark night, and a 2-inch glass will reveal a hazy patch of light with a brighter center. With a 6-inch glass the central nucleus appears prominently and dominates the misty glow of the system. The spiral form may be glimpsed, under the best conditions, and with some uncertainty, in an 8-inch telescope. In a 10-inch it may be held unmistakably when atmospheric conditions allow, and in a 12-inch the spiral coils begin to resemble the familiar photographs which have graced countless astronomy texts. The greatest telescopes resolve the spiral arms into a vast complex of star clouds, bright and dark nebulosity, individual stars, and nebulous "knots" which may be star groups and clusters. The entire spiral pattern is dominated - in fact it is defined - by the narrow dust lanes which may be traced deep into the nuclear region on short exposures.

SPIRAL GALAXY M51. The Whirlpool Galaxy, photographed
with the 200-inch reflector at Palomar Observatory.

The two principal dust lanes lie on the inner edges of the two major spiral arms. Their structure is very complex with many branching filaments which often cross the associated spiral arms at nearly right angles. The arms themselves can be traced for about 1½ turns. The spiral pattern is evident to within 15" of the nucleus. Within this radius, the central mass has a mottled structure, and appears to break up into a number of separate cloudlets, divided by thin dust lanes. The actual nucleus is about 2.7" in diameter, and appears nearly stellar; the true diameter must be about 450 light years.

A study of the radial velocities at various locations in M51 has been made by E.M. and G.R.Burbidge (1964); they derive a total mass of about 60 billion solar masses for the galaxy, out to the visible radius of some 18,000 light years. E.Holmberg, using the newer distance determination, finds the total mass to be about 160 billion suns, and the true diameter to be slightly over 100,000 light years. M51 now appears to be more or less the equal of the Andromeda System M31; the total luminosity is about 10 billion suns.

F.Zwicky (1955) has experimented with a technique of superimposing negatives taken in different colors, revealing many interesting details in the structure of M51. In the examples shown on page 372, the upper print was made by superimposing a blue negative on a yellow positive. The details of the spiral arms are strikingly brought out by this technique, which accentuates the "blue" details of the galaxy. On the lower print, a yellow negative and a blue positive have been superimposed; the distribution of the red and yellow stars is shown by this method. A very interesting feature is the radically different appearance of the satellite galaxy, NGC 5195, on the two prints.

Conspicuous in the small telescope, this satellite system gives the appearance of being attached to the north end of the spiral arm of M51. Evidently it does not lie exactly in the plane of the big spiral, since dust lanes of the M51 arm may be seen crossing in front of it. There are also some dust patches on the opposite side, believed to be directly associated with the smaller galaxy itself. The classification of this peculiar system is uncertain. In the Hubble Atlas of Galaxies, A.Sandage (1961) refers to it as an irregular galaxy of the M82 type. Some long-

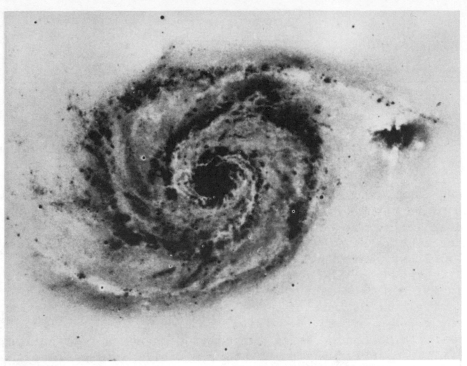

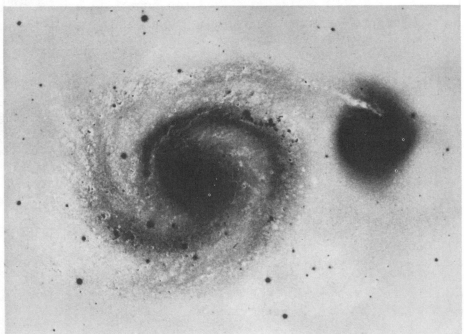

SPIRAL GALAXY M51. Differences in the structure as shown by a technique of superimposing negatives. Top: The blue details. Below: The red details. Palomar Observatory.

DESCRIPTIVE NOTES (Cont'd)

exposure photographs show faint outer filaments which seem
to suggest the structure of an incipient barred spiral. On
the other hand, E.M. and G.R.Burbidge (1964) have classed
it as an S0 system. If the superimposed dust clouds and the
obscuring matter connected with the arm of M51 were removed, the system would probably resemble an elliptical galaxy. Its light is much redder than that of M51, and true
resolution does not appear to have been achieved with any
present telescope. The corrected radial velocities of the
two objects are fairly comparable; 340 and 390 miles per
second in recession.

M63   (NGC 5055) Position 13135n4217. A bright oval
spiral galaxy of about the 10th magnitude, 9' x
4' in size, located some $5\frac{1}{2}°$ southwest of the Whirlpool.
It is easily found by sweeping an area about midway between
Cor Caroli and the end star of the Great Dipper's handle.
M63 was discovered by Mechain in 1779. The 8^m star shown
on the photographs lies 3.6' west and slightly north.
M63 is a very fine Sb type spiral, oriented about 30°
from the edge-on position. It has a very bright central
condensation measuring about 6" in diameter. This nucleus
is encircled by a bright, tightly coiled system of spiral
arms out to a radius of about 50". Here there is a sudden
drop in the surface brightness, and a second pattern of
spiral arms continues to sweep outward in a series of magnificent sprays of star clouds. The outer arms are rather
reminiscent of showers of sparks thrown out by a rotating
fiery pinwheel. To others, the structure apparently resembles some vast celestial flower, since the galaxy has received the popular name of the "Sunflower". The sudden
discontinuity in the brightness between the inner and outer
spiral features is the chief characteristic of the M63 type
of galaxies, which are known as "multiple-arm spirals". The
very beautiful spiral NGC 2841 in Ursa Major would have a
very similar appearance if its outer arms were a little
more loose in structure. Many of the cloudlets and condensations in the arms have been identified as regions of
bright nebulosity.
The distance of M63 is not well determined, but the
red-shift of 345 miles per second (corrected for the solar
motion) suggests a distance in the vicinity of 35 million

SPIRAL GALAXY M63. Two exposures with the 60-inch and the 100-inch reflectors, showing the inner and outer spiral features.   Mt.Wilson and Palomar Observatories.

SPIRAL GALAXY NGC 4258. A bright Sb type spiral in Canes Venatici; photographed with the 200-inch reflector at Palomar Observatory.

SPIRAL GALAXY M94. A very bright, compact spiral in Canes
Venatici; photographed with the 200-inch reflector at
Palomar Observatory.

light years. The actual diameter may be about 90,000 light years, and the total luminosity equal to some 10 billion suns. Radial velocity measurements across the body of the galaxy have been made with the 82-inch reflector of the McDonald Observatory by E.M. and G.R.Burbidge and K.H. Prendergast (1960). The resulting rotation curve implies a total mass of about 115 billion solar masses for the new revised distance of 10.7 megaparsecs. The absolute magnitude for this same distance is close to -20.

M94 (NGC 4736) Position 12486n4123. A bright, very compact and nearly circular spiral galaxy of the 9th magnitude, discovered by Mechain in 1781, and easily found in small telescopes since it forms an isosceles triangle with Alpha and Beta Canum, and lies about $1\frac{1}{2}°$ above a line joining them. Admiral Smyth called M94 a "comet-like nebula; a fine pale-white object" and thought it might be a compressed cluster of small stars. This galaxy is indeed remarkable for the intense brilliance of the central core which measures about 30" in diameter. At the edge of this featureless central hub, a pattern of tightly wound spiral arms emerges, continuing out to a radius of about 60". In these closely packed whorls appear many irregular dust patches which are not shown on most photographs due to very strong over-exposure. At the outer edge of the first spiral zone, a second system of arms begins, continuing outward through a region of greatly decreased luminosity. In the second zone, the pattern of spiral arms is less well-defined than in the region near the hub, but is still quite compact and tight when compared to such "open" spirals as M33 (The Pinwheel) in Triangulum. At the outer edge of the second zone, a little more than 3' from the nucleus, the spiral pattern fades away and the outer boundary appears to have been reached. However, a very faint outer ring may be seen on the best photographs, beginning at 4.3' out from the center. This peculiar feature is not entirely detached from the main body, but contacts it on the west side.

M94 shows a red-shift (corrected) of about 210 miles per second, less than M51 or M63, and suggesting a distance of close to 20 million light years. The resulting diameter is then about 33,000 light years and the total luminosity about 8 billion times the light of the sun.

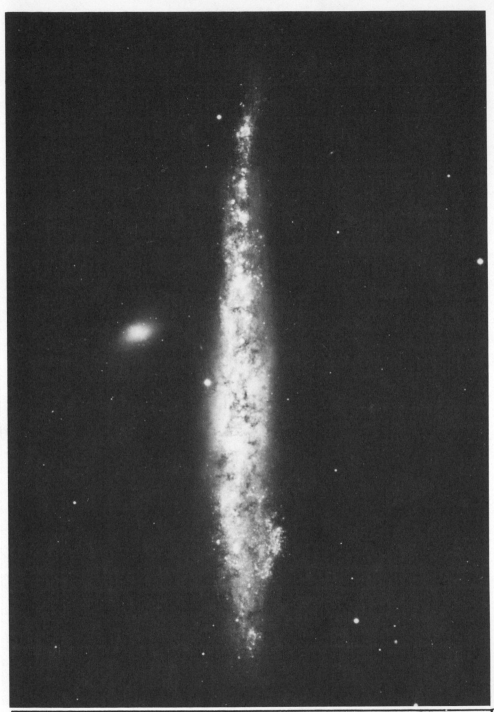

NGC 4631. One of the largest of the edge-on galaxies, believed to be an Sc type spiral. Palomar Observatory 200-inch reflector photograph.

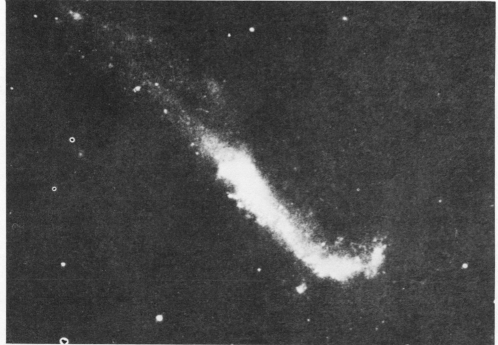

GALAXIES in CANES VENATICI. Top: The fine spiral NGC 5005, photographed with the 100-inch telescope. Below: Irregular system NGC 4656, photographed with the 60-inch telescope.

Mt. Wilson Observatory

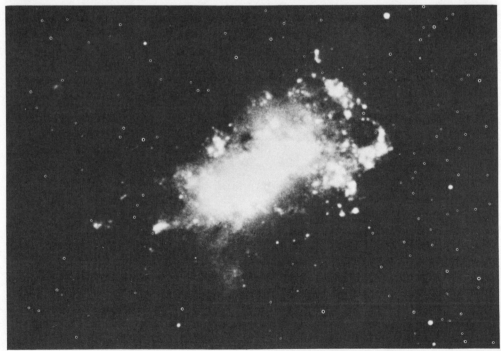

GALAXIES IN CANES VENATICI. Top: The edge-on spiral NGC 4244, photographed with the 200-inch telescope. Below: The irregular system NGC 4449, photographed with the 100-inch telescope.   Mt.Wilson and Palomar Observatories

CANIS MAJOR

LIST OF DOUBLE AND MULTIPLE STARS

NAME	DIST	PA	YR	MAGS	NOTES	RA & DEC
I 349	6.3	43	33	7 - 11	Spect A0	06126s2935
β18	1.5	284	66	7½- 9	PA inc? Spect F2	06144s1202
h3845	35.5	19	07	6 - 10	optical, spect dG0	06149s2242
B110	6.7	49	32	7½- 11	spect F5	06156s2425
Σ3116	4.2	23	35	6 - 10	relfix, cpm pair; spect B2e	06191s1145
β568	0.8	155	66	7 - 7½	relfix, spect B8	06216s1945
Σ903	23.2	295	09	6- 10½	relfix, cpm; spect B8	06220s1256
S518	16.6	88	17	8 - 10	relfix, spect A3	06221s1612
A3029	4.9	197	26	7½-12½	spect B8	06253s1516
β753	1.3	45	42	6 - 7½	relfix, spect B3	06268s3220
h3863	2.6	119	59	7 - 9	relfix, spect A2	06273s2233
h3864	21.6	44	20	7½-10½	relfix, spect B8	06282s1455
Gal 242	36.6	66	02	6 - 11	spect B2	06283s1306
h3866	4.3	138	33	8½- 10	perhaps slight PA inc; spect F5	06283s2406
h2321	5.6	307	46	8 - 9	relfix, spect F2	06292s2035
h3869	24.9	258	30	5½- 7½	relfix, spect B3, A0	06308s3159
Ho 234	0.3	298	62	8 - 8	binary, 170 yrs; spect F0	06322s1111
h3871	7.6	354	55	7½- 8	relfix, spect A2	06322s2935
I 479	1.1	49	42	8½- 9	spect F8	06336s2600
Hwe 13	11.2	300	00	8 - 9	spect B9	06336s1605
ν¹	17.5	262	26	6½- 8	(Hh 239) relfix; cpm; spect G8, G0	06342s1837
h3876	9.1	336	33	7 - 10	relfix, spect B8	06346s2234
Hn 80	4.1	131	46	9 - 9	relfix, spect F8	06349s1413
β319	3.6	166	56	6½- 9	relfix, spect B9	06397s1557
β195	5.8	215	33	7 - 11	spect B5	06404s2311
	34.9	178	17	- 12		
S534	18.2	143	51	6½- 8	relfix, spect F0	06408s2224
α	10.8	72	68	-1; 8	SIRIUS, binary; 50 yrs; spect A1, DA (*)	06430s1639
I 179	4.5	222	33	6½- 10	relfix, spect B8	06431s3032
h3891	4.9	223	59	6 - 8½	relfix, spect B3	06436s3054
Stn 15	2.5	142	38	9 - 9	relfix, spect A0; in cluster M41	06441s2042

CANIS MAJOR

LIST OF DOUBLE AND MULTIPLE STARS   (Cont'd)

NAME	DIST	PA	YR	MAGS	NOTES	RA & DEC
Σ969	7.2	317	23	7 - 10	relfix, spect B9	06456s1103
Σ971	1.2	325	67	8 - 8	slight PA dec; spect G0	06461s1323
β20	3.1	32	43	7½- 11	relfix, spect K0	06466s1609
AC 4	0.9	307	56	6 - 8½	PA inc, spect B6	06467s1505
B120	2.0	278	43	7½-10½	spect K2	06471s2600
β324	1.8	206	42	7 - 8	relfix, spect A1	06477s2401
	30.5	281	33	- 9		
Δ36	42.9	66	19	5½ -7½	relfix, spect B8, A3	06485s3139
β325	1.7	36	35	8 - 9	spect B5	06498s2631
λ71	10.4	99	00	6 - 13	spect M4	06510s2654
Σ990	3.3	275	38	8½- 9½	relfix, spect B9	06521s1411
17	44.4	147	15	6 - 9	relfix, spect A2, K5	06529s2020
	50.5	184	15	- 9		
19	11.6	18	33	5½- 9½	relfix, spect F2; cpm (π)	06535s2004
μ	3.0	340	44	4½- 8	(Σ997) relfix; spect G5, A2	06538s1359
Stn 16	3.7	97	33	7½-10½	relfix, spect F0	06538s2527
S541	23.2	44	29	8 - 9	relfix, spect K0, F	06545s2235
Ho 517	2.9	319	59	7- 13	PA dec, spect F5	06548s1922
I 432	1.3	203	55	8½- 8½	PA dec, spect G0	06549s2838
B122	1.0	264	59	5½- 7½	cpm; spect dF0	06555s2434
B707	0.2	20	59	6½- 7½	spect B3	06561s2706
ε	7.5	161	51	1½- 8	relfix, spect B2 (*)	06567s2854
Hd 198	35.0	315	00	6½- 9	spect B8	06568s3056
	70.0	320	01	- 10		
B708	3.3	34	44	7½-11½	spect B8	06570s2624
λ73	0.3	3	44	8 - 8	slow PA inc; spect A2	06576s2749
λ74	0.1	126	59	7 - 7	spect B5	06582s2203
	13.2	231	00	- 13		
β572	5.2	143	33	7 - 11	relfix, spect A3	06584s2034
Σ1011	4.3	298	15	8 - 8½	relfix, spect A2	06586s1515
I 183	3.4	144	33	7½- 10	relfix, spect B5	06587s2534
Hu 112	0.6	189	46	7½- 8	relfix, spect B3	06595s1114
ADS 5736	8.7	87	08	7 - 12	spect K0	07006s1637

CANIS MAJOR

LIST OF DOUBLE AND MULTIPLE STARS (Cont'd)

NAME	DIST	PA	YR	MAGS	NOTES	RA & DEC
β328	0.6	116	50	6 - 7	PA dec,spect B0	07043s1113
	17.8	350	58	- 9	(Σ1026)	
A3043	0.1	13	52	7½- 8½	PA inc, spect G0	07056s1537
Σ1031	3.7	249	15	8½- 9	relfix, spect A2	07063s1355
	12.0	351	08	- 12		
β329	32.5	100	15	6½-11½	spect B1	07073s1609
h3934	13.6	236	37	7 - 8½	relfix, spect B5	07092s2143
h3934b	0.6	273	49	9 - 11	(RST 4840)spect B9	
h3938	19.7	250	39	7 - 9½	relfix, spect B3, A5	07117s2249
B131	2.8	39	28	7½- 14	spect K0	07119s2326
β575	0.7	75	59	8 - 8	(Σ1057) PA inc,	07125s1523
	15.6	2	19	- 10	spect F8	
Cor 48	7.8	10	40	8 - 10	spect B9	07127s3124
h3945	26.6	55	59	5 - 7	slight PA & dist dec; spect M0, F0; splendid colors	07145s2313
Hu 113	1.8	54	23	8 -12½	spect B8	07148s1354
	8.6	125	23			
A2123	0.3	30	58	7½- 7½	PA inc, spect F5	07148s1157
	15.6	240	30	- 9½	(Σ1064)	
Brs 2	37.9	182	28	6½- 8	spect A5	07150s3048
τ	8.2	90	57	4½- 10	(30) relfix, in	07166s2452
	14.5	79	00	- 11	cluster NGC 2362; spect 09 (*)	
h3950	3.9	346	49	8 - 8	relfix, spect A2	07172s2157
λ76	8.8	217	33	6 - 13	spect B3	07189s2652
λ78	0.6	200	59	7½- 9	(B719) PA inc;	07209s2540
	2.8	291	36	- 11	spect M0	
Kui 29	2.7	310	37	6½- 11	spect A2	07226s1855
Δ47	1.9	309	49	5½-11½	spect gK2	07228s3143
	99.2	342	22	- 7½		
β199	1.8	22	57	7 - 8	relfix, spect B2	07230s2104
	6.7	117	59	- 13		
β578	1.8	46	42	6 - 11	PA & dist dec; spect F0	07249s1746
Σ1097	0.7	175	58	6 - 8	PA slow inc; spect G8, B, B; (β332)	07255s1127
	20.0	313	58	-8½		
	23.2	157	58	-9½		

CANIS MAJOR

LIST OF VARIABLE STARS

NAME	MagVar	PER	NOTES	RA & DEC
β	2.0---	.250	Spectrum variable (*)	06205s1756
ξ¹	4.3--4.36	.2096	Spect B1; Beta Canis type	06298s2323
15	4.66--	.1846	Spect B1; Beta Canis type	06514s2010
R	5.9---6.6	1.136	Ecl.bin.; spect F1, G	07172s1618
T	8.7--11..	310	Semi-reg	07194s2521
U	9.0--11..	305	Semi-reg	06168s2609
V	9.5--12.0	243	LPV.	06416s3144
W	7.0--8..	Irr	Spect N	07057s1151
X	8.5--10..	107	Semi-reg; spect M6	06547s2354
Z	8.9--11.3	Irr	Erratic, spect Bep	07014s1129
RV	9.0--10..	Irr	Spect M6	06584s1417
RY	7.8---8.7	4.678	Cepheid; spect F6--G2	07143s1124
RZ	9.5--10.3	4.255	Cepheid; spect F6	07193s1635
SS	9.5--10.4	12.362	Cepheid; spect F6--G2	07241s2509
SU	9.9--13..	248	LPV. Spect M6	06592s1859
SW	9.4--9.9	10.092	Ecl.bin.; spect A8	07061s2221
SY	9.1--13.0	219	LPV.	07084s1945
TU	9.7--10.7	1.1278	Ecl.bin.;	06296s2408
TW	9.0--10.3	6.994	Cepheid; spect F5--F8	07198s1413
TX	9.6--10.8	2.397	Ecl.bin.; spect A	06113s2232
TZ	9.8--10.5	1.911	Ecl.bin.; spect A0	06396s1937
UV	9.5--13..	340	LPV.	07032s2814
UW	4.7---5.0	4.393	Supergiant Ecl.bin. (*)	07166s2428
VW	9.0---9.2	.7208	Ecl.bin.; lyrid	07104s2526
VY	8.8---9.3	Irr	Spect M3e	07209s2540
VZ	9.2---9.7	3.126	Cepheid	07244s2549
WW	9.0--9.5	.499?	class uncertain; spect F6---G1 Cl.Var?	06177s2138
CO	8.7---9.5	80:	Semi-reg; spect M3	07120s2601
CW	8.7-- 9.4	1.059	Ecl.bin.; lyrid	07198s2342
CX	9.9--10.6	.9546	Ecl.bin.; lyrid	07200s2547
EH	9.5--12..	290	LPV.	06153s3101
EW	4.3--4.6	Irr	(27 Canis) shell star, spect B4p + B8	07122s2616
EZ	6.9--7.0	1.01	Ecl.bin; spect W5	06521s2351
FF	7.9--8.1	.5474	Ecl.bin; spect B3	07086s3035

LIST OF STAR CLUSTERS, NEBULAE, AND GALAXIES

NGC	OTH	TYPE	SUMMARY DESCRIPTION	RA & DEC
2204	13[7]		L,pRi,1C; diam 10', about 20 faint stars, class E	06135s1835
2207			Sc; 12.3; 2.5' x 1.5' pB,pL,mE,pslbM, BN; double galaxy or interacting pair	06143s2121
2217			SBa; 12.0; 4.0' x 3.0' vB,S,1E,psbM	06187s2714
----	I.2165		Mag 12.5; diam 8"; nearly stellar	06196s1257
2223			SBb; 12.7; 3.5' x 2.5' F,pL,1E,vg1bM	06225s2249
2243			pB,S,R; diam 4'; about 50 faint stars; class F	06276s3115
2280			Sb; 12.7; 2.0' x 1.0' pF,pL,E,gbM	06428s2735
2283	271[3]		Sc; 12.8; 1.0' x 1.0' S,F,bM	06438s1809
2287	M41		vL,B,1C, mag 6; diam 30'; 50 stars mags 7... class E fine object (*)	06449s2042
2325			E4; 12.9; 1.1' x 0.7' pB,pL,E,g1bM	07007s2838
2327	25[4]		pB,L,Irr; 20' diam with L, F extensions N-S	07019s1114
2354	16[7]		L,C,cRi; diam 25'; 60 faint stars; class E	07122s2538
2359	21[5]		vF,vvL, 6' x 8' with curved filaments; central star is Wolf-Rayet type, mag 11	07154s1307
2360	12[7]		vL,Ri,pC; diam 10', mag 9; 50 stars mags 9....12; class G (*)	07154s1533
2362	17[7]		pL,Ri,cC; diam 6'; 40 stars incl 4^m star 30 Canis (*)	07166s2452
2374	35[8]		vL,1C,diam 15'; scattered field of bright stars	07217s1309
2383			pS,pmC, diam 2'; about 50 stars mags 12...	07226s2050

"......*shining forth amid the host of stars in the darkness of the night, the star whose name men call Orion's Dog*....."

DESCRIPTIVE NOTES

ALPHA Name- SIRIUS, "The Sparkling One" or the
"Scorching One", also called the "Dog Star"
and the "Nile Star". Position 06430s1639. This is the
brightest of the fixed stars, the "leader of the host of
heaven", and a splendid object throughout the winter months
for observers in the northern hemisphere. To Americans the
coming of Sirius heralds the approach of the Christmas
season and conjures up visions of sparkling frosty nights
and snow-laden fir trees; at Thanksgiving Week the star
rises at about 9:00 pm, but by Christmas Eve he may be seen
coming over the eastern horizon by 7:00. On New Year's Eve
he dominates the southern sky, reaching culmination just at
midnight.

Sirius is 9 times more brilliant than a standard first
magnitude star. A magnitude of -1.58 has been quoted for
years in standard texts, but it now seems that the figure
is somewhat in error, due to the lack of any comparison
stars of comparable splendor. The most accurate of modern
observations indicate a magnitude of -1.42. T.W.Webb states
that Sirius has been observed at noon with an aperture of
one-half inch, and that Hevelius and Bond both perceived it
by day. In any good telescope, Sirius is a truly dazzling
object; to the Herschels the approach of the star to the
field of their great reflectors was heralded by a glow
resembling a coming dawn, and its actual entrance was almost
intolerable to the eye. In color the star is a brilliant

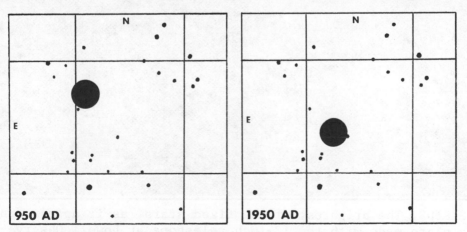

THE PROPER MOTION OF SIRIUS over a period of 1000 years
is illustrated here. Grid squares are 1° on a side.

SIRIUS. The brightest of the fixed stars, as it appears on a plate made with the 13-inch telescope at Lowell Observatory.

white with a definite tinge of blue, but in its rapid
scintillation it often seems to flicker with all the colors
of the rainbow. This is a purely atmospheric phenomenon,
of course, and is most noticeable when the star is at a low
altitude. *"He comes richly dight in many colors,"* wrote
Martha E. Martin (1907) in her charming and informal book
"The Friendly Stars", *"twinkling fast and changing with
each motion from tints of ruby to sapphire and emerald and
amethyst. As he rises higher and higher in the sky he gains
composure and his beams now sparkle like the most brilliant
diamond- not a pure white, but slightly tinged with irid-
escence."*

Sirius is an A1 type main sequence star about 23
times the luminosity of the Sun, 1.8 times the diameter,
and 2.35 times the mass. The surface temperature is about
10,000°K and the central temperature is computed to be over
20 million degrees. The star has a yearly proper motion of
1.324" in PA 204°; in the last 2000 years it has thus
changed its position by 44' or about 1½ times the apparent
width of the Moon. This motion was first detected by Halley
and was announced in 1718; he found that the positions of
Sirius, Arcturus, and Aldebaran were clearly different
from those given in the catalogues of Ptolemy and other
ancient records. Thus was "proper motion" discovered. The
motion of Sirius itself in the last 1000 years is illustrated on page 387. The radial velocity of the star is 4½
miles per second in approach. Midnight culmination, or
date of opposition, is January 1.

At a distance of 8.7 light years, Sirius is the 5th
nearest star known. Among the naked-eye stars it is the
nearest of all, with the exception of Alpha Centauri. The
vastness of space is dramatically illustrated by the fact
that even such a "nearby" star is fully 550,000 times more
distant than the Sun.

Sirius is a member of a moving group of stars often
called the Ursa Major Stream, with members scattered all
over the sky. This widely-dispersed stream shows very nearly the same space motion as the Ursa Major cluster, but it
is not definitely known if the association is real; among
prominent members are Alpha Ophiuchi, Beta Aurigae, Delta
Leonis, and Alpha Corona Borealis. Rather curiously, a
similar "stream" may or may not be associated with the

DESCRIPTIVE NOTES   (Cont'd)

well known Hyades Cluster in Taurus; one of the presumed
members is the 1st magnitude star Capella.

Sirius has been throughout human history the most
brilliant of the permanent fixed stars, and was an object
of wonder and veneration to all ancient peoples. Richard H.
Allen, author of the classic "Star Names and Their Meanings", devotes some 10 pages to a discussion of the various
titles and mythological references concerning Sirius. The
name appears to be derived directly from the Greek word for
"sparkling" or "scorching", though some connection with the
Greek name for the Egyptian god Osiris has also been suggested. The Arabic name "Al Shi'ra" resembles the Greek,
Roman, and Egyptian names, and suggests a common origin
from an older tongue, possibly Sanskrit, in which the name
"Surya", the Sun God, simply means "The Shining One".   In
the ancient Vedas the star is called "Tishiya" or "Tishtrya
-the Chieftain's Star"; in other Hindu writings it is referred to as "Sukra", the rain god or rain star, or sometimes as the Hunter. "Sivanam", the Dog, in the *Rig-Veda*,
is described as "he who awakens the gods of the air and
summons them to their office of bringing the rain.."; this
appears to be another reference to Sirius.

Plutarch calls the star "The Leader", while in the
time of Homer it seems to have been known as the "Star of
Autumn". In late Persian times it was called "Tir", the
Arrow. In Chaldea the star was honored with such titles as
"Kak-shisha", the "Dog Star That Leads", and "Du-shisha",
"The Director". Another Babylonian name was "Kakkab-lik-ku"
or the "Star of the Dog", perhaps derived from the Assyrian
"Kal-bu-sa mas", the "Dog of the Sun". An older Akkadian
name, "Mul-lik-ud" has been translated "The Dog Star of the
Sun".   The association of Sirius with a celestial Dog seems
to have been very nearly universal throughout the classical
world; in fact even in remote China the star was identified
as a "Heavenly Wolf". The Australian aborigines, however,·
regarded it as an Eagle.

The identification of Sirius with the Biblical star
"Mazzaroth" of the Book of Job is probably uncertain.
Isaac Asimov, in his "Guide to the Bible", speculates that
the word possibly referred to the whole cycle of the zodiac
or to the planets. The Hebrews, in any case, seem to have
known Sirius under the Egyptian name of Sihor; the Semitic

name "Hasil" probably also refers to Sirius. According to R.H.Allen, the Phoenicians are said to have known it as "Hannabeah", "the Barker".

Sirius was the revered "Nile Star" or "Star of Isis" to the ancient Egyptians; its annual appearance just before dawn at the summer solstice heralded the coming rise of the Nile, upon which Egyptian agriculture- and in fact all life in Egypt- depended. In about 3000 BC this "heliacal rising" occurred about June 25, and is referred to in many temple inscriptions where the star is called the "Divine Sepet" (or Sopet or Sothis) and is identified with the soul of Isis. In the temple of Isis-Hathor at Denderah appears the inscription: *"Her Majesty Isis shines into the temple on New Year's Day, and she mingles her light with that of her father Ra on the horizon."* Sir Norman Lockyer in his book "The Dawn of Astronomy" (1894) states that this temple, dating from the time of the Ptolemies (3rd - 1st centuries BC) was oriented to the rising of Sirius. At least two other temples at Karnak were similarly oriented, dating from the time of the 18th Dynasty, and probably begun in the days of the Pharaoh Thutmose III, about 1500 - 1450 BC.

Sirius is referred to in a striking passage from the *Iliad*, where King Priam, from the walls of Troy, sees the wrathful Achilles advancing across the Trojan plain..... *"blazing as the star that cometh forth at Harvest-time, shining forth amid the host of stars in the darkness of the night, the star whose name men call Orion's Dog. Brightest of all is he, yet for an evil sign is he set, and bringeth much fever upon hapless men..."*

In the ancient Greek and Roman world, the influence of Sirius was regarded as extremely unfortunate, as the allusion to the wrathful Achilles in the *Iliad* would seem to suggest. In Virgil's *Aeneid* we read of the *"Dog Star, that burning constellation, when he brings drought and diseases on sickly mortals, rises and saddens the sky with inauspicious light"*. The scorching heat of July and August occurs when Sirius rises with the Sun, and was attributed to the dire influence of the blazing star, bringing forth fever in men and madness in dogs. These ideas prevailed well up into the time of the Renaissance, as we find Dante speaking of *"the great scourge of days canicular"*. A more sensible view, however, was taken by Geminus (about 70 BC)

DESCRIPTIVE NOTES   (Cont'd)

when he wrote *"It is generally believed that Sirius produces the heat of the Dog Days, but this is an error, for
the star merely marks a season of the year when the Sun's
heat is the greatest"*.

But, says R.H.Allen, "he was an astronomer".
WAS SIRIUS A RED STAR IN ANCIENT TIMES? This question was
first brought to the attention of the astronomical world
by Thomas Barker who published a paper called "On the
Mutations of the Stars" in the *Philosophical Transactions*
for 1760. Citing the testimony of Aratus, Cicero, Horace,
Seneca, and Ptolemy, he pointed out that all these ancient
writers described Sirius with terms that can only be trans-
lated as "ruddy", "reddish", "blazing as fire", etc. More
than a century later the indefatigable and reportedly
somewhat eccentric T.J.J.See made a thorough study of the
ancient records: "therefore to satisfy my own curiosity
I undertook a critical investigation of all of the ancient
authors hitherto examined, and a great many others with a
view of deciding definitely whether in antiquity Sirius was
really red." After a series of articles and notes totalling
29 pages, published in 1892 in *Astronomy and Astrophysics*,
Professor See concluded that "the results of this research
seem to establish beyond doubt the ancient redness of the
Star".

Among the more convincing statements were those made
by Cicero, Horace, Ptolemy, and Seneca. Homer, in the *Iliad*
seems to compare the gleam of Achilles' copper shield to the
light of Sirius. In a Babylonian cuneiform text the star,
called "Kak-si-di", is described also as "shining like
copper". Aratus describes the star with the term ποιχίλος
which is usually translated as "ruddy". In the 1st century
BC, Cicero refers to Sirius with the term "rutilo cum lum-
ine" or "with a ruddy light". Horace, only a few decades
later, calls it the "rubra Canicula" or "ruddy Dog-star".
Seneca, in the days of Nero, definitely speaks of it as
redder than Mars, whereas Jupiter "is not at all red". The
poet Columella, a contemporary of Seneca, compares the hues
of roses to Tyrian purple, the rising sun, Sirius and Mars.
Pliny, Ovid, and S.Pompeius Festus state that "ruddy dogs"
were sacrificed at the ancient Floralia festival in honor
of the Dog Star; this celebration was instituted at Rome
in 238 BC in accordance with a decree of the oracle of the

Sibyl. Ptolemy, in about 140 AD, refers to Arcturus, Aldebaran, Pollux, Betelgeuse, Sirius, and Antares as "fiery red". However, Al Sufi, in the 10th century, does not mention Sirius among stars which he classes as red. Presumably by that time the star was no longer the "rubra canicula" of ancient times.

What conclusions can be drawn from this impressive collection of statements by the classical writers? The whole question was revived in modern times by recent obser- vations of the faint companion to Sirius (now a white dwarf star) which seem to show that this star may be one of the hottest and therefore newest of all degenerate stars. Measurements by K.Rakos with the one-meter telescope at La Silla in Chile have been cited in support of the hypothesis that Sirius B might possibly have been in the red giant stage as recently as 2000 years ago. Stephen P.Maran, in the July-August issue of *Natural History* discusses this problem and concludes that "this explanation sounds logi- cal, but unfortunately it contradicts much of what we know- or think we know- about the life cycles of the stars". The most serious objection, of course, is that the time-scale seems unacceptably short; the expected time from the red giant stage to the white dwarf stage is about 100,000 years rather than a mere 2000. Yet, some astronomers have found theoretical reasons for supposing that, in certain types of stars at least, the final transformation from the giant to the degenerate dwarf might happen with great rapidity. Sirius B has a present mass of nearly 1 sun, and in its red giant stage would not have been a supergiant like Antares or Betelgeuse, but might have been bright enough at least to equal Sirius A and affect the naked-eye color of the system. The fairly impressive testimony of ancient writers at least suggests that this idea should be seriously con- sidered. But in addition to the various possible explana- tions connected with stellar evolution, it seems to the author of this book that there is another possibility which might be considered, and which has nothing to do with the stars, namely the suggestion that the color-sensitivity or color balance of the average human eye has changed or evolved somewhat in the last few thousand years, and that the ancient peoples did not see colors quite the same as we do today.

In support of this hypothesis, one might consider the fact that Ptolemy also classes Arcturus and Pollux among the "fiery red" stars, and that Capella was called red by ancient writers. All these stars today are yellowish; the term "topaz" is often used to describe Arcturus, but no honest observer today would call it "fiery red". There are other odd color phrases used in ancient writings; consider Homer's repeated use of the term "wine-dark sea". It is true that Homer is possibly semi-legendary, and was also traditionally blind, but the authors, whoever they may have been, still employed the phrase as an appropriate metaphor for the normal color of the sea. Until more conclusive evidence is available, it seems unwise to state dogmatically that the ancient redness of Sirius must be dismissed as an impossibility. *"It is always a capital mistake to theorize before you have all the evidence"*, stated Sherlock Holmes. *"It biases the judgment."*

THE COMPANION TO SIRIUS. In the years between 1834 and 1844 the astronomer and mathematician F.W.Bessel found that Sirius had wavy irregularities in its motion through space, and came to the conclusion that the star had an invisible companion revolving about it in a period of about 50 years. The theoretical orbit of this unseen body was actually calculated in 1851 by C.H.F.Peters, but the expected companion persistently refused to show itself, despite the

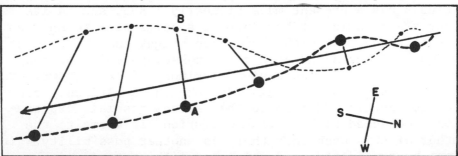

The apparent proper motion of Sirius (heavy dashed line) showing how the irregular curved path is caused by the attraction of the massive companion. The center of gravity of the system moves along the solid line in the direction of the arrow.

DESCRIPTIVE NOTES   (Cont'd)

careful searches made by many experienced observers. Then, in January 1862, the prediction was fulfilled by the discovery of the companion near its expected place, by Alvan G.Clark, with an 18½-inch refracting telescope, then the largest refractor in the world. This instrument is still in use, at the Dearborn Observatory of Northwestern University in Illinois. The companion to Sirius has a magnitude of about 8.65, the distance from Sirius varying from about 3" to 11½" in a period of 49.98 years. Widest separation occurs in 1975, 2025, etc. The companion, usually called Sirius B, or "The Pup", is an extremely difficult object in small and moderate size telescopes unless atmospheric conditions are very good. Usually it is completely lost in the overpowering glare of the brilliant Sirius. In the winter of 1962, during a period of exceptionally good seeing, the visibility of the companion was studied at Lowell Observatory with the 24-inch refractor (another superb Clark telescope) using an adjustable iris diaphragm over the objective. It was found that the faint star was most conspicuous with the aperture reduced to 18 inches, which helped to reduce some of the dazzling glare of the primary; it was still very definite at 12 inches, difficult at 9 inches, and detectable at 6 inches only because its exact position was known. The tests were made with magnifications ranging from 200 to 900. With the higher powers, it was possible to view the companion with Sirius itself placed entirely outside the field!

Although the view of the mysterious companion through the Lowell telescope was undoubtedly the finest that the author of this book has ever experienced, he has since observed the star on many occasions with a 10-inch reflector, and no longer considers it an exceptionally difficult object. The air, however, must be very steady. With reflectors of the usual 4-vane diagonal-holder type, the image of the companion may also fall on one of the diffraction rays, where it is totally lost. The observer should determine the expected PA beforehand, and the telescope should be oriented so that the companion will fall between the diffraction spikes. This is the technique used to photograph the pair, as illustrated in the plate on page 398.

The possible duplicity of Sirius B is an unsolved question. Philip Fox in 1920 reported the image to be

"persistently double" in 231°, separation 0.8". Since Fox
was an experienced observer and was using the same 18½-inch
telescope with which the companion was originally detected,
his observations should carry some weight. The suspected
third star has also been seen by R.T.Innes in South Africa
and by the well known double star expert van den Bos. Due
to the great difficulty in the observations, it has not
been possible to verify these reports. A third star in the
system might explain reported slight irregularities in the
orbits of the visible pair. In 1973, however, a thorough
study by I.W.Lindenblad at the U.S.Naval Observatory concluded that there is no astrometric evidence for the
existence of a third body in the Sirius system. And there,
for the present, the matter rests.

MONTHLY NOTICES

OF THE

ROYAL ASTRONOMICAL SOCIETY.

Vol. XXII.	March 14, 1862.	No. 5.

Discovery of a Companion of Sirius.

In the *Ast. Nach.*, No. 1353, Prof. Bond communicates the
discovery of a Companion of *Sirius*, made on the evening of
Jan. 31 by Mr. Clark, with his new object-glass of 18½ inches
aperture.   Prof. Bond was able to observe it with the Refractor
of 15 inches, at the Observatory of Harvard College, as follows:

$$1862,\ \text{Feb. 10, Angle of Position}\quad 85°\ 15\quad \pm 1°\ 1$$
$$\text{Distance}\quad ...\quad 10''\cdot 37\quad \pm 0''\cdot 2$$

when the images were tranquil, the Companion was seen dis-
tinctly enough, but on account of the atmospheric disturbances
these moments are quite rare.

The Companion was seen at Paris by M. Chacornac, the
20th March, with the telescope with silvered mirror of 80
centimeters, constructed according to the plans and under the
direction of M. Foucault.

It appears from the *Cosmos* of 28th March, that Dr. Peters
does not accept the identity of the Companion thus discovered
with that which he had calculated.

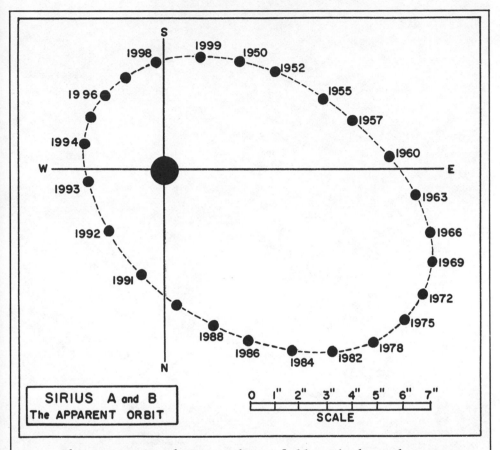

SIRIUS A and B
The APPARENT ORBIT

SCALE

The companion has now been followed through two complete revolutions of the system and the orbital elements
are well known. The orbit has a semi-major axis of 7.62"
and an eccentricity of 0.58; the mean separation of the 2
stars is 24 AU. Periastron occurs in 1944 and 1994. From
the observed orbit the masses of the stars are 2.35 and
0.98 the solar mass.

THE NATURE OF SIRIUS B. The companion to Sirius has
been an object of greatest interest ever since its discovery. The mass is nearly equal to that of the Sun. The
luminosity, however, is less than 1/400 that of the Sun.
The abnormally low luminosity might be explained in two
ways: either by an extremely low temperature which would
imply a very feeble surface brightness, or by an unusually
small diameter. The spectrum of the star was difficult to

SIRIUS and the COMPANION. An unusual photograph made with the 24-inch refractor at Sproul Observatory. A hexagonal diaphragm was used over the objective to create the 6-rayed pattern so that the elusive companion might be recorded.

DESCRIPTIVE NOTES (Cont'd)

determine due to the overpowering glare of the primary,
but was finally obtained by W.Adams at Mt.Wilson in 1915.
It was found to be class A or early F, probably about A5.
The corresponding temperature is 8500° or 9000°K, several
thousand degrees hotter than the Sun, and not much cooler
than Sirius itself. The surface brightness is thus about 4
times greater than the Sun's, and the low total luminosity
implies an exceedingly small diameter of roughly 2% that
of the Sun.

From this result it is an obvious step to the final,
most amazing characteristic of the "Sirius B" stars, now
to be called "white dwarfs". With a mass nearly equal to
that of the Sun, but a diameter some 40 or 50 times smaller, the typical white dwarf must have an incredibly high
density. According to the most recently determined values
for mass and radius (table below) Sirius B has a density
of about 90,000 times that of the Sun, or 125,000 times the
density of water. A cubic inch of this star material weighs
about $2\frac{1}{4}$ tons.

SIRIUS A and B							
	Mag	Abs.Mag.	Spect.	Diam.	Lum.	Mass.	Density
A	-1.42	1.45	A1 V	1.8	23	2.35	0.4
B	8.65	11.4	DA5:	0.022	0.002	0.98	92,000

The exact value computed for the density is dependent
on the value accepted for the surface temperature, which
determines the surface brightness per unit area. As we have
seen, this is difficult to determine due to strong interference from the light of Sirius, computed values varying
between 25,000 and 100,000 times the solar density. But
though the exact figure is uncertain, there is no doubt
whatever about the general order of magnitude, since the
temperature would have to be comparable to the coolest
stars known in order to reduce the density to anything
approaching "normal" conditions.

The statement often appears that Sirius B was the first
white dwarf known. This is not strictly true, however,
since the star 40 Eridani B was recognized as an A-type
star of very low luminosity as early as 1910. Although the
full implications of this were not immediately realized,
it was evident that there was a strange peculiarity in the

DESCRIPTIVE NOTES (Cont'd)

combination of high temperature and low luminosity. In a
discussion of the problem, Professor H.N.Russell pointed
out that all the known stars of very low absolute magnitude
were class M, and the case of 40 Eridani B presented "an
exception to what looked like a very pretty rule of stellar characteristics. I knew enough about it, even in those
paleozoic days, to realize at once that there was an extreme inconsistency between what we would have then called
'possible' values of the surface brightness and density".
With characteristic optimism, W.H.Pickering observed that
"It is just these exceptions which lead to an advance in
our knowledge". And A.Eddington stated that "Strange objects which persist in showing a type of spectrum entirely
out of keeping with their luminosity may ultimately teach
us more than a host which radiate according to rule".
     Sirius B is still the brightest and nearest of all the
white dwarfs, and still remains the most famous member of
this strange and wonderful class of stars. For the owner
of a small telescope, 40 Eridani B is the most easily observed white dwarf; although a member of a triple system,
it is sufficiently far from the 4th magnitude primary so
that there is no interference from the bright star. This
first known white dwarf is also called "Omicron 2 Eridani"
and is described under the constellation Eridanus.

A SUMMARY OF WHITE DWARF CHARACTERISTICS. The main facts
about the white dwarf stars are presented in the brief review which follows.
     DIAMETERS are very small, comparable to the sizes of
the planets, and averaging about 1/50 the diameter of the
Sun. Sirius B is approximately 19,000 miles in diameter
and the computed size of 40 Eridani B is 17,000 miles. Van
Maanen's Star in Pisces is believed to be slightly smaller
 than the Earth; the estimated size is 7800 miles. For
Wolf 219 the computed figure is 5600 miles. Until rather
recently, the smallest known white dwarf was AC +70°8247
in Draco, about half the size of the Earth. In 1962 and
1963, however, two faint stars were identified as probably
the tiniest white dwarfs yet found. The first of these,
discovered by W.J.Luyten, and designated LP 357-186, is
located in Taurus and has an apparent magnitude of 18.3.
The computed diameter is about 1200 miles. Even more fan-

tastic is LP 768-500 in Cetus, discovered also by Luyten and announced in November 1963. It is a star of magnitude 18.3 with an annual proper motion of 1.18". The computed diameter is about 1/1000 that of the Sun, or probably about 900 miles. These results are somewhat provisional since the exact distances are not well determined. Luyten has introduced the term "pigmy" as a designation for stars of unusually small size and exceptionally great density. Only a few stars are known with diameters less than Earth's Moon.

LUMINOSITIES are very low. Sirius B is 10,000 times fainter than its primary, and 1/435 the brightness of our Sun. Sirius B and Procyon B have absolute magnitudes of 11.4 and 13.1 respectively; Van Maanen's Star is still fainter with an absolute magnitude of 14.2. In general, the values for white dwarfs range from the 9th to the 16th absolute magnitudes, with only a few stars exceeding these limits.

HZ 29 in Canes Venatici appears to be one of the most luminous known white dwarfs; it has about 1/40 the luminosity of the Sun. The computed absolute magnitude is +8.9. At the other extreme, only a few white dwarfs are recognized with absolute magnitudes below 15. Wolf 457 in Virgo and Wolf 489 (also in Virgo) both have luminosities 15,000 times less than the Sun's; the absolute magnitudes in each case are about +15.4. The star LFT 555 in Volans appears to be about 35,000 times fainter than the Sun, with a computed absolute magnitude of 16.2. A recently discovered star, HL4 in Orion, seems to be comparable to W489 in luminosity. Finally, the three new Luyten stars may claim the record for low-luminosity white dwarfs. LP 357-186 may be about 16½ absolute, and LP 768-500 is probably fainter than 17th. The star LP9-231, once thought to be among the least luminous of all stars, now seems to be more distant than originally estimated, and is therefore not as intrinsically faint as was first believed.

TEMPERATURES are high for most of the stars of the white dwarf class. More than half of the well observed examples fall into spectral class A, and have surface temperatures ranging from 8000° to 10,000° K. The few stars known of class B are somewhat hotter still. White dwarfs of class F (Ross 627 and Ross 640) are rather scarce, and those of later types are still scarcer. Van Maanen's star seems to

be of type G, and W489 is type K. The newly discovered HL4
is apparently similar to W489. As of 1975, no M-type white
dwarf is known with certainty; the two stars G5-28 and
G7-17 had been tentatively placed in this class, but it
now seems more likely that these stars are red sub-dwarfs
and not truly degenerate stars. According to J.L.Green-
stein at Palomar (1974) G5-28 is definitely a sub-dwarf.

MASSES are known accurately for only three white dwarf
stars which are members of well-observed binary systems.
These are Sirius B, 40 Eridani B, and Procyon B, with masses of 0.98, 0.44, and about 0.65 the solar mass. Future
additions to this short list will be the stars G175-34 in
Camelopardalis, and G107-70 in Lynx, discovered during the
Lowell Observatory proper motion survey. The first of these
forms a binary with a red dwarf companion and an orbital
motion of about 1° per year; the other is a close double DC
pair of about 0.7" separation; a preliminary estimate of
the period is about 16½ years. The total mass of each pair
appears to be close to one solar mass.

The masses of all other white dwarfs are derived from
theoretical calculations, and range from 0.2 to about 1.25
the solar mass. The larger value approaches the "Chandrasekhar Limit", beyond which contraction into a stable white
dwarf is not possible; the stars AC +70°8247, LP 357-186,
and LP 768-500 are believed to possess masses which are
near this theoretical limit. The majority of white dwarf
masses seem to be below the mass of the Sun. In general,
the stars of greater mass have smaller radii, but there is
no obvious correlation between radius and spectral type.

DENSITIES are incredibly high, averaging several tons
to the cubic inch. Sirius B is approximately 125,000 times
denser than water. Van Maanen's Star is some 10 times denser yet, and weighs about 20 tons to the cubic inch. Wolf
219, smaller than the Earth, has a computed density of 4½
million times that of the Sun, or roughly 105 tons to the
cubic inch. One of the densest stars known must be the
object AC +70°8247, 12 million times denser than the Sun,
and weighing a calculated 295 tons to the cubic inch.

The record for density may eventually be claimed by
the two new Luyten stars previously mentioned. LP357-186
is estimated to be nearly 500 million times denser than the
Sun, weighing about 11,000 tons to the cubic inch. On the

DESCRIPTIVE NOTES (Cont'd)

assumption that the diameter of LP 768-500 is about 1/1000 that of the Sun, Luyten finds a density of about 1 billion times that of water for this star, equivalent to 18,000 tons to the cubic inch!

SUB-DWARF STARS. In addition to the "classical" white dwarf stars, there exist a number of semi-degenerate or "intermediate" stars concerning which very little is known. When plotted on the H-R diagram they are found to lie between the main sequence and the realm of the true white dwarfs. The bluish companion to Mira (Omicron Ceti) seems to be a star of this class; it has a B-type spectrum but the absolute magnitude is only about +6. The pre-nova and post-nova stars also appear to be members of this rare sub-dwarf class, as well as the SS Cygni stars and the blue components of the "symbiotic stars" such as R Aquarii. Much remains to be learned about these peculiar objects.

EXPLAINING THE DENSITIES OF THE WHITE DWARFS. Without our present knowledge of the nature of matter, such amazing densities would seem completely unbelievable. The density of the Sun is scarcely over ½ ounce to the cubic inch, and the densest substances known on Earth are only some 20 times heavier than water. However, it must be remembered that all normal matter on Earth consists mainly of empty space, and that even our densest metals are composed of atoms separated from one another by relatively enormous distances. In addition, the atoms themselves are exceedingly "open-work" structures in which the nuclei and electrons could be represented on a scale model by a few gnats flying about in Grand Central Station. Could we fill in all these spaces and pack the atomic particles tightly against each other, we would have a density comparable to that of the white dwarf stars. The material of such stars is tremendously compressed, the atoms having been more or less broken down and the constituent nuclei and electrons packed together, forming so-called "degenerate matter". The cause of this compression is the star's own gravitational field, which brings up the logical question: Why then does not the Sun collapse into the white dwarf state, and why does not gravitation produce this super-dense condition in every star? The answer is found in the nuclear energy supply of the Sun and the other stars; gravitation cannot bring about the collapse of the star as long as the interior energy

DESCRIPTIVE   NOTES   (Cont'd)

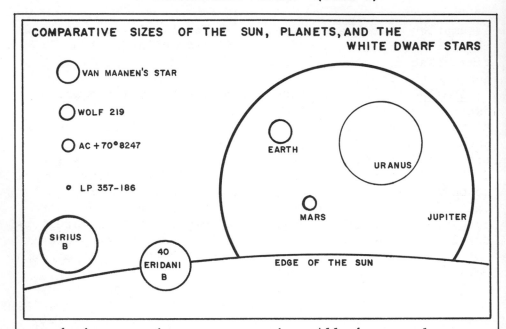

COMPARATIVE SIZES OF THE SUN, PLANETS, AND THE
WHITE DWARF STARS

VAN MAANEN'S STAR

WOLF 219

AC + 70° 8247

LP 357-186

SIRIUS B

40 ERIDANI B

EARTH

URANUS

MARS

JUPITER

EDGE OF THE SUN

-producing reactions are operating. All the normal stars can be regarded as natural and perpetual nuclear furnaces in which energy production is the result of the nuclear conversion of hydrogen into helium at temperatures of many million of degrees. While this continuous chain reaction is operating, the energy supply prevents gravitational contraction. But once the hydrogen "fuel" has been consumed, the star will begin to contract, and the density will grow greater and greater.

White dwarfs can therefore be considered "dead stars" in the sense that they have exhausted their hydrogen supply, and their evolution is at an end. Though no longer producing nuclear energy, a white dwarf is still visible because the contraction has left the star at a high temperature. Eventually even this energy source must come to an end, and the ultimate fate of the star seems to be the "black dwarf" state, in which the star has cooled to a non-luminous planet-like body of incredible density. It is not known whether there are any such objects in existence at the present time; naturally we should never be able to observe them directly.

The cooling and fading of a white dwarf requires a fantastic length of time, and suggests that some of these objects, particularly the cooler and redder ones, must be among the most ancient stars known. It is thought that a white dwarf requires some 3 billion years to cool from a bluish A-type star to a yellowish F-type star, and another 5 billion years to cool to type K. Van Maanen's Star has presumably been a white dwarf for 4 or 5 billion years at least, and the reddish W489, HL4, R193b, and G7-17 may be the oldest individual stars yet identified. It is thought that the majority of the degenerate stars in our galaxy are the remains of rather massive stars which consumed their hydrogen supply at a rapid rate. The same fate possibly awaits all present high-luminosity giants in the astronomically near future. Such giants as Rigel, Beta Lyrae, and S Doradus have life expectancies tremendously shorter than that of the Sun. Rigel, for instance, is at present about 60,000 times more luminous than the Sun, but its glory will necessarily be short-lived. It is of interest to reflect that when Rigel has ended its career, our Sun will still be shining with much of its present brightness.

SPECTROSCOPIC STUDIES OF THE WHITE DWARFS. The peculiar physical conditions of the white dwarfs make spectroscopic studies rather difficult. The dense mass of the star is covered by a relatively thin layer of non-degenerate material which may be 50 or 60 miles deep; above this is the atmosphere of the star which in typical cases is less than 100 feet thick. This atmosphere is the only portion of the star accessible to spectroscopic study. Because of the huge surface gravity, often exceeding 50,000 times that of the Earth, the white dwarf atmosphere is extremely compressed. The atmospheric pressure on Van Maanen's Star, for example, is estimated to be about 2000 times that of the Earth. Such extraordinary conditions produce spectral peculiarities which are difficult to interpret.

A few white dwarfs appear to show virtually continuous spectra, with no detectable lines at all. These are classified as type "DC", though it is possible that more sensitive equipment will eventually reveal faint lines in some of these spectra. It has been suggested that the disappearance of the lines may in some cases be due to extreme widening caused by high pressure.

Type "DB" white dwarfs such as HZ29 are among the hottest known and the spectra show strong lines of helium,

DESCRIPTIVE NOTES (Cont'd)

apparently indicating unusually high abundances of this element. The commonest type "DA" white dwarfs show only hydrogen lines, suggesting that all heavier elements have been forced into the dense mass of the star, and that the tiny amount of remaining hydrogen has been squeezed to the surface. In cooler stars of type "DF" hydrogen lines and calcium lines often appear together, as in Ross 627; but when the temperature falls below 8000° K the hydrogen lines vanish entirely, and the spectrum shows only a few lines of metallic elements such as calcium and magnesium. Ross 640 is such a star. In these objects it may be that the heavier elements were produced from helium, after the exhaustion of hydrogen.

In a very late "DG" star, such as Van Maanen's, the prominent lines are those of calcium and iron, but hydrogen lines are totally absent. In addition to these chief types, there are a few white dwarfs which show individual peculiarities. AC +70°8247 has an unidentified band at $\lambda 4135$, and several stars are known with a single broad band at λ 4670, now attributed to the carbon molecule C_2. Wolf 219 is the standard star of this type. HZ9 and WZ Sagittae are two unusual white dwarfs whose spectra show emission lines. The latter star is of special interest since it is a well known recurrent nova which underwent outbursts in 1913 and 1946.

WHITE DWARFS AND THE NOVAE. The case of WZ Sagittae raises some interesting questions. It has long been recognized that there is some connection between white dwarfs and the exploding stars called novae. All known post-nova stars are hot dwarfs which appear to be at least partially degenerate, and may be intermediate between normal stars and the true white dwarfs. It is thought that the contraction toward the white dwarf state causes one or more periods of stellar instability, particularly in the more massive stars. Remembering that Chandrasekhar's Limit is about 1.25 solar mass, it seems reasonable that stars of greater mass would become violently unstable during contraction, and that the nova outbursts can be explained in this way. In some cases the presence of a close companion star could be a contributing factor. Several of the nova-like variable stars (AE Aquarii, SS Cygni) are known to be close binaries in which one star may be partially degenerate. Studies of

these stars may provide valuable evidence toward a solution
of the nova mystery. (Refer also to Nova Aquilae 1918, WZ
Sagittae, Nova Cygni 1975, and SS Cygni.)

WHITE DWARFS AS DOUBLE STARS. A considerable number of the
known degenerate stars are members of double and triple
star systems. At about the time of the prediction and subsequent discovery of the companion to Sirius, it was shown
that the bright star Procyon (Alpha Canis Minoris) was a
very similar system with an orbital period of about 40
years. The faint companion, discovered visually at the Lick
Observatory in 1896, is a white dwarf with a mass of about
0.65 sun. The "easiest" white dwarf, Omicron 2 Eridani B,
forms a binary with a red dwarf companion "C" in a period
of about 248 years; both stars in turn are in orbit about
a normal K-star in a period of at least several thousand
years. The system of Zeta Cancri contains an unseen star
which is probably a white dwarf. In G175-34 in Camelopardalis, we have another red dwarf- white dwarf combination
with a period probably exceeding 3 centuries. W.J.Luyten in
1974 has published a catalogue of the known binaries with
white dwarf components, a total of 320 systems, a number of
which, however, await spectroscopic confirmation.

Among the interesting objects listed in the Luyten
census is a faint yellowish star which appears to share the
proper motion of Xi Draconis, but is 25,000 times fainter
than the 4th magnitude K-type primary. If the two really
form a common motion pair, then the companion, at magnitude
15.6, must definitely be a degenerate star. The separation,
according to Luyten, is 316" in PA 290°. Observers who wish
to try for another white dwarf which is not generally known
might turn their telescopes on GC 21205 in Lupus, position
15442s3745. It is the 6th magnitude star shown on the
Skalnate-Pleso atlas about 1/3° SE of the globular cluster
NGC 5986. The primary is a G6 main sequence star; the white
dwarf companion is 15.2" distant in PA 131°; it is magnitude about 13.3, and spectral type DA.

Double stars with white dwarf components are of very
great interest to the theoretical astrophysicists, since
they often present the strange circumstance that the more
massive star is still a main sequence object, while the
less massive star has reached the white dwarf stage. If
the two stars are of the same age, and have always been a

physical pair, then the more massive star should evolve
faster than the other. Both Sirius and Procyon present us
with this puzzle. If the two components of a pair are very
close together, then some form of mass-exchange may occur,
which may alter the normal pattern and rate of evolution.
Both Procyon and Sirius, however, are rather wide pairs,
and this explanation does not appear feasible unless the
separation was much less in the past.

There are a number of odd variable stars known in
which one component appears to be a white dwarf. In addition to WZ Sagittae and the U Geminorum stars, the following deserve special mention: BD+16°516, a white dwarf in
the Hyades cluster in Taurus, appears to be a member of a
close eclipsing binary system with a period of 0.52118 day.
According to B.Nelson and A.Young (1970) the primary
eclipse is total and lasts 47 minutes with a drop of about
0.09 magnitude (photographic); the eclipsing body appears
to be a normal K0 dwarf star. HZ9, another member of the
Hyades, is a peculiar emission star with large changes in
radial velocity and seems a probable binary with a period
of about half a day. In Canes Venatici is the star called
HZ22 or UX Canum, believed to be a highly evolved but low-
mass binary; the secondary is thought to be a white dwarf.
J.L.Greenstein (1971) gives the orbital period as 0.5737
day, and finds the primary to be a blue B-type star of the
13th magnitude; the light of the system varies continually
with an amplitude of about 0.3 magnitude. Another object
in this same class is G61-29, near Alpha Comae; it is a
white dwarf of type B with emission lines, and shows an
eclipse type light curve resembling that of U Geminorum;
the orbital period is $6^h 17^m$.

Among the rarest objects in space, however, are double
star systems where BOTH components are white dwarfs. Up to
1976, only three such pairs have been positively identified
though a number of excellent suspects appear in the Luyten
catalogues. The pair LDS 275 in Antlia, discovered by
Luyten, consists of two 15th magnitude DC stars, separated
by 3.6"; the orbital period may be about 700 years. Another
such pair is G107-70 in Lynx, discovered in 1962 by N.G.
Thomas and the author of this book during the course of the
Lowell Observatory Proper Motion Survey directed by H.L.
Giclas. G107-70 is a double DC star, separation about 0.7"

and a period, according to preliminary measurements, of probably less than 20 years. A much wider pair is G206-17 and G206-18 in Hercules, separated by 55" and with a common proper motion of about 0.27" per year in PA 220°. Spectral types are DA (weak-lined) and DC (late type). The best of the Luyten suspects for double white dwarf systems are the following three objects:

LP370-50/51, at 09423n2341, about 19' south of the bright star Epsilon Leonis. Both stars are magnitude 15.8 according to Luyten, and share a common motion of about 0.22" per year in PA 244°; the separation of the pair is 13". Color measurements show that both stars are bluish, though no spectra have yet been reported.

LP332-27/28, at 17276n2919 in Hercules. Luyten gives the magnitudes as 17.1 and 17.9 with a separation of 16". Both stars have colors equivalent to A-type stars, and show the same proper motion of 0.25" per year in PA 217°. No spectra have been reported.

LP77-24/25, at 23193n6910 in Cepheus. Luyten gives the magnitudes as 14.6 and 17.2, with colors equivalent to types B and A, respectively. The separation is 29" and the components show the same proper motion of 0.13" per year in PA 271°. No spectra have been reported.

These pairs, of course, are relatively wide systems, which are not expected to show any sign of orbital motion until observations have been carried out for centuries! Very little is known about the probable frequency of very close degenerate binary systems in space. Imagine, for example, a hypothetical system consisting of two white dwarf stars, each about the size of the Earth, separated by about 10,000 miles, and with a total mass of 1 sun. The orbital period would be about 35 seconds! The eclipses of such a pair, if the system was suitably oriented, would last a few seconds, and would be very difficult to detect by conventional photometric techniques. Observers with large telescopes might find it interesting to make occasional visual checks of known white dwarfs in the hopes of detecting such rapid variations.

THE SEARCH FOR WHITE DWARF STARS. Until very recently, only a few hundred white dwarfs were known; because of their intrinsic faintness we can expect to see only the relatively nearby specimens. But from their comparative abundance

DESCRIPTIVE NOTES (Cont'd)

in our own region of the Galaxy, it appears that these
stars are by no means as rare as was once believed. J.L.
Greenstein (1959) estimated that they comprise some 3% of
the population of the Galaxy. Other estimates have ranged
up to 10%. A list of white dwarfs and semi-degenerate stars
published by Luyten in 1949 contained 96 objects. A total
of 267 white dwarfs was reached in 1953 according to E.
Schatzman, but the figure is uncertain since a number of
these stars have not been studied spectroscopically, and
may be subdwarfs or intermediates. Several thousand new
suspects are now known from the systematic searches of
Luyten, the Lowell Observatory proper motion survey, and
other sky survey programs. "It would be almost correct to
state", wrote Luyten in 1963, "that we are now finding
white dwarfs wholesale. I once identified seventeen of them
in one afternoon". The method of search is suggested by the
fact that a white dwarf is an intrinsically faint object,
and must be relatively nearby in order to be seen at all.
The simplest method of detecting nearby stars is by means
of a "proper motion survey" in which plates are made some
years apart and compared. Any stars which have shown a
change in position are known to be fairly close. The next
step is to secure color measurements or spectra. The great
majority of faint proper motion stars are then found to be
normal low-luminosity stars of the main sequence, but
occasionally the searcher is rewarded by the finding of a
white dwarf. A number of proper motion stars recorded by
F.E.Ross, M.Wolf, and other early investigators were later
identified as white dwarfs. The record for discoveries is
indisputably held by W.J.Luyten, who is credited with the
finding of more than half of the known stars of the type.
WHITE DWARFS AND THE RELATIVITY THEORY. The discovery of
the white dwarfs has provided astronomers with a rare
opportunity to make an observational check of Einstein's
Relativity theory. According to one provision of the theory
all radiant energy may be regarded as possessing a certain
amount of mass, and will be subject to gravitational force
as would any material body. Specifically, Einstein predicted that the vibrational frequency of light should decrease in a strong gravitational field, increasing the
apparent wavelength and shifting all the spectral lines
toward the red. The amount of this shift is given by the

formula M/R, where mass and radius are in terms of the Sun. Since the Sun's gravitational red shift is 0.6 km/sec, that of Sirius B should be 0.98/0.022= 44.5 times greater or about 26.5 km/sec. The actual measured red shift is in close agreement with the predicted value, thus verifying one of the provisions of the relativity theory by actual observation.

NEUTRON STARS. After the white dwarfs or "stellar bankrupts" became accepted members of the stellar community, astronomers went on to speculate on the possibility of stars existing in a vastly greater degree of density. The Russian physicist L.Landau in 1932 postulated the existence of "neutron stars" and F.Zwicky in 1934 analysed the conditions under which a stellar body might contract to a single mass of nuclear material with a density several hundred million times that of a typical white dwarf. Such a super white dwarf might be no more than 6 to 10 miles in diameter, but would contain all the mass of a normal solar type star. J.Robert Oppenheimer, in 1939, studied the theoretical properties of such objects and determined that a newly formed neutron star should be a strong source of X-rays; at the same time G.Gamow, W.Baade, F.Zwicky and others theorized that such objects might be formed in the core of a gravitationally collapsing star as it approaches the supernova state.

By the 1960's it was known that a very remarkable stellar remnant of some sort existed in the heart of the famous "Crab Nebula" M1 in Taurus, the accepted remnant of the brilliant supernova of 1054 AD, and a strong source of both X-rays and radio energy. This object, which presumably supplied the energy to keep the entire nebula radiating, appeared as a star of the 16th magnitude, and seemed to be at a fantastically high temperature.

The possible identification of this object as a true neutron star was still being debated when, in 1968, a very remarkable radio source was detected by A.Hewish, J.Bell, and their associates at the Cambridge University Observatory. The new object was located in the northern Milky Way at 19196n2147, about 1.5° SE of the star 2 Vulpeculae; the radio energy was coming in short pulses following each other with remarkable regularity, at intervals of 1.337301 seconds. This object, promptly dubbed a "pulsar" was given

DESCRIPTIVE NOTES   (Cont'd)

the identifying number CP1919; more recently a standard
numbering system for all pulsars has come into use and the
official designation is now PSR1919+21; the designation
thus gives the approximate coordinates in the sky. When
the discovery of this first pulsar was announced in February 1968, three other similar objects had already been
located by the Cambridge team, and within two years the
number had grown to 40. By 1976 more than a hundred were
known, none of them coinciding with any visible object,
with one important exception. That faint hot star at the
center of the Crab Nebula was shown to be emitting pulses
of both radio energy and visible light, in the amazingly
short period of 0.033089 seconds. The identification of
this object, PSR0531+21, as a neutron star is now regarded
as definite. Not even the smallest and densest known white
dwarfs could revolve about each other or rotate with periods as short as a fraction of a second. (The Crab Pulsar
is also the X-ray source Tau X-1, and the radio source
Taurus A or 3C144.)

     According to the best current evidence, a pulsar may
be regarded as a rapidly rotating neutron star with a remarkably intense magnetic field; the magnetic poles not
coinciding with the axis of rotation. Radio energy is
emitted in a comparatively narrow beam at the magnetic
poles; as the neutron star rotates a radio pulse is detected twice in each revolution as the beam sweeps across our
position. Only a small percent of the neutron stars are so
oriented that such radio pulses can be detected. It is not
surprising, therefore, that no pulsar has been detected at
the site of Tycho's Supernova of 1572 in Cassiopeia, or at
the center of the Veil Nebula in Cygnus. On the other hand
the pulsar PSR0833-45 is very probably the remnant of the
ancient supernova associated with the strong X-ray source
Vela X. It must be remembered that neutron stars are
incredibly tiny objects by astronomical standards; a star
6 to 10 miles in diameter has an extremely small radiating
surface and cannot be detected at great distances even
though the surface temperature is extremely high. The Crab
Pulsar may be visible only because of its extreme youth;
the explosion occurred only 900 years ago, and the neutron
star remnant is still at a fantastically high temperature.

DESCRIPTIVE NOTES   (Cont'd)

At the present time (1976) it is not known if all supernovae produce a neutron star, or if all neutron stars and pulsars are supernova remnants. The exact chain of events evidently depends on the type of star and the mass. In general, it seems reasonably certain that stars of less than 1.25 solar mass may contract to the white dwarf state directly without going through an explosion. Stars of greater mass are presumably fated to end their careers as supernovae, leaving a neutron star core in many cases, after the outer portions of the star are blasted away. The supernova phenomenon may explain the odd fact that in the Sirius and Procyon systems the less massive stars seem to be more evolved than the primaries. Possibly these stars were once objects of much greater mass, losing a large portion of their material in the supernova outburst which reduced them to degenerate stars.

There is, however, a theoretical upper limit to the mass of a neutron star, now generally agreed to be about 3.2 solar masses. Any stellar core whose mass exceeds this limit will continue to contract indefinitely, ultimately producing a gravitational field so intense that nothing – not even the star's radiation – can escape.

THE BLACK HOLE. The concept of the "Black Hole" is not exactly ultra-modern; the idea was probably first mentioned in 1798 by Pierre Laplace who showed that a body of sufficient mass and density would be invisible since not even light could escape the gravitational field. In 1916 the German physicist K.Schwarzschild computed the radius required to satisfy this condition for any given mass:

$$R_s = \frac{2GM}{C^2}$$

where R is in meters, the mass is in kilograms, time in seconds, C = the velocity of light in meters per second, and G= the gravitational constant of 6.7×10^{-11}. For the Sun the Schwarzschild radius is found to be about 3 km; for the Earth it would be slightly less than 1 cm. An object of one solar mass, if squeezed down to a diameter of about 3½ miles, would become a "collapsar" or "black hole" since all the matter would be inside a "Schwarzschild Singularity" in which the escape velocity exceeds the velocity of light. Such a body would be effectively sealed off from the rest of the Universe by its "event horizon", defined by the critical radius; it could neither be seen nor detected by any conventional techniques. Yet a black hole is not entirely unobservable; evidence for its exis- tence might be found by its effect on nearby stars.

The black hole hypothesis has been occasionally advanced as a possible explanation of phenomena observed in a few odd binary systems such as Beta Lyrae, Epsilon Auri- gae, and SV Centauri; these systems appear to contain very massive objects which are not visible stars, or are at least severely under-luminous. At present, however, the most convincing candidate is the strange X-ray source called Cygnus X-1 or HDE 226868, located at 19565n3504, about ½° NE from Eta Cygni. The visible 9th magnitude star appears to be a B0 supergiant of about 22 solar masses; it is a single-line spectroscopic binary with a period of 5.59982 days. The X-ray energy seems to originate in a tremendously heated stream of gas which is passing from the B-star to an unseen companion; rapid variations in the X-ray intensity suggest that the source is of very small dimensions. The mass of the unseen body is in the range of 15 to 20 solar masses, much too great for a white dwarf or neutron star. An object of this mass, if a normal star, should contribute nearly half the total light of the system but no evidence of its existence whatever can be detected spectroscopically. The large mass, lack of visible light energy, and the computed small size of the X-ray source all suggest that the unseen secondary is a collapsar or black hole.

Another possible object in this class is the X-ray source recently discovered near Theta Orionis in the Orion Nebula. Although black holes may possibly exist in fair numbers throughout the Universe, it seems possible to detect only those which are members of close double star systems. Particularly promising suspects are binaries with massive unseen companions, and which are also strong emitters of X-ray energy. In the case of the strange star SV Centauri, the evidence is of a different kind. The star is an eclipsing binary of the Beta Lyrae type with a 1.661 day period and computed masses of 9.4 and 11.1; both stars seem to be of type B. The odd feature of this binary is that the orbital period has been changing with abnormal rapidity over the years; since 1900 it has decreased by

DESCRIPTIVE NOTES (Cont'd)

0.14%, and is still decreasing. In a study of the star in 1971, J.B.Irwin and A.U.Landolt found evidence in the light curve for the existence of a third component, but the third body could not be detected spectroscopically, nor was any evidence found for the presence of gas streams between the components or gaseous shells surrounding them. In any case, the hypothetical third body would be a very massive object indeed to produce the steady change in the period of the visible pair; something in the nature of a black hole seems suggested by the evidence though no final conclusion can yet be made. The famous unseen companion to Epsilon Aurigae is another possible black hole candidate though the evidence is still circumstantial.

ANOTHER SIRIUS MYSTERY? In a recent book called "The Sirius Mystery" by Robert K.Temple (1975) the claim is made that the existence of the white dwarf companion to Sirius was known to the members of the Dogon tribe of Mali in Africa, a people whose religion and culture involves unusually sophisticated concepts concerning the stars and planets. According to the two French anthropologists M. Griaule and G.Dieterlen of the Societe des Africanistes in Paris, the Dogon have long had a tradition of an unseen companion to Sirius, with an orbital period of 50 years, and consisting of a material called "sagala" ("strong") which is said to be vastly heavier than any metal on the Earth, "so heavy that all earthly beings combined cannot lift it". According to Temple, the Dogon also accept a heliocentric theory of the Earth's motion, are familiar with the four large satellites of Jupiter, and know that the planet Saturn is surrounded by a ring which "is different from the ring sometimes seen around the Moon". Evidently, as Temple says, "the obvious parallels between this tribal information and the known facts concerning the true Sirius B are too elaborate and precise to be ignored". He suggests that the Dogon reverence for Sirius may have been inherited from ancestors who once lived in Egypt, but admits, of course, that this would not explain their seemingly scientific knowledge of the nature of the star's faint companion.
      If these are indeed genuine tribal traditions, we would seem to be faced here with a truly inexplicable

enigma. The "traditions" concerning the Jovian moons and
Saturn's rings, however, would seem to lead us to the
virtually unavoidable conclusion that these people have at
some time in the past been in contact with travelers from
the western world who had some astronomical knowledge.
"Surely the most reasonable hypothesis", state I.W.Roxburgh
and I.P.Williams of Queen Mary College in London, "is that
fairly soon after the discovery of Sirius B, a missionary,
explorer or French administrator, by accident or design,
comes across this tribe of Sirius-worshippers and decides
to give them new information about their god. He may even
have had a telescope with him (a very popular piece of
hand luggage in Victorian times) which he used to demonstrate his knowledge of the heavens, showing Jupiter's
satellites and Saturn's rings. The Dogon would rapidly
absorb such information into their religion so that by the
thirties, when they were anthropologically investigated,
the knowledge about Sirius B had become firmly part of
their traditional beliefs."

A REPRESENTATIVE LIST OF WHITE DWARF STARS. The following
table (pages 417--425) contains the chief information concerning the well-observed white dwarfs. The information
was compiled from the lists and catalogues of Greenstein,
Schatzman, Luyten, Giclas, Wolf, Ross, Eggen, and others;
most of the stars listed here have been definitely classed
as white dwarfs either by actual spectra or by proper
motion data combined with three-color photometry. Spectra
are given when known.
    The first column of the table gives the usual desig-
nation or discoverer's number. LTT, LP, L, LDS, and BPM
numbers are from the many lists published by W.J.Luyten.
R= F.E.Ross; W= M.Wolf; F= J.Feige; HZ= M.Humason and F.
Zwicky; Ton= Tonantzintla Observatory. G and GD objects
were discovered by the author of this book with H.L.Giclas
and N.G.Thomas, during the Lowell Observatory proper motion
survey. LFT numbers are from Luyten's catalogue of stars
with motions exceeding 0.5" annually. Values given for
magnitudes, motion and PA, etc., are in many cases preliminary measurements and are subject to future refinement.
Magnitudes are photographic. Numbers in the last column
refer to notes following the list.

A REPRESENTATIVE LIST OF WHITE DWARF STARS								
STAR	LFT	CON	MU"	PA°	MAG	SP	RA & DEC	N
GD-2		Andr	0.08	250	14.5	DA	00050n3301	
G130-49		Andr	0.55	220	16.8		00070n3052	1
W1		Pisc	0.52	115	15.3	DA	00111n0003	
LB 433		Pisc	0.04	78	15.2	DB	00174n1336	
GD-603		Scul	0.17	100	14.5	DA	00187s3359	
G158-78		Ceti	0.27	160	16.2	DA	00235s1054	2
LTT 300		Ceti	0.28	138	15.7	DC?	00314s1224	
L 1011-71		Ceti	0.44	200	15.5	DA	00331n0137	
LP12-438		Cass	0.08	100	16.2	DA	00332n7707	
G132-12		Andr	0.28	123	16.2	DA	00364n3115	
GD-8		Andr	0.07	130	14.0	DA	00373n3116	
Van Maanen	76	Pisc	2.98	156	12.4	DG	00465n0509	3
G69-31		Andr	0.54	109	15.5	DA	00521n2240	
L796-10		Ceti	0.47	350	15.3	DA	00533s1145	
G1-45		Pisc	0.43	52	14.1	DA	01012n0448	
GD-273		Cass	0.15	115	15.5	DA	01033n5549	
GD-11		Andr	0.15	100	14.5	DA	01066n3717	
L 1373-25		Pisc	0.25	122	14.9	DA	01075n2645	
W1516	122	Pisc	0.65	180	13.8	DC	01154n1556	
GD-13		Andr	0.08	115	14.1	DA	01268n4213	
GD-14		Pisc	0.16	100	15.6	DA	01273n2701	
R548		Ceti	0.44	99	14.1	DA	01337s1136	
L870-2	142	Ceti	0.67	121	12.8	DA	01354s0514	
LP13-249		Cass	0.15	135	15.2	DA	01367n7654	
G34-49		Pisc	0.28	109	17.4	DC	01425n2303	4
W82		Arie	0.27	240	15.0	DA	01439n2140	
LP768-500		Ceti	1.18	188	18.3	DC	01458s1726	5
GD-279		Andr	0.17	5	13.0	DA	01489n4645	
G71-B5b		Pisc	0.10	154	17.2		01505n0857	6
F17		Pisc	0.07	300	14.6	DA	01551n0658	
Oxf+25°6725		Arie	0.41	106	13.2	DA	02059n2500	
G74-7		Andr	1.12	117	14.3	DA	02082n3942	
GD-25		Andr	0.17	245	14.4	DA	02132n3938	
G134-22		Andr	1.07	127	16.2	DC	02138n4244	
h Per 1166		Pers	0.16	94	13.7	DA	02140n5653	7
G94-B5b		Arie	0.15	113	15.8	DA	02207n2214	8
F22		Ceti	0.10	101	12.7	DA	02277n0503	
GD-30		Tria	0.24	90	16.2	DA	02306n3420	
PHL 1358		Ceti	0.16	70	14.0	DA	02316s0525	
F24		Ceti	0.08	83	12.3	DA	02325n0331	9
G4-34		Arie	0.36	152	16.1	DC	02394n1100	
G76-48		Ceti	0.54	100	16.0	DA	02573n0800	

STAR	LFT	CON	MU"	PA°	MAG	SP	RA & DEC	N
A REPRESENTATIVE LIST OF WHITE DWARF STARS								
GD−40		Ceti	0.09	250	15.0	DBp	03003s0120	
F31		Ceti	0.08	210	14.0	DA	03020n0245	
LB 3303		Hydi	0.08	97	11.2	DA	03100s6848	10
G5−28		Arie	0.30	181	15.5	DM?	03157n1502	11
BPM 85584		Pers	0.16	158	13.3	DA	03166n3432	
L587−77a	286	Forn	0.80	63	14.0	DA	03268s2734	12
G37−44		Pers	0.47	138	15.5	DA	03322n3202	
W219	306	Taur	1.21	159	15.2	Dλ	03416n1818	13
BD+16°516		Taur	0.09	99	10.4	DA?	03476n1705	14
LB 1497		Taur	0.06	144	16.5	DA	03491n2447	15
HZ4		Taur	0.15	85	14.5	DA	03521n0937	16
G7−17		Taur	1.20	166	15.5	DM?	03587n1836	17
LB 1240		Taur	0.28	146	13.8	DA	04016n2501	
LB 227		Taur	0.12	102	15.4	DA	04066n1659	18
HZ10		Taur	0.10	137	14.1	DA	04073n1754	19
GD−56		Erid	0.23	170	15.0	DA	04086s0406	
LP357−186		Taur	0.45	144	18.3		04094n2347	20
HZ2		Taur	0.10	149	13.9	DA	04101n1145	
HG7−138		Taur	0.11	99	15.7	DK?	04124n1457	21
40 Eridani B	339	Erid	4.08	213	9.7	DA	04130s0744	22
GD−60		Pers	0.24	150	15.2	DA	04169n3329	
G38−29		Pers	0.29	170	15.6	DA	04170n3609	
LB 212		Taur	0.02	259	16.6	DA	04189n1522	
VR−7		Taur	0.12	97	14.3	DA	04210n1614	23
LB 1320		Taur	0.29	206	15.7	DC	04231n1205	
VR−16		Taur	0.12	98	14.0	DA	04257n1652	24
G174−34b		Caml	2.38	145	12.8	DC	04268n5853	25
HZ9 = L1239−16		Taur	0.12	108	13.9	DA	04294n1738	26
HZ7		Taur	0.12	90	14.2	DA	04310n1235	27
G39−27		Taur	0.36	116	15.9	DC	04336n2704	28
GD−61		Pers	0.17	165	15.3	DB	04352n4104	
L879−14	367	Erid	1.50	171	13.7	Dλ	04354s0853	29
HZ14		Taur	0.10	83	13.8	DA	04382n1053	30
G83−43		Orio	0.36	131	16.6	DA	04522n1023	
GD−64		Auri	0.25	180	15.0	DA	04538n4152	
G86−B1b		Auri	0.18	113	16.2	DA	05184n3319	31
G98−18		Auri	0.44	141	16.3	DA	05339n3214	
G102−39		Orio	0.30	181	15.9	DC	05511n1224	32
HL4 = LP658−2		Orio	2.38	167	14.5	DK	05527s0409	33
G99−47		Orio	1.06	207	14.9	DC	05538n0522	34
GD−72		Auri	0.23	215	15.0	DA	06068n2815	
L1244−26		Orio	0.40	188	13.4	DA	06124n1745	

A REPRESENTATIVE LIST OF WHITE DWARF STARS								
STAR	LFT	CON	MU" PA°		MAG	SP	RA & DEC	N
G105–B2b		Mono	0.10	151	16.6	DF	06250n1002	35
GD–77		Auri	0.23	190	15.0	DA	06374n4747	
SA26–82		Auri	0.04		16.7	DA	06396n4443	
Sirius B	486	CMaj	1.32	204	8.7	DA	06430s1639	36
He 3	487	Auri	0.95	191	12.1	DA	06443n3736	
LP58–53		Caml	0.48	199	16.5	DA	06487n6408	
GD–80		Mono	0.15	140	14.5	DA	06517s0205	
G108–42		Mono	0.27	161	15.9	DC	06548n0245	
L886–6	507	Mono	0.82	185	16.0	DA	06596s0623	
G87–29		Auri	0.44	222	15.8	Dλ	07069n3745	37
GD–83		Gemi	0.24	220	15.1	DA	07104n2139	
GD–294		Lynx	0.08	140	12.0	DA	07133n5830	
GD–84		Lynx	0.24	205	15.5	DC:	07144n4553	
G89–10		Gemi	0.30	225	16.3	DA	07154n1235	
GD–85		Auri	0.17	120	14.5	DB	07165n4027	
G107–70		Lynx	1.30	190	14.9	DC	07271n4817	38
GD–86		Lynx	0.25	230	15.0	DA	07307n4848	
L384–24	533	Pupp	0.66	5	13.5	DA	07321s4247	
Procyon B	541	CMin	1.25	214	10.8	D?	07367n0521	39
L745–46a	543	Pupp	1.25	117	13.0	DF	07380s1717	40
GD–89		Lynx	0.17	210	15.5	DA	07439n4416	
G193–74		Lynx	0.27	195	15.7	DC	07496n5237	
NGC 2477–116		Pupp	0.06	108	13.7		07498s3821	41
G90–28		Lynx	0.49	196	16.0	DA	07522n3630	
L97–12	555	Voln	2.04	136	14.5	DC	07528s6738	42
L817–13		Pupp	0.32	204	13.6	DA	07528s1438	
G111–54		Lynx	0.31	187	15.6	DFp	08023n3841	
GD–90		Lynx	0.08	210	15.5	DAp	08165n3741	43
G111–71		Lynx	0.38	212	16.4	DA	08167n3844	44
W309		Lynx	0.18	220	16.0	DA	08267n4531	
G51–16		Canc	0.64	192	15.7	DA	08275n3252	
LB 390		Canc	0.04	252	17.8		08369n2011	45
LB 1847		Canc	0.03	242	18.2		08369n1957	45
LB 393		Canc	0.04	249	17.7		08376n1954	45
L532–81	600	Pyxi	1.71	322	12.0	DA	08397s3247	
LDS 235b		Hyda	0.13	247	15.6	DB	08453s1848	46
GD–98		Lynx	0.17	170	15.0	DA	08543n4028	
LP90–70		UMaj	0.53	216	15.5	DC:	08557n6029	47
G47–18		Lynx	0.34	271	14.7	Dλ	08562n3309	
GD–99		Lynx	0.25	200	15.0	DA	08587n3619	
LP36–115		Caml	0.60	210	17.2	DC	09004n7327	
G195–19		UMaj	1.55	223	14.8	DCp	09125n5339	48

STAR	LFT	CON	MU"	PA°	MAG	SP	RA & DEC	N
G116-16		Lynx	0.27	176	15.5	DA	09134n4412	49
G117-B15a		LeoM	0.14	264	15.5	DA	09212n3530	50
LDS 275a		Antl	0.37	295	14.7	DC	09350s3707	51
LDS 275b		Antl	0.37	295	15.0	DC	09350s3707	51
LP370-50		Leon	0.22	244	15.8		09423n2341	52
LP370-51		Leon	0.22	244	15.8		09423n2341	52
G117-B11b		LeoM	0.14	133	17.2		09434n3305	53
SA29-130		UMaj	0.31	360	13.3	DA	09435n4408	
G49-33		Leon	0.38	218	15.1	DA	09550n2447	
G48-57		Leon	0.34	223	17.2	DC	09557n0901	
G42-33		Leon	0.34	276	15.1	DC	09591n1456	
GD-111		UMaj	0.08	260	16.0	DA	10028n4303	
LP92-65		UMaj	0.15	125	15.5	DB	10113n5704	
G43-38		Leon	0.36	304	16.2	DF	10124n0821	
G235-67		UMaj	0.38	49	14.9	DC	10196n6343	
GD-117		UMaj	0.16	290	16.5	DA	10196n4616	
G43-54		Leon	0.62	127	16.1	DC	10266n1143	
GD-122		LeoM	0.17	215	15.5	DC	10293n3256	
L825-14		Hyda	0.33	260	13.0	DA	10314s1129	
F34		UMaj	0.04	125	11.1	DO	10367n4322	54
G44-32		Leon	0.31	149	16.6		10392n1432	55
GD-124		Sext	0.16	195	16.2	DB	10460s0145	
Ton 547		LeoM			15.4	DA	10468n2810	
Ton 556		LeoM	0.16	270	14.2	DA	10520n2723	
L898-25	753	Crat	0.80	275	14.3	DA	10551s0715	
LB 253		UMaj	0.25	228	13.8	DA	11048n6014	
L970-30		Leon	0.43	180	12.9	DA	11055s0453	56
Ton 573		Leon	0.17	140	15.0	DB	11073n2635	
LP 129-10		UMaj	0.14	140	16.5	DA	11082n5621	
L971-14	792	Leon	0.54	293	15.3	DC	11158s0258	57
R627	800	Leon	1.05	270	14.2	DF	11217n2139	
LB 2012		UMaj	0.16	240	15.5	DA	11225n5436	
F43		UMaj	0.09	238	14.9	DA	11265n3828	
GD-140		UMaj	0.15	260	12.5	DA	11345n3005	
F46		Leon	0.07	309	13.2	DB	11349n1427	58
L 145-141	844	Musc	2.68	97	11.4	Dλ	11429s6434	59
G237-28		UMaj	0.42	201	16.3	DC	11431n6323	
G148-7		UMaj	0.28	203	13.6	DA	11434n3206	60
L 1405-40		Leon	0.35	259	15.6	DA	11477n2535	61
GD-312		UMaj	0.15	300	15.4	DA	11491n4104	
L 1261-24		Leon	0.32	274	15.5	DC	11542n1839	
LP7-200		Caml	0.28	263	15.9	DA	11593n8022	

A REPRESENTATIVE LIST OF WHITE DWARF STARS

STAR	LFT	CON	MU" PA°	MAG	SP	RA & DEC	N
HZ20		CVen	0.01	15.0		12102n4257	
HZ21		Coma	0.10 294	14.2	D0	12114n3312	62
HZ22 (UX Canum)		CVen	0.01	12.7	DB?	12123n3657	63
C 1		UMaj	0.12 206	13.3	DA	12133n5247	64
L 1046-18b	892	Virg	0.70 292	14.9	DA	12143n0314	65
LP39-76		Drac	0.51 271	16.6	DC	12147n6906	
G148-B4b		Coma	0.20 238	16.9		12150n3222	66
HZ28		CVen	0.12 277	15.7	DA	12300n4146	
HZ29		CVen	0.04 53	14.2	DB	12324n3755	67
GD-148		CVen	0.23 145	14.5	DA	12326n4754	
LP321-98		Coma	0.55 234	19.5		12398n3014	
GD-479		Drac	0.15 210	16.2	DB	12412n6509	
G61-17		Coma	0.44 300	15.9	DA	12447n1459	68
HZ32		CVen	0.02	15.7		12487n3727	
GD-151		Coma	0.08 320	15.4	DA	12499n1813	
HZ34		CVen	0.06 214	15.7	D0	12530n3749	69
GD-320		CVen	0.08 250	16.5	DA	12532n4814	
LTT 13724		Coma	0.16 185	13.0	DA	12546n2218	
L 1408-19		Coma	0.28 305	15.4	DA	12575n2750	
W457	960	Virg	1.00 210	15.9	DC	12577n0346	70
LB 2520		UMaj	0.07 30	15.2	DA	12585n5920	
R974		Virg	0.47 175	12.8	DK	12595s0149	71
HZ39		Coma	0.01 150	15.4	DA	13024n2823	
LB 2539		UMaj	0.16 320	14.6	DBp	13025n5944	
G61-29		Coma	0.35 280	16.0	DBe	13033n1817	72
HZ43		Coma	0.17 245	12.9	DA	13140n2922	73
G177-31		CVen	0.53 289	13.9	DA	13171n4521	
G199-71		UMaj	0.49 301	16.1	DF	13258n5810	
W485	1014	Virg	1.19 248	12.3	DA	13277s0819	74
G165-7		CVen	0.58 263	16.0	DK	13287n3045	
GD-325		CVen	0.17 265	14.0	DB	13340n4844	75
W489	1023	Virg	3.87 254	14.7	DK	13343n0358	76
LDS 455a		Virg	0.09 253	15.3	DA	13343s1604	77
AC+70°5824		UMin	0.40 267	12.8	DA	13378n7033	
AC+58°43662		UMaj	0.29 318	16.7	DA	13406n5715	
LTT 14019		UMaj	0.28 315	14.0	DA	13443n5715	
G63-54		Boot	0.87 265	15.7	DA	13450n1037	
LP380-5		Boot	1.47 276	16.7	DK	13458n2350	78
L619-50		Hyda	0.23 164	15.0	DA	13481s2718	79
G165-B5b		CVen	0.14 275	16.3	DA	13549n3403	80
G64-43		Virg	0.29 259	16.0	DB	14038s0105	
LTT 14141		Boot	0.25 175	13.9	DA	14083n3223	

STAR	LFT	CON	MU" PA°	MAG	SP	RA & DEC	N
LP439-354		Boot	0.68 270	18.1		14095n1547	
LP439-356		Boot	0.28 259	16.2	DA	14115n1544	
G166-14		Boot	0.27 283	15.9	DA	14135n2311	
F93		Boot		15.3	DA	14153n1316	
GD-335		Boot	0.08 150	16.0	DC	14199n3509	
Ton 197		Boot		15.3	DA	14215n3150	
LTT 14236		Boot	0.24 230	14.5	DA	14222n0931	
Ton 202		Boot		15.7	DC	14253n2646	81
L 19-2		Cham	0.45 208	13.0	DA	14254s8107	
LP 98-57		Drac	0.30 265	16.8	DF	14262n6124	
GD-336		Boot	0.15 290	14.5	DA	14299n3720	
GD-337		Boot	0.08 295	16.0	DA	14331n5349	82
L 1126-68	1146	Virg	0.91 243	15.5	DA	14484n0747	
GD-173		Virg	0.17 260	15.0	DA	14513n0038	
GD-175		Libr	0.16 260	15.5	DA	15032s0703	
GD-340		Drac	0.07 210	15.0	DA	15087n6344	
GD-178		Boot	0.16 300	14.9	DA	15094n3216	
LP 135-155		Drac	0.38 218	16.2		15106n5637	83
GD-344		Boot	0.16 170	15.0	DA	15255n4324	
BPM 77964		Serp	0.15 200	14.5	DA	15315s0217	
GD-347		Boot	0.16 150	15.0	DA	15348n5024	
GD-189		Libr	0.15 110	15.5	DA	15396s0332	
GD-190		Serp	0.08 165	15.0	DB	15421n1816	
CPD-37°6571b		Lupi	0.48 243	13.3	DA	15442s3745	84
BD+1°3129b		Serp		15.3	DA	15442n0053	85
LTT 14705		Serp	0.18 310	14.5	DA	15502n1819	
G152-B4a		Libr	0.12 198	15.7	DA	15554s0900	86
R808	1242	CorB	0.57 167	14.4	DA	15596n3658	
C 2		Herc		13.8	DA	16067n4215	
G138-8		Herc	0.59 181	15.1	DA	16091n1330	
BPM 91358		Herc	0.24 225	15.0	DA	16108n1640	
GD-198		Scor	0.09 135	15.5	DB	16126s1111	
LTT 6494		Scor	0.26 200	16.1	DA	16147s1250	
L 770-3		Scor	0.25 223	12.4	DA	16151s1528	
Ton 816		Herc	0.07 225	15.0	DA	16205n2602	
R640	1280	Herc	0.87 328	13.9	DF	16268n3651	
GD-354		Drac		15.5	DA	16304n6149	
LP 101-16		Drac	1.62 316	15.5	DC	16335n5716	87
G138-47		Herc	0.27 186	16.5	DC	16354n1347	88
G138-49		Herc	0.59 215	16.0	DA	16365n0547	
L 1491-27		Herc	0.43 185	14.6	DA	16374n3332	
G138-56		Herc	0.72 179	15.7	DA	16393n1519	

A REPRESENTATIVE LIST OF WHITE DWARF STARS

A REPRESENTATIVE LIST OF WHITE DWARF STARS

STAR	LFT	CON	MU"	PA°	MAG	SP	RA & DEC	N
LP43–146		Drac	0.32	340	16.3	DF	16413n7316	
Ton 261		Herc			16.1	D0	16445n2643	89
GD–358		Herc	0.15	295	14.1	DB	16454n3234	
LP 101–148		Drac	0.33	153	12.6	DA	16476n5909	
G169–34		Herc	0.61	176	14.5	DA	16550n2132	
G181–B5b		Herc	0.14	170	16.1	DA	17070n3317	90
L845–70		Ophi	0.36	132	14.3	DC	17085s1445	91
BPM 92077		Herc	0.14	170	15.1	DB	17098n2305	
LP7–172		Drac	0.27	349	17.1	DF	17101n6823	
G240–51		Drac	0.36	191	13.4	DA	17135n6935	
W672b	1339	Ophi	0.62	234	14.3	DA	17162n0200	92
LP332–27		Herc	0.25	217	17.1		17276n2919	93
LP332–28		Herc	0.25	217	17.9		17276n2919	93
G140–2		Ophi	0.28	187	15.5	DA	17362n0518	
GD–363		Herc	0.15	345	15.0	DA	17370n4154	
G154–B5b		Serp	0.09	25	15.0	DA	17431s1317	94
LP44–113		Drac	1.65	311	14.6	DC	17489n7053	95
G140–B1b		Ophi	0.10	174	15.8	DC	17505n0949	96
Xi Draconis B		Drac	0.14	56	15.6		17527n5653	97
LP9–231		Drac	3.59	337	15.4		17570n8244	98
BPM 92960		Herc	0.15	160	15.0	DA	18097n2829	
G206–17		Herc	0.27	220	16.2	DA	18117n3248	99
G206–18		Herc	0.27	220	17.0	DC	18118n3248	99
G155–15		Scut	0.62	196	15.8	DA	18213s1310	
GD–378		Lyra	0.15	360	14.0	DB	18220n4102	
R137		Ophi	0.37	218	13.9	DA	18248n0402	
L993–18		Scut	0.30	175	14.5	DA	18265s0431	
G155–19		Scut	0.32	140	15.9	DA	18279s1039	
G227–35		Drac	0.36	318	16.1	DC	18294n5445	100
L849–15		Scut	0.40	224	15.1		18402s1112	
GD–215		Serp	0.16	300	15.0	DA	18409n0417	
L 1498–127		Lyra	0.34	7	14.6	DA	18557n3352	
G141–54		Aqil	0.27	42	15.5	DA	18575n1154	
AC+70°8247	1446	Drac	0.52	12	13.2	Dp	19006n7036	101
G142–B2b		Aqil	0.10	180	14.2	DA	19113n1331	102
G125–3		Lyra	0.28	179	15.0	DC	19173n3838	
LDS 678a		Aqil	0.20	198	12.2	DA	19179s0746	103
GD–218		Aqil	0.08	150	16.0	DA	19182n1105	
BPM 94172		Aqil	0.08	200	13.0	DA	19194n1435	
LDS 683b		Sgtr	0.14	188	15.9	DA	19329s1336	104
G185–32		Cygn	0.46	85	13.4	DA	19352n2737	
BPM 94484		Cygn	0.16	200	13.5	DA	19366n3247	

A REPRESENTATIVE LIST OF WHITE DWARF STARS

STAR	LFT	CON	MU"	PA°	MAG	SP	RA & DEC	N
L 1573-31		Cygn	0.20	353	14.5	DB	19404n3724	
L 1140-73		Aqil	0.22	209	15.0	DA	19419n0847	
G142-50		Aqil	0.28	182	14.5	DA	19433n1620	
L997-21	1503	Aqil	0.80	213	13.7	DA	19540s0109	
WZ Sagittae		Sgte	0.08	100	15.2	DA	20053n1733	105
L 710-30		Capr	0.29	155	14.2	DA	20073s2155	
G230-30		Cygn	0.27	204	16.0	DA	20082n5103	
GD-229		Cygn	0.08	20	14.5	D:	20104n3105	106
G24-9		Aqil	0.71	206	15.7	DC	20115n0634	107
GD-391		Cygn	0.15	60	13.5	DA	20281n3903	
L 116-79		Pavo	0.02	140	13.2	DA	20306s6816	
W1346	1554	Vulp	0.66	217	11.5	DA	20323n2453	108
L 711-10		Capr	0.33	106	12.0	DA	20397s2016	
G210-36		Cygn	0.27	50	13.4	DA	20472n3717	
R193b = VB11		Aqar	0.82	105	17.8	DC	20541s0502	109
GD-393		Cygn	0.16	230	15.0	DA	20589n5039	
GD-232		Delp	0.08	100	15.6	DA	20590n1809	
G187-15		Cygn	0.48	212	14.5	DC	20597n3137	
G144-51		Delp	0.31	200	15.8	DA	20598n1901	
L24-52		Octn	0.37	167	13.5	DA	21052s8201	
GD-394		Cygn	0.14	90	13.1	DA	21111n4954	
G231-40		Cygn	0.27	333	12.5	DA	21174n5400	
R198		Cygn	0.30	42	14.7	DA	21248n5459	
AC+73°8031		Ceph	0.32	172	12.9	DA	21266n7325	
LDS 749b		Aqar	0.35	87	14.7	DB	21296n0002	110
L 1002-62		Aqar	0.24	85	14.0	DB	21310s0446	
GD-234		Pegs	0.16	285	14.5	DA	21343n2151	
G126-18		Pegs	0.29	70	15.7	DA	21365n2258	
AC+82°3818	1649	Ceph	0.64	29	13.0	DA	21367n8249	
L 1363-3	1655	Pegs	0.72	201	13.2	DC	21403n2045	111
GD-396		Cygn	0.23	60	15.6	DA	21433n3518	
L930-80		Aqar	0.30	114	14.8	DB	21449s0758	
G93-48		Pegs	0.33	182	12.8	DA	21499n0209	
L 1003-16		Aqar	0.27	177	14.4	DA	21515s0131	
LDS 766a		Grus	0.22	144	14.0	DA	21549s4342	112
G18-34		Pegs	0.40	48	16.0	DC:	22074n1415	113
LDS 785a		PscA	0.21	94	14.5	DB	22246s3427	114
GD-236		Pegs	0.23	240	15.5	DA	22266n0607	
G233-19		Lacr	0.29	226	16.6	DC	22346n5248	
F106 = PHL380		Aqar	0.24	80	16.0	DA	22402s0430	
G28-13		Aqar	0.34	209	16.2	DA	22405s0143	
G67-23		Pegs	0.53	85	14.4	DA	22466n2221	

A REPRESENTATIVE LIST OF WHITE DWARF STARS

STAR	LFT	CON	MU"	PA°	MAG	SP	RA & DEC	N
G128-7		Pegs	1.27	83	16.1	DA:	22490n2924	
G156-64		Aqar	0.59	91	16.4	DA	22532s0806	115
GD-243		Aqar	0.08	90	15.5	DB	22532s0617	
GD-244		Pegs	0.24	80	16.0	DA	22543n1237	
G28-27		Pisc	0.31	220	17.2	Dλ	22549n0740	116
GD-245		Pegs	0.16	115	13.5	DA	22564n2500	
BPM 97895		Pegs	0.15	90	12.5	DA	23098n1031	
F108		Pisc	0.05	260	12.9	DA	23136s0207	117
F110		Aqar	0.01	135	11.5	D0	23174s0526	117
LP77-24		Ceph	0.13	271	14.6		23193n6910	118
LP77-25		Ceph	0.13	271	17.2		23193n6910	118
G128-62		Pegs	0.29	154	16.6	DA:	23235n2536	
GD-248		Pegs	0.08	210	14.5	DC	23236n1544	
G29-38		Pisc	0.56	237	13.3	DA	23263n0458	
C 3		Andr	0.30	116	13.8	DA	23292n4045	
G128-72		Pegs	0.52	87	15.3	DA	23295n2642	119
LB 1526		Phoe	0.03		13.4	DA	23316s4730	
GD-251		Pegs	0.17	215	15.5	DA	23319n2902	
L 1512-34		Pegs	0.25	256	12.9	DA	23414n3215	120
GD-561		Ceph			14.2	DA	23429n8040	
LDS 826a	1835	Scul	0.50	217	14.6	DA	23515s3333	121
L505-42	1837	Scul	0.68	178	14.7	DA	23517s3650	
L362-81	1849	Phoe	0.90	138	13.1	DA	23596s4325	

NOTES TO THE TABLE. Numbers in the last column refer to notes, as follows:

1. G130-49. Double star with G130-50; sep= 50". Primary is magnitude 13.2, spectrum dK.
2. G158-78. Double star with G158-77 (BPM 70193). Primary is magnitude 14.6, spectrum probably dM. Separation 90".
3. Van Maanen's Star. Probably the nearest of the white dwarf stars with the exception of Sirius B and Procyon B. Description and chart in constellation Pisces.
4. G34-49. Double star with G34-48; separation 24". The primary is magnitude 13.1, spectrum sub-dwarf M3.
5. LP768-500. One of the smallest and densest of all the degenerate stars, with a diameter of possibly about 900 miles. (See note on page 401)

6. G71-B5b. Double star with G71-B5a; separation 13". The primary is magnitude 14.3, spectrum about dM0.

7. h Per 1166. This star appears in the field of NGC 869, one-half of the famous Double Cluster in Perseus. The designation is perhaps somewhat unfortunate, since the star is obviously not a true cluster member at all, but merely a foreground object. The computed distance is about 130 light years.

8. G94-B5b. Double star with G94-B5a; separation 26". The primary star is GC 2869, magnitude 8.3, spectrum dG.

9. F24. J.L.Greenstein and O.J.Eggen (1965) report that the spectrum of this star resembles a post-nova.

10. LB 3303. The star has a red companion of about the 14th magnitude, 8" distant.

11. G5-28. This star was given a tentative spectral class of DM by Greenstein and Eggen (1965) but with a note of caution; it now appears that the star is probably an unusual red sub-dwarf instead. See Note 17.

12. L587-77a. Double star with L587-77b; separation 8". The companion is a 15th magnitude red dwarf.

13. W219. The spectrum has been occasionally classed as DC, but is peculiar for the presence of a single broad band at $\lambda 4670$, attributed to C2. The color index is equivalent to class F. According to R.A.Bell (1962) the star is a member of a group of seven objects which appear to show the same space motion toward a convergent in the southern sky at 14440s5900. Although the reality of the group is somewhat questionable, it would appear to be more than a coincidence that four of the supposed members should be white dwarfs of the rare $\lambda 4670$ type: W219, L879-14, L145-141, and G28-27. Two other members are DC stars (L97-12 and L1363-3) and the last probable member is G7-17 which is either a red degenerate or a late-type subdwarf. The computed space velocity of the group is about 90 miles per second. See also Note 17.

14. BD+16°516. This star appears to be a member of a close eclipsing binary system. See note on page 408.

15. LB 1497. Possibly a member of the Pleiades star cluster in Taurus; if so, the absolute magnitude is about +11.

16. HZ4. Probably a member of the Hyades star cluster.

17. G7-17. When this star was discovered at Lowell Observatory, it was noticed that the proper motion is nearly the same as W219, which is about 4° distant. In 1962, R.A.Bell announced that several other white dwarfs show the same motion toward a common convergent; seven possible members are now known. (Refer to note 13). G7-17 itself has a very peculiar spectrum, tentatively

classed as "DM" by Greenstein & Eggen (1965) but with some uncertainty: "The spectrum is so peculiar that it probably cannot be confused with those of ordinary dM stars, but not enough is known about the spectra of either low-mass, partially degenerate stars or extremely metal-deficient dM stars to eliminate these stars as possible explanations of the observed spectrum".

18. LB 227. Probably a member of the Hyades star cluster.
19. HZ10. Probably a member of the Hyades.
20. LP357-186. One of the smallest and densest of all known degenerate stars. Refer to note on pages 400-402.
21. HG7-138. Probably a member of the Hyades. The spectral type is uncertain; Greenstein in 1974 classed it as a sub-dwarf G, but not a true white dwarf.
22. 40 Eridani B, also called "Omicron 2 Eridani B". The first known white dwarf and an easy object for the small telescope. A member of a triple star system. See detailed description in the constellation Eridanus.
23. VR-7. Probably a member of the Hyades.
24. VR-16. Probably a member of the Hyades.
25. G174-34. This system is a binary, also known by the double star number given at the Vatican Observatory in 1908, Stein 2051. The two stars are magnitudes 11.2 and 13.0 (visual) and the separation is presently 6.5" with orbital motion of about 1° per year. The primary is a red dwarf of class dM4; the degenerate star is presently the eastern component of the pair and has a spectral type of DC. This is one of the few systems known which permits an accurate determination of the mass of a white dwarf star. See also note on page 402.
26. HZ9. Probably a member of the Hyades. The spectrum is peculiar for the presence of emission lines and large changes in radial velocity. Luyten suggests that HZ9 is a binary with a period of about half a day; the companion may be an M-type red dwarf.
27. HZ7. Probably a member of the Hyades star cluster.
28. G39-27. Double star with G39-28, separation about 2.1'. The bright primary is magnitude 8.42, spectrum dK5e. The space motion appears to class this pair as an out-lying member of the Hyades.
29. L879-14. One of the rare "λ4670" stars, and apparently a member of the W219 group. Refer to note 13.
30. HZ14. Probably a member of the Hyades.
31. G86-B1b. Double star with G86-B1a; separation 9". The primary is magnitude 14.2, spectrum near dM0.
32. G102-39. This star may be a distant companion to the 7th magnitude star GC 7413, which is located about 1.5'

north-following. The bright star is magnitude 7.7, spectral type F8

33. HL4. Announced by Haro and Luyten in the Bulletin of the Tonantzintla and Tacubaya Observatories in January 1960, with a note that the color corresponds to class G5 or K, and the star may be similar to W489. G.Herbig in 1963 obtained a spectral type of DK, and a parallax by Luyten gives the distance as about 22 light years. The absolute magnitude is thus about +15.4 visual. This is one of the reddest white dwarfs, with a color index (B-V) of +1.0 magnitude. See also note 76.

34. G99-47. One of the few stars known in which the light shows optical polarization. G195-19 and AC+70°8247 are two other examples; very strong magnetic fields are believed to be the explanation of this effect.

35. G105-B2b. Double star with G105-B2a; separation about 127". The primary is magnitude 13.2, spectrum dM2.

36. Sirius B. The famous "Pup", best known of the white dwarfs. Refer to text, pages 394 ff.

37. G87-29. Double star with G87-28; separation 15". The primary is magnitude 14.6, spectrum sdM6. The spectrum of the white dwarf shows the λ4670 band weakly.

38. G107-70. Double star with G107-69; separation 106". The companion is magnitude 13.5, spectrum sdM5. The white dwarf is the southern star; it is a close binary (about 0.7"); both spectra probably late DC, period estimated at less than 20 years, distance about 30 light years. See also notes on pages 402 and 408.

39. Procyon B. No spectrum available, due to the strong light of the 1st magnitude primary. The separation is about 4", the period about 40 years. For details refer to the constellation section Canis Minor.

40. L745-46a. Double star with L745-46b; separation 21". The companion is a 17th magnitude M-type red dwarf.

41. NGC 2477-116. The star is not a true member of the star cluster, but merely a foreground object at an estimated distance of about 130 light years. The color indicates a spectral type of about DA.

42. L97-12. Possibly a member of the W219 moving group. Refer to note 13.

43. GD-90. This star has a unique spectrum, and appears to be a strongly magnetic DA star.

44. G111-71. Double star with G111-72; separation 34". The primary is magnitude 13.2, spectrum dM2.

45. LB 390, LB 1847, and LB 393. These stars are possibly members of the Praesepe star cluster M44 in Cancer.

46. LDS 235b. Double star with LDS 235a; separation 30". The primary is magnitude 11.6, spectrum dK3.
47. LP90-70. Double star with LP90-71; separation 44". The companion is a dM2 star, photographic magnitude about 16.2. Eggen and Greenstein (1965) state that the spectrum of the white dwarf may be composite, possibly a combination of DC and dKe.
48. G195-19. One of the few stars showing optical polarization, resembling G99-47 and AC+70°8247. An additional peculiarity of G195-19, reported in 1972, is that the polarization varies in a cycle of 1.33 day.
49. G116-16. This star may be a distant proper motion companion to G116-14, also known as LTT 12432. The separation is 17.2' and the bright primary is magnitude 9.0, spectrum dG5.
50. G117-B15a. Double star with G117-B15b. Separation 15", the companion is magnitude 16.1, spectrum dM2.
51. LDS 275. This interesting pair appears to be one of the few binaries known in which BOTH components are white dwarfs. Both spectra are DC; separation 3.6". Orbital motion is quite slow, suggesting a period of about 700 years; on this assumption the total mass would be 1.4 solar masses. See also the note on page 408.
52. LP370-50 and LP370-51. Another pair resembling the system described above. On color measurements alone, this common motion pair appears to consist of two degenerate stars. See note on page 409.
53. G117-B11b. Double star with G117-B11a; separation 14". The primary is magnitude 15.2; the color indicates a spectral type of about dM3.
54. F34. The spectrum is peculiar; the star may be an O-type subdwarf, rather than a true white dwarf.
55. G44-32. A slight variability of this star was suspected at Lowell Observatory, and confirmed at Cerro Tololo in 1969, where variations of about 2% were detected in periods ranging from 10 to about 27 minutes.
56. LP970-30. Common motion with LP970-27; separation about 280". The companion is magnitude 12.6, spectrum dM6.
57. L971-14. The spectrum appears to be composite, and the star may be a binary resembling SS Cygni. Eggen and Greenstein (1965) suggest a subdwarf O-star with a red companion, both of low luminosity. The color of the system is similar to that of the nova WZ Sagittae.
58. F46. Spectral class somewhat uncertain; may be only a subdwarf of type O.
59. L 145-141. One of the rare "λ4670" white dwarfs, and a

probable member of the W219 moving group. Refer to note 13.

60. G148-7. Double star with G148-6; separation 11". The companion is magnitude 14.1, spectrum dM4.
61. L 1405-40. Double star with L 1405-41; separation 36". The companion is magnitude 15.5, spectrum dM2e.
62. HZ21. Spectral type uncertain; Humason and Zwicky in 1946 classed the star as type B0. Greenstein regards it as either a D0 or extreme subdwarf 0.
63. HZ22, also called UX Canum Venaticorum. The star is a short period variable with a range of about 0.3 mag. Modern studies suggest that the star is a low-mass binary; at least one component is a white dwarf or extreme subdwarf. See note on page 408.
64. C 1. The spectrum seems to be composite, combining the features of a DA white dwarf with a dMe star.
65. L 1046-18b. Double star with L 1046-18a; separation 2.6". The companion is an M-type red dwarf of magnitude 14.7, photographic.
66. G148-B4b. Double star with G148-B4a; separation 8". The primary is magnitude 15.1, spectrum early dM.
67. HZ29. Probably one of the hottest and most luminous white dwarfs known, with a computed absolute magnitude of about +8.9. The spectrum is classed as DB, and shows broad shallow spectral features of helium.
68. G61-17. Double star with G61-16; separation 25". The primary is magnitude 13.5, spectrum sdM2.
69. HZ34. Classed as either D0 or extremely hot sd0.
70. W457. A trigonometric parallax indicates a distance of about 42 light years; the computed absolute magnitude is then +15.4. This is one of the lowest luminosities known for any white dwarf.
71. Classed as "DK" with some uncertainty; Greenstein and Eggen state that the spectrum is that of a very late type subdwarf with very weak lines. A direct parallax leads to a distance of about 45 light years and an absolute magnitude of +12.2; this would place the star definitely among the degenerate objects.
72. G61-29. An unusual spectrum with emission lines, resembling HZ9. The star shows light variations in a period of about 6.28 hours; the light curve suggests an eclipsing binary of the U Geminorum type.
73. HZ43. Double star; the companion is magnitude 14.7, spectral type dM; the separation is 3".
74. W485. The star R476 is located about 8.4' distant; it shows nearly the same proper motion and the two stars may form a wide common motion pair. R476 is magnitude

14.3 and spectral type dM5.

75. GD-325. The star shows a composite spectrum: DB + dM.

76. W489. This remarkable object was the first of the late type white dwarfs to be discovered, and is still one of the few known with a color index (B-V) as great as 1.0 magnitude. The only other degenerate stars of an equal redness known in 1975 are: HL4, R193b, LP380-5, G107-70, and possibly G7-17 if that strange object is truly a degenerate star. The distance of W489, from a direct parallax, is about 25 light years; this gives the absolute magnitude as +10.8. See also note 109.

77. LDS 455a. Double star with LDS 455b; separation 14". The companion is an M-type red dwarf, visual magnitude about 13.9, photographic about 15.5.

78. LP380-5. Double star with LP380-6; separation 187". According to photoelectric measurements at the U.S. Naval Observatory in 1975, the photographic magnitudes are 16.72 and 17.26; the brighter star is the white dwarf. This is one of the few degenerate stars which is redder than the classic W489; the color index (B-V) of LP380-5 is +1.09 mag. The faint companion is a late M-type red dwarf. See also note 76.

79. L619-50. Double star with L619-49; separation 8". The primary is magnitude 13, spectral type dK.

80. G165-B5b. Double star with G165-B5a. Separation 58". The bright primary is BD+34°2473, magnitude 9.6.

81. Ton 202. Greenstein and Eggen (1965) state that the spectrum resembles that of an old nova, with very weak absorption and emission lines.

82. GD-337. The spectrum is composite: DA + dK.

83. LP 135-155. Double star with LP 135-154; separation 18". The primary is magnitude 15.7 with a color index equal to that of a late dM star. This is one of the pairs where the photographically brighter star is visually fainter than the other star; this has caused the designations "primary" and "secondary" to be reversed in some lists, and the numbers to be reversed as well! The white dwarf, in any case, is the western member of the pair, and photographically the brighter of the two stars although usually called the "secondary".

84. CPD-37°6571b. This is the companion to the 6th magnitude star GC 21205 or λ249 near NGC 5986 in Lupus. The primary is also designated DM-37°10500 and BS 5864; the white dwarf is 15.2" distant and is magnitude 13.3. See note on page 407.

85. BD+1°3129b. The primary is magnitude 9.8, spectrum dG0. Separation 16". Proper motion very slight, if any.

86. G152-B4a. Double star with G152-B4b; separation 10". The white dwarf is the eastern star of the pair, and is photographically the brighter of the two stars. The companion is a red M-type dwarf.
87. LP 101-16. Double star with LP 101-15, separation 25". The primary is a dM5e star of visual magnitude 12.9.
88. G138-47. Double star with G138-46, separation about 178". The eastern star of the pair is the white dwarf; the other star has a spectral type of about dM2, and a visual magnitude of 14.0.
89. Ton 261. Spectrum either D0 or sd0.
90. G181-B5b. Double star with G181-B5a, separation 36". The primary is BD+33°2834, visual magnitude 8.7.
91. L845-70. Usually classed as "DC" though Greenstein states that the λ4670 band may show faintly.
92. W672b. Double star with W672a, separation 13". The primary is magnitude 14.0, spectrum sdM6.
93. LP332-27 & 28. This common proper motion pair, on color measurements alone, appears to be a double white dwarf system. Refer to note on page 409.
94. G154-B5b. Double star with G154-B5a, separation 32". The primary is type dM3, visual magnitude 11.9.
95. LP44-113. Another of the rare white dwarfs which shows optical polarization. G195-19 and AC+70°8247 are two other examples.
96. G140-B1b. Double star with G140-B1a, separation 27". The bright primary is magnitude 9.5 visual, spectral type dK2.
97. Xi Draconis B. The primary is magnitude 3.76, spectrum K2 III; the white dwarf is magnitude 15.6 and 316" distant in PA 290° according to Luyten. This is one of the very few cases known in which a giant star of any type has a white dwarf companion. No spectrum of the faint star appears to be available at present (1976) but the color measurements appear to make the identification certain. See also the note on page 407.
98. LP9-231. Discovered by Luyten in 1965, and originally thought to be one of the smallest and least luminous of all white dwarf stars. Based on a preliminary trigonometrical parallax, Luyten estimated the distance to be about 10 light years, which gave an absolute magnitude of +17.9. Newer measurements have not confirmed these results, and the star now seems to be at least several times more distant than was originally thought, and the absolute magnitude now appears to be about +13.

99. G206-17 and G206-18. One of the rare double star systems known in which both components are white dwarf stars. Separation 55", magnitudes 16.2 and 17.0. See note on page 409.

100. G227-35. The light of the star shows some polarization.

101. AC+70°8247. The spectrum is unique, showing an unidentified band at λ4135; the color index is closely comparable to class A. This star is believed to be one of the smallest white dwarfs, probably about half the size of the Earth. This is also one of the magnetic white dwarfs, showing optical polarization.

102. G142-B2b. Double star with G142-B2a, separation 19". The primary is magnitude 12.7, spectrum dM2.

103. LDS 678a. Double star with LDS 678b, separation 27". The designations "a" and "b" are somewhat confused since the two stars are nearly equal in the visual. The white dwarf, which is designated "a" in this list, has a visual magnitude of 12.2. The other star is 12.1 visual, spectrum dM5.

104. LDS 683b. Double star with LDS 683a, separation 28". The primary is magnitude 13.6, spectrum sdM1.

105. WZ Sagittae. Famous recurrent nova with outbursts in 1913 and 1946. The star is an extremely close and rapid binary with a period of 81.6 minutes, and the spectrum of at least one component is definitely that of a white dwarf, but with superimposed emission lines. Refer to the constellation section Sagitta for additional description and charts.

106. GD-229. The star shows optical polarization which varies in at least two different time scales; one on the order of a few minutes, and the other of about 1 day.

107. G24-9. Double star with G24-10, separation 102". The primary is LFT 1534, magnitude 13.2, spectrum dM5.

108. W1346. Possibly a spectroscopic binary, based on reported variations in radial velocity.

109. R193b. Double star with R193, separation 13". The primary is magnitude 13.3, spectrum dM4. The faint star is one of the reddest of the degenerate stars with a color index (B-V) of +1.13 magnitude. As of 1975 this appears to exceed the color index of any other degenerate star known, including HL4, W489, LP380-5, and G107-70. Refer to note 76.

110. LDS 749b. Double star with LDS 749a, separation 133". The bright primary is magnitude 9.9, spectrum sdK4.

111. L 1363-3. This star is one of the presumed members of the W219 moving group; the spectral type is DC. Refer to note 13.

112. LDS 766a. Double star with LDS 766b, separation 9". Luyten gives the photographic magnitude of the faint companion as 15.8; the color index classes it as a dM star.

113. G18-34. The star is usually classed as "DC" though it has been reported to show very weak and broad lines of hydrogen. Greenstein classes it as a DA with weak lines.

114. LDS 785a. Double star with LDS 785b, separation 9". The companion is a dM star. Luyten gives the magnitudes as 13.9 and 14.1 (pg).

115. G156-64. Double star with G156-65, separation 43". The bright primary is LFT 1749, magnitude 8.7, spectrum G6.

116. G28-27. One of the rare $\lambda4670$ stars, and probably a member of the W219 moving group. Refer to note 13. Greenstein and Eggen (1965) state that the spectrum may be composite (DK+DA?) and is possibly variable.

117. F108 and F110. Spectral types uncertain; may be hot subdwarfs. F108 shows only hydrogen lines; F110 has additional sharp lines of helium.

118. LP77-24 and LP77-25. A possible new addition to the list of binaries in which both components are white dwarfs. See note on page 409.

119. G128-72. This is also L 1440-18, or LTT 16922.

120. L 1512-34. Double star with L 1512-35, separation 174". The primary is magnitude 11.7, spectrum dM5. This pair appears in some lists under the designation L 1512-34 A&B.

121. LDS 826a. Double star with LDS 826b, separation 7". The companion is a red dwarf of photographic magnitude 15.0.

DESCRIPTIVE NOTES (Cont'd)

BETA   Name- MURZIM or MURZAM, "The Announcer", so called from the fact that it rises just before Sirius and therefore heralds the appearance of the great Dog Star. Murzim is magnitude 1.98, spectrum B1 II, color white. Position 06205s1756. This is the standard example of a "Beta Canis Majoris type" of variable, a rare group of of pulsating B-type giants distinguished by ultra-short periods and small amplitudes. Otto Struve (1955) referred to them as "quasi-cepheids". The first star of the type to be recognized was Beta Cephei, whose variable radial velo- city was discovered by E.B.Frost in 1902. Beta Canis itself was identified as a similar type of object by S.Albrecht in 1908; the period was found to be almost exactly 6 hours.

The small variations in light, about 0.03 magnitude, were first detected by J.Stebbins in 1928. In addition to the periodic changes in light and radial velocity, F.Henroteau found in 1918 that the appearance and width of the spectral lines also changes periodically, in a cycle about 2 minutes longer than the radial velocity variations. In a detailed analysis of the star in 1934, W.F.Meyer found that the ob- served velocity curve indicates that the star is oscillat- ing in two superimposed periods of $6^h 00^m$ and $6^h 02^m$. The two interfering pulsations produce a secondary harmonic cycle or "beat period" of about 49 days, the longest known for stars of this type.

A direct parallax obtained at McCormick gave a distance of about 465 light years, but recent studies of the spec- tral features suggest a somewhat greater distance, probably about 750 light years. The computed absolute magnitude is -4.8, and the actual luminosity is 7600 times that of the Sun. The mean radial velocity is 20 miles per second in recession; the very small proper motion has been measured at 0.004" annually.

The Beta Canis Majoris variables are all B-type stars of high luminosity, apparently restricted to spectral types B1, B2, and B3. The spectral changes are accompanied in some cases by slight variations in light, amounting to 0.2 magnitude in the most extreme case, that of BW Vulpeculae. When plotted on the H-R diagram, the stars form a well- marked group merging with the main sequence at B3, and ly- ing about 1 magnitude above it at B1. According to a study

by D.H.McNamara (1953) the stars show a period-luminosity
relation similar to that displayed by the better known
cepheids. Stars near the top of the sequence (B1) have absolute magnitudes of about -4.9 and periods of about six
hours; for those near the lower end (B3) the figures are
-2.8, and about $3\frac{1}{2}$ hours. O.Struve (1962) suggests that
Beta Canis itself has a mass of about 10 solar masses and
a diameter in the range of 10 times that of the Sun. For
the less luminous stars of the class, the derived masses
and diameters appear to be about half the values assigned
to Beta Canis. As in the case of the cepheids, there is a
correlation of the periods with the mass, radius, and the
density.

It is of interest to note that several Beta Canis stars
exist in relatively young star groups such as the Scorpio-
Centaurus Association. Hence it seems that these objects
are young, rather massive stars, which are beginning to
evolve away from the main sequence. It is believed that a
slow expansion and decrease in density occurs during this
stage, which must result in a gradual lengthening of the
pulsation period. Here is perhaps one of the rare chances
to observe an actual change in the characteristics of a
star due to its rapid evolution. Several stars of the type
show slight increases in period; that of BW Vulpeculae is
the greatest known, at about 3 seconds per century. This
is, however, a much larger increase than is suggested by
theory, and implies a faster evolution than expected. The
interpretation of such period changes therefore remains
uncertain.

Beta Canis Majoris stars are now often referred to in
modern literature as "Beta Cephei stars" since that star
was the first example known. They may not be truly rare
stars, but only a limited number are known, since the
small light variations make them difficult to detect. In
1967, only 33 were known, but several dozen additional
specimens have since been discovered, including a dozen
belonging to the Perseus I association which surrounds
the great Double Cluster. Among the closer and brighter
members of the class are: Beta Cephei, Sigma Scorpii,
Gamma Pegasi, Delta Ceti, Beta Crucis, Theta Ophiuchi,
Nu Eridani, Tau-1 Lupi, Xi-1 Canis Majoris, 15 Canis
Majoris, 12 Lacertae, 16 Lacertae, and 53 Arietis.

GAMMA   Mag 4.10, spectrum B8 II. Position 07015s1533.
The estimated distance is about 1250 light
years, and the absolute magnitude about -3.8. The star is
a giant with about 2700 times the luminosity of the Sun.
The proper motion of Gamma Canis is very slight, about
0.01" per year; the radial velocity (somewhat variable)
is about 18 miles per second in recession.

Gamma Canis presents us with the interesting and
unsolved problem of the supposed "secular variations" in
the light of certain stars; the question of whether the
star's light has perceptibly changed over many hundreds
of years. Although labeled Gamma by Bayer, this star is
much fainter than the stars designated Delta, Epsilon,
Zeta, Eta, and even Omicron! According to R.H.Allen,
"Montanari said that it entirely disappeared in 1670, and
was not observed again for twenty-three years, when it
reappeared to Miraldi, and since has maintained a steady
lustre, although faint for its lettering." Beta Librae
is another star which presents a similar puzzle.

DELTA   Name- WESEN. Mag 1.82, spectrum F8 Ia. Position
07064s2619. The distance of this star is too
great for accurate parallax measurements, but indirect
calculations give about 2100 light years. The spectral
characteristics are those of a supergiant of absolute magnitude -7.0, and the actual luminosity must be about 60,000
times that of the Sun. There is no measurable proper motion. The radial velocity is 21 miles per second in recession.

EPSILON   Name- ADHARA. Mag 1.49, spectrum B2 II. The
position is 06567s2854. This is the 22nd star
in order of brightness in the heavens, and should really
be included in lists of the 1st magnitude stars. It lies
at a distance of about 680 light years, and has an absolute
magnitude of about -5.0 (luminosity = 9000 suns). The
annual proper motion is immeasurably small; the radial
velocity is 16 miles per second in recession.

The companion star, of the 8th magnitude, was discovered at the Cape Observatory in 1850. No change in PA
or separation has been noted in over a century. The projected separation of the pair is about 1600 AU.

ZETA Mag 3.02; spectrum B2.5 V. The position is 06184s3002. The computed distance is about 390 light years; the absolute magnitude about -2.4, and the actual luminosity about 750 times that of the Sun. The star is a spectroscopic binary with a period of 685 days and an orbit which has the rather high eccentricity of 0.57. The very slight annual proper motion is less than 0.005"; the radial velocity is about 19 miles per second in recession.

ETA Name- ALUDRA. Mag 2.41; spectrum B5 Ia. The position is 07221s2912. This is another of the highly luminous supergiants which seem to be unusually plentiful in the rich Orion-Canis Major section of the sky. The computed distance is about 2700 light years, giving the true luminosity as about 55,000 times that of the Sun. The absolute magnitude is probably about equal to that of Rigel (about -7.1). The annual proper motion is again very slight, less than 0.01"; the radial velocity is 24 miles per second in recession.

A distant companion of the 7th magnitude may be located in binoculars, at 169" in PA 285°. The two stars, however, do not form a true physical pair. There are two interesting groupings of small stars south of Eta Canis; one about 2° south, and the other about 1½° southwest. Not true clusters, these objects are suitable for low powers only.

OMICRON 2 Mag 3.02; spectrum B3 Ia. Position 07009s 2346. Another highly luminous supergiant, probably the equal of Rigel in luminosity. The estimated distance is 3400 light years, and the resulting absolute magnitude about -7.1. Proper motion is negligible; the radial velocity is 29 miles per second in recession.

UW (29 Canis Majoris) Mag 4.95 (variable); Spectrum 07. Position 07166s2428. A super- giant binary star, undoubtedly one of the most massive and luminous systems known in our Galaxy. It lies in the field of the great cluster NGC 2362, about 24' to the north, but the difference in radial velocities appears to rule out any

THE MILKY WAY IN CANIS MAJOR. This field is south and east
of Sirius. Star Cluster M41 is at upper right; Eta Canis is
at lower left. Lowell Observatory photograph.

STAR CLUSTER NGC 2360 in CANIS MAJOR. This object lies
about 3½° east of Gamma Canis Majoris. Lowell Observatory
photograph with the 13-inch telescope.

DESCRIPTIVE NOTES   (Cont'd)

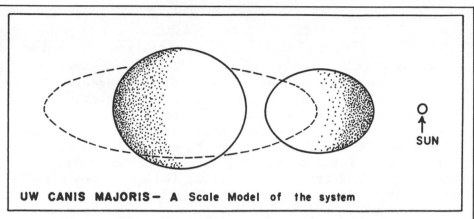

UW CANIS MAJORIS— A Scale Model of the system

possibility of true cluster membership. The radial velocity
of the star is about 5½ miles per second in approach, but
the brightest member of the cluster shows a recession velocity of about 24 miles per second.

UW Canis Majoris is one of those giant systems of the
Beta Lyrae and A0 Cassiopeiae class, revolving in a period
of 4.39351 days, with a center-to-center separation just
under 17 million miles. The computed orbit is nearly circular, with an eccentricity of 0.06. Both components are
distorted into flattened ellipsoids by rapid rotation and
tidal effects. The eclipses of the system are partial, resulting in a light decrease of about 1/4 magnitude.

Spectroscopic studies show that the primary (type 07)
is ejecting material into space, resulting in a gaseous
stream from the larger star toward the smaller. Some of
this material, lost to both stars, goes into an expanding
cloud which surrounds the system. Radial velocity measurements are distorted by these moving gas streams, complica-
ting the problem of calculating accurate orbital elements.
Computed masses of 40 and 30 suns are now believed to be
spuriously high. In his model of the system, J.Sahade (1959)
adopted masses of 19 and 23 suns, and diameters of 18.6 and
14.8 suns. The primary component is thus somewhat less massive than the secondary, which possibly indicates that it is
still in the early stages of gravitational contraction. UW
Canis may be an unusually "young" binary star.

The computed distance is about 3600 light years, leading to a total luminosity of about 16,000 suns. The total
absolute magnitude may be near -5.7.

M41   (NGC 2287) Position 06449s2042. A fine bright galactic star cluster, visible to the naked eye and partially resolvable in field glasses. It is easily located, about 4° south of Sirius. M41 is a beautiful object in low power instruments, and is a favorite of deep-sky observers. It contains about 25 bright stars and many fainter ones scattered over a field as large as that covered by the Moon. There is a bright reddish star near the center; many of the other stars seem to be arranged in curving rows and groups, a peculiar feature noted also in other open clusters such as M35 (Gemini) and M37 (Auriga).

M41 was stated by C.E.Barns to be "possibly the faintest object recorded in classical antiquity"; it was mentioned by Aristotle about 325 B.C. as one of the mysterious "cloudy spots" then known in the sky.

Approximately 100 stars are now recognized as true members of this cluster, ranging in brightness from 7th to 13th magnitudes. The 10 brightest members have the following magnitudes and spectra:

1. Mag 6.93; spect K3 II	6. Mag 7.87; spect K4 II
2. " 7.30 " G + B9	7. " 8.29 " B9 V
3. " 7.46 " K1 II	8. " 8.39 " B8 V
4. " 7.79 " G8 II	9. " 8.42 " B8 V
5. " 7.82 " K0 II	10. " 8.50 " B9 V

Star #1 is the central reddish star, a K-type giant with about 700 times the luminosity of the Sun. Its absolute magnitude may be about -2.4. Star #2 has a composite spectrum and is undoubtedly a close binary. Several other K-type giants are known in the cluster; most of the other prominent members are bright blue giants of types B8 and B9. According to studies by A.N.Cox (1954) the distance of M41 is about 2350 light years, giving the actual extent of the group as about 20 light years. Cox suggests that the total membership may be about 150 stars, which would imply a space density of about 1.1 star per cubic parsec. The total luminosity of all the members would add up to about 1500 times the light of the Sun. Radial velocity measurements show a speed of about 20 miles per second in recession.

STAR CLUSTER M41. A bright galactic star cluster in Canis
Major, a few degrees south of Sirius. Lowell Observatory
13-inch telescope photograph.

STAR CLUSTER NGC 2362. This fine group surrounds the 4th magnitude star Tau Canis Majoris.

100-inch reflector, Mt.Wilson Observatory

NGC 2362   Position 07166s2452. An unusually attractive and interesting cluster of stars surrounding the 4th magnitude star Tau or 30 Canis Majoris, just 24' south of the giant eclipsing binary UW. In very small telescopes the object may at first present the appearance of a nebulosity about the star, but any good 2-inch telescope should resolve it easily into a rich little cluster of some 40 stars. The apparent diameter is about 6', and the magnitudes of the members range from $7\frac{1}{2}$ to about 13. All the brighter stars are O and B-type giants of great size and luminosity. According to a study by H.L.Johnson (1950) the distance of the group is about 4600 light years; our Sun at that distance would appear as a star of magnitude $15\frac{1}{2}$. The true diameter of the cluster is about 8 light years.

NGC 2362 has become an object of great interest in recent years, since it seems to be one of the youngest of all known star clusters. The peculiar feature of the H-R diagram is the considerable displacement of many of the members to the right of the normal main sequence. From a comparative study of many clusters, this feature evidently means that many of the stars have not yet reached the main sequence state, and are still in the process of gravitational contraction. The age of the group can scarcely be as much as 1 million years. NGC 2362 thus resembles the wonderful Double Cluster in Perseus and the great NGC 2264 in Monoceros. And like the Double Cluster (but unlike NGC 2264) there is no evident nebulosity in the immediate vicinity. One might speculate that the formation of the cluster has exhausted the supply of interstellar gas in the region. Or has the nebulosity simply been "blown away" by the intense radiation of the newly formed giant stars? Evidence for such a process appears to exist in some other clusters, as in the group which illuminates the Rosette Nebula in Monoceros. The British astronomer Fred Hoyle suggests that the actual formation of a star cluster is unlikely to be directly observable, since it must occur deep in the heart of thick obscuring nebulous clouds. The cluster is revealed when the stellar radiation disperses the nebulosity.

Nearly central in the cluster is the bright star 30 or Tau Canis Majoris, in all probability an actual member of the group. A slight variability has been suspected,

with a possible range of about 0.10 magnitude. The mean of
several modern catalog values is 4.44; the spectral class
is 09 III. O.Struve and A.Pogo (1928) found the star to
be a spectroscopic binary with a rather long period of
154.8 days. Only one star is observed spectroscopically.
The eccentricity of the computed orbit is 0.36, and the
estimated separation of the components is about 2 AU. The
total mass of the system appears to be in the range of 40
to 50 solar masses. If accepted as a member of NGC 2362,
this star is one of the most luminous supergiants known,
with an absolute magnitude of about -7 and an actual luminosity of over 50,000 suns. R.J.Trumpler (1935) found the
radial velocity of the star to be somewhat higher than the
mean value derived from 7 cluster members; he attributed
the difference to a gravitational red-shift, which gave a
very large mass (about 300 suns) for Tau itself. Later
studies have made such abnormally large masses seem quite
unlikely. The most massive binary known appears to be
Plaskett's Star in Monoceros, where the total mass may be
about 100 times that of the Sun. (Refer also to NGC 2264,
and the Double Cluster in Perseus. See M13 in Hercules
for a discussion of cluster age-dating)

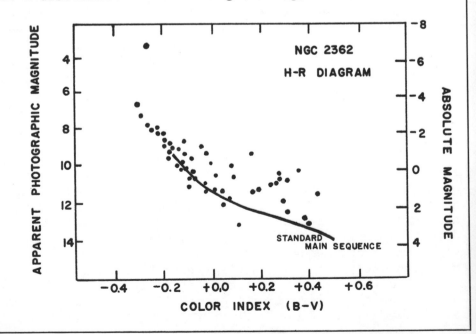

CANIS MINOR

LIST OF DOUBLE AND MULTIPLE STARS

NAME	DIST	PA	YR	MAGS	NOTES	RA & DEC
0Σ170	1.3	94	62	7½- 7½	PA dec, spect G0	07149n0923
Σ1074	0.6	161	58	7½- 8	PA inc, spect B9	07180n0030
	12.8	101	24	-12½	(β3577)	
	15.3	11	22	- 13		
	53.6	278	19	- 11		
A2939	3.7	244	21	7½-13½	spect F5	07181n0945
Σ1073	8.8	67	27	8 - 10	relfix, spect A0	07183n1017
Σ1076	2.9	110	62	8½- 8½	relfix, spect A0	07184n0409
Σ1082	19.9	326	27	8 - 8½	relfix, spect A3	07210n1048
	15.6	22	04	- 13		
Σ1095	10.1	78	32	8½- 9	relfix, spect A2	07247n0851
η	4.0	25	34	5½- 11	(β21) relfix spect F0	07253n0703
γ	30.0	240	11	4 - 13	(Lam 4) spect K3	07254n0902
A2739	3.8	225	21	8 - 12	spect A5	07267n0329
Σ1099	4.0	343	27	8½- 9	relfix, spect A0	07267n1138
A2869	0.3	59	58	8½- 8½	PA dec, spect A5	07278n0750
Σ1103	4.4	243	49	7 - 8½	relfix, spect B9	07279n0522
Σ1114	6.5	54	21	8½- 9	relfix, spect G0	07310n0924
0Σ176	1.3	217	66	7½- 9½	relfix, spect B9	07359n0037
	3.1	335	20	- 13		
α	3.9	113	62	1 - 13	PROCYON. binary,	07367n0521
	119	13	58	- 12	40 yrs (*)	
Σ1126	0.9	161	68	7- 7½	PA inc, spect A0	07375n0521
	44.4	249	14	- 10½		
Σ1130	0.3	258	66	8½- 9	PA inc, dist dec; spect G0	07390n0949
A2534	0.1	360	62	7 - 7	PA inc, spect G5; triple system, AC PA inc	07405n0019
	0.7	225	62	- 8		
Σ1134	10.2	147	28	8 - 11	spect F8	07409n0337
	85.5	347	27	- 10½		
A2879	1.2	158	35	7½-11½	spect A0	07440n0159
Σ1137	2.9	132	40	8 - 9	relfix, spect F5	07440n0415
Σ1149	21.7	41	27	7½- 9	relfix, spect G0	07469n0321
A2880	0.2	237	59	7½- 7½	PA inc, spect K1	07482n0324
0Σ182	1.0	21	58	7 - 7½	PA dec, spect A2	07501n0331
0Σ185	0.4	16	62	7 - 7	binary, 57 yrs; spect F6	07547n0116
Σ1168	6.2	220	24	8 - 12	slight PA inc; spect B9	07561n0546

LIST OF DOUBLE AND MULTIPLE STARS   (Cont'd)

NAME	DIST	PA	YR	MAGS	NOTES	RA & DEC
Σ1175	1.2	258	66	8 - 10	PA inc, dist dec, Spect G5	07598n0418
Σ1182	4.5	73	44	7 - 9	relfix, spect B9	08028n0558

LIST OF VARIABLE STARS

NAME	MagVar	PER	NOTES	RA & DEC
R	7.3--11..	338	LPV. Spect Se	07060n1006
S	7.0--13.2	332	LPV. Spect M6e--M8e	07300n0826
T	9.3--14.7	319	LPV. Spect M5e	07312n1151
U	8.3--13.5	410	LPV. Spect M4e	07386n0830
V	7.8--15..	366	LPV. Spect M6e	07043n0858
W	9.9--11..	Irr	Spect R6	07461n0531
UX	9.0--10..	150	Semi-reg; spect M5	07429n0519
YY	8.5-- 9.1	1.094	Ecl.bin.; spect F5	08040n0205
ZZ	9.3--11..	500:	Semi-reg; spect M6e	07215n0900
AD	9.0-- 9.5	.1230	Spect F2; Delta Scuti type ?	07502n0144
AI	7.5--9..	Irr		07331n0022
BC	6.15-- 6.4	35:	Semi-reg; spect gM4	07495n0325

DESCRIPTIVE   NOTES

ALPHA   Name- PROCYON. Mag 0.35; spectrum F5 IV or V.
Position 07367n0521.   Procyon is the 8th brightest star in the sky. Among all the naked eye stars it is
the 5th nearest; probably only Alpha Centauri, Sirius,
Epsilon Eridani, and 61 Cygni are closer. The value of
0.288" has been obtained for the parallax of the star,
giving a distance of 11.3 light years. Procyon has a rather large annual proper motion of 1.25" in PA 214°. The

radial velocity is about 1.8 miles per second in approach.
Procyon is about 6 times the luminosity of our sun, and
slightly over twice the diameter. The actual size cannot
be measured directly, but may be computed from the known
spectral type and total luminosity. The surface temperature
is close to 7000°; the absolute magnitude is +2.6.

The name Procyon has been in use since the days of
ancient Greece, and is the equivalent of the Latin word
"Antecanis" or "Before the Dog", an allusion to the fact
that Procyon rises immediately preceding Sirius, and thus
heralds the appearance of the great Dog Star. In Arabian
records it appears as "Al Shi'ra al Shamiyyah" and Riccioli
designated it as "Siair Siami"; both names might be trans-
lated as "The Northern Sirius". Babylonian records designate it as "Kakkab Paldara", the "Star of the Crossing of
the Water Dog", while in China it was "Nan Ho", or the
"Southern River". Today the popular name most often used
is simply "The Little Dog Star". In an almanac for the
year 1553, published by Leonard Digges, the star is quaint-
ly referred to:

*"Who, learned in matters astronomical, noteth not the
great effects at the rising of the starre called the
Litel Dogge..."*

THE COMPANION TO PROCYON. The fact that the Little Dog Star
is not a single object has been known for over a century.

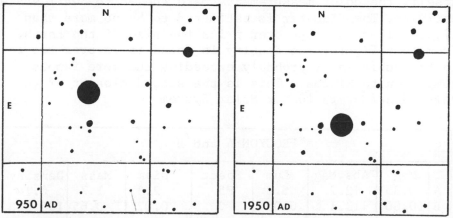

THE PROPER MOTION OF PROCYON over a period of 1000 years
is illustrated here. Grid squares are 1° on a side.

DESCRIPTIVE NOTES (Cont'd)

From observed irregularities in the proper motion, A.Auwers deduced the existence of a faint but massive companion, and in 1861 published a computed period of 40 years. The hypothetical companion was searched for many times by O.Struve, S.W.Burnham, and others, but without success. Finally, in 1896, it was detected visually with the 36-inch refractor at Lick Observatory by J.M.Schaeberle; the position was at 4.6" from the primary in PA 320°. Because of its extreme faintness and its proximity to the brilliant 1st magnitude primary, the faint star is a very difficult object for observers, and can be seen only in great telescopes. The magnitude of the star was estimated to be about 13 at the time of discovery, but it appears likely that the brilliancy of Procyon causes the small star to appear fainter than it actually is; the true magnitude may be about 11. The orbital motion is direct, with a period of 40.65 years. According to computations by K.A.Strand (1949) the semi-major axis of the orbit is 4.55", giving the mean separation of the components as about 15 AU. The orbit has the moderate eccentricity of 0.40, and periastron occurs in 1968. The apparent separation of the pair varies from 2.2" (1968) up to about 5.0" (1990).

The companion, usually called Procyon B, is a very remarkable example of a white dwarf star. It is at least 15,000 times fainter than Procyon, but has a mass of 65% that of the Sun. Although no spectrum of the companion has been obtained, due to the overpowering glare of Procyon itself, the color measurements make the white dwarf status definite. The diameter is estimated to be no more than 17,000 miles, or just over twice the size of the Earth. From this figure, the density is found to be over two tons to the cubic inch, probably exceeding the more famous companion to Sirius. This is the second closest of the white dwarf stars to our Solar System.

PROCYON A and B							
	Mag	Abs.Mag	Lum	Spect	Diam	Mass	Density
A	0.35	+2.7	5.8	F5	2.3	1.7	0.14
B	10.80	+13.1	.0005	?	0.02	0.65	100,000

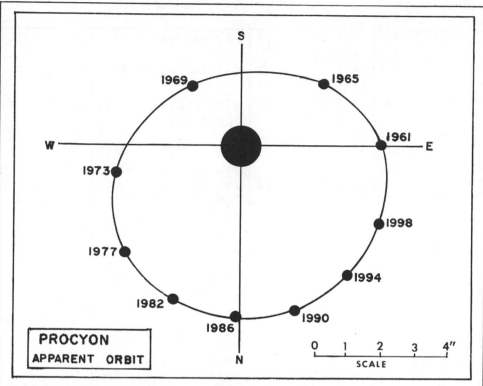

PROCYON
APPARENT ORBIT

Procyon also has several faint optical companions, but none of these share the proper motion of the A-B pair. The brightest of these, Procyon C, was 56.6" distant when first measured in 1836, but had increased to 119" by 1958, and the separation will continue to widen. A minor mystery concerns a reported 9th magnitude star discovered by Smyth in 1833, at 145" from Procyon in PA 85°; this object could not be found by Bond at Harvard in 1848, but was seen again by Fletcher in 1850 for the last time. Possibly this object was some odd variety of variable which is only occasionally bright.

BETA   Name- GOMEISA. Position 07244n0824. Mag 2.84, Spectrum B7 V.   Located about 4.3° NW from Procyon. The computed distance is about 210 light years; the absolute magnitude about -1.1, and the actual luminosity about 230 times that of the Sun. Beta Canis Minoris shows an annual proper motion of about 0.07"; the radial velocity (variable) is 15 miles per second in recession.

CAPRICORNUS

LIST OF DOUBLE AND MULTIPLE STARS

NAME	DIST	PA	YR	MAGS	NOTES	RA & DEC
Hd156	6.2	88	46	7 - 10	relfix, spect F0	20042s0904
	8.4	58	46	- 12		
3	26.9	34	12	5½- 13	(β294) spect B9	20136s1230
3b	7.5	180	00	13- 13		
α	376	291	25	3½- 4	wide optical pair, spect G9, G3 (*)	20153s1242
α²	6.6	172	59	3½- 11	optical, PA inc; spect G9; B = AGC 12	20153s1242
AGC 12	1.2	240	59	11- 11	relfix	
α¹	45.4	221	32	4 - 9	(β295) relfix; spect G3	20149s1240
	44.3	182	60	- 13		
σ	55.9	179	13	5½- 10	relfix, spect K3	20165s1917
β	205	267	25	3 - 6	wide cpm pair; spect F8, B9 (*)	20182s1456
βᵇ	0.8	89	52	6 - 10	PA dec, spect B9	
Ho456	14.1	208	28	8 - 13	slight PA dec, spect K0	20205s2654
π	3.2	148	55	5- 8½	(β60) cpm; relfix spect B8	20245s1822
Σ2683	22.8	67	15	9 - 9	relfix, spect both F8	20255s1318
ρ	0.5	158	58	5 - 9	dist dec, PA dec, spect F2 (β61)	20260s1759
	55.2	151	01	- 13		
o	21.9	239	55	6 - 6½	relfix, cpm; spect A2, A3	20270s1845
β668	2.9	29	62	6 - 11	relfix, spect G3, cpm pair	20297s1002
λ--	0.1	345	62	8 - 8	PA dec, spect F0	20297s1647
h2975	9.8	23	55	7½- 11	relfix, spect F8, cpm pair	20306s2224
Σ2699	9.5	195	47	8 - 9	relfix, spect F0	20342s1255
h1537	3.3	22	59	8½- 8½	relfix, spect F8	20335s1529
τ	0.4	107	62	5½- 7	(Hu 200) binary, 95 yrs; spect B6	20365s1508
Hn 40	5.2	356	51	8½- 9	relfix, spect F0	20406s1940
β674	1.6	99	42	8- 10½	slight PA dec, spect K0	20419s2104
h5220	18.0	354	51	7½- 9	relfix, spect F2	20435s2703
β153	1.6	265	47	7½- 8½	PA dec, spect A2	20444s2636

LIST OF DOUBLE AND MULTIPLE STARS   (Cont'd)

NAME	DIST	PA	YR	MAGS	NOTES	RA & DEC
S 763	15.8	294	50	6½- 7	relfix, spect G5	20456s1823
h2998	5.9	141	37	8½- 9	relfix, spect G5	20460s2048
h5226	18.6	68	50	7- 8½	relfix, spect K0	20471s2733
h3003	1.7	206	59	6 - 8	cpm; slight PA dec spect sgG5	20501s2358
λ436	0.2	76	36	8- 8½	PA dec, spect F5	21009s2431
HI 47	3.7	313	57	8 - 8	PA dec, spect G5	21096s1512
β271	3.2	255	59	7 - 9½	PA inc, spect G4,	21169s2633
	81.7	72	09	- 12	K4, cpm.	
β1262	2.0	112	62	8½- 9½	relfix, spect G8	21195s1507
ζ	21.3	13	33	4 - 13	(λ446) Spect G5p no certain change, possibly cpm.	21238s2238
β683	2.9	195	44	8½- 11	spect K0, relfix	21247s2026
λ449	2.0	190	43	7- 13½	no certain change, spect F0	21286s1927
A2096	0.8	38	59	7 -10½	PA dec, spect F5	21309s1625
Ho464	17.5	104	08	7 - 11	relfix, spect F0	21388s1505
41	5.5	205	54	5½- 12	(λ454) relfix, cpm; spect gG9; near cluster M30	21392s2329
β1036	4.7	206	44	8 - 11	relfix, spect B9	21448s1732
β168	5.7	72	53	8 - 9½	relfix, spect F8	21510s2015
β168b	0.4	56	01	10- 10	(Hu 380) binary, about 106 yrs.	
h3071	18.0	323	59	7 - 11	dist dec, spect A2	21551s1523

LIST OF VARIABLE STARS

NAME	MagVar	PER	NOTES	RA & DEC
δ	2.9--3.1	1.023	Ecl.bin.; spect A7 (*)	21443s1621
R	9.4--14..	345	LPV. Spect Ne	20085s1425
T	8.5--14.3	269	LPV. Spect M2e	21193s1523
V	8.2--14.4	276	LPV. Spect M5e	21047s2407
X	9.9--14..	218	LPV.	21057s2133
Y	9.9--15.0	206	LPV.	21316s1412
Z	8.6--15.0	181	LPV. Spect M2e	21078s1623
RR	7.8--14.6	277	LPV. Spect M5e	20594s2717
RS	7.0--9..	340	Semi-reg; spect M4	21045s1637
RT	6.5---8.1	395	Semi-reg; spect N3	20141s2129
RU	9.2--15.2	347	LPV. Spect Me	20296s2152
RW	9.8--11.0	3.392	Ecl.bin.; spect A2+A4	20151s1750
TU	9.6--11..	88	Semi-reg; spect K4	20114s1603
TW	9.7--10.5	28.557	Cepheid; W Virginis type Spect F8e	20117s1400
TX	9.5--14..	129	LPV. Spect M4e	20377s1715
UU	8.7--10..	100	Semi-reg; spect M4	21335s1406
AB	9.1--10.0	Irr	Spect M3	20567s1500
AD	8.7---9.2	6.118	Ecl.bin.; spect K0	21371s1614
AE	7.9--9..	180	Semi-reg; spect M4	20161s1601

LIST OF STAR CLUSTERS, NEBULAE, AND GALAXIES

NGC	OTH	TYPE	SUMMARY DESCRIPTION	RA & DEC
6907	141[3]	⊘	SBb; 12.1; 2.5' x 2.0' cF,cL,1E,vglbM; S-shape spiral	20221s2458
7099	M30	⊕	Mag 8, diam 6', class V ! B,L,1E; vRi; stars mags 12.... (*)	21375s2325

DESCRIPTIVE NOTES

ALPHA   Name- AL GIEDI, "The Goat". Position 20153s 1242. A wide naked-eye double with a separation of 376". The brighter star is Alpha 2; magnitude 3.56, spectrum G9 III. The fainter is Alpha 1, magnitude 4.24, spectrum G3 Ib. Both stars have small companions; details are given in the list of double and multiple stars.

The two bright stars do not form a true physical pair. Alpha 2 is about 100 light years distant; it shows an annual proper motion of about 0.06" and a radial velocity of about zero. The other star appears to be about 5 times as distant, and shows a radial velocity of 14 miles per second in recession.

BETA   Name- DABIH. Mag 3.08; spectrum composite, F8 V + A0. Position 20182s1456. A wide and easy pair, offering a fine color contrast for the small telescope. The two stars share a common proper motion of about 0.04" per year; the projected separation is 9400 AU. Dabih is estimated to be about 150 light years distant; the resulting luminosity of the primary is about 100 times that of the Sun. The radial velocity is 11 miles per second in approach. Beta B is a close double star itself.

The bright star is a spectroscopic triple, with periods of 8.678 days and 1374 days. T.W.Webb also mentions a tiny pair between the wide components, with a separation of 6.4" in PA 322°, both stars being of the 13th magnitude.

DELTA   Name- DENEB ALGIEDI. Mag 2.82; spectrum A7 V. Position 21443s1621. The distance is about 50 light years; the actual luminosity about 25 times that of the Sun. The star is an eclipsing binary system of small range, with a period of 1.023 day. The spectral type of the secondary remains unknown. From the radial velocity measurements, the orbit is found to be very nearly circular, with a computed separation of the components of 1 to 2 million miles The star shows an annual proper motion of 0.39" in PA 138; the radial velocity is about 2 miles per second in approach.

About 4° NE from Delta, near Mu Capricorni, is the spot where the planet Neptune was first detected by J. Galle at the Berlin Observatory, Sept.25, 1846, the result of predictions by J.C.Adams and U.J.Leverrier.

M30   (NGC 7099) Position 21375s2325. Globular
star cluster, located in the eastern part
of the constellation, about 6½° south of Gamma Capricorni,
and some 25' west and slightly north from 41 Capricorni.
M30 is one of Messier's discoveries, found in August 1764,
and described as a "nebula....seen with difficulty in an
ordinary telescope of 3½ feet...It is round and I saw no
star there, having observed it with a good Gregorian telescope of 104X". The cluster was probably first resolved by
Sir William Herschel in 1783; he found it "brilliant...with
two rows of stars, 4 or 5 in a line which probably belong
to it". According to John Herschel, M30 has a noticeably
elliptical shape, about 4' X 3', and this effect was noted
also by Admiral Smyth who called it a "fine, pale white
cluster...bright and from the straggling streams of stars
on its N. edge has an elliptical aspect with a central
blaze; few other stars in the field". Lord Rosse, as in the
case of several other globulars, thought to discern some
hint of spiral arrangement in the outer streams of stars.
E.J.Hartung in his "Astronomical Objects for Southern Telescopes" (1968) seems to confirm the Rosse observation: "the
well-resolved centre is compressed and two short straight
rays of stars emerge Np. while from the N edge irregular
streams of stars come out almost spirally.."
        The central nucleus of M30 is fairly dense, about
1.5' in size, and the extreme diameter on photographs is
about 9'. According to the catalogue published by H.B.S.
Hogg (First Supplement 1963) M30 has a total integrated
magnitude of 8.58 (pg) and an integrated spectral type of
F3. From the color-magnitude diagram the distance must be
close to 40,000 light years, giving the extreme diameter as
about 100 light years. Three short-period variable stars
are known in the cluster, and a fourth object which seems
to be an eruptive variable perhaps resembling U Geminorum.
Eight additional variables have been detected up to 1973,
but their classification is yet uncertain. M30, like many
of the globulars, shows a very high radial velocity, about
108 miles per second in approach.
        The nearby bright star 41 Capricorni is a common
proper motion double; magnitudes 5½ and about 12, separation
5.5", with perhaps a slight PA increase since the early
observations of T.J.J.See in 1897.

GLOBULAR STAR CLUSTER M30 in CAPRICORNUS. The bright star in the field is 41 Capricorni. Lowell Observatory photograph made with the 13-inch telescope.

CARINA

LIST OF DOUBLE AND MULTIPLE STARS

NAME	DIST	PA	YR	MAGS	NOTES	RA & DEC
h3884	25.6	282	18	7 - 11	spect K0	06366s5518
I 157	1.6	347	43	6½- 9½	spect G5	06456s5438
h3921	6.0	272	12	7½- 11	spect A0	06598s5819
△ 39	1.7	82	52	6 - 7	slight PA inc, spect B9	07025s5906
I 184	0.5	165	51	8 - 8½	PA dec, spect F0	07085s6030
	24.5	340	00	- 12		
h3941	0.6	291	51	7½- 8	PA dec, spect K0	07087s6018
△41	7.0	225	54	8 - 8	(Rmk 5) relfix, spect K0	07094s5531
Hd 199	0.3	132	51	6½- 7	PA dec, spect A0	07116s6306
Hd 304	15.0	350	00	7 - 10	spect A0	07119s5724
h3952	16.2	277	17	7 - 11	spect K0	07151s5357
I 485	3.1	356	26	8 - 10	spect F5	07178s5227
△44	9.5	23	55	6½- 7	(Rmk 6) PA inc, both spectra F2	07192s5213
h3958	29.7	281	55	7½- 8½	slight dist dec; spect A0, F0	07194s5206
h3962	8.6	105	15	7½- 8½	relfix, spect B9	07200s5641
h3962b	1.1	257	30	8½- 12	(Rst 247)	
I 777	1.3	47	42	8½- 10	spect K5	07351s5343
h4000	1.6	241	42	7 - 9½	slight PA inc, spect B9	07414s5833
h4005	36.2	218	15	6½- 11	relfix, spect F0	07446s5636
Hu1585	1.6	68	41	7½- 11½	spect K5	07446s5850
Cor 58	22.9	43	17	8½- 8½	both spectra G5	07454s5941
h4008	20.8	230	17	7 -12½	spect B9	07462s5312
Cor 60	4.0	55	33	7½- 9	spect A0	07474s5457
λ 88	6.9	183	33	5½- 11	spect K0	07481s5613
h4014	11.1	154	17	8 - 9	spect A0	07481s6334
h4018	5.1	327	33	7½- 9	relfix, spect B9	07513s5929
	60.4	259	00	- 11		
h4021	8.3	307	33	7½- 13	PA inc, spect G0	07523s5818
Gls 77	34.8	341	28	8 - 8½	spect K0	07541s5328
h4027	9.4	115	17	8½- 9	relfix, in cluster NGC 2516	07559s6041
LDS 198	48.0			5½- 11	cpm; spect G2	07569s6010
h4031	5.5	357	21	8 - 8½	relfix, spect B8, in NGC 2516	07574s6043
I 1104	8.5	243	34	8 - 8	relfix, spect A0, in NGC 2516	07580s6030

LIST OF DOUBLE AND MULTIPLE STARS   (Cont'd)

NAME	DIST	PA	YR	MAGS	NOTES	RA & DEC
△60	40.4	161	38	6 - 8	spect B5, B8	08002s5422
Hrg 12	4.6	38	37	8 - 10½	spect A0	08002s6123
I 190	2.3	118	43	7½- 10	spect A2	08036s5450
M1b 2	1.6	351	44	8 - 8½	relfix, spect A0	08042s6015
h4053	11.6	98	17	8 - 10	spect K0	08075s6056
Rmk 8	3.9	66	43	5½- 8	(c Carinae) cpm;	08145s6246
					relfix, spect A2	
h4079	30.4	171	13	7- 11½	spect K0	08165s5543
Hu1438	8.4	201	33	7- 11½	spect A2	08241s5535
h4125	7.6	237	18	5½- 10	relfix, spect G5	08364s6241
h4128	1.4	210	46	6½- 7	PA & dist dec,	08382s6009
					spect A0	
Hu1443	0.6	260	13	8 - 8½		08395s5600
h4130	3.8	238	55	6½- 9	PA inc, spect A2	08395s5722
Rmk 9	4.1	292	51	7 - 7	relfix, spect B8	08439s5832
	50.9	359	13	- 11		
	61.4	222	13	- 11		
φ 37	0.9	46	43	7 - 10	Spect A0	08511s5509
△74	40.4	75	17	5½- 7	(b¹) relfix,	08557s5902
					spect B3, B8	
h4175	19.9	132	17	7½- 10	spect F0	09021s6209
S1r 16	1.0	310	47	7½- 9½	spect B9	09029s6429
h4178	3.3	162	34	6½- 10	spect A3	09034s5739
h4190	8.2	22	32	6½-10½	relfix, spect B3	09102s5745
Hd 207	0.2	219	59	7½- 7½	PA inc, spect A0	09116s6042
Cp 43	4.7	338	40	8 - 8½	spect G5	09170s7011
Rmk 10	10.4	18	26	7½- 7½	relfix, spect A0	09173s6936
L3846	0.3	254	35	6 - 6½	(h4206) (I 12)	09175s7441
	7.1	343	35	- 10	PA dec, spect A0	
	48.2	353	19	-10½		
I 494	1.9	238	34	7½- 10	spect A0	09203s6037
h4213	8.7	327	32	6 - 9½	relfix, spect A2	09242s6144
I 830	17.9	58	11	7½- 10	spect M0	09265s5707
I 200	1.1	355	38	7 - 9	spect A0	09308s6101
I 201	1.9	144	36	7½- 9½	PA dec, spect F0	09312s7331
R 123	1.9	34	47	7½- 7½	relfix, spect B9	09318s5744
I 359	1.6	169	45	7½-10½	PA dec, spect A0	09348s6759
h4232	10.9	302	51	7½- 8	relfix, spect A0	09369s5718
I 203	0.3	311	47	8 - 8½	relfix, spect B9	09371s6219

LIST OF DOUBLE AND MULTIPLE STARS   (Cont'd)

NAME	DIST	PA	YR	MAGS	NOTES	RA & DEC
B 780	0.2	238	59	6 - 6½	binary, 10.7 yrs; spect A2	09392s5745
h4241	34.2	304	18	6½- 11	spect A0	09411s6641
h4240	12.4	57	17	7½- 9½	relfix, spect B8	09418s5948
Hd 208	3.1	140	32	7 - 11	spect A0	09442s6932
I 204	1.7	126	11	7½- 10	spect B8	09448s6953
	3.7	320	01	- 12		
ʊ	5.0	127	43	3 - 6	(Rmk 11) relfix; fine object, cpm; spect A9, F0	09459s6450
Rmk 12	9.2	213	18	7 - 8½	spect B9	09541s6857
I 291	1.2	302	38	7½- 10	PA dec, spect F2	10003s7029
Hrg 47	1.2	350	47	6½- 8	relfix, spect B8	10020s6136
h4292	60.2	123	19	5½- 9	optical, spect G9	10073s6534
I 13	0.7	131	47	7 - 7	PA dec, spect A0	10083s6827
	26.0	40	33	- 11½		
S1r 17	3.8	341	44	7 - 11	relfix, spect B9	10115s6455
S1r 17b	4.5	287	31	11- 13		
Hu1597	0.4	260	60	7 - 7	PA inc, spect A2	10143s5939
h4306	2.1	134	41	7 - 7	relfix, spect A0	10175s6425
R 141	1.9	44	47	7½- 8½	PA inc, spect B9	10187s6655
R 151	3.4	192	13	7½- 9½	spect A0	10297s6838
h4333	32.0	101	17	5 - 12	spect gK5	10299s7258
h4335	7.9	219	29	8 - 8½	spect F5	10306s6949
△ 93	25.0	39	39	8 - 8½	relfix, spect A0	10332s6353
△ 93b	3.0	232	39	9 -10½	(I 74)	
△ 94	14.5	21	32	4½- 8	relfix, cpm; spect K5, A	10369s5855
G1s152	23.5	78	59	6 - 8½	(R153) dist inc; optical, spect M1 + B	10371s5834
R 152	2.1	16	47	7 - 9	PA inc, spect B9	10372s6415
△ 97	12.4	174	50	7½- 8½	relfix, spect B5	10413s6054
Cor112	6.6	241	34	8½- 11		10419s5918
h4356	2.4	149	34	7½- 9½	relfix, in great nebula NGC 3372	10420s5917
h4360	2.0	115	39	8 - 8	relfix, spect A3; in NGC 3372	10422s5919
	12.6	288	34	- 8		
δ 8	7.3	278	34	6½-12½	spect B0	10424s5944

LIST OF DOUBLE AND MULTIPLE STARS (Cont'd)

NAME	DIST	PA	YR	MAGS	NOTES	RA & DEC
△99	63.0	75	17	6½- 6½	spect A4, A3	10428s7035
	39.9	42	17	- 10		
R 161	1.0	284	60	6½- 7½	PA inc, spect A0	10474s5904
h4378	31.0	345	18	7 - 10	relfix, spect A0	10490s5941
I 418	2.3	202	43	8 - 8	spect A3	10511s6249
h4383	1.5	285	46	6½- 7	relfix, spect B8	10521s7027
h4387	23.6	152	13	7½- 10	spect B0	10542s5717
h4393	8.4	131	18	7 - 9½	relfix, spect B8	10556s6846
R 164	4.0	78	39	6½- 10	PA dec, spect B9	10572s6103
Gls159	17.6	274	20	8 - 9	spect B0	11029s5926
Hd 210	12.0	230	00	6- 11½	spect gG8; in NGC	11043s5824
	15.0	180	00	- 12½	3532	
I 874	0.5	80	41	8 - 8½	spect F8	11097s7411
h4416	14.7	170	17	9½- 12	A is suspected variable	11102s7109
I 231	2.6	2	41	7½- 10	spect B9	11103s7056
R 163	1.6	58	59	7 - 7½	relfix, spect A0	11152s5850
R 163b	0.4	24	59	8 - 8½	(Rst 4472)	

LIST OF VARIABLE STARS

NAME	MagVar	PER	NOTES	RA & DEC
η	-0.8---8	---	Nova-like irregular; in great nebula NGC 3372 (*)	10431s5925
l	4.3---5.1	35.56	Cepheid, spect F8--K0	09439s6217
R	4.0--10.0	309	LPV. Spect M4e--M7e	09310s6234
S	5.4---9.5	149	LPV. Spect K7e--M4e	10078s6118
U	5.7---7.1	38.76	Cepheid; spect G0--K2	10558s5928
V	7.4---8.1	6.697	Cepheid; spect G2--K0	08277s5957
X	8.0---8.7	1.083	Ecl.bin.; both spect A0	08302s5904
Y	7.7---8.3	3.640	Cepheid; spect F5	10313s5814
Z	10--15.2	384	LPV. Spect M6e	10122s5836
RR	7.4---8.6	109	Semi-reg; spect M6	09565s5837

CARINA

LIST OF VARIABLE STARS (Cont'd)

NAME	MagVar	PER	NOTES	RA & DEC
RS	7.0----	---	Nova 1895	11060s6140
RT	9.2--10.7	Irr	Z Andromedae type? spect M3	10428s5909
RU	8.4--10..	Irr	Spect N3	09144s6601
RV	9.9--14..	367	LPV. Spect M6e	09570s6339
RW	8.5--13..	318	LPV. Spect M4e	09189s6833
RX	10..12..	333	LPV.	10350s6203
RY	9.8--13..	422	LPV.	11179s6136
RZ	9.0--14..	273	LPV. Spect M4e	10342s7027
ST	9.1--10.0	.9016	Ecl.bin.; spect A0+F6	10142s5958
SU	8.6--15..	231	LPV. Spect Me	10118s6038
SV	8.7--12..	298	LPV.	09468s5918
SX	9.0--9.9	4.860	Cepheid, spect F5--G8	10441s5717
SY	8.8--10.5	Irr	Spect N3	11134s5739
SZ	8.0--10..	126	Semi-reg; spect N3	09583s5959
TY	9.6--10.7	180	Semi-reg; spect M5	10502s7230
TZ	8.4--9.6	70	Semi-reg; spect R5	10443s6521
UU	8.8---12..	198	LPV. Spect M3	09268s7319
UW	9.0---9.8	5.346	Cepheid, spect G0	10251s5925
UX	7.7---8.4	3.682	Cepheid, spect F4--G5	10273s5721
UY	8.4---9.5	5.544	Cepheid, spect G	10303s6131
UZ	8.9---9.7	5.205	Cepheid, spect G0	10345s6045
VY	7.0---8.1	18.93	Cepheid, spect F9--K0	10426s5718
WW	9.6--10.5	4.677	Cepheid, spect F0	10496s5907
WZ	8.7--10.4	23.01	Cepheid, spect F8	10533s6040
XX	8.5--10.2	15.72	Cepheid, spect G0	10552s6452
XY	8.6-- 9.7	12.44	Cepheid, spect G5	11003s6400
XZ	8.0-- 9.2	16.65	Cepheid, spect K5	11022s6043
YY	10--10.8	2.643	Ecl.bin.	10185s6112
YZ	8.3-- 9.5	18.16	Cepheid, spect G5	10264s5906
AC	8.0--9..	100	Semi-reg; spect M	07060s5818
AF	8.5--13..	434	LPV.	10122s5847
AG	7.1--9..	Irr	Erratic, spect A2e--B5e	10542s6011
AQ	8.8---9.3	9.770	Cepheid, spect G0	10197s6049
BO	8.2--9...	Irr	Spect M2	10439s5913
BZ	7.8---9.6	97	Semi-reg; spect M2	10522s6146
CK	8.1---9.4	525	Semi-reg; spect M2	10266s5956
CL	8.6--11..	513	Semi-reg; spect M3	10520s6049
CO	9.6--10.8	8.312	Ecl.bin.	10164s6308
CX	10--10.8	3.347	Ecl.bin.; spect A	10555s5817
CY	9.6--10.3	4.266	Cepheid, spect G0	10558s6039

LIST OF VARIABLE STARS (Cont'd)

NAME	MagVar	PER	NOTES	RA & DEC
DO	9.1---9.3	3.852	Ecl.bin.; spect B8	10116s5858
DV	10--10.3	.8405	Ecl.bin.; spect B8	10382s5956
DW	9.7--10.3	1.328	Ecl.bin.; spect B5	10414s5946
EM	8.9---9.2	3.414	Ecl.bin.; spect B0	11099s6049
ER	7.0---7.5	7.718	Cepheid, spect F8	11075s5834
EV	8.2---9.7	347	Semi-reg; spect M4	10186s6012
EX	9.5--11.0	1.396	Ecl.bin.; spect G0	10234s6323
EY	9.8--10.2	2.876	Cepheid	10405s6054
EZ	9.6--10.0	1.189	Ecl.bin.; spect B8	10411s6208
FR	9.3--9.9	10.72	Cepheid, spect G5	11122s5947
GG	9.1---9.5	62.09	Ecl.bin.; lyrid, spect Be	10540s6008
GH	9.2--9.6	5.726	Cepheid, spect G0	11086s6029
GI	8.0---8.6	4.431	Cepheid, spect F8	11118s5738
GL	8.9---9.6	2.422	Ecl.bin; spect B3	11125s6023
GM	9.0---9.2	1.535	Ecl.bin; spect B9	10354s5859
GV	8.9---9.4	4.295	Ecl.bin; spect A0	11034s5828
GW	9.9--10.8	1.129	Ecl.bin; spect B5	09350s5946
GX	8.6---9.9	7.196	Cepheid, spect K0	09538s5811
GY	9.9--12..	294	LPV.	10004s6235
GZ	9.9--10.4	4.159	Cepheid, spect G5	10186s5907
HK	9.7--10.3	6.696	Cepheid, spect G5	11017s6022
HP	8.5-- 8.8	1.600	Ecl.bin.; spect B2	10178s5709
HR	8.2-- 9.6	Irr	Spect B2e, perhaps XX Ophiuchi type	10211s5922
HW	8.0---8.8	9.200	Cepheid, spect K0	10375s6053
IT	7.7---8.1	7.536	Cepheid, spect F7--K2	11100s6129
IW	7.6---9.3	67.5	RV Tauri type, spect G0	09257s6325
IX	8.0--9..	400	Semi-reg; spect M1	10485s5943
KV	8.2--9..	150:	Semi-reg; spect M4	11011s6651

LIST OF STAR CLUSTERS, NEBULAE, AND GALAXIES

NGC	OTH	TYPE	SUMMARY DESCRIPTION	RA & DEC
2516		⬡	vL,B,pRi; diam 60'; about 100 stars mags 7...13; class G (*)	07597s6044
----	I.2448	◎	Mag 12, diam 8"; vS,R, appearance nearly stellar	09066s6944
2808	△265	⊕	! Mag 6.0, diam 7', class I, L,vB,eRi,eCM, stars mags 13 ...	09109s6439
2867		◎	Mag 9.7; diam 12", vS,B,R	09200s5806
----	I.2501	◎	Mag 11, diam 2", stellar	09374s5952
3059		⊖	SBb; 12.2; 2.9' x 2.6' F,L,iR, glbM	09495s7341
3114	△297	⬡	L,B,lC, diam 40', about 100 stars mags 9...13; class E	10011s5953
3136		⊖	E4; 12.4; 1.4' x 0.7' pB,pS,lE, gbM, BN	10045s6708
----	I.2553	◎	Mag 13, diam 4", nearly stellar	10078s6222
3199		☐	B,L, irregular neby	10151s5743
3211		◎	Mag 12, diam 14", F,S,R	10162s6226
----	I.2581	⬡	pL,B, mag 5, diam 6'; about 35 stars with central 5^m star.	10254s5723
3293		⬡	B,Ri, diam 8', about 50 stars mags 6...13, class D; dark neby to south	10315s5758
3324	△322	☐	pB,vL, 15' diam with 8^m star spect type 05e	10355s5822
----	Mel 101	⬡	Diam 15', 40 faint stars; class E	10404s6450
----	I.2602	⬡	vvL,B, scattered group incl 3^m star Theta Carinae; diam 70', about 30 stars mags 5..	10410s6408
3372	△309	☐	! B,eL,irregular, with dark lanes, diam 80' x 85'. "Keyhole Nebula", contains nova-lie variable η Carinae (*)	10431s5925
----	I.2621	◎	Mag 10.5; diam 2", stellar	10584s6458

LIST OF STAR CLUSTERS, NEBULAE, AND GALAXIES   (Cont'd)

NGC	OTH	TYPE	SUMMARY DESCRIPTION	RA & DEC
3503		☐	F,S, diffuse neby, several faint stars involved	10593s6027
3532	△323	⬡	! eL,Ri,1C, diam 60', 150 stars mags 8....12; class F, fine object (*)	11034s5824
3572		⬡	pL,B,diam 6'; about 40 stars mags 7...14, class D	11073s5958
3590		⬡	pRi,C,E, diam 4'; about 25 stars mags 9.... class F	11098s6032
3581		☐	F,S, fan-shaped neby; many faint nebulous patches in field, incl 3582, 3584, 3579, 3586	11101s6100
----	I.2714	⬡	diam 12', pC, 150 faint stars; class E	11152s6226

DESCRIPTIVE NOTES

ALPHA   Name- CANOPUS. Mag -0.72, spectrum F0 Ib or F0 II.   Position 06228s5240. This is the second brightest star in the sky, exceeded only by Sirius. Canopus is the Great Star of the South - a name and a legend only to many North American observers, but a dazzling gem to our more fortunately situated neighbors to the south. From the southern half of the United States it may be glimpsed during the winter months, low on the southern horizon, and culminating about 20 minutes before Sirius. The low altitude is evidently the cause of the widespread impression that Canopus is golden or orange in color; the true tint is nearly white. Opposition date (midnight culmination)is December 27.

Probably because of the inaccessibility to the great telescopes of the northern hemisphere, Canopus had not been adequately observed until recently, and very discordant estimates of distance, size, and brightness have appeared in astronomical catalogs. A distance of over 600 light years has been quoted in many observing lists, and the luminosity has been thought to be as high as 60,000 times that of the Sun. Modern studies do not support these large estimates, yet there is no doubt that Canopus is actually a very large and brilliant star, at least when compared with our Sun. According to a trigonometric parallax obtained at the Cape Observatory in South Africa, the distance is in the range of 100 to 120 light years. This gives Canopus an absolute magnitude of about -3.1, in good agreement with the luminosity computed from the spectral features. The diameter may be about 30 times that of the Sun, and the true brightness about 1400 times the Sun's. The annual proper motion is 0.025"; the radial velocity is 12 miles per second in recession.

BETA Name- MIAPLACIDUS. Mag 1.67, spectrum A1 IV.
Position 09127s6931. The distance is about 85 light years, the actual brightness about 110 times that of the Sun, and the absolute magnitude about -0.4. The annual proper motion is 0.18"; the radial velocity is about 3 miles per second in approach.

EPSILON Name- AVIOR. Mag 1.86; spectrum composite,
K0 II and B. Position 08215s5921. The computed distance is 340 light years, the actual luminosity about 1400 times that of the Sun. The annual proper motion is 0.03"; the radial velocity is 7 miles per second in recession.

ETA Position 10431s5925. A very remarkable nebular
variable star which should perhaps be classed with the novae. It was first recorded by Halley in 1677 as a 4th magnitude star. For the next century it varied in an irregular manner, reaching 2nd magnitude in 1730, falling to 4th magnitude about 1782, brightening again about 1801, and fading again to 4th magnitude in 1811. In 1820 the star began to brighten steadily, rising to 2nd magnitude in 1822 and attaining 1st magnitude in 1827. The first maximum was only a preliminary; the star faded back to 2nd magnitude for about 5 years, then rose again to become as bright as Rigel. After a second slight decline it increased once more

and in April 1843 it reached its maximum brilliancy of
about -0.8 when it outshone every star in the sky with the
exception of Sirius. After this final flare-up the star
faded slowly, becoming invisible to the naked eye in 1868.

The variations since 1870 have been comparatively un-
spectacular. A rise of about a magnitude occurred in the
1890s, but by 1900 the brightness had faded to 8th magnitude where it remained for a number of years. In 1941 Eta
Carinae brightened again, and in 1953 was about 7th magnitude. The future activity of the star is quite unpredictable, but it seems possible that it may rise to great brilliancy again.

Eta Carinae is located in one of the most splendid regions of the southern Milky Way, the great diffuse nebulosity NGC 3372, often called the "Key-hole Nebula", remarkable both for its great size and the complexity of its
structure. Sir John Herschel found words inadequate "to
convey a full impression of the beauty and sublimity of
the spectacle offered by this nebula, when viewed in a
sweep, ushered in as it is by a glorious and innumerable
procession of stars, to which it forms a sort of climax.
Situated in one of those rich and brilliant masses, a suc-
cession of which, curiously contrasted with dark adjacent
spaces, constitute the Milky Way between Centaur and Argo,
its branches with their included vacuities cover more than
a square degree, and are strewn by above 1200 stars".

Dark lanes divide the nebulosity into several separate
islands of glowing light; the brightest of these contains

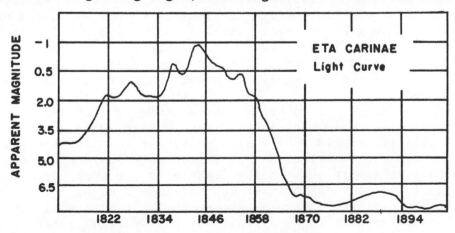

an irregular dark elongated mass- the "key-hole" itself -
from which the nebula derives its name. In addition to
this nebulosity, which forms a brilliant setting for Eta
Carinae, the star itself is surrounded by a much smaller
nebulous shell which is expanding at a rate of about 4"
per century; presumably this shell is connected with the
last bright outburst of the star in 1843, and resembles
the gaseous shells ejected by some of the classical novae.
Bright nebulous condensatioñs in the shell were detected
visually by R.T.Innes in 1914, and were at first recorded
as faint "companion stars".

The first spectrum of Eta Carinae was obtained in 1891
at the time of a minor increase in light of the star, and
was classified as type F5 with sharp lines and some superimposed emission features. Shortly afterwards, as the star
faded back to 8th magnitude, the spectrum showed remarkable
changes, developing a unique pattern of bright emission features. One of the strongest features is the line called H-
alpha, produced by glowing hydrogen, causing the star to
appear reddish in the telescope. The color was compared to
Aldebaran at the great maximum of 1843. The spectrum is
also characterized by bright lines of ionized iron and
other metals; some of these features have been observed
for a short time in nova spectra, but their persistence
over a period of years in the Eta Carinae spectrum is a
unique and unexplained phenomenon. Equally surprising is
the large velocity of expansion measured for the nebulous
shell as late as 1952, a value of about 270 miles per second. This value is confirmed, however, by the expansion of
the visible gaseous shell, which is now about 20" in size.

Eta Carinae, as is evident from the preceding account,
shows some resemblance to the orthodox novae, yet the many
differences are striking. The total range of about 9 magnitudes is quite normal, but the star was bright for more
than a century before the great maximum, in contrast to
the typical nova which shows a single sharp rise and slower decline. Regarded as a slow nova, the star would still
be unique, the maximum having lasted some 35 years. But
perhaps the most outstanding feature of the star was its
high luminosity at maximum. From a comparison of the radial
velocity and the observed expansion of the nova shell, the
distance appears to be approximately 3700 light years. In

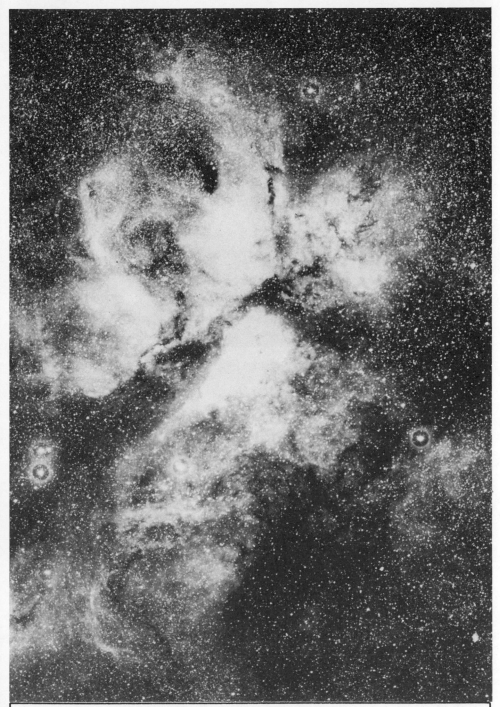

NGC 3372. The great Key-hole Nebula, photographed with the 60-inch reflector at the southern station of Harvard Observatory in South Africa.

DETAILS IN NGC 3372. Top: A view centered on the dark "Keyhole". (Cerro Tololo Observatory) Below: Eta Carinae and the surrounding nebulosity. (Radcliffe Observatory)

a study by A.D.Thackeray (1953) a distance of about 1200
parsecs was derived; approximately the same distance had
been determined earlier by B.J.Bok (1930). From these results, the peak luminosity of Eta Carinae is found to be
over a million times that of the Sun; the computed absolute
magnitude is near -11. The star thus appears to have been
intermediate in brilliance between the ordinary novae which
rarely exceed -9, and the supernovae which range from -13
to -19. In 1843, Eta Carinae was probably the most lumin-
our object in our Galaxy, and in its present 7th magnitude
state is still a giant star some 1600 times brighter than
our Sun. E.Hubble and A.Sandage (1953) found that similar
high-luminosity variables exist in the nearer external
galaxies (M31 and M33) with average absolute magnitudes of
about -8. F.Zwicky (1965) has compiled evidence to show
that these stars may be regarded as a variety of supernovae; in addition to the well known types I and II he has
identified three additional types and classes the Eta Carinae stars as members of type V. These objects, which may
 also be called "high luminosity ejection variables", are
the faintest of the five types, and are characterized by
slow and irregular changes, rather than by sudden outbursts
as shown by types I and II. But whatever the exact classification of Eta Carinae, it seems unlikely that the history of this strange star is ended. Astronomers will watch
its future activities with great interest. (For a review
of supernovae, refer to"Tycho's Star" B Cassiopeiae)

THETA   Mag 2.74; spectrum 09.5 V. Position 10412s6408.
          This is the central star of the large scattered
galactic cluster IC 2602, containing 30 stars brighter than
9th magnitude, and an indeterminate number of fainter members. Of the bright stars, 23 have spectral types of B and
A; the remainder range from F0 to K5. The entire group is
more than a degree in apparent diameter, requiring wide-
angle low power telescopes. Although relatively little-
studied due to its far southern position, this may be one
of the nearest galactic star clusters. Theta itself has a
computed distance of about 700 light years, and an actual
luminosity of about 3300 suns. The annual proper motion is
0.02"; the radial velocity is 14½ miles per second in
recession.

IOTA   Mag 2.25; spectrum F0 I. The computed distance is about 750 light years, and the actual luminosity about 5200 times that of the Sun. The absolute magnitude is about -4.5. The star shows an annual proper motion of 0.02"; the radial velocity is 8 miles per second in recession.   Position 09158s5904.

UPSILON   Mag 2.96; spectrum A9 II. Position 09459s6450. The computed distance is about 340 light years which leads to an actual luminosity of 630 times the Sun. The annual proper motion is 0.01"; the radial velocity is 8 miles per second in recession.

Upsilon Carinae is a fine double star for the small telescope; the 6th magnitude companion is 5.0" away and has a spectral class of F0. Although the two stars undoubtedly form a physical pair, there has been no definite change in either the separation or the angle in 150 years. The projected separation is about 520 AU.

CHI   Mag 3.48; spectrum B2 IV. Position 07555s5251. The distance is computed to be about 430 light years, and the actual luminosity about 600 times that of the Sun. The annual proper motion is 0.04"; the radial velocity is 11½ miles per second in recession.

OMEGA   Mag 3.38; spectrum B7 IV. Position 10126s6947. The distance is about 300 light years, and the actual luminosity about 275 times that of the Sun. The annual proper motion is 0.03"; the radial velocity is 2½ miles per second in recession.

a   (Not to be confused with Alpha) Mag 3.43; spectrum B2 IV. Position 09097s5846. The star is estimated to be nearly 600 light years distant, giving the true luminosity as equal to 1200 suns. The annual proper motion is 0.03"; the radial velocity is 14 miles per second in recession. The star is a spectroscopic binary with a period of 6.744 days. The two stars appear to be nearly equal in mass and luminosity, and the eccentricity of the computed orbit is 0.18.

P Mag 3.30; spectrum B5e V. Position 10302s6126.
  The distance is about 430 light years, leading
to an actual luminosity of about 700 suns. The annual proper motion is 0.02"; the radial velocity is 15½ miles per
second in recession.
  The spectrum is somewhat peculiar, classifying
the star as a B-type giant with emission lines, and indicating the presence of a surrounding gaseous shell. A
slight variability has also been detected; the maximum recorded range is from 3.22 to 3.39. No regular periodicity
is evident.

q Mag 3.41; spectrum K5 Ib. Position 10154s6105.
  At a distance of about 1300 light years, the
computed luminosity of this star is about 5800 times that
of the Sun. The absolute magnitude may be about -4.6. The
annual proper motion is 0.02"; the radial velocity is 5
miles per second in recession. Recent studies reveal a
slight variability with no regular period; the range seems
to be about 0.06 magnitude.

l (Variable) position 09439s6217. One of the
  brightest of the pulsating cepheid variable
stars, visible without optical aid throughout its cycle,
but unfortunately too far south to be observed from the
latitude of the United States. The star has an unusually
long period of 35.556 days and a visual range of about 0.8
magnitude. The Moscow "General Catalogue" (1958) gives the
photographic range as 5.0 to 6.0, with a spectral change
of F8 to K0. This is undoubtedly one of the largest and
most luminous of all known cepheids, a supergiant whose
diameter may average about 200 times that of the Sun. The
photographic absolute magnitude is given by the well known
period-luminosity relation (refer to Delta Cephei) and is
found to be about -4.6. The peak visual luminosity may be
over 12,000 times the light of the Sun. From the distance
modulus method, the distance is estimated to be slightly
over 3000 light years. S.Gaposchkin (1958) finds evidence
for a seconday variation in the light of this star, an
effect which may raise or lower an individual cycle nearly
0.1 magnitude. (Refer also to Delta Cephei).

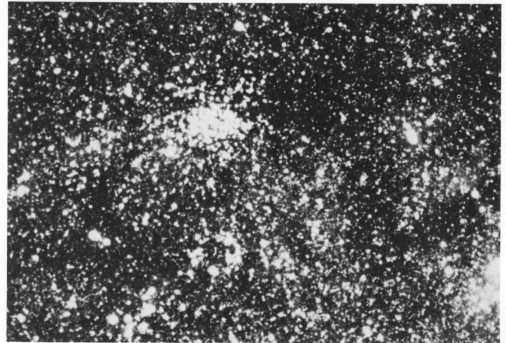

STAR CLUSTERS IN CARINA. Top: The bright group NGC 2516.
Below: The large cluster NGC 3532 in the Carina Milky Way.
Radcliffe Observatory

NGC 2516   Position 07597s6044. A fine open star cluster on the edge of the Carina Milky Way, about 15° SE from Canopus. This is a large and brilliant group, easily visible to the naked eye, with more than 100 stars scattered over a field 1° in diameter. The bright red giant star near the center is very obvious in small telescopes; T.W.Webb called it orange. E.J.Hartung in his "Astronomical Objects for Southern Telescopes" found the cluster "a glorious sight with its scattered groups and irregular sprays of stars, effective for small apertures." The cluster is some 1200 light years distant, and about 20 light years in diameter. For details of the three double stars in this cluster, refer to page 458.

NGC 3532   Star Cluster in the Carina Milky Way at 11034s5824. A superb galactic star cluster situated in a rich field about 3° WNW from the Eta Carinae Nebula. Although one of the finest open clusters in the sky NGC 3532 is almost unknown to observers in the northern hemisphere, owing to its far southern declination. Pickering found it by far the finest irregular cluster in the sky while Sir John Herschel considered it the "most brilliant he had ever seen". The group is very large and much elongated, requiring a wide-field telescope; it measures about 60' x 30' and contains at least 150 stars down to magnitude 12. Possibly some 400 stars are known to be true members. E.J.Hartung (1968) refers to it as a "magnificent cluster ....numerous bright scattered stars....small straight and curved lines of stars are very evident...A number of bright orange stars will be noted..."

The cluster is unusually rich in bright A-type stars; H.Shapley in 1930 reported that fully 93% of the brighter stars (out of 204 measured) were class A. Seven G stars and eight K stars were noted, but no members of type M. S.Raab in 1922 derived a distance of about 1550 light years; the mean of several newer catalogue values is about 1300, which gives the true diameter of the group as about 25 light years. Our sun at that distance would appear of magnitude 12.8. The cluster seems to be intermediate in age between the Pleiades and the Perseus Double Cluster.

The cluster includes the triple star Hd 210; for data refer to page 461.

LIST OF DOUBLE AND MULTIPLE STARS

NAME	DIST	PA	YR	MAGS	NOTES	RA & DEC
Σ3053	15.2	70	58	6 - 7½	relfix, cpm pair; spect K1, A2;fine color contrast	00000n6549
Σ3057	3.9	298	67	7 - 9	relfix, spect B3; cpm pair	00023n5815
Hu502	2.4	113	35	7- 10½	spect A3	00034n4921
Σ3062	1.3	259	67	7 - 8	binary, 107 yrs; PA inc, spect G4, G8	00035n5809
A1253	3.0	87	43	7½-12½	spect G5	00050n5230
Σ7	1.3	212	55	8- 8½	relfix, spect B8	00090n5541
0Σ1	1.6	210	47	7½- 10	slight PA inc, spect A0	00092n6551
β1026	0.3	334	61	8 - 9	binary, 72 yrs; spect F0	00095n5321
Σ10	17.6	176	50	7½- 8	relfix, spect A0	00121n6234
Σ16	5.8	40	62	7½- 9	relfix, cpm pair; spect A3	00140n5423
β392	19.4	69	56	6 - 12	relfix, spect G4	00142n6115
Es 41	6.0	218	24	8 - 11	spect F8	00157n4914
Hu506	0.2	17	62	6- 8½	spect B5	00216n5145
0Σ9	1.9	52	50	7 - 10	slow PA dec,	00235n5630
	22.9	4	10	- 10	spect G0	
Σ30	15.3	307	62	7 - 9	Dist dec, PA inc;	00245n4942
	64.0	98	10	- 12½	spect B9	
β1094	0.7	254	48	6 - 9	PA inc, cpm; B9	00276n5942
β394	0.8	286	66	8½- 8½	PA inc, spect G0	00280n4715
λ	0.5	179	66	5½- 6	(0Σ12) Binary, PA inc, spect B8	00290n5415
β1227	2.9	202	36	7½-11½	spect F2	00296n5804
	22.5	86	36	- 11½		
0Σ15	0.2	314	60	7½- 8½	PA inc, spect A2 + K	00331n4845
β1097	0.5	72	57	8½- 8½	relfix, spect B9	00344n5745
h1989	32.3	51	08	7½-11½	spect B3	00346n7237
h1989b	10.8	343	08	12-12		
Σ45	14.5	88	30	7 - 10	dist inc, spect G5	00360n4641
0Σ16	13.3	23	55	6 - 10	optical, dist dec; spect K5	00364n4905
β257	0.6	240	54	8 - 9	spect G0	00374n4658

LIST OF DOUBLE AND MULTIPLE STARS (Cont'd)

NAME	DIST	PA	YR	MAGS	NOTES	RA & DEC
α	64.4	280	13	$2\frac{1}{2}$- 9	(h1993) optical;	00376n5616
	38.3	105	08	- 13	spect K0 (*)	
Σ48	5.4	334	62	7 - 7	relfix, cpm pair;	00395n7106
					spect A0	
O	33.6	302	60	4 - 11	(β231) (22 Cass)	00419n4801
					cpm; spect B2	
21	36.0	160	61	6- $9\frac{1}{2}$	YZ Cass. A = ecl.	00423n7443
					bin; spect A2	
β492	2.2	152	33	6 - 12	cpm; spect A2	00424n5457
	88.3	24	09	- 11		
Σ59	2.1	148	61	7 - 8	relfix, spect A0	00452n5110
η	11.0	297	61	$3\frac{1}{2}$- $7\frac{1}{2}$	(Σ60) very fine	00461n5733
					binary, PA inc;	
					color contrast (*)	
h1054	8.6	180	13	8 - 10	spect F5	00467n6029
	93.0	62	09	- 10		
β232	0.7	228	61	$8\frac{1}{2}$- 9	Binary, ±150 yrs,	00476n5022
	26.1	296	61	$-10\frac{1}{2}$	PA inc, G4+G6	
β781	0.7	22	67	8 - $8\frac{1}{2}$	PA dec, spect A2	00484n6843
A812	1.8	323	45	7 - 11	spect K0	00490n4747
Σ65	3.1	40	62	8 - 8	slow PA inc,	00496n6836
					spect A2	
β1	1.4	82	36	8 - 10	spect B2, multiple	00499n5621
	3.8	33	36	- 9˙	group, in nebula	
	8.9	194	36	- $9\frac{1}{2}$	NGC 281	
	15.7	332	15	$-12\frac{1}{2}$		
β497	130	171	23	6 - 9	spect F8, dist inc	00501n6051
Σ70	8.0	245	17	$6\frac{1}{2}$- $9\frac{1}{2}$	relfix, cpm pair;	00509n5225
	78.5	148	13	- 10	spect A0, G4	
Σ70c	1.7	88	09	10-$10\frac{1}{2}$		
Hu802	0.4	209	55	7- $7\frac{1}{2}$	spect A0	00520n4908
γ	2.3	252	61	$2\frac{1}{2}$- 11	(β1028) primary	00537n6027
	52.7	347	61	- 13	variable (*)	
β1099	0.2	197	61	6 - 7	binary, about 85	00538n6006
	41.4	156	59	- 13	yrs; spect B9	
Es 940	62.6	356	10	7 - 9	spect K2	00544n5158
Es 940b	7.0	33	10	9 - 13		
A2901	0.3	49	58	$7\frac{1}{2}$- $7\frac{1}{2}$	PA inc, spect B9	00582n6905
Es 45	7.7	243	38	7 - 11	spect G5	00586n4917
β1161	0.4	353	59	7 - $7\frac{1}{2}$	PA inc, spect B5	01000n5132

LIST OF DOUBLE AND MULTIPLE STARS (Cont'd)

NAME	DIST	PA	YR	MAGS	NOTES	RA & DEC
β396	1.2	66	45	6 - 9	relfix, spect A9	01005n6048
HIV 66	21.7	75	35	6½- 10	spect K0	01042n5314
OΣ23	14.6	192	50	7½- 8	cpm; spect F8	01072n5129
	50.0	94	09	- 12		
β235	1.1	118	61	7½- 7½	PA inc, spect F5;	01076n5045
	43.4	285	53	-10½	multiple group;	
	59.4	68	53	- 9	D = 8" pair	
Σ96	0.8	283	67	8 - 9	Dist dec, spect F0	01094n6445
β258	1.3	267	62	6 - 9	PA slow inc, B9	01100n6127
β1100	0.4	230	61	8 - 8	binary, 75 yrs;	01116n6041
					spect F5, PA dec.	
φ	48.6	208	25	5 - 12	spect F0, B6; on	01169n5758
	134	231	56	- 7	edge of cluster	
					NGC 457 (*)	
Hu523	0.4	94	48	6½- 10	spect B9	01176n5120
35	53.8	347	19	6 - 8	dist inc, spect A0	01176n6424
Σ115	0.7	145	62	7½- 7½	cpm; slow PA dec,	01201n5753
	45.5	280	25	- 13	spect F4	
Σ114	3.7	356	26	7½-10½	relfix, spect A0	01203n7235
ψ	2.5	45	63	4½- 13	AB cpm; spect K0	01224n6752
	23.2	118	63	- 9		
ψ c	2.9	254	63	9 - 9½	(Σ117) relfix	
Σ131	13.8	142	56	6 - 9	in cluster M103;	01299n6026
	28.2	145	56	- 10½	relfix, spect B3	
M1b 80	58.4	79	23	6½- 11	spect K0	01320n6250
M1b 80b	6.4	45	23	11-11½		
OΣ33	25.4	76	24	7 - 8	relfix, spect B8	01341n5823
A1267	0.3	357	58	8 - 8½	PA inc, spect A0	01374n5441
h1088	19.6	168	33	7 - 9½	spect B9	01390n5823
44	1.6	358	34	6 - 12	(β1103) slow PA	01399n6018
	66.0	310	12	- 10	dec, spect B9	
OΣ35	12.8	99	58	7 - 10	PA dec, dist inc,	01405n5538
					spect A2	
β870	1.1	19	61	7 - 8½	PA dec, spect A2	01410n5717
Σ151	7.1	38	02	9½- 10	in NGC 663	01425n6058
Σ152	9.3	105	22	9 -10½	spect B2, in NGC	01426n6059
					663	
Σ153	7.6	69	18	8½- 10	spect B3, in NGC	01431n6101
					663	
Σ163	34.8	35	36	6½- 8½	relfix, spect K5;	01476n6436
					color contrast	

LIST OF DOUBLE AND MULTIPLE STARS (Cont'd)

NAME	DIST	PA	YR	MAGS	NOTES	RA & DEC
Σ170	3.3	246	55	6½- 7½	relfix, spect A4	01506n7559
Σ182	3.6	123	53	7 - 7	relfix, cpm pair;	01529n6102
	29.9	71	53	- 13	AC PA and dist inc spect A0	
h1100	41.5	310	59	6 - 10	optical, dist inc, spect A0	01558n6423
Σ185	1.0	10	66	7 - 8½	PA dec, cpm; spect A0	01573n7516
48	0.3	47	62	5 - 7	(β513) binary,	01578n7040
	23.7	51	23	- 13	60 yrs; PA inc, spect A4; all cpm	
Σ191	5.5	195	62	6 - 8½	relfix, spect A3	01586n7337
49	5.4	246	11	6 - 13	(β785) cpm pair	02006n7553
	28.0	128	11	- 13	spect G8	
Σ216	0.3	246	53	7½- 8½	PA dec, spect F0	02076n6207
Σ234	0.9	246	61	8½- 9	binary, 150 yrs; spect G0	02137n6107
St 356	5.8	68	05	7½-10½	spect B5	02149n6412
0ΣΣ26	63.3	200	25	6½- 7	wide pair, spect A2, G5; relfix	02160n5948
Σ257	0.3	14	61	7½- 8	binary, about 280 yrs; PA inc, spect B8	02219n6120
ι	2.2	241	66	4 - 7	(Σ262) triple	02249n6711
	7.3	114	61	- 8	system (*)	
A823	0.5	256	25	7½-11½	spect A0	02295n5947
Σ277	3.1	137	15	7½- 11	spect A0	02332n5940
Σ283	1.8	207	59	8 - 8½	relfix, spect G5	02367n6116
	18.3	15	04	- 13½		
A970	5.2	100	29	7 - 13	spect A0	02394n5841
Σ302	5.1	167	13	8 -10½	spect B9	02458n6425
Σ306	2.1	93	38	7 - 9	relfix, spect B0	02472n6013
	27.4	157	01	-11½		
β--	15.6	194	12	8 - 12	spect A3	02491n6010
	75.5	203	12	- 12		
Σ312	2.2	32	66	7 - 8	slow PA inc;	02510n7241
	42.9	129	57	- 9	spect G0	
Σ329	16.0	273	15	7½- 9	relfix, spect A2	02572n5850
Σ349	6.1	321	62	7½- 8	relfix, spect F8	03066n6336
0Σ50	1.1	175	66	7½- 7½	PA dec, spect F8	03076n7122
	27.2	299	31	- 13		

LIST OF DOUBLE AND MULTIPLE STARS (Cont'd)

NAME	DIST	PA	YR	MAGS	NOTES	RA & DEC
0 Σ485	20.1	50	58	6 - 9	spect B9	23006n5457
	56.6	260	24	- 9½		
0 Σ490	1.3	299	55	7 - 9	relfix, spect G5	23079n5710
β229	17.5	36	15	7 -11½	spect K2	23176n5657
Es220	36.9	83	22	8- 11½	spect A0	23187n6208
Es220b	6.0	76	21	-12		
0 Σ495	0.2	111	60	7½- 7½	PA dec, spect B5	23218n5716
AR	1.1	347	47	5 - 10	(1 Cass) (0 Σ496)	23277n5816
	75.7	269	22	- 8	AB cpm; primary is	
	43.4	114	18	- 9	ecl.bin.; spect B3	
	67.3	338	05	- 9	C = 1.4" pair	
0 Σ498	17.2	244	23	7 - 10	spect F5	23289n5208
0 Σ499	9.5	78	29	7 - 9	spect G5	23309n5708
0 Σ499b	0.3	160	58	9 - 11	(A641)	
h1896	16.3	116	12	7 - 11	spect A2	23364n6151
0 Σ 502	3.6	223	16	7- 10½	relfix, spect A2	23375n6327
β993	2.7	275	55	7- 11½	slight PA dec, spect M0	23400n6414
0ΣΣ248	52.7	140	23	7 - 9	spect K0	23435n5024
	23.3	339	15	- 12		
Σ3037	2.7	213	56	7- 8½	relfix, spect K0	23436n6012
	29.2	186	56	- 9		
	52.6	229	56	- 9½		
β390	15.4	231	58	8½- 12	dist dec, spect B9	23450n4902
0 Σ507	0.7	296	61	7- 7½	binary, PA inc;	23462n6436
	50.4	351	59	- 8	spect A0, C is optical	
6	1.5	199	61	6 - 8	(0 Σ508) relfix,	23464n6157
	62.4	309	12	-10½	spect A3	
Es 700	14.6	35	08	6½- 10	spect F5	23481n5356
0 Σ511	10.5	35	13	7 - 11	spect K5	23506n6026
β1224	4.0	203	51	6½- 13	relfix	23544n5534
	77.2	357	12	-12		
0 Σ 512	3.0	293	36	6½- 11	dist dec, spect M	23548n6046
Arg 99	4.8	318	36	9½- 10	6' east from 0Σ512	23549n6046
Σ3047	1.1	70	62	8½- 8½	relfix, spect B9	23553n5707
	8.2	189	17	- 12		
R	14.0	273	10	var-14	(Es 37) LPV. (*)	23559n5107
	27.8	331	10	- 10		
σ	3.0	326	58	5½- 7½	(Σ3049) relfix; spect B1, B3	23564n5529

LIST OF DOUBLE AND MULTIPLE STARS   (Cont'd)

NAME	DIST	PA	YR	MAGS	NOTES	RA & DEC
WZ	58.1	89	53	7½- 8	(0ΣΣ254) relfix; fine colors; WZ = N type variable; spect N and A	23587n6005

LIST OF VARIABLE STARS

NAME	MagVar	PER	NOTES	RA & DEC
α	2.1---2.6?	---	Spect K0; variability uncertain (*)	00376n5616
β	2.25 ±0.04	.1043	Spect F2; Delta Scuti type (*)	00065n5852
γ	1.6---3.0	Irr	Erratic, spect B0e (*)	00537n6027
δ	2.7---2.8	759	Ecl.bin.; spect A5 (*)	01225n5959
ι	4.5 ±0.03	1.740	Spect A5p; Alpha Canum type. Also visual triple star (*)	02249n6711
ρ	4.1---6.2	Irr	Semi-reg; spect F8 (*)	23519n5713
B	-4.5.....	---	"Tycho's Star"; supernova of 1572 (*)	00220n6352
R	5.4--13.0	431	LPV. Spect M6e--M8e (*)	23559n5107
S	7.8--15.1	611	LPV. Spect S4e (*)	01159n7221
T	7.1--12.4	445	LPV. Spect M6e--M8e	00205n5531
U	8.0--15.4	278	LPV. Spect S5e--S8e	00436n4758
V	7.3--12.8	228	LPV. Spect M5e--M7e	23095n5925
W	8.3--12.4	405	LPV. Spect M3	00519n5818
X	9.5--13..	423	LPV. Spect Ne	01532n5901
Y	9.0--15.3	414	LPV. Spect M6e--M8e	00008n5524
Z	9.0--15.0	496	LPV. Spect M7e	23421n5618
RR	9.8--13.9	301	LPV. Spect M5e	23533n5327
RS	9.9--10.7	6.296	Cepheid; spect G5	23349n6209
RU	5.5---?		(32 Cass) Spect B8 variability unconfirmed	01084n6444

LIST OF VARIABLE STARS   (Cont'd)

NAME	MagVar	PER	NOTES	RA & DEC
RV	7.6--15.5	331	LPV. Spect M6e--M7e	00499n4709
RW	9.2--10.7	14.80	Cepheid, spect G2--K8	01339n5730
RX	8.5--9.2	32.32	Ecl.Bin.; spect gG3+gA5e	03032n6723
RY	9.5--11.0	12.135	Cepheid; spect G2	23496n5828
RZ	6.4---7.8	1.195	Ecl.bin. (*)	02443n6926
SS	8.9--13.2	141	LPV. Spect M3e	00070n5117
ST	9.0--10.5	Irr	Spect N	00149n5001
SU	5.8---6.2	1.949	Cepheid; spect F5--F7	02475n6841
SV	8.1--11.5	276	Semi-reg; spect M6	23366n5159
SW	9.3--10.0	5.441	Cepheid; spect F6--G4	23050n5817
SX	9.1--10.2	36.57	Ecl.bin.; spect gA6+gG6	00081n5437
SY	9.4--10.2	4.071	Cepheid; spect F5--G2	00125n5809
SZ	9.6--10.1	13.62	Cepheid, spect F6--G4; W Virginis type	02235n5914
TU	7.2---8.1	2.139	Cepheid; spect F3--F5 W Virginis type	00236n5100
TV	7.3---8.4	1.813	Ecl.bin.; spect A0	00166n5852
TW	8.3---8.9	1.428	Ecl.bin.; spect B9+A0	02417n6531
TX	9.2---9.8	2.927	Ecl.bin.; spect B1	02482n6235
TZ	9.0--10.5	Irr	Spect M2	23504n6043
UW	9.7--16..	291	LPV. Spect M8	00333n5740
UY	9.8--11..	103	Semi-reg; spect M3e	22599n5722
VY	9.0--10.2	100:	Semi-reg; spect M6	00484n6239
VZ	8.3--12.3	169	LPV. Spect M0e	01133n5608
WW	9.1--11.7	Irr	Spect N	01303n5730
WX	9.8--10.6	Irr	Spect M1	01505n6052
WY	9.5--14..	478	LPV. Spect Spe	23554n5613
WZ	7.4--10..	186	Semi-reg; spect Np; with 8^m comp at 58", forms wide double 0ΣΣ254.	23587n6005
XX	9.0--10.4	3.067	Ecl.bin.; spect B4+B6	01263n6043
XZ	9.0-- 9.6	Irr	Spect M0	01196n6055
YZ	5.6---6.1	4.467	(21 Cass) Ecl.bin.; spect A2; also visual double	00423n7443
AA	8.5---9.7	Irr	Spect gM6	01163n5604
A0	6.0---6.2	3.523	supergiant Ecl.bin.; spect 08 (*)	00151n5109
AR	4.7---4.9	6.066	Ecl.bin.; spect B3; also visual double 0 Σ 496.	23277n5816

LIST OF VARIABLE STARS   (Cont'd)

NAME	MagVar	PER	NOTES	RA & DEC
BM	9.4--10.2	197.3	Ecl.bin.; spect F0	00517n6349
CC	7.5-- 7.6	3.369	Ecl.bin.; spect both 08	03101n5923
CQ	9.9--11.4	2300	Semi-reg; spect M6	02426n6248
DL	9.5--10.2	8.000	Cepheid; spect K3	00272n5957
DN	9.9--10.2	1.155	Ecl.bin.; spect 08	02194n6036
DO	8.6---9.2	.6847	Ecl.bin.; spect A2	02375n6020
FM	9.2---9.9	5.809	Cepheid; spect G0	00118n5559
GG	9.5---9.9	3.759	Ecl.bin.; spect B5+K0	01131n5604
HS	9.5--11..	Irr	Spect M3	01051n6319
IM	9.0--10.0	323	Semi-reg; spect M2	01286n6204
KN	9.5--10.1	Irr	Spect M1+B	00069n6223
KR	9.7--11.0	4.904	Ecl.bin.	00511n5415
MN	8.8--9.4	1.917	type uncertain	01388n5442
MZ	9.6--10.8	Irr	Spect M5	00186n5940
NQ	9.0-- 9.9	Irr	Spect R5	00218n5400
PV	9.9--11.0	.8752	Ecl.bin. ? spect B8	23079n5856
PZ	8.5--11..	900	Semi-reg; spect M3	23416n6131
V368	8.5-- 9.2	4.4516	Ecl.bin; spect B8	03086n5945
V373	6.0--6.1	13.4187	Ecl.Bin; spect B0	23536n5708
V377	7.8--8.3	Irr	Spect F0	00165n5927
V391	7.9---8.8	Irr	Spect M4	01525n6957
V436	7.56 ±0.06	160:	Type uncertain, spect A5	23304n5738

LIST OF STAR CLUSTERS, NEBULAE, AND GALAXIES

NGC	OTH	TYPE	SUMMARY DESCRIPTION	RA & DEC
103		(cluster)	pS,pC, diam 7'; about 35 stars mags 11...18; class D	00226n6103
129	79[8]	(cluster)	vL,pRi,lC, diam 14'; 50 stars mags 9...13; class E	00270n5957
133		(cluster)	diam 7'; about 40 stars mags 10.... class E	00284n6204
136	35[6]	(cluster)	S,F,vmC, diam 1'; 30 faint stars	00287n6114
146		(cluster)	pL,lC, diam 6'; 50 stars mags 11...15; class E	00303n6301
147		(galaxy)	E4; 12.1; 6.5' x 3.8' vF,pL,E; pair with NGC 185; distant companions to M31 in Andromeda (*)	00304n4814

LIST OF STAR CLUSTERS, NEBULAE, AND GALAXIES   (Cont'd)

NGC	OTH	TYPE	SUMMARY DESCRIPTION	RA & DEC
185	707[2]	⊘	E1; 11.8; 3.5' x 2.8' pB,pL,1E; pair with NGC 147; companions to Andromeda Galaxy M31 (*)	00361n4804
225	78[8]	⣿	L, scattered group, W-shape 14' diam; about 20 stars mags 9...	00406n6131
278	159[1]	⊘	Sc; 11.6; 1.2' x 1.2' (*) cB,pL,R, very compact spiral	00492n4718
281		□	F,vL, Irr; 23' x 27'; incl 8½ mag 05e star, also double star β1 (*)	00504n5619
---	I.59	□	pF,L, nebulous patch, 30' north of Gamma Cass.	00537n6048
---	I.63	□	pF,L, fan-shaped neby; 20' north-following Gamma Cass	00570n6035
358		⣿	S,pRi,C, diam 3', about 25 faint stars	01020n6146
381	64[8]	⣿	pS,Ri,cC; diam 5', about 40 faint stars	01052n6119
436	45[7]	⣿	pS,pC, diam 4'; 40 stars mags 10.... class D	01124n5833
457	42[7]	⣿	B,L,pRi, 10' diam; 100 stars mags 8.... class E. Phi Cass on edge of group (*)	01159n5804
559	48[7]	⣿	B,pL,pRi, diam 7'; mC; 50 stars mags 10.... class E	01261n6302
581	M103	⣿	pL,B,Ri, mag 8, diam 8'; 40 stars mags 8....12; class D; incl Σ131 (*)	01299n6027
---	Tr.1	⣿	S,pC, diam 4'; about 25 stars mags 10...15	01323n6102
609		⣿	vS,Ri,vC; 3' diam; 80 vF stars	01337n6418
637	49[7]	⣿	pS,C, diam 3'; about 20 stars mags 10... class D	01383n6347
654	46[7]	⣿	pS,vRi,vC; diam 5', about 50 stars mags 11...14; class D; in field with NGC 663	01405n6139

LIST OF STAR CLUSTERS, NEBULAE, AND GALAXIES (Cont'd)

NGC	OTH	TYPE	SUMMARY DESCRIPTION	RA & DEC
659	65[8]	⦂⦂	S,pRi,cC; diam 4'; about 30 stars mags 12... class D; in field with NGC 663	01408n6028
663	31[6]	⦂⦂	B,L,vRi; mag 7; diam 11'; about 80 stars mags 9.... class E. Incl double stars Σ151, Σ152, Σ153. (*)	01426n6101
---	I.1747	◎	Mag 13½, diam 12"	01538n6304
---	I.1795	▢	vL,F, 13' x 27'; brightest portion of vast neby I.1805 which encloses cluster Mel 15	02210n6140
---	Mel 15	⦂⦂	coarse group, 20' diam with 20 stars mags 7... class D; surrounded by neby I.1805, eF,Irr, loop 90' diam.	02287n6113
1027	66[8]	⦂⦂	L, 8' diam, scattered group of 12 stars, mags 8.... class D	02388n6120
---	I.1848	▢	eL,vF,Irr; 90' x 45' with 7th mag 07 star.	02474n6013
---	I.289	◎	Mag 12; 45" x 30" with 15^m central star	03062n6108
---	H1	⦂⦂	L,nC, 15' diam, 30 stars; class E	03072n6303
7635	52[4]	▢	vL,F, diam 205" x 180"; 07 star of 8^m inv. Contains great nebulous shell or ring = "Bubble Nebula" (*)	23185n6054
7654	M52	⦂⦂	L,Ri,Irr;mCM, mag 7, diam 12'; 120 stars mags 9... class E (*)	23220n6120
---	H21	⦂⦂	pRi, C; diam 5'; 20 F stars class D	23518n6129
7788		⦂⦂	S,pRi,vC, diam 3', mag 10; stars 10...13; class E	23542n6107
7789	30[6]	⦂⦂	vL,eRi,vmC; mag 10, diam 20'; 900 stars mags 11..(*)	23545n5626
7790	56[7]	⦂⦂	pRi,pC,diam 5', 25 stars mags 11.... class D	23545n6056

DESCRIPTIVE NOTES

ALPHA   Name- SCHEDAR. Mag 2.23, spectrum K0II or III. Position 00376n5616. Alpha Cassiopeiae has been suspected of light variations by various observers. Birt in 1831 found a range of 2.2 to 2.8 with no regular period. Sir John Herschel confirmed the variability, and Argelander thought the period to be about 80 days with considerable uncertainty. Chandler noted that the variability is only occasionally evident. No definite changes have been detected in recent years, and the variability is now considered doubtful. The star is listed as "constant" in the Moscow "General Catalogue" (1958).

The 9th magnitude companion at 63" was first seen by Sir William Herschel in 1781, and is an easy object for small telescopes; its color usually appears to be bluish or pale white, contrasting finely with the bright orange of the primary. A second, fainter companion at 38" is also mentioned in the list of double stars. Both companions are optical attendants only. The separation of the brighter companion has been increasing from Herschel's measured value of 56" in 1781; the change is due to the proper motion of the primary.

In addition, a third fainter component of the 14th magnitude was detected by S.W.Burnham in 1889; the separation was then 17.6" in PA 272°. This star also appears to be an optical attendant only, and the apparent separation has now increased to about 20" (1960).

Parallax measurements of Alpha Cassiopeiae have been somewhat discordant , but suggest a distance in the range of 150 to 200 light years. The smaller distance seems to be supported by the spectroscopic features of the star, which indicate an absolute magnitude of about -1.1 and an actual luminosity of about 230 suns. The annual proper motion is 0.06"; the radial velocity is about 2½ miles per second in approach.

BETA   Name-CAPH. Mag 2.25; spectrum F2 IV. Position 00065n5852. The distance of this star is about 45 light years; the actual luminosity about 19 times that of the Sun. (Absolute magnitude +1.6) The annual proper motion is 0.56" in PA 109°; the radial velocity is 7 miles per second in recession.

THE MILKY WAY IN CASSIOPEIA. Gamma Cass is at center; Beta is near bottom center with NGC 7789 near lower edge. Near the top edge is Epsilon Cass. Lowell Observatory photograph

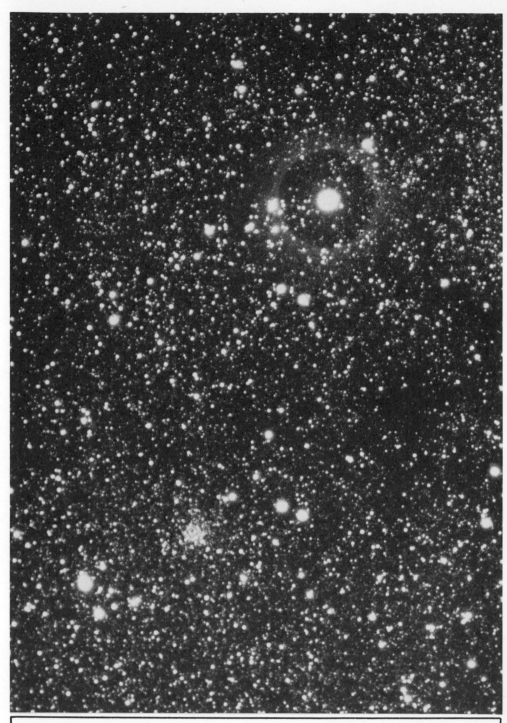

BETA CASSIOPEIAE. The star is the brightest object in this field. The star cluster in the lower portion of the print is NGC 7789. Lowell Observatory 13-inch telescope photo.

DESCRIPTIVE NOTES (Cont'd)

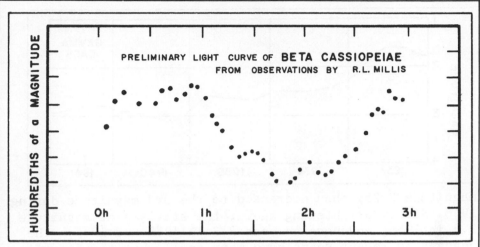

PRELIMINARY LIGHT CURVE OF **BETA CASSIOPEIAE**
FROM OBSERVATIONS BY R.L. MILLIS

HUNDREDTHS of a MAGNITUDE

Oh   1h   2h   3h

A 14th magnitude optical companion at 23" was discovered by A.G.Clark in 1889; the separation is increasing since the star does not share the proper motion of the primary. According to a note in the ADS Catalogue, Beta Cassiopeiae is a spectroscopic binary with a period of about 27 days.

In addition, the star is slightly variable in light. R.L.Millis (1964) found a very rapid variation with an amplitude of about 0.04 magnitude, in a period of 0.1043 day. The variations appear to class the star as a member of the Delta Scuti group, a classification which appears to be supported by the spectral type and position on the H-R diagram. Beta Cassiopeiae is the brightest, and evidently the nearest, of the stars which have been assigned to this rare class of pulsating variable. (Refer also to Delta Scuti).

GAMMA   Mag 2.40, spectrum B0 IV e. Position 00537n 6027. Gamma Cassiopeiae is the central star of the large "W-shaped" figure which identifies the constellation. It is a peculiar variable star which - during the last half century - has shown puzzling and unpredictable variations in its light. Before the year 1910, the star appeared constant at magnitude 2.25. It appears to have slowly risen a half magnitude by 1936, then increased rapidly during the next year to a maximum of about 1.6 in April 1937. Toward the end of that year it returned to

DESCRIPTIVE NOTES (Cont'd)

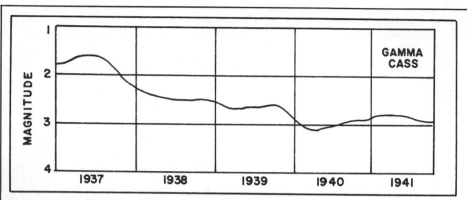

magnitude 2.25, then decreased to the 3rd magnitude during 1940. Slowly brightening again, the star was at magnitude 2.5 in 1954, and hovered near 2.2 in 1975 and 1976. The changes in recent years have been small, but the future activity of the star is totally unpredictable.

Gamma Cassiopeiae is a B0 subgiant with bright hydrogen lines, a peculiarity first noticed by Father Secchi in 1866, and said to have been the first such case known. Spectroscopic studies seem to show that the star is subject to periods of violent change and fluctuation, during which the magnitude, spectrum, color, temperature and diameter all change. The spectroscopic variations began about 1927, some years before any light changes were detected. The maximum of 1937 was accompanied by a drop in temperature, from about 12,000°K to about 8500°; this large change was completed in the space of a few months, in early 1937. And from spectroscopic measurements, it appears that the star at this time ejected a gaseous shell which grew from its original size of about 8 solar diameters, to about 18. The shell activity resulted in peculiar changes in both the absorption and emission features of the spectrum; some of these changes have not yet been successfully interpreted. There is no evidence, however, that the star has any connection with the novae, a suggestion that has occasionally been made. All known novae were dense bluish dwarfs when at minimum light, in no way resembling the giant Gamma Cass. From observations made with the SAS-3 satellite in 1976, the star is known to be a weak source of X-ray energy.

The exact distance is still uncertain, but has been estimated to be close to 100 light years. The corresponding actual luminosity at the present time is just over 100

GAMMA CASSIOPEIAE. The erratic variable, and the two nebulosities IC 59 and IC 63, which appear to be associated with the star.   Haute Provence Observatory.

times the Sun's. The annual proper motion is 0.025"; the radial velocity is 2½ miles per second in approach.

Gamma Cassiopeiae is also a visual double star, but a very difficult one due to the great difference in the magnitudes of the two stars. The companion was found by S.W. Burnham with the 36-inch refractor at Lick Observatory in 1888. It is 2.3" distant from the primary, and is estimated to be about 11th magnitude. The faint star probably shares the proper motion of the primary, but there is no evidence for orbital motion. The projected separation of the two stars is about 70 AU.

DELTA   Name- RUCHBAH. Mag 2.68; spectrum A5 V. The position is 01225n5959. Delta Cassiopeiae is computed to be about 45 light years distant, giving the actual luminosity as about 12 times that of the Sun. The annual proper motion is 0.30" in PA 99°; the radial velocity is about 4 miles per second in recession. According to a note in the Yale "Catalogue of Bright Stars" (1964) the space motion identifies the star as an outlying member of the moving Taurus group associated with the Hyades cluster.

The star shows a slight brightness variation of 0.1 magnitude in the long period of 759 days; usually attributed to the partial eclipse of the star by a revolving companion. Additional studies are needed to establish the exact nature of the light curve and the elements of the system.

EPSILON   Mag 3.38; spectrum B3 IV. Position 01508n6325. The star is about 520 light years distant, and must have an actual luminosity of about 1000 times that of the Sun. The annual proper motion is 0.04"; the radial velocity is 5 miles per second in approach.

ETA   Mag 3.47; spectrum G0 V. Position 00461n5733. Possibly one of the best known binary stars, discovered by Sir William Herschel in August 1779. The two stars are magnitudes 3.5 and 7.2, and their separation varies from 5" (1890) to about 16" (2150). The period is approximately 500 years. In an analysis made in 1937, K.A. Strand obtained a period of 526 years, but in a more recent computation he has revised this to 480 years. The apparent orbit is very nearly circular, but has the primary star considerably displaced from the center. The true orbit has a semi-major axis of 12", and an eccentricity of 0.50. Periastron was in 1889. The mean separation of the two stars is about 68 AU.

Eta Cassiopeiae has an especially beautiful contrast in colors. Some observers have seen the components as gold and purple, some as yellow and red, and others as "topaz and garnet". Facts about the two stars are given here:

	Spect	Mag	Abs.Mag.	Diam.	Lum.	Mass
A	G0 V	3.47	+4.5	0.8	1.2	1.1
B	dM0	7.22	+8.3	0.5	0.04	0.6

The spectral class of the fainter star is still somewhat uncertain, and is given by various authorities as K3, K5, M0, or M1. According to A. Slettebak (1963) the star should be classed as a late-type metal-poor dwarf on the basis of the peculiarities of its absorption spectrum. P.C.Keenan suggests a type near K3. The system is only 18 light years distant, and shows a large annual proper motion of 1.22" in PA 115°. The radial velocity is 5½ miles per second in recession.

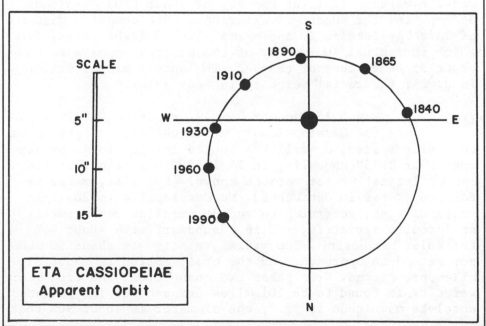

ETA CASSIOPEIAE
Apparent Orbit

IOTA Mag 4.51; spectrum A5p. Position 02249n6711.
One of the finest triple stars in the sky, resolvable in a good three-inch telescope when the seeing conditions permit. The primary appears yellowish to most observers, and the companions are usually described as bluish. These colors, like those of many double stars, may be illusionary. It is interesting to note that the three stars have the spectral classes A5, F5, and G4; thus the primary is actually bluer than either of the companions.

The close pair form a binary in slow retrograde motion with a period of about 840 years, according to a recent orbit computation by Heintz (1962). The semi-major axis of the orbit is 2.3", and the eccentricity is 0.40. Astrometric measurements show that another much closer unseen star is present, with a period of about 52 years. Finally, the third visible component at 7" is a physical member of the system, but has shown no definite relative motion since discovery by F.G.W.Struve in 1829. The period must be at least several thousand years. The projected separations are: AB = 115 AU; AC = 350 AU.

The primary of this system is a spectrum variable of the Alpha Canum Venaticorum type, with a period of 1.74 days. A very small light change of about 0.03 magnitude accompanies the spectrum variations. The computed distance of Iota Cassiopeiae is approximately 160 light years, from which the actual luminosity of the primary appears to be about 35 times that of the Sun. The annual proper motion is 0.02"; the radial velocity is near zero.

MU Mag 5.15; spectrum G5 V. Position 01049n5441.
(30 Cassiopeiae) One of the near neighbors of the solar system, a small star noted for its large proper motion of 3.75" annually, in PA 115°. This is one of the twenty largest proper motions known. (For list, refer to Barnard's Star in Ophiuchus). Mu Cassiopeiae is 26 light years distant, according to recent parallax measurements at Sproul Observatory, and is a subdwarf with about 40% the solar luminosity. The radial velocity is about 60 miles per second in approach, and the cross-motion is about 81 miles per second. From these two quantities, the true space velocity is found to be 101 miles per second. The computed absolute magnitude is +5.7, the diameter is about 90% that

of the Sun, and the mass is estimated to be about 75% the mass of the Sun.

Mu Cassiopeiae has long been known as an astrometric binary, a system in which only one component is actually seen, but the presence of a small companion is proved by periodic variations in the proper motion. From measurements made at Sproul, the period of the companion is known to be about 18½ years, with periastron in 1956. In 1966 the faint companion was detected visually for the first time by P.A.Wehinger with the 84-inch reflector at Kitt Peak National Observatory. The two stars differ by about 3 magnitudes, and the separation at discovery was near 0.8". Evidently the small star is a red dwarf, and its mass appears to have the unusually small value of about 0.2. The average separation is about 7 AU.

RHO   (Variable) Spect F8 Ia; position 23519n5713.
A peculiar irregular variable star, showing slow and unpredictable changes in both its light and spectrum. It has a normal range of magnitude 4.4 to about 5.1, but on occasion has faded to 6th magnitude. Although no real periodicity is evident, the interval between some maxima has been measured at about 100 days.

When near maximum the spectral type is classified as F8, although the light is redder than normal for an F-type star. During the variations the spectral type fluctuates between F8 and K5, and has reached M5 on at least one occasion, in June 1946. Studies of the star at Harvard have shown that the spectrum changes do not always follow the light variations. The star was once observed to be type K when near maximum. Another peculiar feature is that the color does not alter as much as the spectral changes would seem to require. Even when at type M, the star does not become as red as a normal M-type star.

The distance and true luminosity of this star are not definitely known, and widely different results are obtained by various methods. Trigonometrical parallaxes have been measured at Allegheny, McCormick, and Mt.Wilson, and agree in giving a distance of about 200 light years. This makes the peak absolute magnitude about +0.4. The spectrum appears to be that of a supergiant, however, suggesting a luminosity about 100 times greater. An absolute magnitude

of about -4.5 would be normal for a star of type F8 Ia,
but this in turn would imply that the true distance must be
something close to 3000 light years! The answer to this
puzzle may lie in certain spectral characteristics which
have caused the star to be erroneously identified as a
supergiant. On the other hand, L.W.L.Sargent (1961) found
evidence for an absolute magnitude brighter than -8, and
derives a mass of about 25 suns for the star. His studies
of the spectrum show that the star is surrounded by an
expanding gaseous shell which is moving outward at about
25 miles per second; the mass loss is estimated to be one-
millionth of a solar mass per year. Although this star has
occasionally been classed among the variables of the R
Coronae Borealis type, it is definitely not a typical member. The amplitude of the variations is much less, and the
spectrum does not show the strong carbon features which are
so typical of the R Coronae stars.

The very small annual proper motion has been measured
at about 0.005"; the radial velocity is 26 miles per second
in approach. (Refer also to R Coronae Borealis)

R   (Variable) Position 23559n5107. The brightest
           of the long-period variable stars in Cassio-
peia, and the first to be discovered, found by N.Pogson
in 1853. Although it often reaches naked-eye visibility
at maximum, R Cass is not one of the easier variables to
locate; it is situated in a field richly sprinkled with
multitudes of faint and distant stars, without any bright
objects nearby to serve as guideposts. As a rough aid to
memory, the field is located approximately halfway between

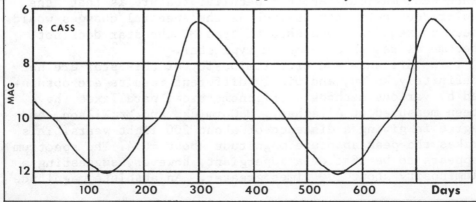

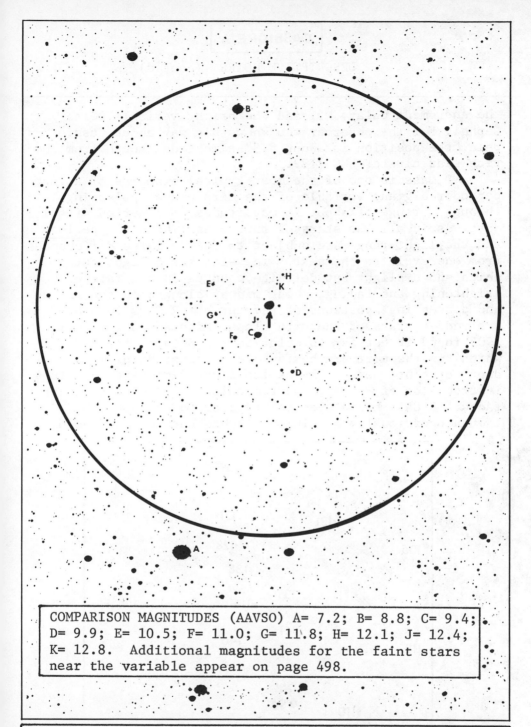

COMPARISON MAGNITUDES (AAVSO) A= 7.2; B= 8.8; C= 9.4; D= 9.9; E= 10.5; F= 11.0; G= 11.8; H= 12.1; J= 12.4; K= 12.8. Additional magnitudes for the faint stars near the variable appear on page 498.

R CASSIOPEIAE. Finder chart made from a 13-inch telescope plate at Lowell Observatory. Circle diameter = 1° with north at the top. Limiting magnitude about 15.

the Andromeda Galaxy M31 and the bright variable Delta
Cephei. For the observer who knows his way about Cassiopeia, the position is about 5.3° almost due south from the
rich star cluster NGC 7789.

R Cass is one of the most typical examples of a long
period red giant variable of the Mira class, pulsating
through a range of about 7½ magnitudes in a period of 431
days. As in all the stars of this class, both the period
and the light range are subject to fairly large differences
from one cycle to the next. The star has a light curve
which shows slight but definite changes in its rate of
brightening and fading, about midway along both the rising
and descending portions of the curve. R Cass is one of
the redder stars of the Mira class; Miss Agnes Clerke in
1905 found it not far inferior to the N-type star V Cygni
which "in the northern hemisphere...bears the palm for
depth of tint, especially as its light diminishes.." This
is true also of R Cass; the red color deepens as the star
fades. R Cass has a spectral type of M6 to about M8e,
though occasionally at minimum it has been classed as M10.

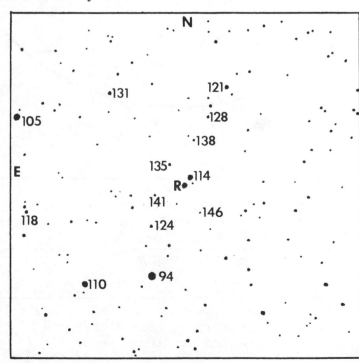

R CASS
15'
FIELD

As in all the Mira-type stars, the enormous change in the
light output is not exactly real; as the star appears to
fade, the total energy emitted drops by only a factor of
about two, but the radiation shifts over into the invisible
infrared portion of the spectrum. R Cass is an especially
remarkable example of this effect, showing a difference of
nearly 10 magnitudes between the visual and the infrared
radiation; it would appear as a 1st magnitude star if the
human eye was sensitive to radiation at all wavelengths.

Distances of the Mira-type stars are chiefly determined by statistical methods, as no stars of the class are
near enough to permit an accurate trigonometrical parallax.
The spectral features suggest a peak absolute magnitude of
about -1 (visual) and the resulting distance is close to
800 light years. R Cass shows an annual proper motion of
0.08"; the radial velocity is about 12 miles per second in
recession.

In attempting to identify R Cass, particularly when
the star is faint, observers should remember that the star
has a companion of magnitude 11.4 (AAVSO mag) some 28"
distant in PA 331° and an even closer companion of the 14th
magnitude almost due west. T.E.Espin gave the separation as
14" in 1910, but according to the current AAVSO chart it is
now 11". These stars do not appear to be true physical
companions to R Cass, which may explain the slow change in
the separation. The chart on the opposite page will assist
in identifying the star when near minimum, and shows a
field 15' in diameter. Star magnitudes are given according
to the AAVSO, but with decimal points omitted to avoid
confusion with star images; thus "131" = magnitude 13.1.

S   (Variable) Position 01159n7221, about 12½°
           north from Delta Cassiopeiae. Discovered at
Bonn, Germany in 1861. S Cass is a long-period red variable
star which sometimes rises above the 8th magnitude at maximum, but is usually fainter than 14th at minimum. Cycles
of the star average about 611 days, which is unusually long
for stars of the type; the time from minimum to maximum is
about 275 days and the star often shows a temporary slow-
down on the ascending branch of the light curve. S Cass
is one of the best known variables of type S; spectral
features resemble the M-type stars but show lines of

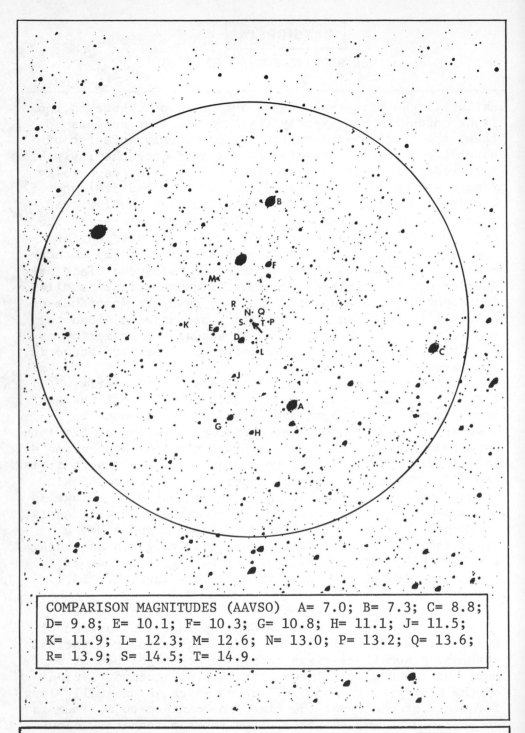

COMPARISON MAGNITUDES (AAVSO)   A= 7.0; B= 7.3; C= 8.8;
D= 9.8; E= 10.1; F= 10.3; G= 10.8; H= 11.1; J= 11.5;
K= 11.9; L= 12.3; M= 12.6; N= 13.0; P= 13.2; Q= 13.6;
R= 13.9; S= 14.5; T= 14.9.

S CASSIOPEIAE. Finder chart made from a 13-inch telescope
plate at Lowell Observatory. Circle diameter = 1° with
north at the top. Limiting magnitude about 15.

zirconium oxide instead of the usual titanium oxide. In a
few stars, as R Andromedae, the lines of both compounds
appear in the spectrum. S-type stars also have a somewhat
lower temperature than M-type; S Cass itself shows a range
of about 2500° to about 1900°K, one of the coolest stars
known. A computed absolute magnitude of about -1 (maximum)
suggests a distance of about 2000 light years.

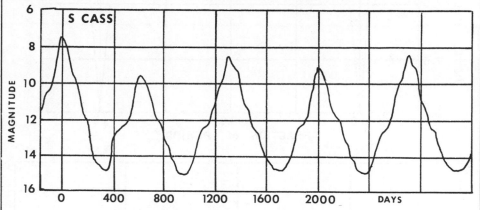

RZ   (Variable) Spectrum A0. Position 02443n6926.
              An interesting eclipsing variable star, located
near Iota Cassiopeiae, and discovered by G.Muller in 1906.
It is one of the most suitable objects of its class for
observation by amateurs, and can be studied in very small
telescopes. RZ Cassiopeiae is normally magnitude 6.4. When
primary eclipse begins, the star requires only 2 hours to
fade to magnitude 7.8. The brightening then begins immediately, and in another two hours the star is at normal magnitude. The period of the system is 1.195252 days, or more
usefully 1d 4h 41m. The primary star is type A0; the spectral class of the companion remains as yet undetermined.
              An intriguing feature of this eclipsing binary is
a gradual decrease in the length of the period. In 1960 the
minima were occurring nearly half an hour earlier than the
predictions made from a formula that was correct in 1953.
Reliable observations of the star are needed in order to
determine the nature of this change. Visual magnitude estimates may be made by comparing RZ with the nearby field
stars on the chart (Page 502). Accurate times should be
recorded with each observation.

DESCRIPTIVE NOTES (Cont'd)

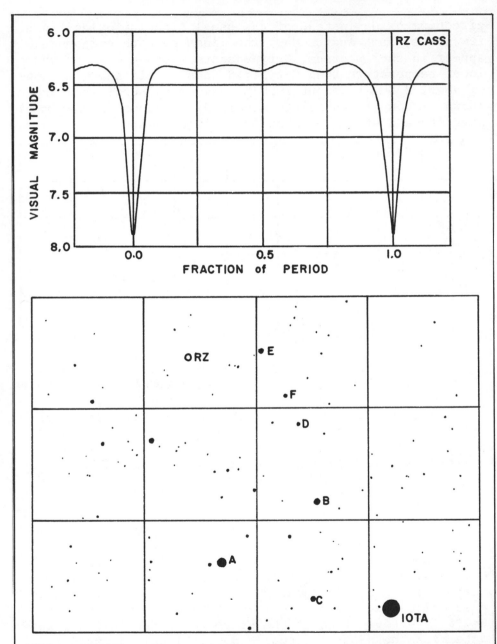

RZ CASSIOPEIAE Chart. Grid squares are 1 degree on a side and north is at the top. Comparison star magnitudes are: A = 6.0; B = 6.8; C = 7.3; D = 7.4; E = 7.7; F = 8.0.

DESCRIPTIVE NOTES (Cont'd)

AO   (HD 1337) (Boss 46) (Pearce's Star). Magnitude
6.05 (variable); spectrum O8 or O9 III. The
position is 00151n5109. This is a noted binary star, one
of the most massive systems known, and undoubtedly among
the most luminous objects in our Galaxy. The two components are giant O-type stars revolving almost in contact in
a period of 3.52355 days. The computed separation is some
15 million miles, which means that their surfaces must be
nearly touching. The orbit is nearly circular, and the two
stars form an eclipsing system with the small amplitude of
0.2 magnitude.

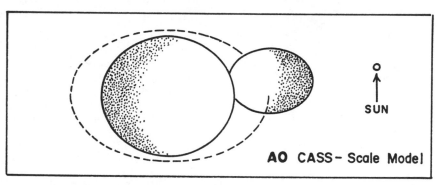

AO CASS- Scale Model

Both components are giants of type O8 or O9, with calculated surface temperatures of about 28,000°K, and computed
masses of 32 and 30 times the solar mass. The diameters
appear to be about 23 and 15 times that of the Sun. For
the larger star, O.Struve suggests a total radiation of
about 300,000 times the Sun's, only a fraction of which
appears as visible light. The total absolute magnitude may
be about -6, which suggests a distance of about 7000 light
years. Systems of this type are extremely rare in space,
but can be seen and identified at vast distances because
of the enormous energy output. UW Canis Majoris appears
to be a very similar object. (Refer also to Beta Lyrae,
and Plaskett's Star in Monoceros)

TYCHO'S STAR   (B Cassiopeia). The great supernova of
1572, the most brilliant nova recorded
during the past half millennium, and one of the four known
supernovae observed in our Galaxy. This famous "new star"
appears to have first been seen by W.Schuler on Nov. 6, 1572

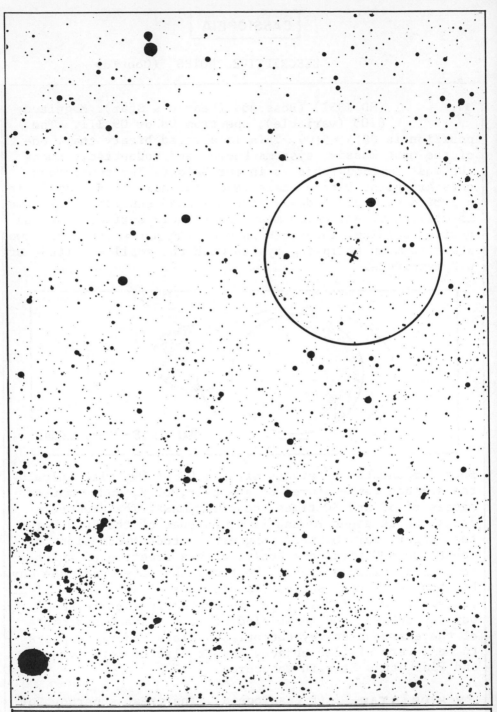

THE FIELD OF TYCHO'S STAR, from a Lowell Observatory 13-inch telescope plate. Circle diameter = ½°; north is at the top. Limiting magnitude about 16. Kappa Cass (mag 4.2) is at lower left. Cross indicates computed nova position.

and was probably observed by several others in the next three days, including Haintzel, Chytraeus, Maurolycus, and Cornelius Gemma. The early observers make no definite statement about its brightness, but when accidentally and independently discovered by Tycho Brahe on November 11 the star was more brilliant than Jupiter, and soon became the equal of Venus. A rather free translation of Tycho's own account of the discovery reads as follows:

"On the eleventh day of November in the evening after sunset... I was contemplating the stars in a clear sky... I noticed that a new and unusual star, surpassing the other stars in brilliancy, was shining almost directly above my head; and since I had, from boyhood, known all the stars of the heavens perfectly, it was quite evident to me that there had never been any star in that place in the sky, even the smallest, to say nothing of a star so conspicuous and bright as this. I was so astonished at this sight that I was not ashamed to doubt the trustworthiness of my own eyes. But when I observed that others, on having the place pointed out to them, could see that there was really a star there, I had no further doubts. A miracle indeed, one that has never been previously seen before our time, in any age since the beginning of the world."

For about two weeks the nova outshone every star in the sky, and could even be seen in full daylight. At the end of November it began to fade and change color; from brilliant white it turned yellowish, then orange, and finally reddish, fading from sight in March of 1574, having been visible to the naked eye for about 16 months.

Tycho Brahe, fascinated by this miracle in the supposedly changeless heavens, made a special study of the new star. There were no telescopes then, of course. Nevertheless, his account of the light changes and his position measurements form a valuable record for the modern astronomer, and in his honor the nova is generally referred to as "Tycho's Star". The visual light curve is shown on page 506, and is compared with the light curves of two other known supernovae, Kepler's Star of 1604 in Ophiuchus, and the supernova which appeared in the faint external galaxy IC 4182 in August 1937.

It has long been debated whether this star, or the remnant of it, is still visible at the present time. Tycho's

instruments were sufficiently accurate to permit a determination of the position to within about 30". His results
(precessed to 1950 coordinates) are:

     RA = 0h 22m 00.2s   Dec = +63° 52' 12"

No star exists near this position which can be identified
as a probable nova-remnant, although the field has been
thoroughly studied with large reflectors, and any typical
post-nova star as bright as 19th magnitude would have been
detected. Faint shreds of nebulosity have been found on
plates made at Palomar, however, and radio studies have
made the identification with Tycho's star virtually certain. These nebulous remnants are in no way comparable to
the vast expanding nebulosity resulting from the supernova
of 1054 A.D. in Taurus. However, the brightness of the
Taurus cloud, known as the "Crab Nebula" (NGC 1952) seems
to be attributable to the fantastic "synchrotron process",
the radiation of high speed electrons being accelerated in
a magnetic field. In all probability the Cassiopeia supernova is surrounded by an equally extensive nebulosity, but
the conditions may be different and the cloud remains all
but invisible from a lack of illumination. From recent
radio studies (1966) the distance of Tycho's Star appears
to be somewhat over 10,000 light years which implies that
the star at maximum had an actual luminosity of about 300
million times that of the Sun, and an absolute magnitude
of about -16.5. The expanding shell of gases is now about

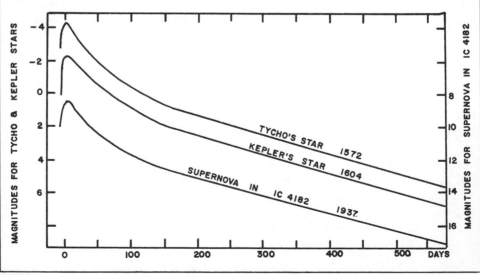

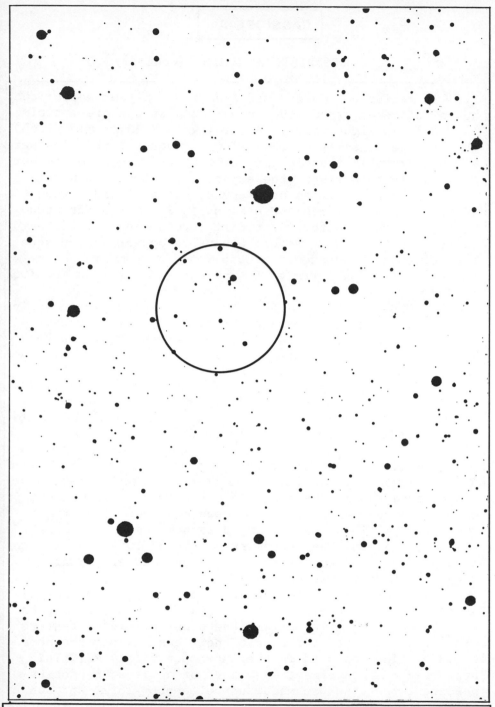

THE FIELD OF TYCHO'S STAR, enlarged from the chart on page
504. The circle here is 10' in diameter, centered on the
computed position. Chart made from a Lowell Observatory
plate obtained with the 13-inch telescope.

3.7' in radius, as determined from radio measurements; the
actual diameter of the supernova cloud at the present time
is nearly 20 light years. According to R.Minkowski (1966)
the average velocity of expansion is nearly 5600 miles per
second, probably the highest velocity ever measured in our
Galaxy. In comparison, the expansion of the Crab Nebula
is only about 600 miles per second, but as a radio source
it "outshines" the remnant of Tycho's Star by about   ten
times. The reasons for these differences are still rather
obscure, but the Crab Nebula begins to appear as a unique
type of object and should probably not be compared with
the more orthodox variety of supernovae. (Refer to NGC 1952
in Taurus).

THE PHENOMENA OF SUPERNOVAE. These colossal explosions, in
which a giant star appears to be almost completely demol-
ished, are the greatest stellar cataclysms which man has
actually witnessed in the Universe. At the time of such an
outburst, the exploding star may brighten by 20 or more
magnitudes, becoming for a time several hundred million
times brighter than the Sun. A supernova may be more than
50,000 times the brightness of an ordinary nova! But while
ordinary novae appear rather frequently, some 30 or 40 a
year in our Galaxy, the expected frequency of supernovae
is about one every third century in any one galaxy. In the
last thousand years, there have been four such super explosions witnessed and recorded in our Galaxy, though there is
good evidence that some others have occurred. The bright
"new star" of 1006 A.D. in Lupus is now recognized as the
earliest example. The Taurus supernova of 1054 A.D. was the
second, Tycho's Star was the third, and Kepler's Star of
1604 in Ophiuchus was the last.

        Supernovae are also detected at intervals in the gal-
axies beyond our own, sometimes equalling or even surpassing the combined light of all the other billions of stars
composing the system. The best known example was the nova
of 1885 which appeared in the Andromeda Galaxy M31. This
star reached an apparent magnitude of at least 6, corresponding to an absolute magnitude of about -18.2, and an
actual luminosity of 1.6 billion suns. About equal in brilliance was the supernova of August 1937 in the faint galaxy
IC 4182; this star rose to magnitude 8.2, becoming over 100
times brighter than the galaxy in which it appeared.   The

light curve is compared with that of Tycho's Star on page 506. With an absolute magnitude of about -18.4, this was one of the most brilliant supernovae on record. Such a star, in the course of a few days, radiates into space an amount of energy equal to the entire output of the Sun for several million years. The total energy released is about 10^{48} ergs; the total power output at maximum about 10^{35} watts, comparable to the power output of an entire galaxy.

Over 100 supernovae have been recorded up to 1965, and a study of the accumulated data has led to the recognition of at least two main types and possibly several sub-types or minor varieties. Supernovae of Type I appear to be the rarest and most brilliant; the average absolute magnitude at maximum is about -16, equal to 200 million suns. The spectrum is unlike anything else known, showing extremely broad bright bands, even before maximum. The light curve is characterized by a rapid rise to maximum followed by a rapid fading at first, and a slower fading after a month or so. Three to four months after maximum the decrease in brightness becomes linear, with a gradient of about 0.016 magnitude per day. This strictly exponential decline suggests the radioactive decay of an unstable element with a half-life of about 55 days. The heavy element Californium (atomic weight 254) has been considered a possible suspect but the identification now seems unlikely for various theoretical reasons. Such heavy elements are not known naturally on Earth, but have been synthesized in thermo-nuclear reactions. Presumably they could be formed in the cores of super-dense contracting stars. A somewhat different interpretation has been suggested by P.Morrison and L.Sartori (1966); in their picture the expanding "light sphere" from the outburst produces fluorescence in the interstellar gases, and the linear decrease in brightness results from the increasing inefficiency of the illuminating process as the radiation shell expands into space.

Supernovae of Type II seem to be 8 or 10 times more plentiful than those of Type I. They have usually been regarded as being several magnitudes fainter than Type I, but current studies show that the difference is not so great as had been thought. The spectrum is less unusual, however, resembling that of a normal nova on a gigantic scale. The bright bands do not appear until after maximum.

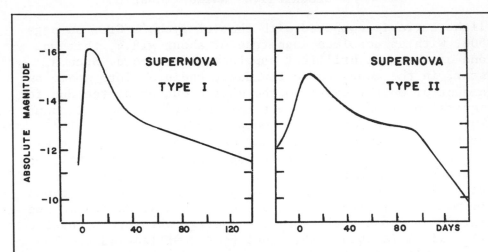

The pre-maximum spectrum is essentially continuous,   and
remains so until 5 or 6 days after peak brilliancy. For
these stars, measured expansion velocities range up to
4000 miles per second, higher than the velocities found
for some of the Type I supernovae. The light curve of Type
II is characterized by a slower rise to maximum, a more
leisurely decline at first, and a more rapid decline later,
beginning about 100 days after maximum. Typical examples
of both types are shown above.

   The distribution of the two types is of much interest,
and must be considered in any attempt to explain supernovae and their relation to stellar populations. All known
type II stars have appeared in spiral galaxies, and seem
to favor the regions of the spiral arms. Not a single example is known in any elliptical galaxy.   Supernovae of
type I have appeared in all classes of galaxies, but if in
spirals, tend to lie between the spiral arms. These facts
appear to identify the type II stars as members of Population I (confusingly enough), while the type I supernovae
are thought to be older stars, of Population II. At least
three additional types of supernovae have been identified
by F.Zwicky (1965) but their relation to types I and II is
still obscure.   Type III may be only a sub-class of Type
II, but show maxima which may persist for many weeks. Only
one star is presently classed as Type IV, the supernova
which appeared in the galaxy NGC 3003 in 1961. It had a
unique light curve, dropping in two major steps rather than

declining steadily. Finally, supernovae of Type V may not
be true supernovae at all, but unusually luminous variable
stars which show occasional slow increases to absolute magnitudes as high as -12. In our own galaxy the peculiar
"nova" Eta Carinae may be such a star. A similar object
has been detected in the external galaxy NGC 1058.

THE CAUSE OF SUPERNOVAE. The search for a possible cause
of supernova outbursts has long been one of the most interesting and exciting problems in astrophysics. The study is
obviously handicapped by the great rarity of the phenomena
and the fact that no cataclysm of this type has been seen
in our own galaxy since the invention of the telescope and
the spectroscope. Thus we are restricted to data obtained
from observations of supernovae in other galaxies, which
are, naturally, inconveniently distant. A number of supernovae remnants have been identified in our galaxy, however,
and much has been learned from studies of these expanding
clouds of debris. The estimated frequency of supernovae
gives us reason to hope that another outburst of the type
will be observed in our galaxy in the near future.

According to the best present evidence, a supernova explosion shows us the sudden (perhaps nearly instantaneous)
collapse or "implosion" of a very massive star. In our review of the white dwarf stars (page 403) the point was made
that a star must contract to amazing density once the hydrogen "fuel" has been consumed and there is no internal
energy supply to counteract the effects of gravitation. It
seems certain, however, that a very massive star cannot
shrink quietly into a stable white dwarf. Considering the
most massive stars known, calculations lead to a peculiar
paradox: the conclusion that the weight of the star's outer layers will be too great to be supported by the inner
regions. The star will thus have a virtually unlimited contraction; mathematically speaking it will be shrinking to
a geometrical point! Obviously, the stage is now set for
catastrophe. The later stages of contraction must result
in inconceivably high internal pressures and temperatures
of billions of degrees. It is the fascinating and frustra-
ting task of the astrophysicist to analyse these conditions
and explain the processes which lead to the eventual destruction of the star. One suggestion is that the core of
the star collapses into a "neutron star"; that the individ-

ual atomic particles are fused into a single gigantic mass of nuclear matter once the pressure passes a certain critical value. The density of such a mass would make even the white dwarf stars seem rarified; it would surpass the density of our heaviest metals by a factor of several hundred billion! Following the sudden collapse of the core, all the outer layers of the star would fall inward under the action of gravitation, and the entire star would be blown apart in a blast of intense radiation. A similar theory attributes the collapse of the core to the mass formation of those mysterious particles called "neutrinos" which have zero charge and zero mass, and astonishing power of penetration. G.Gamow has picturesquely pointed out that a neutrino beam could be stopped only by a layer of lead several light years thick! These particles are thus able to pass right through the body of the star and escape into space, taking most of the energy of the interior with them. If the density and temperature are sufficiently high, the cooling of the interior through neutrino emission will be so great that the internal pressure of the star may be reduced to a small fraction of its former value in a matter of minutes. The collapse of the star would result immediately.

In his review of the supernova problem, Fred Hoyle has shown that the pre-supernova star develops a multi-layered structure rather like an onion, in which various nuclear reactions are proceeding in the different layers according to the temperature required. As the star exhausts each "fuel", the core shrinks and the temperature rises still higher until some new reaction is started. When the central temperature exceeds 2 billion degrees, the reactions in the core result chiefly in the production of the nuclei of the heavier elements. It is at this point also that the energy loss through neutrino emission becomes critical, and the contraction of the star begins at an accelerated rate, resulting in even higher temperatures and pressures. As is now evident, the cycle is a closed circle: increasing temperature causes increased neutrino production; this in turn causes the star to shrink at an ever-increasing rate, and the shrinkage results in a continual rise in the temperature. This cycle continues until a critical temperature - about 5 billion degrees - is reached. "At this temperature", says Hoyle, "an extremely sharp change sets in. Instead of

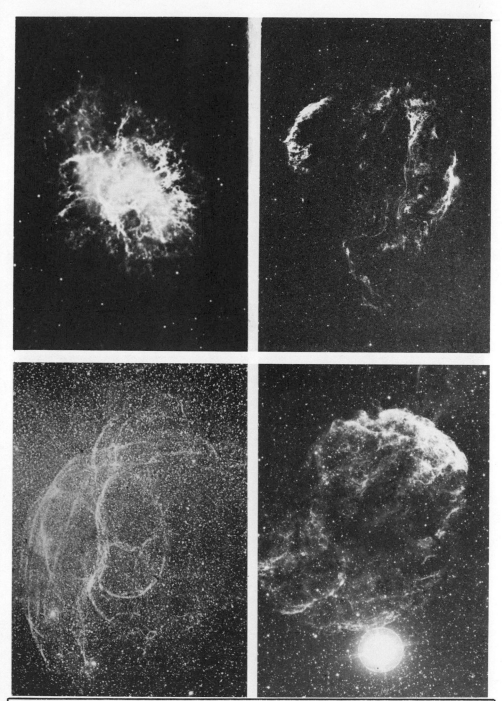

SUPERNOVA REMNANTS IN THE MILKY WAY GALAXY. Top: The Crab
Nebula NGC 1952 in Taurus, and the Veil Nebula NGC 6960-
6992 in Cygnus. Below: S147 in Taurus, and IC 443 in Gemini.

EXAMPLES OF SUPERNOVAE IN OTHER GALAXIES. Top: The bright supernova in NGC 5253 in 1972. Below: Supernova in NGC 7331 in 1959.   Mt. Wilson and Palomar Observatories

the material of the innermost parts of the star continuing
to belong to the iron group (the heavier elements) a dra-
matic change of composition occurs. The material changes
back into helium. Astonishing as this may be, there can be
no doubt at all about its correctness...the material must
change almost entirely into helium if the temperature rises
to a value in the neighborhood of 5,000 million degrees."
    The sudden transformation of the core into helium is
virtually equivalent to removing the central mass bodily;
Hoyle estimates that the resulting collapse of the star
takes place in about 1 second, and gives a graphic description of the explosion which follows immediately.   "The
energy released in only a second of time is as much as the
nuclear reactions inside the sun yield in about 1 billion
years. The amount of energy released is sufficient to endow
the exploding outer parts of the star with velocities of
from 2000 to 3000 kilometers per second, and is sufficient
to enable the star to radiate at 200 million times the rate
of the Sun for a time of about a fortnight."
    CHANDRASEKHAR'S LIMIT. This cryptic term is encountered
frequently in literature relating to stellar explosions.
It defines the mass required in a star to insure unlimited
contraction, and therefore determines how massive a star
must be in order to end its career as a supernova. This
critical mass is given in many texts as about 1.44 times
the solar mass, but recent studies seem to indicate that
about 1.25 is a more accurate figure. This concept leads
to an interesting line of thought. If our interpretation
is correct, it seems that the explosion of a star occurs
when the mass of the degenerate core exceeds the "Limit",
regardless of what the total mass of the star happens to
be. Obviously, a star of small mass will never reach this
state at all, and a star of about 1.5 solar mass will reach
it only when approaching the end of its hydrogen-consuming
life. But a very massive star will reach this stage when
only a fraction of its "fuel" has been exhausted; a star
of mass 15, for example, will reach the critical point when
less than 10% of its mass has gone to make up the degenerate core. Such a star would explode while the normal hydrogen-to-helium reaction was still proceeding in the outer
layers. The pre-explosion collapse would bring the hydrogen into direct contact with the inner core where it would

react with tremendous violence and contribute, probably, to the total effect of the explosion. Such a supernova would be "hydrogen-rich", in contrast to the "hydrogen-poor" supernova which has exhausted its normal nuclear fuel. It is tempting to identify these two types with the major classes I and II which we mentioned previously. Type I supernovae are presumably older stars (Pop. II) whose masses may exceed the critical limit only slightly, and which would be expected to be hydrogen-poor. Type II supernovae must be younger, more massive stars which have "aged" very rapidly; they are commonly found in the spiral arms of the galaxies where star formation is still underway, and massive high-luminosity stars are conspicuously evident.

Both types, then, might be expected to occur in our own galaxy, which is populated by a wide variety of stellar types. The four known examples all seem to have been Type I although the classification of the 1054 supernova is still uncertain and its vast cloud of debris (the Crab Nebula) is more or less unique. The famous Veil Nebula in Cygnus is undoubtedly a supernova remnant, though the explosion must have occurred many thousands of years ago. A very similar filamentary nebula in Taurus (S147) is another object which can hardly have originated in any other way. A great ring-shaped cloud over 400 light years in diameter exists in the Large Magellanic Cloud, and must have had its origin in a supernova outburst, many centuries ago.

The remnant of a more recent supernova in the Milky Way was identified in 1958, the discovery resulting from a fascinating piece of astronomical detective work. The first clue was the finding, in 1944, of an unusually strong radio source called "Cassiopeia A". One of the most intense in the sky, it is located at 23h 21m; +58°32'. After the position had been accurately measured, direct photographs were made with the 200-inch reflector at Palomar. A peculiar field of nebulous shreds and filaments was discovered, covering an area of 4'. These nebulous fragments showed a large proper motion of nearly 0.5" annually, outward from the center. Radial velocity measurements reveal that some of the filaments are moving with speeds of more than 3600 miles per second. The identification of such an object as a supernova cloud may be regarded as certain. The date of the outburst, computed from the enormous expansion rate,

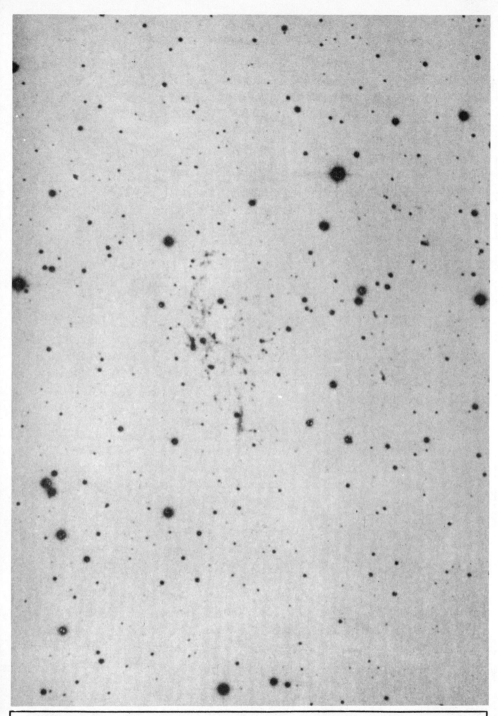

CASSIOPEIA A. The field of the Cassiopeia A radio source, photographed with red-sensitive plates at Palomar, with the 200-inch telescope.　　(negative print)

turns out to be fairly recently, probably around the year
1680. The distance of Cassiopeia A is some 11,000 light
years, and the star should have appeared as an object of
apparent magnitude 0 or -1. There are no records of such a
star having been seen, but the explanation is obvious. The
nova appeared in a portion of the sky in which thick dark
nebulosity produces an estimated 6 magnitudes of absorption. Thus the star probably appeared about 5th magnitude,
still within naked-eye range, but too faint to attract
attention at the time. As more radio sources are studied,
other such objects will undoubtedly be identified.

The most notable supernova remnant known, however, is
the famous Crab Nebula M1 or NGC 1952 in Taurus, the result
of the brilliant exploding star seen in the summer of 1054
AD. Identified as a strong radio source (Taurus A or 3C144)
and an X-ray source (Tau X-1) this vast cloud of stellar
debris is now some 5 to 6 light years in diameter and still
expanding at the rate of about 600 miles per second. Near
the center of the cloud lies a faint but extremely hot star
of the 16th magnitude whose identification as a white dwarf
or possible neutron star has been debated for some years.
In 1968, however, the discovery of a remarkable radio
source, now called a "pulsar" was announced by A.Hewish,
J.Bell, and their research group at Cambridge University
Observatory. Now called PSR1919+21, the new object is located near the star 2 Vulpeculae, and shows remarkably regular radio pulses occurring at intervals of 1.337301 second.
The identification of this object as a neutron star seems
definite, as the period is much too short to be attributed
to the pulsation, rotation, or orbital revolution of even
the smallest and densest white dwarf. With one such object
identified, radio astronomers went on to discover more than
a hundred others, including that enigmatic star in the
heart of the Crab Nebula; it is now known to be a pulsar
with the extremely short period of 0.033089 second. So in
this one case at least, the neutron star hypothesis has
been triumphantly verified. It is not thought, however,
that every supernova leaves a neutron star remnant. From
the analysis of various theoretical models it seems likely
that some stars leave only a white dwarf remnant, while
still others may be totally destroyed, leaving only a huge
expanding gas cloud.

What are the chances of observing a supernova in our own Galaxy in the near future? According to F.Zwicky, who discovered 122 extra-galactic supernovae during his life-time, the expected frequency is about 1 in three centuries per galaxy, but the figure is highly uncertain, and may depend critically on the type of galaxy. M83 in Hydra has shown four supernovae in only 45 years, while NGC 6946 in Cepheus has had four in 51 years. Both galaxies are Sc-type spirals. Two supernova appeared in a single year (1921) in the Sc-spiral NGC 3184 in Ursa Major, followed by a third outburst in 1937! Three supernovae each have been recorded in NGC 2841 in Ursa Major, and in M61 and M100 in the Coma-Virgo Galaxy Group.

It seems unlikely that any type I supernova has been missed in our Galaxy in modern times. At a distance of 5000 light years such a star appears as brilliant as Venus; at 30 light years (about the distance of Vega) it would shine with 40 times the light of a full moon! Even on the oppos-ite rim of the Galaxy, the apparent magnitude would still be about +2. Much of the Galaxy is, of course, hidden from our view by cosmic dust clouds, which might totally obscure even the most brilliant exploding star.

Type II supernovae are supposedly more frequent than those of Type I, and it is possible that some of the known bright novae were actually stars of this type at great distances. A check of the records, however, reveals no very convincing suspects at all. Nova Aquilae 1918 was very brilliant for an ordinary nova, but comparatively feeble for a supernova; the spectrum and light curve were also normal. Nova Puppis 1942 and Nova Cygni 1975 were somewhat more peculiar. The great light ranges are the outstanding features of these stars; 18 and 19 magnitudes respectively. This alone seems to set these two stars apart from all other novae, but the light curves and spectra were normal in all other respects, as were the expansion velocities. It is now thought that these stars were probably "virgin novae" or stars undergoing the nova process for the first time. Finally, the strange star Eta Carinae is sometimes classed among the supernovae as a member of the rare "Type V", though it is not certain that these high-luminosity variable stars should be included among the true supernova at all. See also M1 in Taurus, and Kepler's Star in Ophiuchus.

M52   (NGC 7654) Position 23220n6120. A fine star
                         cluster of the "open" or "galactic" type,
located in a rich Milky Way field on the western edge of
the constellation, near the Cepheus border. To locate, draw
a line from Alpha Cass through Beta, and continue it out
for a distance slightly more than the separation of the two
bright stars. M52 is one of Messier's discoveries, found on
Sept.7, 1774, while observing the comet of that year. The
discoverer described it as a cluster of very small stars
mingled with nebulosity. On this point, Messier was in
error, as there is no nebulosity in or near the cluster,
though the diffuse nebulosity NGC 7635 lies about 36'
distant toward the SW.

John Herschel described M52 as large, rich, round and
much compressed, whereas Admiral Smyth saw it as "irregular
and of a somewhat triangular form with an orange-tinted 8th
magnitude star at the vertex, giving the resemblance of a
bird with out-stretched wings. It is preceded by two stars
of 7 – 8 mag, and followed by another of similar magnitude,
and the field is one of singular beauty..." Lord Rosse
thought that M52 might contain about 200 stars, an estimate
which appears to be closely confirmed by modern star counts
as A.Wallenquist (1959) found 193 probable members out to a
radius of 9'. He derived a distance of 924 parsecs or about
3000 light years for the cluster; studies at Yerkes in 1960
gave a somewhat larger distance of about 1660 parsecs. The
true diameter is in the range of 10--15 light years.

M52 is one of the richer and more compressed clusters
with a computed density of somewhat over 3 stars per cubic
parsec, rising to more than 50 stars per cubic parsec near
the cluster center. In terms of age, M52 appears to be
among the younger open clusters, probably comparable in age
and type to the Pleiades. The brightest main sequence stars
are blue giants of spectral type B7. The two apparently
brightest members of the group are yellow giants of types
F9 (mag 7.77) and G8 (mag 8.22)

The faint nebulosity NGC 7635, located about 36' distant toward the SW, shows dimly near the lower right edge
of the photograph on page 521. The most curious feature of
this nebulosity is a faint ovoid arc of gas about 3' in
size, resembling a great ghostly bubble (photograph on page
522. This object is often classed as a planetary nebula,

DEEP SKY OBJECTS IN CASSIOPEIA. The galactic star cluster M52, located in a rich region of the Milky Way. Lowell Observatory 13-inch telescope plate.

NEBULA NGC 7635 in CASSIOPEIA. The peculiar "Bubble Nebula" is a vast sphere of tenuous gas, sometimes classified as a planetary nebula.

200-inch telescope, Palomar Observatory

but can hardly be considered a typical member of that odd
family of objects; it may be an ancient nova remnant. Some
2° to the SW lies another large field of faint nebulosity,
unnumbered on standard star atlases, but measuring more
than 1° in diameter. This field lies on the Cassiopeia-
Cepheus border, near the flattened galactic star cluster
NGC 7510.

M103   (NGC 581) Position 01299n6027. Galactic star
             cluster located in a rich Milky Way field about
1° NE from Delta Cass. This is the last object in the
original Messier Catalogue, though many modern versions
include at least one additional object, the Sombrero Galaxy
in Virgo (M104), while other versions propose additions up
to M109. The cluster, however, is not one of Messier's
original discoveries; it was first seen by M.Mechain in
1781. The best description of M103 is still that given by
the tireless Admiral Smyth in his "Cycle of Celestial
Objects"; he found the cluster to be "a fan-shaped group
diverging from a sharp star in the N.F. quadrant, brilliant
from the flash of a score of its larger members, four
principal ones of which are from 7 to 9 magnitude. Under
the largest in the S.F.quadrant is a red star of mag 8".
Smyth also called attention to the "neat double star" Σ131
on the NW point of the cluster, giving its colors as "straw
and dusky blue" and the PA as 141°, separation 14.4" (1832).
Very little change, if any, has occurred in the relative
alignment of the two stars since Smyth's day; the spectrum
of the brighter component is about B3. D'Arrest, in his
catalogue published in 1867, also mentioned this double
star, and described the cluster as "an irregular cluster
of 9- 10- 11 mag stars, size approximately 9'; a beautiful
10 mag reddish star prominent, its color is rose-tinted."
        Modern catalogues give the apparent diameter as
about 6.5' and the total integrated magnitude as about 7.
According to a study by A.Wallenquist (1959) the distance
is probably somewhat over 8000 light years, and the true
diameter about 15 light years. At least 40 stars seem to
be true members, the brightest ones of which are giants of
spectral type B3. This is not one of the richer clusters
but is a fairly compact group, with a much-flattened or
wedge-shaped outline, and easily identified when sweeping

STAR CLUSTER M103 in CASSIOPEIA. This compact group lies
about 1° from Delta Cassiopeiae. Lowell Observatory photograph made with the 13-inch telescope.

DIFFUSE NEBULA NGC 281 in CASSIOPEIA. This nebulous cloud
lies about 1.5° east of Alpha Cass. Lowell Observatory
photograph in red light with the 13-inch telescope.

A FIELD OF STAR CLUSTERS IN CASSIOPEIA. NGC 654 is at top, NGC 663 just below center, and NGC 659 at lower right. Lowell Observatory photograph made with the 13-inch telescope.

the area with low powers. As do many open clusters, M103 contains a single red giant star, type gM6, magnitude about 10.8. About 1.5° to the east, and extending north- ward will be found the trio of star clusters illustrated on page 526. The brightest of these, NGC 663, is probably also the nearest, at about 2600 light years. NGC 654 is believed to be about the same distance as M103, while the other cluster, NGC 659, has a computed distance of about 6000 light years. The four clusters, apparently, do not form a real group in space.

NGC 185 and NGC 147   These two miniature elliptical galaxies appear to be distant companion of the Great Andromeda Galaxy M31. They are some 7° north of it in the sky, and are approximately the same distance from us, about 2.2 million light years. With an apparent separation of 58' they may be viewed together in the field of a wide-angle eyepiece. The true separation from the Andromeda Galaxy appears to be about a quarter of a million light years.

NGC 185 is the brighter of the two, and may be seen in a good 6-inch telescope when its position is accurately known. No smaller glass is recommended. The object is an elliptical galaxy of dwarf characteristics, about 2300 light years in diameter. It has been well resolved into stars with a 4-hour exposure in red light with the 100- inch telescope. Appearing like a gigantic globular cluster, it must contain many millions of faint stars. The apparent magnitude is about 11.8, and the total luminosity some 8 million times that of the Sun. An unusual feature of this Population II system is a small irregular dust patch which often disappears on photographs due to over-exposure of the bright central mass. It may be seen on the photograph on page 148, made with the 200-inch telescope.

NGC 147 is a more difficult object for the amateur telescope, detectable with a 6-inch on the best of nights, but requiring something considerably larger (or more use of the imagination) to view it with any degree of certain- ty. The total brightness is about 12th magnitude, and the true diameter may be about 4400 light years across the longer dimension. Resolution into stars was accomplished with red-sensitive plates used on the 100-inch and 200-inch

NGC 185. A 4-hour exposure in red light with the 100-inch reflector; the first photograph which resolved this small galaxy into stars.

Mt.Wilson Observatory

NGC 147. A distant dwarf companion to the Great Andromeda
Galaxy M31. Resolution into millions of stars is shown on
this 200-inch telescope photograph. Palomar Observatory

telescopes. The total luminosity is only about 6 million
times that of the Sun, which places this dwarf system among
the intrinsically faintest galaxies known. The Andromeda
Galaxy itself is about 2000 times more luminous! (Refer
also to M31 in Andromeda)

NGC 457   Position 01159n5804. A bright galactic star
cluster located in the rich star fields of
the Cassiopeia Milky Way, about 4° southeast of Gamma Cass.
It is a rich scattered group of stellar points, some 10' in
apparent diameter, containing about 100 stars brighter than
13th magnitude. Some 60 of these are presently identified
as true cluster members. The stellar population of the
group resembles that of the Perseus Double Cluster, implying that NGC 457 is a rather young star group. In the main
mass of the cluster, the brightest star is a red supergiant
of type M0, of apparent magnitude 8.6 and absolute magnitude -5.2. The true luminosity is about 10,000 times that
of the Sun.

The bright star Phi Cassiopeiae, magnitude 5.0, spectrum F0, is of special interest from its position on the
southeast edge of the cluster. If actually a member, this
star is at a distance of about 9300 light years, and must
be one of the most luminous of all known stars, exceeding
even Rigel. The absolute magnitude, after correcting for
space absorption, would be about -8.8, or about 275,000
times the light of the Sun. While not definitely proven,
membership in the cluster seems supported by radial velocity measurements, studies of polarization in the cluster,
lack of measurable proper motion, and the spectrum, which
is that of a highly luminous supergiant. Another possible
supergiant member is the star HD 7902 (magnitude 7.0, spectrum B6) located near Phi on the edge of the cluster. If
proven to be a member, its absolute magnitude is -6.8. As
a standard of comparison, our sun at a distance of 9300
light years would appear as a star of magnitude 17.3! Such
a consideration may help the observer to realize- in some
degree- the true splendor of some of these distant groups
of giant suns. The true diameter of such a group cannot be
much less than 30 light years, and the total population,
allowing for the presence of the common low-luminosity
stars, may easily be several thousand stars.

STAR CLUSTER NGC 457 in CASSIOPEIA. A splendid group for
amateur telescopes. This photograph was made with an 8-
inch reflector.   Photograph by Kent de Groff.

STAR CLUSTER NGC 7789 in CASSIOPEIA. An exceptionally
rich galactic star cluster in the Cassiopeia Milky Way.
Lowell Observatory 13-inch telescope photograph.

NGC 7789   Position 23545n5626. An unusually rich
galactic star cluster located in a splendid
Milky Way field between the stars Rho and Sigma Cass, discovered by Caroline Herschel in the 18th century. To the
observer with binoculars, it is a hazy patch of unresolv-
able star dust; in a good 3-inch glass a rich sprinkling
of star points begins to appear across its surface, and the
view grows steadily more impressive with every increase in
the size and quality of the telescope. Sir John Herschel
described it as a most superb cluster which fills the field
and is full of stars, gradually brighter in the middle but
without a nuclear condensation. T.W.Webb refers to it as a
"large faint cloud of minute stars" and Smyth speaks of the
surrounding area as "a vast region of inexpressible splendor". The whole group covers an area nearly half a degree
in diameter, and the stars range from the 11th to the 18th
magnitudes.

At least 1000 stars are probably actual members of
this cluster, and the distance of the group, according to
recent studies by H.Arp (1962) is close to 6000 light years.
The true diameter is about 50 light years, and the total
light of all the stars may be something like 3000 times the
light of the Sun.

NGC 7789 has been considered by some observers to be
of a type intermediate between the true galactic clusters
and the less condensed globulars. A study by A.Sandage and
E.M.Burbidge (1958) has shown that the cluster is indeed a
galactic type, but a rather unusual one. It appears to be
much older than most of the well known galactic clusters,
and the stars seem to be well advanced in stellar evolution. The brightest members are orange giants of type K4 III
with absolute magnitudes of about -2.3; the majority of the
other bright members are giants and subgiants. All stars
brighter than absolute magnitude +2 appear to have evolved
away from the main sequence, and the resulting population
of stars resembles NGC 752 in Andromeda which has been
classed as an "intermediate-age cluster" by Arp (1962).
The computed ages of both clusters are in the range of 1.0
to 1.5 billion years, older than most galactic clusters
but not so ancient as the globulars. (Refer also to M13 in
Hercules, NGC 752 in Andromeda, M67 in Cancer, and NGC 188
in Cepheus)

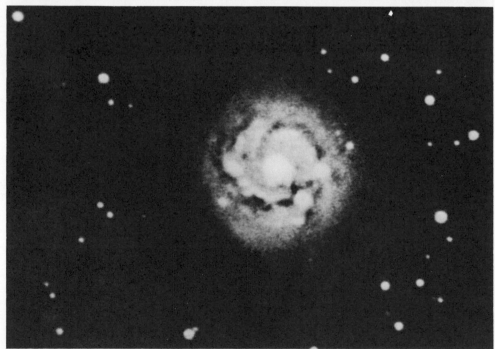

DEEP SKY FIELDS IN CASSIOPEIA. Top: Star field surrounding
Gamma Cassiopeiae, photographed at Lowell Observatory.
Below: The compact spiral galaxy NGC 278, photographed at
Mt.Wilson with the 100-inch reflector.

LIST OF DOUBLE AND MULTIPLE STARS

NAME	DIST	PA	YR	MAGS	NOTES	RA & DEC
h4405	19.2	48	17	7½- 9½	spect B8	11040s5300
h4409	1.4	260	60	5½- 8½	cpm; PA & dist dec; spect A5p	11050s4222
R165	3.4	67	52	7½- 7½	slow PA inc, spect F8	11108s4647
Rst 4940	0.4	114	49	7½- 8½	spect A2	11110s4951
λ128	0.1	85	59	7½- 7½	PA inc, spect A0	11128s3912
h4421	23.3	67	33	7 - 10	spect A0	11136s4739
h4423	2.4	276	59	7- 7½	relfix, spect F2	11141s4537
λ129	6.7	7	33	7½- 12	spect A2	11175s5312
π	0.2	111	40	4½- 5½	(I 879) binary, 43 yrs; PA inc; spect B5	11187s5413
h4426	13.3	174	47	7- 11	relfix, spect K0	11193s4317
Brs 5	1.6	131	27	7½- 9½	binary, about 265 yrs; PA inc; spect K7	11226s6123
I 1542	1.3	120	27	5½- 10	spect gK0	11231s3547
B 796	5.2	60	40	6 - 11	relfix, spect gM3	11231s3728
I 883	0.2	315	60	6½- 7	PA inc; spect G0, A2	11244s5253
h4438	23.0	196	20	7½- 12	relfix, spect A0	11251s3936
I 885	1.0	176	44	7½- 10	PA dec, spect G0	11262s4452
Brs 6	13.1	167	47	5½- 8	(△109) relfix; spect B9	11262s4224
Cor 125	6.6	82	28	7½-12½	(Fin 186) spect A0	11268s5831
	20.0	123	20	- 9½		
I 1544	2.2	86	36	6½- 11	spect G0	11304s3556
h4448	7.5	89	48	8 - 10	relfix, spect A0	11305s4325
I 78	1.0	97	59	6½- 6½	PA slow inc; spect A2	11312s4019
B 797	4.0	23	52	6 - 13	spect A0	11333s6101
I 421	1.5	117	44	7 - 10	in cluster NGC 3766; spect B5	11351s6113
Cor 126	11.0	271	20	7½- 12	spect B3	11358s6256
	13.5	203	20	- 11		
I 422	0.4	103	46	7½- 7½	PA inc, spect B2; in cluster I.2948	11360s6306
	1.7	3	43	- 10		
	9.6	323	43	- 12		

LIST OF DOUBLE AND MULTIPLE STARS   (Cont'd)

NAME	DIST	PA	YR	MAGS	NOTES	RA & DEC
h4460	8.6	176	32	8 - 9	relfix, spect A0	11368s5728
Hwe 70	3.3	104	54	8 - 8½	relfix, spect F8	11370s3709
△114	17.0	95	34	6½- 8	spect G5	11375s3750
Rst 4949	1.1	25	49	7 - 11	PA inc; spect A0	11375s5012
Hd 212	1.0	325	46	7½- 8	spect K0	11388s6052
I 890	2.0	304	59	7½-10½	slight PA dec; spect A0	11424s5511
Gls 169	4.3	228	54	7½- 8½	relfix, spect B8	11493s6419
I 892	54.6	122	60	5½- 10	dist inc; spect A2	11497s5643
I 892b	5.1	122	60	10- 11	slight PA dec	
Hwe 71	1.3	275	56	6½- 9	cpm; spect dF8, no certain change	11519s3729
Hld 114	3.2	180	56	7½- 7½	PA dec, dist inc; spect G0	11525s5549
I 80	1.3	99	54	8 - 8	PA dec, spect A2	11529s4138
h4482	0.1		52	8 - 8	(B 1203) PA dec;	11553s4326
	23.3	289	29	- 12	spect A0	
h4484	3.1	310	43	7- 9½	relfix, spect K0	11558s4040
I 895	0.9	312	47	8 - 8	spect A2	11592s4133
λ143	0.6	173	27	7- 7½	binary, 115 yrs; PA dec, spect F5	12011s3844
h4492	15.8	273	17	7½- 11	relfix, spect F2	12012s5426
h4491	23.4	41	33	8½- 8½	relfix, spect both F8	12012s4351
h4500	50.3	31	20	7 - 9	spect K0	12041s3735
I 216	1.1	27	51	7½- 9½	PA dec, spect F8	12072s5130
λ146	0.9	58	59	8 - 10	relfix, spect A2	12080s3633
I 423	2.6	166	35	6½- 11	spect K0	12084s4509
D	2.9	244	54	5½- 6½	(Rmk 14) relfix; cpm; spect gM0	12114s4527
λ147	0.3	334	60	7½- 8	PA inc, spect A2	12115s3627
	29.5	303	28	- 12		
Slr 10	1.9	245	53	7½- 9½	relfix, spect A2	12124s3557
R193	0.6	164	54	7 - 7	dist dec, spect A0	12152s3549
I 1220	0.1		59	7½- 8	spect B9	12155s5202
I 1221	1.7	83	59	7- 10	spect A0	12185s4217
Hd 216	4.6	54	60	7 - 13	spect A5	12215s3839
h4518	10.0	208	59	6½- 8½	relfix, spect K0	12221s4106
B 1718	1.0		29	6 - 10	spect G5	12320s4423

LIST OF DOUBLE AND MULTIPLE STARS   (Cont'd)

NAME	DIST	PA	YR	MAGS	NOTES	RA & DEC
λ157	7.7	121	00	6½- 12	spect A0	12340s5004
γ	0.3	302	27	3 - 3	(h4539) binary; 84½ yrs; PA dec; spect A0 (*)	12388s4841
h4546	15.0	223	17	7½- 9	relfix, spect F2	12420s5229
I 1559	6.6	309	42	7½-12½	spect G0	12428s3711
h4554	31.6	25	20	7½- 11	spect M0	12490s3049
Hd 220	6.4	210	47	6½- 12	spect K0 (L5299)	12495s5333
λ165	3.5	167	36	7- 13	spect K2	12504s3907
	20.6	233	20	- 14		
h4555	22.8	304	19	7- 12	relfix, spect A5	12508s4248
I 83	0.2	322	28	7½- 7½	binary, about 295 yrs; PA inc; spect F5	12538s4725
Cor 143	16.5	112	38	8- 9½	spect A0	12552s5355
△127	16.8	126	38	8½- 8½	relfix, spect B9	12568s5538
I 1224	0.3	353	56	8 - 8	PA inc, spect F2	12572s4107
Cp 13	5.1	67	47	7 - 9	relfix, spect K0	12574s4820
h4563	6.4	237	52	7- 8½	relfix, spect G5	12583s3321
B 238	5.9	292	34	7½- 12	(Daw 164) spect K0	12591s3034
λ169	11.2	235	42	7- 11	spect G5	13010s3359
Cor 148	4.8	99	43	8½- 9	relfix, spect F2	13022s4618
h4567	11.4	79	33	5- 11	(f Centauri) spect B5; relfix	13034s4812
I 917	1.2	289	59	8 - 8	cpm; spect F2; PA dec	13037s4546
ξ²	25.1	100	33	4½- 9½	(△128) cpm; spect B2	13040s4938
R213	0.8	24	56	7 - 7	relfix, spect B9	13042s5936
h4569	4.8	242	34	7 - 9	relfix, spect A0	13050s5625
I 1565	1.2	316	27	7½- 11	spect A3	13056s3351
Cor 149	12.7	256	20	8 - 10	spect A0	13063s5125
h4571	23.5	267	30	6½- 9	relfix, spect K0	13087s3452
I 424	0.3	113	27	5½- 6	(λ170) both PA inc, spect B8	13091s5939
	1.7	2	60	- 8		
Cor 152	25.8	147	20	6½- 8½	spect dG0	13098s5933
I 1227	0.4	69	60	7 - 7	PA dec, spect A0	13104s5026
I 399	8.2	138	31	7- 10½	spect F5	13108s6223
Fin 205	2.7	341	28	5- 10	spect F8	13111s5850
Mlb 3	1.7	40	59	7 - 9	(Hrg 83) spect B3	13114s6319

LIST OF DOUBLE AND MULTIPLE STARS   (Cont'd)

NAME	DIST	PA	YR	MAGS	NOTES	RA & DEC
λ171	0.4	296	00	8 - 8	spect K0	13121s3422
Cor 153	4.6	350	42	7½- 11	(λ172) relfix;	13125s3353
	20.0	227	00	- 14	spect A0	
h4576	5.6	128	33	7½- 10	(R216) relfix;	13131s5648
					spect F0	
I 233	3.4	110	53	7½-10½	(λ173) perhaps	13139s4101
	44.0	4	00	-10½	slight PA dec;	
					spect G5	
h4578	8.6	150	40	7- 10½	relfix, spect A2	13148s3645
J	0.1	168	59	5½- 5½	(△133) (Fin 208)	13194s6044
	60.0	343	00	- 6½	spect B5,B4; A-C	
					optical	
Slr 18	0.6	239	60	7 - 7	PA inc, spect A2	13199s4741
I 1231	0.1	290	53	8 - 8	spect A2	13216s5115
B249	0.4	315	59	7 - 8½	PA dec, spect F0	13233s3217
R218	2.4	170	43	7½- 10	relfix, spect K0	13251s4331
λ179	0.2	64	26	4½- 5	(d Centauri)	13281s3910
					binary, 62 yrs;	
					spect G8	
λ180	3.7	231	42	6½- 9	relfix, cpm pair;	13284s4212
					spect K0	
Hrg 86	1.6	239	55	7½- 8	relfix, spect B9	13291s6206
Hrg 87	3.1	336	56	9- 9½	relfix, spect A2	13296s6208
Rst 4985	0.5	166	49	6½- 8½	PA inc, spect A0	13315s4801
I 365	0.4	105	26	6- 6½	binary; 34.6 yrs;	13337s6126
	45.0	229	26	- 12	PA dec, spect dF8	
I 221	0.5	132	59	8 - 8½	PA dec, spect F2	13339s3209
R223	2.5	23	60	6½- 11	relfix, spect K0	13348s5809
λ184	2.4	303	59	7½- 9½	dist inc, PA dec;	13349s3449
					spect G0	
h4600	16.7	119	33	8 - 9	relfix; spect K0,	13362s4845
					G5	
Q	5.3	163	56	5½- 7	(△141) relfix;	13385s5418
					spect B8	
h4608	4.2	185	52	7½- 7½	PA inc, spect F5	13395s3344
△ 142	32.8	90	00	6½- 8½	both spectra B9	13406s5859
Fin 353	0.1	48	60	7 - 7	PA inc; spect B8	13407s4149
	0.8	71	60	- 9	(Rst 1741)	
Hwe 23	1.6	187	47	7½- 8	(I 222) relfix;	13409s3955
					spect A5	

LIST OF DOUBLE AND MULTIPLE STARS   (Cont'd)

NAME	DIST	PA	YR	MAGS	NOTES	RA & DEC
Cor 157	9.4	318	33	6 - 10	spect gG5	13437s6220
△ 143	12.5	36	33	8 - 9	relfix, spect G5, B8, color contrast	13457s6151
Hwe 24	11.6	355	38	6½- 9½	relfix, cpm pair; spect G3	13461s3527
△ 144	9.2	256	35	8 - 9	relfix, spect F8	13465s4707
N	18.0	289	54	5½- 7½	(Rmk 18) (△ 147) relfix, cpm pair; spect B9, A3	13488s5234
h4619	23.6	199	13	7½- 8½	spect G0	13489s4737
k	7.9	108	54	4½- 6	(△148) (3 Cent) fine cpm pair; spect B5, B8	13489s3245
β343	0.7	66	59	6½- 7½	PA dec, spect dF8	13491s3122
h	14.9	185	51	5 - 8	(4 Cent) (Hn 51) relfix, cpm pair; spect B5	13503s3141
y	0.2	264	60	6½- 6½	(β1108) (Hwe 28)	13506s3525
	27.7	158	59	- 12	PA inc, dist dec;	
	66.7	4	59	- 8	spect dF6	
h4624	21.4	350	59	6- 10½	relfix, spect B3e	13508s4653
λ190	7.2	222	33	7½-11½	relfix, spect K2	13510s3002
	31.9	144	34	- 14		
Hwe 74	6.2	117	38	7 - 9	relfix, spect G5	13524s3151
I 1238	1.5	225	25	7 - 12	spect F5	13530s3020
R227	1.8	4	53	6½- 8	PA inc, spect A1	13531s5353
	27.9	286	00	- 12		
I 224	2.8	184	44	7½- 10	relfix, spect F2	13533s3825
△ 151	23.2	45	43	8- 9½	(h4634) optical; PA & dist inc; spect G5, A2	13540s5548
I 225	2.4	301	29	7½- 11	spect K0	13578s6243
β1197	1.9	211	47	6½- 8	PA inc, spect gF5	14001s3127
β	1.3	251	60	1 - 4	HADAR (Vou 31) slow PA dec, spect B1 (*)	14003s6008
h4643	22.2	134	20	7½-11½	spect F8	14011s3701
I 941	0.1		60	8 - 8	PA inc, spect F0. difficult, angle uncertain	14017s3524

LIST OF DOUBLE AND MULTIPLE STARS   (Cont'd)

NAME	DIST	PA	YR	MAGS	NOTES	RA & DEC
h4642	9.1	12	18	$7\frac{1}{2}$- 13	Primary spect is	14034s6312
	20.8	104	18	- 13	composite = G0+A3;	
	26.4	335	18	- 10	$10^m = 0.6''$ pair	
Slr 19	1.4	295	59	7- $7\frac{1}{2}$	PA inc, spect G0	14044s4938
λ198	9.2	137	29	$7\frac{1}{2}$-$11\frac{1}{2}$	(I 942) spect B9	14068s4403
λ199	8.3	226	26	7- $12\frac{1}{2}$	spect A2	14075s2951
△158	3.9	55	38	8 - 9	relfix, spect F2	14078s4641
Cor 167	2.8	159	56	$6\frac{1}{2}$- $8\frac{1}{2}$	relfix, spect 05	14114s6128
					(marked "Cor 33"	
					on Norton's Atlas)	
Hd 228	33.9	170	13	5- $10\frac{1}{2}$	spect B3e	14115s5651
Brs 10	30.2	116	13	7 - 8	spect A0	14130s5704
R	28.0	218	00	var-12	(I 1240) LPV. (*)	14129s5941
I 523	0.3	41	59	8 - 8	PA inc, spect B9	14153s5827
β1110	3.9	132	33	7- $11\frac{1}{2}$	relfix, spect M	14167s3638
I 1241	0.3	131	59	$7\frac{1}{2}$- 9	spect A3.	14176s4213
	80.9	211	00	- 8	C= Cor 168	
Cor 168	1.7	205	59	9 - 9	(I 1241c) PA dec	14176s4213
△ 159	9.3	160	31	5 - 7	(L5893) (Rmk 19)	14190s5814
	20.7	255	03	-$13\frac{1}{2}$	relfix, spect gG4	
	45.6	4	02	- 10		
β1112	2.5	12	60	6- $9\frac{1}{2}$	perhaps slight PA	14302s3030
					inc, spect K0	
Hd 234	14.2	158	41	7- 11	relfix, spect K0	14305s3306
η	5.6	270	00	$2\frac{1}{2}$- 13	(λ207) spect B2	14323s4156
					(*)	
Hwe 75	4.1	215	60	8 - $8\frac{1}{2}$	relfix, spect A0	14341s3719
α	8.7	8	44	0- $1\frac{1}{2}$	ALPHA CENTAURI.	14362s6038
					fine binary, 80	
					yrs; spect G2,K1	
					"nearest star" (*)	
Hd 237	8.7	22	32	7- 13	spect B9	14371s4038
β 414	1.0	347	60	7- $7\frac{1}{2}$	relfix, spect B9	14389s3043
I 528	0.1	340	27	8 - 8	PA inc, spect A2	14423s3556
h4702	0.1	35	60	$7\frac{1}{2}$- 8	PA dec, spect K0	14455s3538
	9.8	215	32	- $9\frac{1}{2}$		
I 226	2.8	218	35	7 - 11	spect A0	14513s3356
β347	13.3	319	33	6 - 11	relfix, spect K0	14516s3306
	58.1	243	03	- 10		

LIST OF DOUBLE AND MULTIPLE STARS   (Cont'd)

NAME	DIST	PA	YR	MAGS	NOTES	RA & DEC
I 84	4.6	259	43	7½- 10	(λ 214) relfix;	14518s3613
	39.0	57	00	- 11½	spect A0	
Ho 390	24.3	170	22	5½- 12	spect A0	14527s3339
I 227	0.4	107	27	8 - 8	(λ 215) binary,	14535s3426
	6.5	114	60	- 13	40 yrs; spect F8;	
	48.6	24	00	- 10	ABC all cpm.	
h4718	1.9	63	60	7½- 9	relfix, spect K0	14545s3511
K	3.9	82	60	3½- 11	(I 1260) spect	14559s4154
					B2, probably cpm	
h4722	8.6	337	56	7 - 9	relfix, spect F0	14565s3031
	32.3	295	00	-13		
h4724	15.4	226	30	8- 10½	spect A3	14587s3644

LIST OF VARIABLE STARS

NAME	MagVar	PER	NOTES	RA & DEC
μ	3.1--3.2	Irr	Spect B2e (*)	13466s4214
R	5.4--12..	547	LPV. Spect M4e-M7e (*)	14129s5941
S	6.5--8...	65	Semi-reg; spect Np	12219s4909
T	5.5--10.0	91	Semi-reg; spect K7e-M3e	13389s3321
U	7.5--13..	220	LPV. Spect M3e-M4e	12307s5423
V	6.5--7.2	5.494	Cepheid; spect F5-G5	14289s5640
W	7.6--13..	201	LPV. Spect M3e-M4e	11525s5859
X	7.0--13.9	315	LPV. Spect M5e-M6e	11467s4128
Y	7.5--9..	180:	Semi-reg; spect M4	14280s2953
Z	7.2-----	---	Supernova of 1895 in	13371s3123
			galaxy NGC 5253	
RR	7.1--7.5	.6057	Ecl.bin; Spect F2;	14134s5737
			W Ursae Majoris type	
RS	8.0--13..	164	LPV. Spect M2e--M4e	11183s6136
RT	8.1--13.5	256	LPV.	13454s3637
RU	8.6---9.3	64.78	Ecl.bin; lyrid type;	12068s4509
			spect Fpe	

LIST OF VARIABLE STARS (Cont'd)

NAME	MagVar	PER	NOTES	RA & DEC
RV	7.2--10.8	446	LPV. Spect N3	13343s5613
RW	9.5--11..	Irr	Spect N3	11051s5451
RX	8.7--14..	328	LPV. Spect M5e	13485s3642
RY	9.1--14..	327	LPV. Spect M5e	14467s4219
RZ	9.0-- 9.5	1.876	Ecl.bin; lyrid type; spect B2	12587s6422
SS	9.2--10.8	2.479	Ecl.bin; spect B9	13103s6353
ST	9.9--10.8	1.223	Ecl.bin; spect F8	11077s5213
SU	8.7-- 9.6	5.354	Ecl.bin; spect F5	11088s4734
SV	8.8---9.8	1.661	Ecl.bin; spect B8; lyrid	11455s6017
SW	9.8--11.6	5.220	Ecl.bin; spect A0	12151s4927
SX	9.2--12.1	32.86	RV Tauri type; spect F5	12185s4856
SZ	8.2---8.9	4.108	Ecl.bin; spect A2	13472s5815
TT	9.8--15..	462	LPV. Spect Me	13164s6031
TU	9.2--13..	294	LPV. Spect M4e-M5e	14310s3129
TV	8.0---8.5	175:	Semi-reg; spect Nb	12119s5115
TW	7.5--11..	269	LPV. Spect M4e-M6e	13548s3049
UU	9.0--13..	368	LPV.	13188s6103
UV	8.8--12..	274	LPV.	11386s5723
UW	8.7--12..	---	R Coronae Bor type; spect K	12404s5415
UX	8.5---9.0	122:	Semi-reg; spect Nb	13188s6357
UY	8.5--10.5	115	Semi-reg; spect K5p	13136s4426
UZ	8.1---9.0	3.334	Cepheid; spect G5	11386s6225
VV	8.8--13..	199	LPV. Spect M3e	11441s6137
VX	8.0--11..	308	Semi-reg; spect M4e-M5e	13477s6010
VZ	8.7-- 9.0	4.929	Ecl.bin; spect B1	11500s6115
WW	9.5--11..	304	Semi-reg; spect Mb	13063s5959
XX	7.3---8.1	10.956	Cepheid; spect F6--K0	13370s5722
XY	9.6--10..	Irr	Spect M2	13457s4416
XZ	7.8--10.7	291	LPV. Spect M5	12215s3521
AD	8.5--10..	Irr	Spect M0e---M3e	11507s5902
AF	9.2--14..	284	LPV.	13012s5604
AL	9.0--10..	125	Semi-reg; spect Mb	12333s5319
AM	9.5--10.4	Irr		13441s5306
AO	9.2--11..	189	LPV. Spect M2e	14575s4218
AP	9.2--12..	357	LPV. Spect M5e	14581s3417
AQ	8.5--12..	387	LPV.	14020s3515
AW	9.0--10..	90	Semi-reg; spect Mb	13106s5642
AY	8.5---9.2	5.310	Cepheid; spect G4	11228s6027

CENTAURUS

LIST OF VARIABLE STARS (Cont'd)

NAME	MagVar	PER	NOTES	RA & DEC
AZ	8.4---9.1	3.213	Cepheid; spect F7	11230s6106
BB	9.7--10.2	3.997	Cepheid; spect F5	11510s6235
BE	8.5--12..	202	LPV.	14587s3004
BF	8.5---9.4	3.693	Ecl.bin; spect B7	11339s6111
BK	9.2---9.8	3.174	Cepheid; spect G5	11468s6248
LT	9.2---9.6	1.626	Ecl.bin; spect A0	11267s6036
LW	9.4---9.7	1.003	Ecl.bin; lyrid; spect B1	11352s6304
LZ	8.2---8.8	2.758	Ecl.bin; lyrid; spect B5	11481s6031
MN	8.7---9.1	3.489	Ecl.bin; spect B7	11258s6108
MO	9.8--10.1	9.656	Ecl.bin; spect B8	11396s6124
MP	9.7--10.4	2.993	Ecl.bin; lyrid; spect B8	11413s6128
MQ	9.8--10.2	3.687	Ecl.bin; spect B5	11418s6126
MT	8.5----15	---	Nova 1931	11416s6017
NP	9.6--10.7	2.853	Ecl.bin; spect A5	13134s6157
V339	8.8---9.6	9.467	Cepheid	14180s6119
V346	8.4---8.8	6.322	Ecl.bin; spect B4	11404s6209
V369	8.0--9..	108	Semi-reg; spect Mb	12123s5433
V377	8.3---9.0	8.252	Ecl.bin; spect A3	12544s4748
V378	8.4---9.1	6.459	Cepheid; spect G0; W Virginis type	13157s6207
V379	8.8---9.6	1.875	Ecl.bin; spect B9	13221s5931
V380	9.8--10.4	1.087	Ecl.bin; spect B8	13240s6137
V381	7.2-- 8.1	5.079	Cepheid; spect F6--G7; W Virginis type	13473s5720
V396	8.7--10..	Irr	Spect M	13142s6119
V402	9.8--10.3	3.720	Ecl.bin; spect A0	13250s6319
V412	8.0--9..	Irr	Spect Mb	13540s5728
V418	7.8---8.5	Irr	Spect K5	14174s6401
V419	7.7---8.3	5.507	Cepheid; spect K0	11286s5637
V420	9.1--10.4	24.77	Cepheid; W Virginis type	11374s4741
V499	9.5--10.4	.3467	Cl.Var.	13522s4300
V553	7.0---8.1	2.061	class uncertain; spect G5p; Cepheid?	14435s3158
V621	9.9--10.4	3.684	Ecl.bin; spect B9	13592s6229
V636	7.7---9.2	4.284	Ecl.bin; spect G0	14137s4943
V644	9.7--10.5		Ecl.bin; spect B3	11407s6027
V645	11---12	---	Proxima Centauri (Alpha Centauri C) Spect dM5e; red dwarf flare star (*)	14267s6229
V646	9.0--11.7	2.247	Ecl.bin; spect B9	11346s5256

LIST OF VARIABLE STARS   (Cont'd)

NAME	MagVar	PER	NOTES	RA & DEC
V659	6.6---7.0	5.622	Cepheid; spect F8	13282s6119
V673	9.5---9.7	.9328	Ecl.bin; lyrid	13596s4827
V701	8.4---9.0	.7384	Ecl.Bin; spect A0	13285s5131
V716	6.0---6.5	1.490	Ecl.Bin; spect B5	14102s5424
V745	9.8--10.8	3.025	Ecl.Bin; lyrid, spect B3	14234s6150
V752	9.1--9.75	.3702	Ecl.Bin; W Ursa Maj type spect G0	11403s3532

LIST OF STAR CLUSTERS, NEBULAE, AND GALAXIES

NGC	OTH	TYPE	SUMMARY DESCRIPTION	RA & DEC
3557		⊖	E2; 12.1; 1.5' x 1.3' pB,S,1E, pgmbM	11075s3716
3680	△ 481	⋰	cL,1C; diam 12'; about 25 stars mags 10....14	11233s4258
3699		☐	S neby with few faint stars	11256s5939
----	I.2872	☐	pL,F neby; 2' x 2'; bM	11261s6240
3706		⊖	E4; 12.7; 1.3' x 0.8' pB,cS, 1E	11273s3608
----	I.2944	☐	vL,F neby; diam 60' x 35'; surrounds Lambda Centauri	11335s6244
3766	△ 289	⋰	pL,pRi,C; diam 12'; about 60 stars mags 8...13; incl double star I421	11340s6119
----	I.2948	⋰	Diam 15'; about 25 stars, incl multiple star I422	11362s6315
3783		⊖	SBa; 12.8; 1.0' x 0.9' cB,1E, sbMN	11365s3728
3882		☐	vF neby; 1E, with 8^m star spect A0	11436s5605
3909		⋰	vL,1C, scattered field of stars mags 9....13	11470s4759
3918		◎	Mag 8, diam 10"; vB,S,R, tiny bluish disc "very like Uranus"	11478s5654
3960	Mel 108	⋰	diam 6'; pL,pRi; about 50 stars mags 12....	11484s5525
4219		⊖	Sc; 12.7; 4.0' x 1.3' pF,pL,pmE, vg1bM	12138s4303

LIST OF STAR CLUSTERS, NEBULAE, AND GALAXIES   (Cont'd)

NGC	OTH	TYPE	SUMMARY DESCRIPTION	RA & DEC
----	I.3253	⊖	S ; 12.3; 2.5' x 1.0' eF,L,mE, lbM	12211s3421
4373		⊖	E3; 12.2; 1.5' x 1.0' pB,S,lE, pgvmbM	12227s3928
----	I.3370	⊖	E2; 12.4; 1.4' x 1.2' pB,pL,vlE	12250s3904
4507	New 2	⊖	SB; 12.9; 1.2' x 1.1' pB,S,R,psbM	12329s3938
4603		⊖	Sc; 12.5; 2.5' x 1.2' F,L,lE,gbM; S-shape spiral	12383s4042
4645		⊖	E3; 13.1; 1.0' x 0.6' pB,S,E, psbM	12413s4129
4679		⊖	Sc; 12.9; 2.0' x 0.9' eF,pL,E	12448s3918
4696		⊖	E1; 12.5; 1.7' x 1.2' pB,L,lE,gbM	12461s4102
4767		⊖	E5; 12.8; 0.8' x 0.4' B,pS,lE,bM	12512s3927
----	I.3896	⊖	E1; 13.0; 1.0' x 0.9' S,F,bM	12537s5003
4835		⊖	SBc; 12.5; 2.3' x 0.7' F,pL,mE,vgbM	12553s4559
4852	△311	⦂⦂	Diam 10'; L,pRi, about 40 stars mags 10....	12571s5920
4936		⊖	E0; 12.6; 1.0' x 1.0' pB,S,R,bM	13015s3015
4945	△411	⊖	Sc or SBc; 9.2; 15' x 2.5' B,vL,vmE; nearly edge-on (*)	13024s4913
4947		⊖	SB; 12.6; 2.0' x 1.0' F,pL,E	13026s3504
4976		⊖	E4; 11.6'; 2.8' x 1.5' pB,pL,lE,gmbM	13059s4914
5011		⊖	E1; 12.9; 0.9' x 0.8' pB,cS,R, BN	13100s4250
5064		⊖	Sb; 13.1; 2.0' x 0.8' B,S,E, pslbM	13160s4739
5090		⊖	E2; 12.9; 1.2' x 1.0' pB,pL,R	13183s4328
5102		⊖	S0; 10.8; 6.0' x 2.5' vB,L,E; SBN; 17' n foll Iota Centauri	13191s3623

LIST OF STAR CLUSTERS, NEBULAE, AND GALAXIES (Cont'd)

NGC	OTH	TYPE	SUMMARY DESCRIPTION	RA & DEC
5121		⊖	S0; 12.5; 1.0' x 0.8' cB,S,R, psmbM	13219s3725
5128	△482	⊖	S0/pec; 7.2; 10' x 8' !! vB,vL,1E; dark central band; radio source (*)	13224s4245
5139	ω	⊕	!!! Mag 4; diam 30'; class VIII; eL,B,eRi,vvC; stars mags 11... finest globular, magnificent object (*)	13238s4713
5138		∴	Diam 10', 1C, about 15 stars mags 11....	13241s5845
5156		⊖	SB; 12.9; 1.5' x 1.4' pB,cS,E,g1bM, SN	13257s4839
5161		⊖	S ; 12.5; 4.0' x 1.5' pF,L,vmE, pgbM	13263s3254
5188		⊖	S ; 12.7; 1.0' x 0.8' F,pL,v1E, vg1bM	13286s3432
5193		⊖	E0; 12.6; 1.0' x 1.0' pB,S,R,gpsbM	13291s3258
----	I.4296	⊖	E0; 11.9; 0.6' x 0.6' pF,pS,R	13338s3343
5253	△623	⊖	E/pec; 10.8; 4.0' x 1.5' B,pL,E,psbM. Supernovae in 1895 (Z Cent) and 1972 (*)	13371s3124
5266		⊖	S0; 12.8; 1.5' x 1.0' B,pL,v1E, vg1bM	13399s4756
5286	△388	⊕	Mag 9; diam 4'; class V; vB,pL,R, stars mags 14...	13430s5107
5281	△273	∴	B,S,pC; diam 3'; about 20 stars mags 10...12	13431s6239
----	I.4329	⊖	E7 or S0; 12.8; 1.5' x 0.5' F,cS,E, bM	13462s3003
5299		∴	Bright Milky Way field; diam 30'; probably not a true cluster	13469s5937
5307		◎	Mag 12; diam 15" x 10" pF,S,1E	13479s5058
5316	△282	∴	pL,pC, diam 12'; about 50 stars mags 11....	13504s6137
5357		⊖	E3; 13.2; 0.6' x 0.5' pF,S,1E , 1bM	13531s3006

LIST OF STAR CLUSTERS, NEBULAE, AND GALAXIES (Cont'd)

NGC	OTH	TYPE	SUMMARY DESCRIPTION	RA & DEC
5367		☐	Mag 10; diam 1.3' x 1.0' L,B,1E; double nucleus	13547s3944
5365		⊖	Sa; 13.0; 1.0' x 0.7' pB,cS,pgbM, SBN	13548s4342
5398		⊖	Sp; 13.0; 1.0' x 1.0' pB,pL,R, vgbM	13583s3250
5419		⊖	E3; 12.4; 1.0' x 0.7' pB,pL,R, gpmbM	14007s3344
5460	△ 431	⫶⫶	Diam 30'; vL,1C; about 25 stars mags 8....	14045s4805
5483		⊖	Sc; 12.4; 2.6' x 2.4' pF,vL,R, vgbM	14074s4305
5494		⊖	Sc; 12.6; 1.5' x 1.5' pB,L,R,gbM	14095s3026
5617	△ 302	⫶⫶	Diam 15'; L,pRi,pCM; about 50 stars mags 8.... 80' west of Alpha Centauri	14260s6030
5662	△ 342	⫶⫶	Diam 8'; L,pRi,1C; about 30 stars mags 9.....	14316s5620

ALPHA CENTAURI. The nearest star is the bright central object; its distant "neighbor" Beta Centauri appears near the top of this print. North is at the left.
Georgetown College Observatory

DESCRIPTIVE NOTES

ALPHA   Name- RIGEL KENTAURUS, but more often called
              simply "ALPHA CENTAURI". The third brightest
star in the sky. Magnitude -0.27; spectra G2 V and dK1.
Position 14362s6038. Opposition date (midnight culmination) is May 3.

       Alpha Centauri is a triple star system, famous as
the nearest star to our Sun. It is 4.34 light years away,
or about 25 trillion miles. The distance was first determined by Henderson at the Cape of Good Hope in 1839, only
two months after the first stellar parallax was measured
and announced by F.W.Bessel, for the star 61 Cygni. Among
all the first magnitude stars Alpha Centauri has the greatest known parallax (0.751") and the largest proper motion
(3.68" per year in PA 281°). The radial velocity of the
system is about 14½ miles per second in approach.

       One of the finest visual binaries in the heavens,
the duplicity of the star was discovered by Father Richaud
at Pondicherry, India in 1689, during the course of his
observations of the comet of that year. The orbital motion
of the pair has been followed since the first accurate
measurements of Lacaille at the Cape of Good Hope in 1752.
Modern computations show that the orbital period is very
close to 80 years; whether slightly under or slightly over
it seems impossible to say. W.D.Heintz (1959) obtained a
period of 79.92 years, whereas the Yale "Catalogue of
Bright Stars" (1964) has 80.089 years. The semi-major axis
of the orbit is 17.66" with the apparent separation varying from about 2" to 22". Periastron was in mid-1955. Tilted about 11° from the edge-on position, the apparent orbit
is an eccentric and elongated ellipse. The computed eccentricity of the true orbit is 0.52, and the true distance
between the stars varies from 11 to about 35 AU. Facts of
interest concerning the two stars are given in the brief
table below.

	Mag.	Spect.	Mass	Diam.	Lum.	Abs.Mag.
A	-0.04	G2 V	1.10	1.07	1.5	+4.3
B	1.17	dK1	0.85	1.22	0.4	+5.7

According to R.H.Allen, there is some suspicion that the
K-star has brightened since the time of Richaud; some of

DESCRIPTIVE   NOTES   (Cont'd)

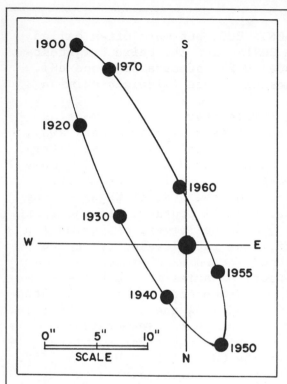

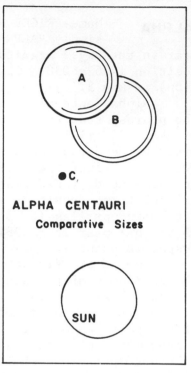

ALPHA   CENTAURI
Comparative Sizes

SUN

the earlier observers refer to it as a 4th magnitude star.
Miss Agnes Clerke (1905) estimated the difference in light
as about 3:1, which agrees very well with modern measurements. There also appears to be some disagreement concerning the precise spectral classes. O.J.Eggen, in his list
of the nearest visual binaries (1956) gives dG4 and dK5;
C.E.Worley (1963) has G2 V and dK5; while the Yale Catalogue of Bright Stars (1964) has G2 V and dK1. The bright
star, in any case, is very near our own Sun in type, size,
and luminosity; Alpha Centaurians (if such exist) would
see our Sun as a star of 1st magnitude near the Cassiopeia-
Perseus border, a few degrees northeast of the Perseus
Double Cluster. The Earth would appear as an infinitesimal
speck 0.75" distant from the Sun at widest separation; it
could not be detected with any telescope in existence.
    PROXIMA CENTAURI. Alpha Centauri is accompanied by
a faint and distant companion star, thought to be slightly
nearer to us than the bright pair. This faint star is
known as Proxima Centauri, and is located 1°51' south and

DESCRIPTIVE NOTES   (Cont'd)

9.9m in RA west of Alpha. It was discovered by R.T.Innes through proper motion measurements in 1915. The measured parallax is 0.762" and the annual proper motion is 3.85", both values being slightly larger than those of Alpha itself. The apparent magnitude is 10.7 and the spectral type is dM5e. Intrinsically one of the least luminous of all known stars, Proxima has an absolute magnitude of about +15.1, and an actual luminosity 13,000 times less than that of the Sun. If such a star were to replace our Sun, it would give as much light as 45 full moons.

There is some evidence that Proxima is in slow orbital revolution about Alpha Centauri, but the period must be extremely long, perhaps in the neighborhood of half a million years. The actual distance between Alpha and Proxima is approximately one trillion miles, or about 1/6 of a light year. This is an immense distance for any physical pair; it is nearly 300 times the greatest separation of the main pair A and B, and more than 400 times their mean separation. It is approximately 10,000 times the distance which separates the Earth and the Sun.

The actual diameter of this miniature star is calculated to be approximately 5% that of the Sun, or about 40,000 miles. Its mass is not known with any real certainty, but is undoubtedly only a fraction of the mass of the Sun. At present, the smallest stellar masses known are those of the binary L726-8 (UV Ceti system) where each star

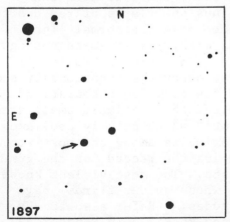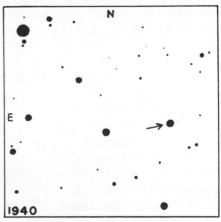

THE PROPER MOTION OF PROXIMA CENTAURI. Total displacement in 43 years = 165"

DESCRIPTIVE NOTES (Contd)

has about 4% the solar mass. Ross 614b has 8% the mass of the Sun, and the figure for Krueger 60b is 14%. Up to 1966 these three are the smallest stellar masses yet determined. From the well known mass-luminosity relation, it appears that the mass of Proxima Centauri is probably between 5% and 15% the mass of the Sun, definitely among the least massive stars known.

It is interesting to note that Proxima Centauri is a "flare star", a red dwarf in which the light may show very sudden changes of as much as a magnitude in only a few minutes. These temporary flashes occur at unpredictable intervals, and the star may be back to normal in less than half an hour. From a study of 592 Harvard plates, Harlow Shapley (1950) has shown that 48 flares of the star were recorded during the interval from 1925 to 1950. A rough estimate of the probable frequency of the outbursts may be made from the fact that 8% of the plates showed the star to be brighter than normal by at least half a magnitude. Shapley finds evidence that the variations are "erratic rapid changes rather than isolated brief outbursts from a normally quiet condition. Probably it is a star on which frequently more than one flare is in operation at the same time." UV Ceti and Krueger 60b are also flare stars.

The cause of such rapid outbursts is not well understood, but the phenomenon is probably not restricted to the faint red dwarfs. A similar flare on a bright star would go undetected, but on a faint dwarf it more than doubles the total radiation. Thus the flares of red dwarfs are probably conspicuous only because the normal light of the star is so feeble to begin with. It is worth noting that the total energy released in a flare on Proxima is comparable to one of the bright outbursts occasionally seen on the Sun, and called "solar flares". The outbursts of Proxima are thus probably limited to relatively small areas on the surface, local "hot spots" which quickly cool and fade to the normal state. Proxima is among the most active of all the flare stars, and holds the record for the greatest number of observed outbursts. The most violent known example, however, is UV Ceti, whose sudden flares have occasionally exceeded 5 magnitudes. (Refer also to UV Ceti, Krueger 60 in Cepheus, and Ross 614 in Monoceros)

BETA   Name- HADAR. Mag 0.66; spectrum B1 II. Position
14003s6008. Hadar is the 10th brightest star in
the sky, forming a wide naked-eye pair of $4\frac{1}{2}°$ separation
with Alpha Centauri. Opposition date (midnight culmination) of the star is about April 23.

Hadar is one of the "Orion type" stars of high temperature and great luminosity. It is located at a distance
of about 490 light years, and has an actual luminosity of
about 10,000 times that of the Sun (absolute magnitude =
-5.2). The star shows an annual proper motion of about
0.03"; the radial velocity is about 7 miles per second in
approach.

The companion, of magnitude 4.1, is only 1.3" away;
a rather difficult object because of the closeness of the
pair and the brilliance of the primary. The two stars
undoubtedly form a real physical pair, with a decrease in
the PA of 8° between 1935 and 1960. The projected separation is about 200 AU; the orbital period is evidently many
centuries. The computed luminosity of the "faint" star is
about 440 times that of the Sun.

GAMMA   Mag 2.17; spectrum A0 III or IV.   Position
12388s4841. The distance of Gamma Centauri is
about 160 light years; the total luminosity is about 275
times that of the Sun (absolute magnitude -1.3). The star
shows an annual proper motion of 0.20" in PA 266°; the
radial velocity is $4\frac{1}{2}$ miles per second in approach.

Gamma Centauri is a fine but close binary star, with
components nearly identical in type, brightness, and size.
The duplicity was first observed by John Herschel at the
Cape of Good Hope in 1835. Owing to the near equality in
brightness, the interpretation of Herschel's position
angles is uncertain. Consequently, two sets of possible
orbital elements have appeared in various catalogs, one
set giving a period of about 85 years and the other about
200 years. Recent measurements show that the shorter period
is undoubtedly the correct one. Van den Bos (1951) obtained 84.5 years, with periastron occurring in 1931. The
orbit has the high eccentricity of 0.79 and the apparent
separation varies from 0.2" to about 1.7"; the computed
semi-major axis is 0.93". The true separation averages
about 50 AU, reduced to about 10 AU at periastron.

DELTA　　Mag 2.59; spectrum B2p. Position 12058s5027.
　　　　　　The computed distance is about 370 light
years; the actual luminosity about 1000 times that of the
sun (absolute magnitude about -2.7). Delta Centauri shows
an annual proper motion of 0.04"; the radial velocity is
5½ miles per second in recession.
　　　The star has a peculiar spectrum resembling those of
Zeta Tauri and Phi Persei. Such objects, called "shell
stars" or "emission stars" appear to display large-scale
atmospheric turbulence, and a really violent example such
as P Cygni may eject material in an almost nova-like fash-
ion. Delta Centauri itself shows slight variations in
light, with the recorded range being about 2.56 to 2.62.
The radial velocity may also be slightly variable.
　　　The two bright stars in the field show the same proper motion, and apparently form a true moving group with
Delta.　These are GC 16576 (mag 4.5; spectrum B6 III) and
GC 16575 (mag 6.4; spectrum B9). The first lies 3.7' north
of Delta; the other is 2.5' south.　All three stars are
members of the widely scattered "Scorpio-Centaurus group"
of early type stars which includes many of the bright
stars in Centaurus, Crux, and Scorpius.　(Refer to the
constellation Scorpius for details concerning this group)

EPSILON　　Mag 2.33; spectrum B1 V. Position 13367s5313.
　　　　　　The computed distance is about 570 light
years, giving an actual luminosity of about 3000 times
that of the Sun (absolute magnitude -3.9). The annual proper motion is 0.03"; the radial velocity is 3.4 miles per
second in recession.
　　　A 13th magnitude companion was discovered in 1948 by
R.A.Rossiter; the separation is 36" in PA 158°. The star
is probably not a true companion to Epsilon; the projected
separation would be approximately 6300 AU.

ZETA　　Mag 2.56; spectrum B2 IV. Position 13524s4703.
　　　　　The star is estimated to be about 520 light
years distant; the actual luminosity is about 1900 times
that of the Sun (absolute magnitude -3.4). Zeta Centauri
shows an annual proper motion of 0.07"; the radial velocity
is 4 miles per second in recession.　The star is a spectroscopic binary with a period of 8.0235 days.

ETA   Mag 2.39 (slightly variable); spectrum B2 V +
A2. Position 14323s4156. The distance is estimated to be about 390 light years; the actual luminosity
about 1300 times that of the Sun (absolute magnitude -3.0).
Eta Centauri shows an annual proper motion of 0.05" and a
radial velocity of about 0.1 mile per second in approach.

A faint companion star of magnitude 13½ was reported
by T.J.J.See in 1897, at a separation of 5.6". The two
stars probably form a common proper motion pair with a
projected separation of about 675 AU. In addition, the
primary star has a composite spectrum, and, according to
the Yale "Catalogue of Bright Stars (1964) has been resolved by Finsen into a pair of 0.1" separation. At least
one of the components is slightly variable; the recorded
range of the system is 2.33 to 2.45.

THETA   Name- MENKENT. Mag 2.04; spectrum K0 III or IV.
Position 14037s3607. The distance is about 55
light years; the actual luminosity some 40 times that of
the Sun, and the absolute magnitude +0.9. The annual
proper motion is 0.74" in PA 225°; the radial velocity is
about 0.8 mile per second in recession.

IOTA   Mag 2.76; spectrum A2 V. Position 13178s3627.
Iota Centauri is approximately 70 light years
distant; the actual luminosity is about 28 times that of
the Sun (absolute magnitude +1.1). The annual proper
motion is 0.35" in PA 255°; the radial velocity is less
than 0.1 mile per second in recession.

The 11th magnitude S0-type galaxy NGC 5102 lies
in the field of Iota Centauri, approximately 17' toward
the north-east.

KAPPA   Mag 3.15; spectrum B2 V. Position 14559s4154.
The computed distance is about 470 light years;
the actual luminosity is about 900 times that of the Sun
(absolute magnitude -2.6). The annual proper motion is
0.03"; the radial velocity is slightly variable but averages 5.4 miles per second in recession. The star is probably a spectroscopic binary.

The faint companion, of the 11th magnitude, was
first recorded by R.T.Innes in 1926, and probably shares

the proper motion of the primary. It is 3.9" distant,
corresponding to a projected separation of 570 AU, and
(if at the same distance as the bright star) has a computed luminosity about equal to that of our Sun. Kappa
Centauri, like many of the bright stars in this region of
the sky, is a member of the large Scorpio-Centaurus moving
group. (For information concerning this association, refer
to the constellation Scorpius)

LAMBDA   Mag 3.15; spectrum B9 II. Position 11335s6245.
Lambda Centauri lies in a rich Milky Way field
about half a degree from the Galactic Equator. The faint
diffuse nebulosity IC 2944 surrounds the star, and the
galactic cluster IC 2948 lies 40' to the southeast. The
computed distance of Lambda is about 370 light years, and
the actual luminosity is about 630 times that of the Sun
(absolute magnitude -2.1). The star shows an annual proper motion of 0.04" and a radial velocity of 4.7 miles per
second in recession.

A companion star of magnitude 11½ was noted by R.A.
Rossiter in 1937, at 16.3" in PA 316°; it is not certain
that the two stars form a true physical pair. The projected separation of the pair is about 1870 AU. The faint
star, if at the same distance as Lambda, has an actual
luminosity of about half that of the Sun.

MU   Mag 3.12; spectrum B2 Vpe. Position 13466s4214.
The computed distance is about 470 light years
and the actual luminosity is about 1000 times that of the
Sun (absolute magnitude -2.7). Mu Centauri shows an annual
proper motion of 0.03"; the radial velocity is 7½ miles
per second in recession. The star is another member of
the large Scorpio-Centaurus association, and is an emission
type B-star which shows irregular light variations of about
a tenth of a magnitude. The recorded range is from 3.08
to 3.17.

A 14th magnitude companion was detected at Harvard
in 1897; it lies 48" from the bright star in PA 128°. This
star is probably not a true physical companion to Mu.
The projected separation would be about 6800 AU, and the
computed luminosity of the companion would be about 1/25
the light of the Sun.

NU   Mag 3.40 (possible slight variability). Spect B2 IV. Position 13465s4126. The computed distance of the star is about 750 light years, leading to an actual luminosity of about 1900 suns (absolute magnitude= -3.4). The annual proper motion is 0.04"; the radial velocity is about 5½ miles per second in recession. Nu Centauri is another member of the Scorpio-Centaurus group of early-type stars.

     The star is a spectroscopic binary, first identified by H.K.Palmer at Lick in 1906. According to an orbit by R.E.Wilson the period is 2.6252 days and the brighter star is about ½ million miles from the center of gravity of the system. The eccentricity of the system is near zero.

R   Variable. Position 14129s5941. A noted long-period red variable star, located about 1½° ENE of Beta Centauri, and discovered by B.A.Gould in 1871. At times it has attained naked-eye visibility with a maximum recorded brightness of magnitude 5.3. In the Harvard "Second Catalogue of Variable Stars" (1907) the period is given as 568.2 days, but now appears to be about 547 days. A gradual decrease of the period has evidently occurred, as in the case of R Aquilae. A changing period seems to imply a rapid change in the star's internal structure, but very little else can be said until the "mechanics" of the pulsating stars is more thoroughly understood.

     The period, in any case, is unusually long for stars of the Mira class. The really peculiar feature, however,

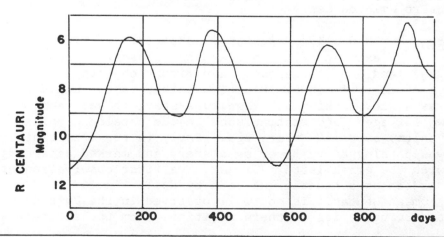

is that the light curve shows double maxima and minima.
The minima alternate quite regularly between 9 and 11, now
and then sinking to an occasional deep minimum of 13th
magnitude. The maxima are nearly equal, but the slightly
higher one systematically follows the shallower minimum.
R Normae and U Canis Minoris are two other stars which
have light curves of this rare type.

Owing to the double maxima and minima, the true period of R Centauri is more or less a matter of arbitrary
definition. The long-period variables display a spectrum-
period correlation, in which the stars of latest type usually show the longest periods. Since R Centauri shows a
spectral class of M4 at maximum, this correlation suggests
that the true period should be regarded as one-half the
long cycle, or about 274 days.

No star of the type is near enough to yield a really
accurate trigonometric parallax. In a statistical study
of a large number of long-period variables, V.Osvalds and
A.M.Risley (1961) derived a maximum absolute magnitude of
about -2.0 for stars of type M4 with periods near 273 days.
If R Centauri is accepted as a member of this class (in
spite of its definite peculiarities) the distance modulus
is found to be about 7½ magnitudes, giving a distance of
1000 light years. At maximum, the light may be about 500
times that of the Sun.

The measured annual proper motion is about 0.03"; the
radial velocity is 12 miles per second in approach. (For
more detailed discussion of the long-period variables,
refer to Omicron Ceti).

OMEGA (NGC 5139) Position 13238s4713. The finest
example of a globular star cluster in the
heavens, and one of the most magnificent objects within
range of the telescope. Plainly visible to the unaided
eye as a hazy looking 4th magnitude star, it has been
known for ages, and was included in the catalog compiled
by Ptolemy over 1800 years ago. Early in the 17th century
it was catalogued by Bayer as a star, and marked according-
ly with the greek letter "Omega". The first observation of
the object as a cluster was made by Halley in 1677.

Omega Centauri is not well observed in the United
States owing to its southern position. From the southern

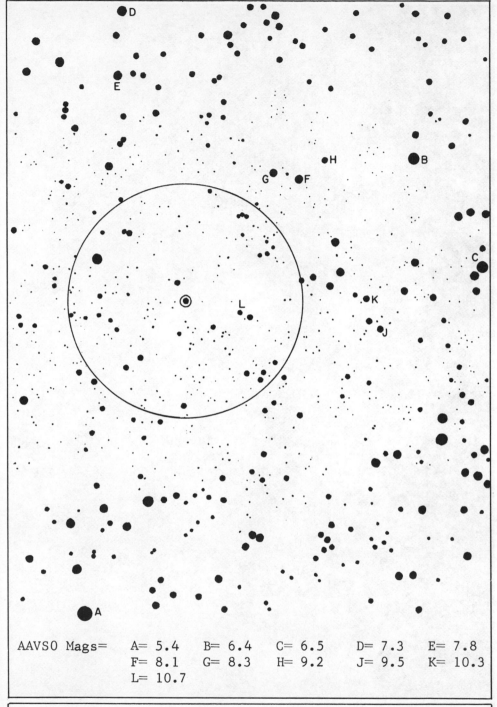

AAVSO Mags=   A= 5.4   B= 6.4   C= 6.5   D= 7.3   E= 7.8
                F= 8.1   G= 8.3   H= 9.2   J= 9.5   K= 10.3
                L= 10.7

R CENTAURI. Identification field, adapted from AAVSO chart. Circle diameter = 1° with north at the top, limiting magnitude about 14.

OMEGA CENTAURI. The greatest of the globular star clusters
is seen at a rather low altitude by observers in the U.S.
This photograph was made with a 5-inch camera at Lowell
Observatory.

half of the country it may be seen low on the horizon in the evenings of spring and summer, and from the latitude of Flagstaff it stands 8° above the southern horizon at culmination. The position is some 36° nearly due south from Spica. Binoculars or even an opera-glass will assist in locating it, and a clear and unobstructed horizon to the south is, of course, a prime requisite. To observers with small telescopes it resembles a "tailless comet" as Norton described it.

Definitely among the nearest of the globulars, Omega Centauri is still probably not THE nearest as has often been claimed. That honor now seems to go instead to the much fainter globular NGC 6397 in Ara, which has a computed distance of only 8200 light years. For Omega Centauri published values for distance range from 15,000 to about 22,000 light years. According to a summary of the present evidence by H.Arp (1965) the best modern value for the distance is probably about 5.2 kiloparsecs, or 17,000 light years. The apparent distance modulus, from a study of the H-R Diagram, is about 14.3 magnitudes, but a correction of nearly a magnitude must be made for absorption since the cluster lies only 15° above the plane of the Galaxy. The true modulus is probably close to 13½ magnitudes.

Visually, the apparent diameter approaches 30', which corresponds to an actual diameter of some 150 light years. On the best photographic plates the full size is not less than 70', or nearly 350 light years. Star counts to the 20th magnitude have been made on plates obtained with the Harvard ADH telescope; the extreme cluster diameter from this method is about 95'. As with all the globulars, the gradual thinning out of the stars around the edges makes it meaningless to attempt any precise determination of diameter. The rich central core is about 100 light years across.

In an early attempt to count the number of stars in this amazing cluster and determine their distribution, 6389 distinct star images were recorded on plates obtained with the 13-inch refractor at Arequipa, Peru, in 1893. The total population is now believed to exceed one million stars, and the mass may be in the neighborhood of half a million solar masses. At the cluster center, the star density is estimated to be about 25,000 times greater than in

DESCRIPTIVE NOTES (Cont'd)

the solar neighborhood, and the average distance between stars must be about 1/10 light year. The outline of the great swarm is not exactly circular, but noticeably elliptical, an effect presumably caused by rotation. This ellipticity is most evident in the outer regions, the distribution of the stars becoming nearly circular toward the center. The total integrated magnitude of all the stars is magnitude 4.25; the integrated spectral class is F7. From the apparent magnitude and known distance the true luminosity is readily calculated; after correction for light loss from absorption it amounts to about 1 million times the light of the Sun; the computed true absolute magnitude is about -10.2. As a standard of comparison, it may be remembered that our Sun, at a distance of 17,000 light years, would appear as a star of magnitude 18.4. From this it can be seen that the brightest members of the cluster outshine the Sun about 1000 times; these stars are red and yellow giants with absolute magnitudes of about -3.

Spectroscopic studies reveal that Omega Centauri is receding from the Sun's region in space at a velocity of 138 miles per second. This motion is the combined result of the Sun's motion and the orbital revolution of Omega Centauri around the center of the Galaxy in a period of some 100 million years.

The task of determining accurate magnitudes and colors for more than 7000 stars in the cluster was begun by astronomers of the Royal Greenwich Observatory in 1961. The observational work was done with the 74-inch reflector at Radcliffe in South Africa, and with the 24-inch and 18-inch refractors at the Cape of Good Hope. As a result of this project, accurate magnitudes and colors of about 7500 stars in the clusters are now known, and are shown on the accompanying H-R Diagram (page 564). Each point represents a star, and the over-all pattern strongly resembles similar graphs constructed for M13, M5, and other bright globular clusters. (For an explanation of the use of H-R diagrams and their significance in the study of stellar evolution and age-dating of clusters, refer to M13 in Hercules).

Omega Centauri contains a rich population of variable stars. As early as 1893, several examples had been discovered at the southern station of Harvard Observatory at

OMEGA CENTAURI. The finest globular cluster in the heavens is resolved into thousands of faint stars on this Harvard Observatory 60-inch telescope photograph.

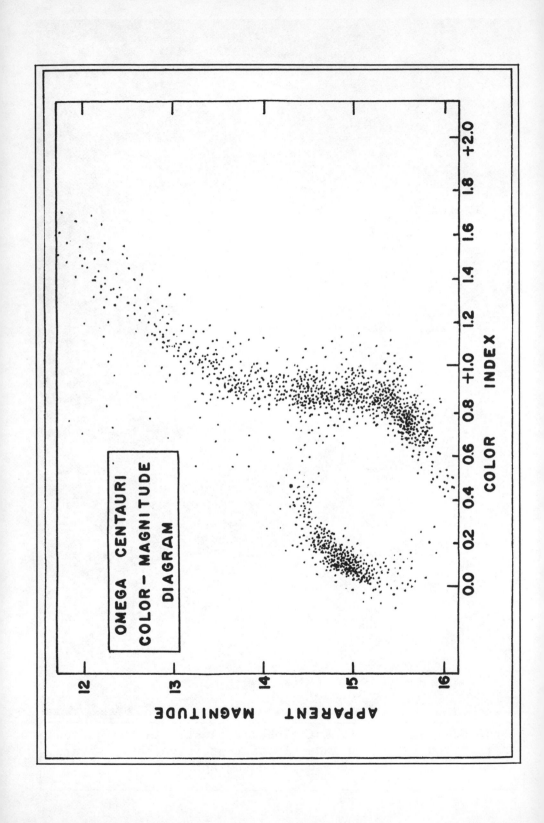

OMEGA CENTAURI
COLOR–MAGNITUDE
DIAGRAM

APPARENT MAGNITUDE

COLOR INDEX

COLOR

DESCRIPTIVE NOTES (Cont'd)

Arequipa, Peru. Less than 10 years later, S.I.Bailey (1902) listed 128 known variables in the cluster. The present total (1965) stands at 165. All but a few of these are the well known "cluster variables" or RR Lyrae type stars, pulsating with periods of less than a day. One of these stars, number 65 in Bailey's list, was shown by H.van Gent and E.Hertzsprung (1933) to have a period of approximately 1½ hours, the shortest period known at that time. Bailey classified the light curves of these stars into three distinct groups, illustrated by the examples shown below: a: rapid rise and large amplitude; b: moderate rise and amplitude; c: symmetrical curve, small amplitude. These divisions are still recognized today, though the classes a and b are often combined into one in modern studies of these stars.

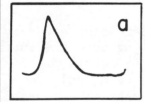 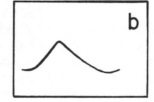 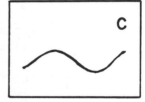

If we arrange the light curves of a number of these stars in order of increasing period, an interesting period-type relationship becomes evident. Stars of type "c" have the shortest periods - from 1½ hours up to about 11½ hours. At about 11½ hours there is a sudden change in the light curve to the sharply peaked type "a". Then, as the period increases from 12 hours to about 1 day, the form of the light curve alters again, and shows a gradual transition to type "b". Stars with periods exceeding about 1¼ days again show a high peaked light curve resembling type a, but these stars are no longer to be classed among the RR Lyrae type variables; they seem instead to be the first of the true cepheids. The radical change in the form of the light curve may in fact be regarded as the "boundary line" between the two classes. Six cepheids have been discovered in the cluster, several long period variables and irregular variables, and one star which appears to be an eclipsing binary. Omega Centauri contains more variable stars than any other known globular cluster, with the single

exception of M3 in Canes Venatici. These stars are of
great interest to astronomers studying the problems of
stellar evolution, since the globular clusters are known
to be the most ancient groups of stars yet identified in
our Galaxy.

"The noble globular cluster Omega Centauri", says Sir
John Herschel, "is beyond all comparison the richest and
largest object of its kind in the heavens. The stars are
literally innumerable, and as their total light affects
the eye hardly more than a star of 4th magnitude, the
minuteness of each star may be imagined.." (Refer also
to M13 in Hercules, M3 in Canes Venatici, M5 in Serpens,
M22 in Sagittarius, 47 Tucanae, and NGC 6397 in Ara. For
an account of the cluster-type variables refer to RR Lyrae
and for a survey of the cepheid variables refer to Delta
Cephei).

NGC 5128   Position 13224s4245. An interesting and very
              peculiar galaxy which has long been the sub-
ject of much controversy. It is located $4\frac{1}{2}°$ north of the
great cluster Omega Centauri, and appears as a luminous
sphere about 10' in diameter, crossed by a prominent dark
obscuring band. Until rather recently the classification
of NGC 5128 was quite uncertain, and various catalogues
listed it either as a diffuse nebula or an external galaxy.
In a Helwan Observatory publication of 1921 it is described as "a large patch of structureless and possibly gaseous
nebulosity, cut in two by a wide belt of absorbing matter,
through which appear several stars and wisps of nebulosity." H.D.Curtis at Lick (1918) classed it among the edge-
wise spirals with dark lanes. In his "Outlines of Astronomy" (1849) Sir John Herschel described it as "two semi-
ovals of elliptically formed nebula appearing to be cut
asunder and separated by a broad obscure band parallel to
the larger axis of the nebula, in the midst of which a
faint streak of light parallel to the sides of the cut
appears." E.Hubble (1922) and J.S.Paraskevopoulos (1935)
have classed it among the local nebulosities, while H.
Shapley and A.Ames (1932) included it in their famous
catalog of galaxies as an irregular system. Modern spectroscopic studies have demonstrated the extra-galactic nature
of NGC 5128 beyond any doubt, but the "pathological" form

GALAXY NGC 5128. One of the most peculiar galaxies known, this strange system seems to be ejecting material with explosive violence. Palomar Observatory 200-inch telescope photograph.

of the system is not yet satisfactorily explained. The
bright central disc appears to be the main body of a huge
elliptical or S0 galaxy, but the wide absorption lane is
a unique feature. The integrated spectrum is type G, but
with emission lines of hydrogen and oxygen. The dark band
is approximately 1' wide where it crosses in front of the
nucleus, widening out to about 2' on the southeast side of
the galaxy. On the northwest the band becomes weaker and
less regular, breaking into a chaotic mass of bright and
dark clouds. The course of the dark lane is from PA 135°
to 315°.

Spectrograms obtained with the 200-inch reflector at
Mt.Palomar show a corrected radial velocity of about 270
miles per second in recession; differences of up to 100
miles per second in the measurements reveal much turbulent
motion in the system. The object is evidently one of the
nearer of the bright galaxies, but the precise distance is
still not well determined. In a study of plates obtained
with the 60-inch reflector at Cordoba Observatory in Arg-
entina, J.L.Sersic (1957) found that the central region
of the band presented "a complex structure formed of chains
of condensations - probably chains of high luminosity O-B
stars... averaging about 18.7 photographic magnitude".
From these observations, Sersic derived a distance modulus
of about 26.6 magnitudes, equivalent to 6.8 million light
years. However, these results do not appear to be confirmed on plates made with the 100 and 200-inch telescopes, on
which no definite resolution into stars can be detected.
These negative observations suggest a distance of 15 to 20
million light years. E.M. and G.R.Burbidge (1958) adopted
a working value of 5 million parsecs in their study of the
system, while the measured red shift suggests a figure of
about 25 million light years.

In the following discussion, a compromise value of 15
million light years will be accepted. This makes the real
luminosity equal to about 20 billion suns, among the highest known for any galaxy. The absolute magnitude is near
-21, perhaps even higher after allowing for the occulting
effect of the dark lane. On the best plates the full diameter is about 25' x 20', giving the actual size as about
100,000 light years. The apparent width of the dark lane
varies from about 4000 light years in front of the nucleus

to 8000 light years at the southeast edge. Spectroscopic measurements show a difference in radial velocity on opposite sides of the galaxy, an effect presumably produced by rotation, and indicating a total mass of about 200 billion suns. This is one of the most luminous and most massive systems known, perhaps comparable to M87 in Virgo.

NGC 5128 is a strong source of radio radiation, over 1000 times as intense as the radiation of our own Galaxy. It is known to radio astronomers as "Centaurus A". The strongest emission comes from the dark band, but weaker radiation can be detected out to a distance of one degree from the center. In addition, radio energy appears to be coming from two large invisible "lobes" about 3/4° from the core of the galaxy, one lobe located north and slightly east of the visible object, the other one to the south and west. Similar lobes have been detected on either side of the intense radio galaxy "Cygnus A".

Baade and Minkowski have suggested that NGC 5128 is actually a colliding system of two galaxies, and that the dark bands belong to a nearly edge-on spiral which is seen in silhouette against the bright disc of a spherical or S0 galaxy. Increasing knowledge from both optical and radio studies now makes this interpretation seem rather unlikely. The radio lobes imply that a strong magnetic field is present, and this is confirmed by radio studies of polarization in the system. Observations made with the 210-foot radio telescope at Parkes, Australia (1962) have shown that the polarization reaches 40% in some regions of the galaxy. It is now thought that gigantic explosions have occurred in the nuclei of such galaxies, involving masses of many millions of suns, and accelerating material out to tremendous distances along magnetic lines of force. In the very peculiar galaxy M82 in Ursa Major such an explosion seems to be in progress at the present time. The giant M87 in Virgo appears to be another case with its strange nuclear "jet". It has even been suggested that the enigmatic Cygnus A, appearing as two objects in contact, is actually a single system with a dark band, similar to NGC 5128. In any case, the collision hypothesis does not appear adequate to explain the structural peculiarities and the enormous radio energy of many of these unusual systems. (Refer also to Cygnus A, M87 in Virgo, M82 in Ursa Major)

SUPERNOVA IN NGC 5253 in CENTAURUS. The supernova is near maximum (top) on a plate made May 16, 1972. The comparison plate was made in 1959. Palomar Observatory photographs made with the 48-inch Schmidt telescope.

NGC 5253   Unusual galaxy, Position 13371s3124, about 1.9° SSE of the great spiral M83 in Hydra. NGC 5253 is a system of uncertain classification, sometimes called irregular, but showing a much more symmetrical outline than most true irregular systems; it has also been classed as an elliptical with definite peculiarities. To the eye and the photographic plate it appears as a fairly regular oval about 4' in length, tilted toward PA 45° and with a brighter central mass. On short exposures, however, the central hub appears roughly rectangular with numerous irregular indentations and extensions; this area contains many condensations which appear to be nebulous aggregates of stars and giant emission regions. Some smaller diffuse spots have been identified as probable globular clusters.

NGC 5253 is accepted as a member of the fairly nearby Centaurus group of galaxies which includes M83, NGC 4945, NGC 5102, and the odd eruptive system NGC 5128 in Centaurus. J.L.Sersic, in the Cordoba "Atlas de Galaxias Australes", adopts a distance of about 12.4 million light years; the apparent size of 4' then corresponds to about 14,000 light years; the absolute magnitude is about -16.4. The observed red shift of about 160 miles per second might imply a slightly larger distance, of 15 or 16 million light years.

NGC 5253 is chiefly noted for its two brilliant supernovae, the brightest extra-galactic supernovae ever observed with the single exception of the star of 1885 in the Andromeda Galaxy M31. The first of these exploding stars, now called by its variable star designation "Z Centauri", was first observed by Miss Fleming at Harvard in 1895; on a plate of July 8 it is magnitude 7.2. The position was 1.28^s east of the center of the galaxy and 23" north. In May 1972 a second brilliant supernova blazed up in the outer environs of the system, 85" south and 56" west of the nucleus; it was found by C.T.Kowal at Palomar about 20 days after maximum. The magnitude was then about 8.5; at maximum about May 4 it had probably reached 7.2. This star, now designated SN 1972e, outshone its entire parent galaxy by a factor of at least 10; at the adopted distance the computed absolute magnitude was about -20.5, close to the light of 13 billion suns. (Refer also to S Andromedae, page 143, and the discussion of supernovae beginning on page 508.)

GALAXY NGC 4945. Possibly a loose-structured spiral seen
nearly edge-on. This photograph was made with the 30-inch
reflector at Mt. Stromlo in Australia.

CEPHEUS

LIST OF DOUBLE AND MULTIPLE STARS

NAME	DIST	PA	YR	MAGS	NOTES	RA & DEC
Σ 3051	16.9	23	25	7½- 9½	relfix, spect F2	00002n8000
Arg 47	10.1	290	17	8½- 9	optical, spect K, A	00002n5941
Σ 2	0.4	33	67	6½- 7	binary, about 300 yrs; PA dec, spect A3	00063n7926
0 Σ Σ 1	76.3	103	23	6½- 7	wide optical pair, spect M	00115n7545
Σ 11	8.0	192	33	8 - 11	relfix, spect K0	00123n7744
Σ 13	0.8	66	61	6½- 7	binary, dist inc; about 1600 yrs. PA dec, spect B9	00134n7640
A 803	0.1	218	58	7½- 7½	PA inc, spect A3	00154n7240
h1018	1.0	89	66	8- 8½	binary, about 160 yrs; PA inc, spect G5	00182n6723
0 Σ 6	0.3	158	61	7½- 8½	binary, about 200 yrs, spect A0	00186n6644
	13.5	114	50	- 9½		
U	13.8	62	00	7 - 11	relfix, spect B8; primary is ecl. binary (*)	00577n8137
	21.2	321	00	- 12		
0 Σ 28	0.9	299	62	7- 8½	PA dec, spect F0	01143n8036
0 Σ 34	0.3	260	61	7½- 8	binary, about 400 yrs; spect A0; PA inc.	01443n8038
0 Σ 37	1.2	209	37	7 - 9	PA dec, spect A3	02041n8115
Σ 223	0.7	46	43	8 - 10½	relfix, spect A0	02147n8030
Σ 320	4.8	232	62	6½- 9½	relfix, cpm, fine colors, spect M1, F7	02592n7913
Σ 327	24.2	283	35	6 - 11	optical, spect A	03038n8117
β1176	1.2	276	00	6- 12½	(48 Ceph) AB cpm; spect A4; AC PA inc, AC optical	03140n7733
	12.0	246	34	- 13		
Σ 343	29.5	325	26	8 - 9	spect G	03188n8353
	57.7	296	22	- 12½		
Σ 460	0.9	100	66	5 - 6	(49H) PA inc, binary, about 420 yrs; spect gG8, A7	04014n8034

CEPHEUS

LIST OF DOUBLE AND MULTIPLE STARS (Cont'd)

NAME	DIST	PA	YR	MAGS	NOTES	RA & DEC
κ	7.4	122	54	4 - 8	(Σ 2675) (1 Ceph) relfix, spect B9	20107n7734
A 731	2.1	214	32	7 - 12	spect A3	20247n6001
A 733	1.3	168	55	8 - 9$\frac{1}{2}$	relfix, spect A3	20274n6005
β152	1.0	89	61	7 - 8	PA dec, spect A2	20410n5712
η	68.4	48	25	3$\frac{1}{2}$- 11	Optical, PA inc, dist dec (*)	20443n6138
Es 134	10.7	265	18	9- 9$\frac{1}{2}$	spect A5	20444n6321
A 751	0.2	113	23	7 - 7$\frac{1}{2}$	binary, 59 yrs; PA dec, spect A8, F2	20525n5907
Es 135	6.6	196	23	7- 11	spect A0	20527n5659
	71.8	117	23	- 11		
A 756	0.5	221	57	7$\frac{1}{2}$- 8$\frac{1}{2}$	relfix, spect A0	20564n5838
	55.1	197	18	- 9		
Es 136	4.3	341	18	9 - 9		20586n5703
β472	0.8	8	54	8 - 8$\frac{1}{2}$	relfix, spect A2	20587n6140
h1607	11.5	82	35	7 - 10	dist inc, PA dec; spect K0	20592n6118
	30.6	325	20	- 13		
β1139	2.0	140	00	6$\frac{1}{2}$- 13	spect B8	21006n5653
Σ2751	1.5	354	62	6 - 7	PA inc, spect B9	21007n5628
Hu 959	1.4	161	62	7$\frac{1}{2}$- 9	slight PA inc, spect A2	21008n6631
Σ2771	2.7	215	62	9 - 9	relfix, spect A0	21041n7034
Σ2764	7.0	301	37	8 - 8$\frac{1}{2}$	relfix, spect A0	21044n6157
Σ2764b	0.6	35	57	8$\frac{1}{2}$- 9	(Hu 765) PA dec	
Σ2766	5.0	249	37	8$\frac{1}{2}$- 8$\frac{1}{2}$	relfix, spect F5	21057n5848
Σ2780	1.0	219	59	6 - 7	PA dec, spect B0	21105n5947
Σ2784	14.2	347	11	8$\frac{1}{2}$- 10	relfix, spect A0	21112n7351
0 Σ 436	11.7	230	00	7- 10$\frac{1}{2}$	spect B9	21126n7606
Σ2783	0.9	18	62	8 - 8	PA dec, spect A0	21127n5805
HI 48	0.7	250	67	7 - 7	binary, 84 yrs; spect gG0	21127n6412
M1b 221	3.6	174	25	9 - 9		21131n6553
Σ2788	8.0	354	54	8$\frac{1}{2}$- 9$\frac{1}{2}$	relfix, spect K0	21150n6708
β1140	4.1	275	46	7- 12	relfix, spect B2	21159n5824
Σ2796	25.9	43	51	7$\frac{1}{2}$- 9	relfix, spect A0; cpm pair	21168n7824
Σ2790	4.5	45	58	5$\frac{1}{2}$- 10	relfix, spect M1p & B3	21178n5825
	74.5	351	00	-10$\frac{1}{2}$		

LIST OF DOUBLE AND MULTIPLE STARS   (Cont'd)

NAME	DIST	PA	YR	MAGS	NOTES	RA & DEC
Es 137	45.5	74	05	$6\frac{1}{2}$- 9	Spect B0	21179n6139
Es 137b	2.7	75	21	9 -$12\frac{1}{2}$		
Es 138	9.0	265	06	$6\frac{1}{2}$- 13	spect F5	21191n6029
Σ2801	1.8	272	62	$7\frac{1}{2}$- 8	cpm; spect F8	21200n8008
Σ2807	1.9	313	62	8 - 8	relfix, spect F8	21205n8218
A 764	0.8	345	61	8 - 9	binary, about 200 yrs; PA inc, spect G5	21209n5721
Σ2798	6.5	126	51	8 - 10	relfix, spect B9	21226n6443
0 Σ440	11.3	179	62	6- $10\frac{1}{2}$	optical, PA dec, spect M1	21260n5932
h1654	4.3	31	20	9 - 9		21274n6124
β	13.6	250	54	$3\frac{1}{2}$- 8	(Σ2806) relfix, spect B2, A3 (*)	21280n7020
Hu964	1.5	277	22	$6\frac{1}{2}$- 12	spect A2	21284n6650
Hu771	2.6	195	21	7- $11\frac{1}{2}$	spect K0	21294n7743
h1659	7.2	298	37	$8\frac{1}{2}$- 9	spect A0	21314n5826
h3044	7.5	77	24	9 - 9		21319n7120
0 Σ442	0.4	348	61	8 - 8	PA dec, spect A2	21326n6144
Σ2810	17.0	290	40	$7\frac{1}{2}$- $8\frac{1}{2}$	relfix, spect F0	21331n5852
Σ2812	2.1	131	62	$8\frac{1}{2}$- 9	relfix, spect G5	21334n5927
Σ2813	10.2	273	16	$8\frac{1}{2}$- 9	relfix, spect F5	21345n5715
β371	8.6	3	17	8- $10\frac{1}{2}$	spect K0	21350n5828
Σ2815	0.9	161	57	$8\frac{1}{2}$- $9\frac{1}{2}$	slow PA inc,	21362n5720
	7.5	82	48	- 10	spect A0	
Σ2816	1.6	324	35	6 - $13\frac{1}{2}$	(β1143) spect O6	21374n5716
	11.7	121	58	- 8	in neby IC 1396	
	19.9	339	58	- 8		
Σ2819	12.4	57	51	$7\frac{1}{2}$- $8\frac{1}{2}$	relfix, spect F5; Σ2816 in field	21388n5721
Hu969	2.6	325	22	$7\frac{1}{2}$-$12\frac{1}{2}$	spect G5	21428n6041
Σ2836	11.8	153	26	7 - 10	relfix, cpm pair, spect F2	21480n6633
0Σ451	4.2	221	62	7 - 8	relfix, spect A2	21495n6122
Σ2840	18.3	196	58	6 - 7	slight dist dec, cpm; spect B6, A1	21503n5533
Σ2843	1.5	144	63	7 - 7	PA inc, spect A2	21504n6531
	56.0	276	39	- 10		
Σ2844	11.6	261	12	8 - 10	spect K0	21505n6439
Σ2845	2.1	172	54	8 - 8	relfix, spect B3	21510n6251

LIST OF DOUBLE AND MULTIPLE STARS (Cont'd)

NAME	DIST	PA	YR	MAGS	NOTES	RA & DEC
O Σ 457	1.2	247	63	6½- 8½	relfix, spect B2	21542n6504
O Σ 458	0.8	349	60	7- 8½	relfix, spect A0	21549n5932
β275	0.4	174	58	7 - 7	PA slow dec, spect B3	21558n6103
Σ2853	3.7	188	25	8 -10½	relfix, spect F5	21586n6744
Kui 110	0.7	225	52	6½- 8½	spect B0	21591n6216
Σ2873	13.7	69	59	6 - 7	PA slow dec, spect dF5, dG5	22004n8238
Σ2860	10.5	256	61	7½- 9½	optical, dist inc, spect K0	22016n6036
ξ	7.6	278	62	4½- 6½	(Σ2863) fine cpm pair, spect A, F7	22022n6423
15	11.1	298	47	6- 10½	(O Σ461) relfix, spect B5	22022n5934
	89.9	39	23	- 9½		
Σ2872	21.6	316	25	6- 7½	relfix, spect A0	22069n5903
Σ2872b	0.8	307	61	8 - 8	PA dec	
Σ2883	14.7	252	63	6 - 8	relfix, cpm pair; spect dF2	22096n6953
Σ2880	4.2	352	41	7½- 9½	relfix, spect K0	22101n5929
β376	3.6	149	31	8 - 11	relfix, spect K5	22104n5951
Σ2893	28.9	348	51	5½- 7½	easy cpm pair; relfix, spect K0, A3	22120n7304
O Σ470	4.2	354	37	7- 9½	relfix, spect A5	22195n6643
Σ2903	4.3	96	55	7 - 8	relfix, spect F5, A2	22202n6627
Krg 60	2.4	59	61	10- 11	famous red dwarf binary system (*)	22262n5727
δ	40.7	192	61	4 - 6½	typical "cepheid" variable; spect F5 (*)	22273n5810
	20.9	284	61	- 13		
O Σ473	14.8	357	22	6½- 10	relfix, spect A2	22284n5658
Hu981	0.3	223	58	7½- 7½	PA dec, spect A0	22288n6122
β706	2.8	15	38	8- 12½	relfix, spect A3	22316n6815
Σ2924	0.5	256	61	7- 7½	binary, 230 yrs; PA dec, spect F2	22316n6939
Σ2923	9.4	46	58	7 - 9	relfix, spect A0	22318n7006
Hu983	0.2	188	60	7½- 7½	PA inc, spect K0	22322n6534
β1092	32.6	261	26	7½- 12	(h3133) dist inc;	22348n7237
	42.2	137	26	- 8½	spect F5. A= 0.3" pair, PA dec.	

LIST OF DOUBLE AND MULTIPLE STARS   (Cont'd)

NAME	DIST	PA	YR	MAGS	NOTES	RA & DEC
OΣ481	2.4	268	54	7½ - 9½	relfix, spect A0	22431n7815
Σ2947	4.2	60	61	7 - 7	PA dec, cpm pair,	22473n6818
					spect F4, F5	
OΣ482	3.5	40	62	5½- 10	(34H) slow PA inc,	22477n8253
					spect K3	
Σ2948	2.6	5	62	7- 8½	relfix, spect B9	22478n6617
Σ2950	1.9	296	61	5½- 7	PA dec, spect G8,	22494n6125
	39.0	354	61	-10	G2	
A 632	0.4	60	21	8½- 9	binary, about 95	22500n5727
					yrs; PA dec,	
					spect K5	
Σ2963	2.3	355	56	8- 8½	relfix, spect A3	22530n7604
OΣ484	0.3	124	62	7½- 8½	binary, 130 yrs;	22546n7234
	31.9	256	11	- 11	PA dec, spect A2	
Σ2971	5.5	4	62	7½- 8½	relfix, spect G5	22557n7813
β851	2.0	160	24	7½- 13	spect G5	22596n7551
OΣ487	0.1	162	55	7- 8½	PA dec, spect A3	23004n8031
OΣ486	33.9	276	20	6- 8½	relfix, spect B5	23013n6010
Σ2977	1.9	350	55	7- 10½	PA inc, spect F5	23044n6110
β180	0.6	155	57	7½- 8	PA dec, spect A2	23050n6033
	34.3	106	15	- 10½		
Σ2984	4.5	295	24	7½- 10	spect K0	23054n7023
Hu994	0.1	306	61	6½- 7	PA dec, spect B3	23058n6322
π	0.7	307	61	5- 7½	(33 Ceph) binary,	23063n7507
	58.6	240	11	-12½	about 150 yrs;	
					PA inc, spect G2	
β992	0.2	68	61	8 - 8	binary, about 480	23137n6350
					yrs; PA dec,	
					spect F0	
O	3.2	214	61	5½- 8	(34 Ceph)(Σ3001)	23164n6750
					binary, about 800	
					yrs; spect K0, F6	
β386	19.9	313	25	6½- 12	spect A2	23241n7023
	46.5	203	10	- 12		
Ho 200	2.3	142	12	6½- 12	spect F0	23245n8608
β1148	2.1	75	48	7 - 13	relfix, spect K0	23251n6520
Σ3017	1.6	28	62	7 - 8	slight PA dec,	23257n7349
					spect F0	
β996	4.6	95	58	7 -11½	PA inc, spect K3	23496n7516
β1154	1.1	322	59	8 - 8	PA inc, spect A3	23566n7434

CEPHEUS

LIST OF VARIABLE STARS

NAME	MagVar	PER	NOTES	RA & DEC
β	3.15--3.2	.1905	Spect B2; Beta Canis Majoris type (*)	21280n7020
δ	3.6---4.3	5.366	Typical "cepheid" star; also visual double (*)	22273n5810
ε	4.20±.03	.0424	Delta Scuti type, spect F0 (*)	22132n5648
μ	3.7---5.0	Irr	Herschel's "Garnet Star" spect M2e (*)	21420n5833
S	7.4--12.9	487	LPV. Spect N8e. Very red star (*)	21359n7824
T	5.3--10.9	390	LPV. Spect M5e--M7e	21089n6817
U	6.8---9.2	2.493	Fine eclipsing binary, algol type, spect B8 + G8 (*)	00577n8137
V	6.5---		Variability doubtful, spect A2	23540n8255
W	7.5---8.5	Irr	Spect K0ep	22346n5810
X	8.1--15..	534	LPV. Spect M5e	21002n8252
Y	8.1--15..	333	LPV. Spect M5e--M6e	00347n8005
Z	9.9--15.4	278	LPV.	02192n8127
RR	9.0--15.5	383	LPV.	02361n8056
RS	9.8--11.5	12.42	Ecl.bin; spect A5+G	04572n8011
RU	8.4---9.4	105	Semi-reg; spect G6--K2	01141n8452
RW	7.6--9...	Irr	Spect M0	22212n5544
RX	7.5---7.8	Irr	Spect G5	00457n8142
RY	9.4--13.6	149	LPV. Spect M0e	23196n7841
RZ	9.0---9.7	.3087	Cl.Var; spect A0--A3 (*)	22375n6436
SS	6.9-- 8.0	90	Semi-reg; spect M5	03413n8010
ST	8.5--10..	Irr	Spect M0	22283n5644
SU	9.8--10.8	.9014	Ecl.Bin; lyrid type; spect B8	21451n5704
SW	9.4--11..	70	Semi-reg; spect M6	21246n6221
TY	9.7--13.3	330	Semi-reg; spect M4	21497n8624
TZ	9.0--11.0	83	Semi-reg; spect G6--K2e	00223n7337
VV	4.9---5.2	7430	Giant eclipsing binary system; spect M2ep + B9 (*)	21552n6323
VW	6.9---7.2	.2783	Ecl.bin; W Ursa Majoris type, spect dG5+dK1	20381n7525
VZ	8.8---9.3	Irr	Spect G0	21494n7112

CEPHEUS

LIST OF VARIABLE STARS   (Cont'd)

NAME	MagVar	PER	NOTES	RA & DEC
WX	9.0--9.7	3.378	Ecl.bin; spect A2+A5	22295n6316
XX	8.4---9.3	2.337	Ecl.bin; spect A8	23360n6403
XZ	8.4---9.2	5.097	Ecl.bin; lyrid type; spect 09	22162n5702
ZZ	9.3--10.0	2.142	Ecl.bin; spect B7+F0	22434n6752
AH	6.9---7.1	1.775	Ecl.bin; lyrid type, spect B0	22461n6448
AI	9.2--10.0	4.225	Ecl.bin; lyrid type, spect B	21448n5641
AN	8.3--11..	250	Semi-reg; spect M8e	23172n8244
AR	7.1---7.8	116	Semi-reg; spect M4	22527n8447
BF	8.7--13..	424	LPV. Spect M7	20305n6247
CQ	8.9---9.4	1.641	Ecl.bin; lyrid type, also Wolf-Rayet star, spect WN6 + 07	22349n5639
CW	7.6---8.0	2.729	Ecl.bin; spect B3+B3	23020n6307
DH	8.9---9.0	2.111	Ecl.bin; spect 06; in cluster NGC 7380	22449n5748
DM	8.0---8.6	Irr	spect M6	22074n7231
DO	9.2--10.5	---	(Krueger 60B) flare star of UV Ceti type, spect dM4e (*)	22262n5727
DQ	7.0--7.1	.0789	Delta Scuti type, spect F1	20564n5517
EK	7.8---9.1	4.428	Ecl.bin; spect A0	21405n6928
BV382	6.9---7.5	.9362	Ecl.bin; spect A0; 20' nf Beta Cephei (GK Ceph)	21304n7036
EM	7.0--7.13	.8062	Ecl.bin; Spect B0+B1	21524n6223
FZ	7.0-- 7.6		Semi-reg; spect M7	21182n5514
GP	9.0---9.1	6.688	Ecl.bin? Spect B0+W6	22167n5553
GQ	8.2 ±0.09	2.036	Alpha Canum type, spect A0p	22383n7524
GT	8.3---8.9	4.909	Ecl.bin; spect B8	22560n6808
IV	7.0--19..	---	Nova 1971	22028n5316
KZ	8.5 ±0.02	.2454	Beta Canis Maj type; B0	22545n6236
MX	7.9 ±0.07	17.20	Alpha Canum type, A2p	22507n5832
NY	7.5 ±0.06	15.27	Ecl.Bin; spect B0+B0	22567n6248

CEPHEUS

LIST OF STAR CLUSTERS, NEBULAE, AND GALAXIES

NGC	OTH	TYPE	SUMMARY DESCRIPTION	RA & DEC
40	58[4]	◎	F,S,1E; mag 10½, diam 60" x 40"; central star mag 11½, spect O	00102n7215
188		⠫	vL,F,Ri,C; diam 15', about 150 stars mags 11...18, class C. "Oldest galactic star cluster" (*)	00394n8503
2276		⦶	Sc; 12.4; 2.5' x 2.0' pL,F,1bM; multiple-arm spiral. 6' pair with 2300	07110n8552
2300		⦶	E2; 12.2; 1.0' x 0.7' pB,pL,1E; bM; NGC 2276 in field, 6' to northwest	07165n8550
6939	42[6]	⠫	vRi,pL,pC; diam 8', about 100 stars mags 12...16, class G. Galaxy 6946 lies 38' to southeast. (*)	20304n6028
6946	76[4]	⦶	Sc; 11.1; 8.0' x 8.0' vF,vL,smbM,sN. One of the nearest spirals (*)	20339n5958
6951		⦶	SB; 12.3; 3.5' x 3.5' pB,pL,1E (*)	20365n6556
7023	74[4]	☐	7th mag B5e star in L,F neby, diam 18', complex structure, bright & dark masses & outer filaments (*)	21014n6758
----	I.1396	☐	E,F,eL neby, 135' x 165'; surrounds O-type double star Σ2816	21375n5714
7129	75[4]	☐	cF,pL, 7' x 7', several F stars inv.	21420n6552
7133		☐	vF, 3' diam; part of 7129	21434n6556
7139	696[3]	◎	vF,cS,R; mag 13½, diam 80"; eF central star	21446n6335
7142	66[7]	⠫	L,vRi,pC; diam 10'; about 50 F stars mags 11...15	21447n6534
7235		⠫	pS,C, diam 4'; about 25 stars mags 10.... Irr.	22108n5702
7261		⠫	pRi, diam 7', about 20 F stars	22186n5750

CEPHEUS

NGC	OTH	TYPE	SUMMARY DESCRIPTION	RA & DEC
7281		⁙	Small scattered group in Milky Way, with three 10^m stars in short row	22229n5735
7354	705^2	◎	B,S,R, mag 13; diam 30" with 16½ mag central star	22384n6101
7380	77^8	⁙	pL,1C, diam 8', 20 stars mags 9.... Class D, with surrounding F neby 20' diam. Association "Ceph I". Incl variable star DH Cephei.	22449n5749
7419	43^7	⁙	pRi,cC, diam 3'; about 30 F stars, class E	22524n6034
----	I.1470	◎	F,S, 12^m 07 star with neby 70" x 45"	23032n5959
7510	44^7	⁙	pRi,mC, diam 3', mE with 30 stars mags 10... class D; L region of F neby to SE (*)	23092n6018
7538	706^2	☐	F,L, 5' x 10', with 10^m star inv; filamentary structure in neby	23115n6114
7762	55^7	⁙	L,pRi,pC; diam 15'; about 70 stars mags 12...15; class D	23475n6744
7822		☐	eeL,eF, portion of 2° arc surrounding nebulous area 1½° to south, which contains star GC 39, mag 5.7, spect gK1	23590n6825

DESCRIPTIVE NOTES

ALPHA   Name- ALDERAMIN. Mag 2.46; spectrum A7 IV or V. Position 21174n6222. The computed distance of the star is about 52 light years, giving an actual luminosity of about 23 suns (absolute magnitude +1.4). The radial velocity is 6 miles per second in approach; the annual proper motion is 0.16".

An optical companion of magnitude $10\frac{1}{2}$ lies at a distance of 207" in PA 22°. About 20" to the south of this star will be found a close and faint pair of 2.6" separation, discovered by S.W.Burnham in 1907. Both components are of the 11th magnitude. None of these stars appear to have any real connection with the bright primary.

Alderamin itself is remarkable for its unusually rapid rotation which causes the spectral lines to become very broad and hazy. It is also interesting to note that the star lies near the path traced by the Earth's axis in space in the course of its 25,800 year precessional cycle. (See Page 54). The star will thus replace Polaris as the Pole Star in the course of time, and will be nearest to the true Pole about 7500 A.D.

BETA   Name- ALFIRK. Mag 3.15 (slightly variable). Spectrum B2 III. Position 21280n7020. The computed distance is about 980 light years and the actual luminosity is about 4000 times that of the Sun (absolute magnitude about -4.2). The star shows an annual proper motion of 0.01"; the radial velocity averages about 5 miles per second in approach.

The 8th magnitude companion at 14" is an easy object for amateur telescopes, and was first recorded by F.G.W.Struve in 1832. There has been no change in separation or PA in more than a century. Despite the lack of definite orbital motion, the two stars probably form a physical pair. The small star has a spectral class of A3, and a true luminosity of about 50 times the Sun. The projected separation of the pair is some 4300 AU.

Beta itself is a variable of extremely short period and small amplitude, belonging to the class sometimes called "quasi-cepheids". The variations, first detected by Dr. E.B.Frost at Yerkes in 1902, consist chiefly of a periodic shifting of the spectral lines with a total range in the radial velocity curve of about 20 miles per second. The

period is 0.19048 day, or about 4 hours and 34 minutes.
In 1913, P.Guthnick detected and measured a slight change
in the light output of about 0.04 magnitude, occurring in
the same period. At first thought by some investigators
to indicate a binary of extraordinarily rapid motion, the
variations are now attributed to rapid pulsations in the
outer regions of a single star. The typical example of
this rather rare class of variable is Beta Canis Majoris,
whose variations were discovered by S.Albrecht in 1908. All
known stars of the type are high-luminosity B giants, of
spectral classes B1, B2, or B3. The periods range from $3\frac{1}{2}$
hours up to about 6 hours, and the stars show a definite
period-luminosity relation. Stars of longest period (such
as Beta Canis itself) have the highest luminosity. These
stars appear to be related to the better known "cepheids"
which show much larger variations in light. (Refer also
to Beta Canis Majoris. Cepheids are described under
Delta Cephei)

GAMMA   Name- ER RAI. Mag 3.21; spectrum K1 IV.
            Position 23373n7721. Er Rai is about 50 light
years distant, and has an actual luminosity of about 11
suns. The computed absolute magnitude is +2.2. The star
shows an annual proper motion of 0.17"; the radial velocity is 25 miles per second in approach. Like Alpha, this
star periodically takes its turn as Pole Star, a position
which it will occupy in about 2000 years.

DELTA   Variable. Position 22273n5810. Spectrum F5 Ib.
            Delta Cephei is one of the most famous of the
variable stars, the typical example of a large number of
short-period pulsating variables whose light changes are
NOT due to eclipse by a revolving companion, but to an
actual pulsation of the star. Stars of this class are
called "cepheids" in honor of Delta Cephei, the first
example to be discovered. The variations were discovered
by John Goodricke in 1784.
        The light changes may be followed by keeping a
careful watch on the star from night to night, and comparing the light with the nearby stars Epsilon and Zeta. The
magnitude range of Delta is from 3.6 to 4.3, with a change
in the spectrum from F5 to about G3. The rise to maximum

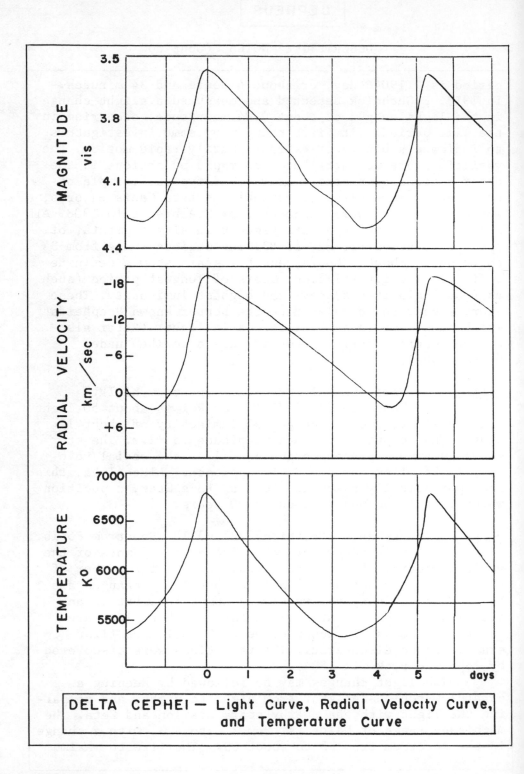

DELTA CEPHEI — Light Curve, Radial Velocity Curve, and Temperature Curve

DESCRIPTIVE NOTES (Cont'd)

requires about 1½ days, and the fall to minimum occupies
about 4 days. The precise period is 5.36634 days, or 5
days 8 hours and 48 minutes. Like all the cepheids, the
star is a supergiant, attaining about 3300 times the luminosity of our Sun at maximum. The diameter, probably
variable by about 6%, may average 25 or 30 times that of
the Sun. The computed distance of the star is slightly
over 1000 light years, the annual proper motion is only
0.01", and the radial velocity (variable) ranges from near
zero up to about 12 miles per second in approach.
CEPHEID CHARACTERISTICS. Since the discovery of Delta
Cephei, over 500 stars of the type have been found, and
these remarkable objects form one of the most important
and interesting class of variables known in the Universe.
Among the brighter examples of the class are such objects
as Eta Aquilae, Zeta Geminorum, Beta Doradus, 48 Aurigae,
and Polaris.

Cepheids are all very luminous white and yellow
giants of spectral types F, G, and K, and may form a connecting link between the red giants and the highly luminous main sequence stars. When plotted by spectral type and
luminosity on the H-R Diagram they occupy a region about
4 magnitudes above the main sequence and some 2 magnitudes
above the red giant region. For reasons not completely
understood, this region of the diagram seems to be characterized by stellar instability, not the violent type which
causes nova outbursts, but a more quiet type which appears
in the form of periodic oscillations. The problem is a
fascinating one; the main points will be discussed after
the following short review of cepheid characteristics.
PERIODS. Cepheids have periods of from a few hours up to
about 50 days. A typical period is from 5 to 8 days; the
longest known at present is 54.3 days for a faint variable
in Vulpecula. Also, in the same constellation, the star
SV has the unusually long period of 45.1 days.

Cepheids of very short period - usually less than a
day - are classified in a separate category, designated
"cluster variables" from their abundance in the globular
star clusters. The shortness of the period is not the only
point of difference between these rapidly fluctuating stars
and the normal or "classical" cepheids. Spectral types
appear to be restricted to A and F, and the stars are much

smaller and less luminous than the true classical cepheids.
A typical example has a period of about half a day, but a
luminosity of less than 100 suns. Cluster variables are
also characterized by unusually high space velocities, in
some cases exceeding 200 miles per second. They are standard Population II members of the Galaxy. The typical star
of the class is RR Lyrae.

A third sub-class must be mentioned briefly, the so-
called "dwarf cepheids" or "ultra-short period cepheids"
as typified by CY Aquarii and SX Phoenicis. These strange
objects have periods of 88 minutes and 79 minutes, respectively. In addition, there are the "Delta Scuti" type
stars, also with extremely short periods but with much
smaller amplitudes. The exact classification and relationship of these sub-types is uncertain. Spectral types seem
confined to A and F, and the luminosities are usually
below those of the cluster variables. (Refer to RR Lyrae
and Delta Scuti)

REGULARITY. The period of any one cepheid variable is
generally as regular as fine clockwork, and in many cases
is known to the fraction of a second. There are only a few
stars known in which the period has changed by as much as
2 or 3 seconds in the last 50 years. A very peculiar fact
about such changes is that they occur very suddenly rather
than cumulatively. Perhaps the best known example is the
peculiar variable RZ Cephei (see page 607) which has shown
several abrupt changes in period. A decrease of about 4
seconds occurred in 1901, an increase of about 4 seconds
in 1916, and another increase of about 2 seconds in 1923.
RZ Cephei is also remarkable for having the highest space
velocity known for any star. But for sheer unpredictabil-
ity the award must be given to the strange cepheid RU
Camelopardi; not only has it shown several sudden changes
in period, but it appears to have ceased operations entirely in 1965 and now shines at an apparently constant magnitude of about 8.5. No other such case is known. (Refer
to page 327)

AMPLITUDE. The light variation of a typical cepheid averages rather less than one magnitude, although there are a
few known that have a range of about $1\frac{1}{2}$ magnitudes. Photographically the range is somewhat greater than when observed visually, due to the change in color (toward the red)
as the star fades. Thus the visual amplitude of Delta

DESCRIPTIVE NOTES (Cont'd)

Cephei is about 0.7 magnitude, but the range in the ultra-violet is 1.48 magnitudes. Conversely, in the infrared the range is only 0.43 magnitude.

As might be expected, the oscillations of these stars are accompanied by changes in temperature and spectral type. The typical star drops about 1500° K from maximum to minimum, while at the same time the spectral type may fall a whole class. The cluster variables range from A to F. Classical cepheids of moderate period (±7 days) are type F at maximum and fall to type G at minimum. With cepheids of long period (about 30 days) the range is from type G to K.

LIGHT CURVES. The light curves of all cepheids show a marked similarity in shape and amplitude. The rise to max-imum is nearly always more rapid than the decline, and in some types, particularly the cluster variables, is accom-plished with almost nova-like rapidity. The ascending part of the curve is usually smooth and steady, while the de-cline is often subject to slight irregularities and tem-porary halts. The chart on page 588 shows typical light curves. They seem to fall into several well-marked groups when the stars are arranged in order of increasing period.

Groups I and II are standard cluster variables. The abrupt change in the form of the light curve at about 0.45 days is well shown on the chart. Groups III, IV, and V are representative of the classical cepheids. Here again there are sudden changes in the form of the light curve at two points, notably at about 2½ days and again at 10 days. A peculiar feature of group IV cepheids is the presence of a conspicuous hump on the descending side of the curve, show-ing a short interval of constant light which interrupts the fading of the star. This hump appears to grow in size as the period increases, so that the later members of the group almost seem to have a double maximum. S Muscae and VX Persei are fine examples of this type of cepheid.

THE PULSATION THEORY OF CEPHEID VARIABLES. The absolute regularity of most cepheids might suggest that we are here dealing with some peculiar type of eclipsing double star. And the fact that the spectroscope reveals alternate velo-cities of approach and recession - as if the star were moving in an orbit - might appear to strengthen this theory. However, it is now definitely established that

TYPICAL CEPHEID LIGHT CURVES WITH PERIODS IN DAYS

GROUP I

RV Arie		.09
AQ CorA		.12
RW Arie		.26
ω Cent v		.30
ω Cent v		.34
ω Cent v		.41
ω Cent v		.47

GROUP II

V467 Sgtr		.43
ω Cent v		.50
ω Cent v		.61
ω Cent v		.73
ω Cent v		.90
DE CorA		1.01
V527 Sgtr		1.26
ω Cent v		1.35

GROUP III

BQ CorA		1.13
XX Virg		1.35
VW Mono		1.53
TU Cass		2.14
AU Peg		2.40
DT Cyg		2.50

GROUP IV

UX Norm		2.4
RT Musc		3.1
T Vulp		4.3
δ Ceph		5.4
CS Car		6.7
η Aqil		7.2
W Gem		7.9
S Sgte		8.4
S Musc		9.7
VX Per		10.9

GROUP V

YZ Sgtr		9.6
Z Lacr		10.9
RX Cass		11.6
U Norm		12.6
AD Pupp		13.6
RW Cass		14.8
X Cyg		16.4
WZ Sgtr		21.9
X Pupp		26.0
U Car		38.8

this idea is totally incorrect. It is true that an eclipsing binary will necessarily show velocities of approach and recession as the components revolve about each other. But this motion will be reduced to zero at the time of eclipse, since both components are then moving across the line of sight and are neither approaching or receding. Now in the cepheids we observe a very different effect. Minimum brightness occurs near the time of maximum recessional velocity; obviously this cannot be explained by orbital motions causing an eclipse. Instead it would seem that we are here dealing with a single star which appears to pulsate in a regular period, alternately expanding and contracting. This basic idea is the pulsation theory of the cepheid variables in its simplest form.

One of the earliest studies of the cepheid problem was made by the noted physicist Sir Arthur Eddington, the first researcher to make a detailed mathematical analysis of the problems of pulsating gas spheres. In Eddington's time it was shown that the pulsations might imply a sort of "contest" between two nearly equal forces - the star's gravitational field and the internal radiation pressure; the first tending to make the star contract, and the second causing it to expand. When these two forces are very nearly equal, some small internal - or possibly external- disturbance could conceivably set the star into oscillation with a period depending upon the mass and density of the star. Such oscillations, however, could maintain themselves for only a limited time since a certain fraction of the pulsation energy is lost during each cycle. Thus it seems clear that the pulsations must be maintained by some type of energy producing mechanism which compensates for the loss during each cycle. S.A.Zhevakin (1953) has shown that energy may be trapped in certain zones of the star which become especially opaque during the maximum compression phase of the cycle. This mechanism is of particular importance in a region of helium ionization which lies a relatively short distance below the star's visible surface. In their analysis of the cepheid problem, N.Baker and R. Kippenhahn (1961) conclude that "this excitation is of such an order of magnitude as to be able to overcome the damping in the interior. Stars which lie to the left of the cepheid region in the H-R diagram no longer pulsate because in them the excitation is smaller than the damping". The cepheid region on the H-R diagram may thus be regarded as a zone of instability, and it seems that a star begins cepheid pulsations whenever its evolution takes it across this zone. This is evidently not a unique event in the history of a star; from theoretical studies of stellar evolution it seems that some stars may pass through the cepheid stage a number of times.

Although the general causes of cepheid pulsation now seem to be fairly well understood, the details remain to be explained, and it must be admitted that the stars still show a number of puzzling features. A peculiar fact, previously mentioned, is that the maximum brightness occurs near the time of most rapid expansion, while minimum brightness coincides with the most rapid contraction. This is contrary to any theory which assumes a simple pulsation of the entire stellar body. It might indeed seem that the star should be brightest and hottest shortly after the contraction has brought it to a state of highest density and pressure. The "time-lag" suggests that the outer layers of the star do not instantaneously follow the pulsations occurring in the unstable zone beneath. Spectroscopic studies show evidence that the various atmospheric layers do not pulsate in phase, and that when the star is near maximum some layers have already begun to contract while others are still expanding. A lag of phase with increasing wavelength is also characteristic of cepheids, and has been demonstrated by observations made in different colors. The time of maximum depends on the color being observed; when the star has begun to fade in the blue it is still brightening in the longer wavelengths. All these facts probably have some bearing on the variety of shapes of cepheid light curves; it seems likely that the form of each curve results from the way in which various pulsating layers interact, either reinforcing or cancelling the total effect.

THE PERIOD-LUMINOSITY RELATION. The most important fact about the cepheids is the discovery that there is a definite relation between the periods and the actual luminosities of these stars. This was first announced by Miss Henrietta Leavitt of Harvard in 1912, as a result of her observations of variable stars in the Small Magellanic

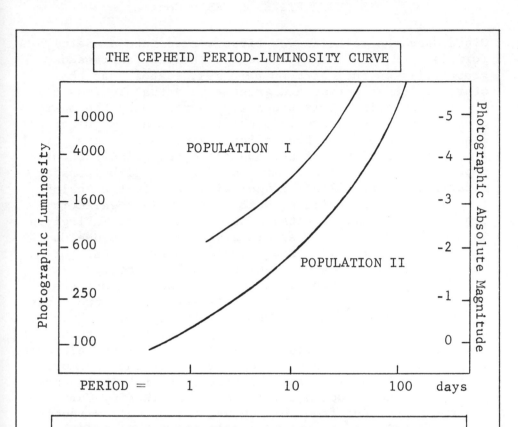

THE CEPHEID PERIOD-LUMINOSITY CURVE

THE PERIOD-LUMINOSITY RELATION OF CEPHEIDS FOR
POPULATION I, FROM THE ABOVE GRAPH

PERIOD = Abs.Pg.Mag.		PERIOD = Abs.Pg.Mag.	
1d	-1.5	13d	-3.6
2	-2.1	14	-3.7
3	-2.4	15	-3.8
4	-2.7	16	-3.9
5	-2.9	18	-4.0
6	-3.0	20	-4.1
7	-3.2	25	-4.3
8	-3.3	30	-4.5
9	-3.4	35	-4.6
10	-3.5	40	-4.8
11	-3.5	45	-4.9
12	-3.6	50	-5.0

Cloud, and the relation was first worked into a useful
formula by Harlow Shapley in 1917. The law, expressed
graphically on the chart on page 591, states that the
stars of longer period are greater in actual luminosity.

At this point it is necessary to recognize the basic
division of these stars into the two stellar "populations"
I and II. Cepheids of Population I are found in the spiral
arms and star clouds of our own and other galaxies, while
cepheids of Population II are found in the globular star
clusters, in the elliptical galaxies, and in the "halo"
components of the spirals. Most of the bright galactic
cepheids are Pop.I objects. The peculiar star W Virginis
is a cepheid of Pop.II, as are also the known cepheids in
globular clusters. The short period RR Lyrae stars or
"cluster variables" are likewise Pop.II objects.

The important point in this discussion is that the
period-luminosity relation is different for the two types,
a Pop.I cepheid being intrinsically 1½ magnitudes brighter
than a Pop.II cepheid of the same period. Before this fact
was realized from studies made with the 200-inch telescope,
many puzzling discrepancies were caused by the assumption
that all cepheids followed the same rule. But it is now
possible to construct a double Period-Luminosity Curve for
the two types. And from this curve, the actual luminosity
of any cepheid can be read directly when the period is
known. In the construction of such curves, photographic
magnitudes are generally used. If visual magnitudes are
desired, a correction may be easily made.

Let us take Delta Cephei itself as an example. The
period is about 5.4 days. Referring to the Pop.I curve, we
see that this period implies an absolute photographic magnitude of about -2.9 at midrange, or an actual luminosity
of about 1200 suns. The visual luminosity would be almost
exactly twice this figure, since the color index at mid-
range is about 0.8 magnitude.

Similarly, SV Vulpeculae, with a 45 day period, is
found to have a median absolute magnitude of about -4.9.
The photographic luminosity is about 6900 times the Sun,
the visual luminosity is about 14,000 times.

The importance of the Period-Luminosity relation lies
in the fact that it provides us with a method of measuring
the vast distances of the universe. For once the actual

luminosity of a cepheid is known, a comparison between the
real and the apparent brightness will quickly give the
distance. Therefore, when cepheids are found in distant
star clusters or galaxies, we are enabled to compute the
distances quite easily. Sometimes called the "measuring
sticks of the Universe", these stars are all intrinsically
very brilliant objects and can be seen and identified at
vast distances, where stars like our Sun would be completely lost to view. It was through the discovery of cepheids
in the so-called "spiral nebulae" that these objects were
finally identified as external galaxies. Some typical
cepheids in the Andromeda Galaxy (M31) are shown in the
photograph on page 136; the period-luminosity relation for
these objects is demonstrated by the light curves on page
135.

Two actual examples of distance determination by the
cepheid rule will now be presented, in order to illustrate
the method.

Let us begin with Delta Cephei itself. The period is
5.366 days; the photographic range is 4.1 to 5.2. What is
the distance?

First we refer to the period-luminosity graph for
Pop.I, and find that the period corresponds to an actual
brightness of -2.9 (photographic absolute magnitude). The
median apparent magnitude is about 4.65. The difference
between the two, which is of course the distance modulus,
is thus 7.55 magnitudes. This corresponds to a difference
of about 1000 times in light intensity (Table III, page
65). The star is thus about 1000 times fainter than it
would be if it was at the standard distance of 10 parsecs
or 32.6 light years. The distance, then, must be:

$$32.6 \times \sqrt{1000} = 1031 \text{ light years}$$

As a second example, let us consider one of the faint
cepheids discovered by Hubble in the Andromeda Galaxy. The
magnitude variation (pg) is 18.8 to 19.8; the period is 42
days. What is the distance?

As before, we find the real absolute magnitude from the
period. The figure is -4.8. The median apparent magnitude
is 19.3. Adding these, the distance modulus is found to
be 24.1 magnitudes, which corresponds to the enormous light
ratio of 4.379 billion times. The square root of this huge

number is about 66,200; multiplying this by 32.6 we now
obtain the distance, about 2.16 million light years.

In actual practice, the calculations with very large
numbers may be avoided by using a table which converts
various distance moduli directly into actual distances.
Such a table will be found on page 67 of this Handbook.
Also, in our sample cases, we have assumed no loss of star
light though absorption by interstellar dust. In many regions of space such a correction may be necessary.

DELTA CEPHEI AS A DOUBLE STAR. Delta Cephei is a well
known double star for the small telescope. The companion,
magnitude 6.3, is located 41" distant from the primary,
and has shown no definite change in separation or angle
since the first measurements of F.G.W.Struve in 1835. The
small star has a spectral class of about B7, and shows a
noticeable color contrast with the yellowish tint of the
primary.

In all probability the two stars form a physical pair.
The measured annual proper motion of Delta itself is about
0.01" and the separation should have increased nearly 1"
if the small star does not share the motion. The radial
velocities are also quite similar, that of the small star
being about 11 miles per second in approach. The computed
luminosity of the B-star is about 250 times that of the
Sun, or about 1/10 the brightness of the giant primary.
At the enormous separation of about 13,000 AU, no sign of
orbital motion is to be expected.

A second companion of the 13th magnitude was discovered
by S.W.Burnham in 1878; the distance from Delta is 20.9".
According to C.E.Worley (1966) this star is probably not
a physical member of the system.

The faint but highly interesting double star Krueger 60
is located near Delta Cephei, approximately 43' to the
south. (See page 598)

TYPICAL CEPHEIDS include the following objects: Eta
Aquilae, RT Aurigae, *l* Carinae, Zeta Geminorum, U Aquilae,
Y Ophiuchi, W Sagittarii, S Sagittae, SU Cassiopeiae, T
Vulpeculae, Kappa Pavonis, T Monocerotis.

EPSILON   Mag 4.20 (slightly variable). Spectrum F0
             IV. Position 22132n5648. Epsilon is a
convenient comparison star for the nearby variable Delta

Cephei. At maximum, Delta is 0.6 magnitude brighter than Epsilon, while at minimum it is 0.1 magnitude fainter than Epsilon. The chief statistics concerning Epsilon are summarized as follows: Distance about 85 light years, luminosity about 11 times the Sun, absolute magnitude near +2.2, annual proper motion 0.45" in PA 84°, radial velocity less than 1 mile per second in approach.

Epsilon Cephei is a variable star of the rare Delta Scuti class, remarkable for its extremely short period of 0.0424 day or about 61 minutes. The variations were first detected photometrically by M.Breger in September 1966, with the 24-inch reflector at Lick Observatory. This is one of the shortest periods known for any pulsating variable star, although the amplitude of the light curve is only about 0.03 magnitude. The very similar star UV or 38 Arietis, has an even shorter period, about 53 minutes. (Refer also to Delta Scuti)

ZETA Mag 3.36; Spectrum K1 Ib. Position 22091n 5757. The computed distance of the star is about 1240 light years; the actual luminosity about 5800 times that of the Sun. The spectral characteristics are those of a supergiant with an absolute magnitude of about -4.6. The annual proper motion is 0.01"; the radial velocity is 11 miles per second in approach.

Just 15' south of Zeta is the faint triple star β436, with magnitudes of 8, 11½, and 13; separations 19.7 and 19.1" in position angles 328° and 100°. No observations of this star are reported in standard catalogs since the measurements of S.W.Burnham in 1903.

ETA Mag 3.43; spectrum K0 IV. Position 20443n 6139. Direct trigonometrical parallaxes give a distance of 46 light years; the actual luminosity is about 7 times that of the Sun and the absolute magnitude is +2.6. Eta Cephei has a fairly large annual proper motion of 0.82" in PA 6°; the radial velocity is about 52 miles per second in approach.

The 11th magnitude companion has no physical connection with the primary, and does not share the large proper motion. The separation is decreasing from 100" in 1879 and will reach a minimum of about 44" around 1990.

DESCRIPTIVE NOTES (Cont'd)

MU (Variable). Spectrum M2 Ia. Position 21420n5833.
The "Garnet Star", so named by Herschel. This
famous and interesting object is perhaps the reddest star
visible to the naked eye in the north half of the sky. The
variability seems to have first been noticed by J.R.Hind in
1848, and was confirmed by Argelander; the visual range is
from 3.7 to about 5.0. The period is irregular, but seems
to average about 755 days, with shorter superimposed oscillations of the order of 100 days or less. In addition, a
long cycle of about 12.8 years has been suspected. In an
analysis of the light curve from 1881 to 1935, V.Balasoglo
(1949) has identified periods of 700, 900, 1100, and 4500
days. Although the reality of these periods has frequently
been questioned, very similar results were obtained by S.
Sharpless, K.Riegel, and J.O.Williams (1966) in a thorough
analysis of the light curves. They conclude that " the
light variations of Mu Cephei are characterized by a much
greater degree of regularity than is generally attributed
to stars classed as semi-regular variables".
The exact distance of the star is uncertain, but is
believed to be in the range of 800 to 1200 light years.
From the spectroscopic parallax method the apparent distance modulus is about 8½ to 9 magnitudes, giving a distance
of about 1800 light years. This result, however, requires
some adjustment for loss of light due to obscuring clouds
in the vicinity, since the star lies near the northern edge
of an extensive nebulosity, IC 1396.
Mu Cephei is a red giant star, evidently of the same
class as the similarly pulsating Betelgeuse in Orion. From
a comparison of the spectra, it seems that Mu probably has

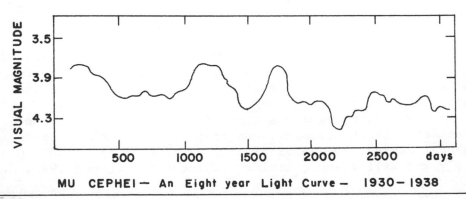

MU CEPHEI— An Eight year Light Curve — 1930—1938

a higher actual luminosity than Betelgeuse, and must certainly rank among the most brilliant of all known red supergiants. The maximum absolute magnitude may be about -5, and the computed diameter is at least several hundred times that of the Sun. Mu Cephei is also one of the few stars known which shows water-vapor bands (steam!) in the spectrum. In a study of the infrared spectrum of the star in 1964, R.E.Danielson, N.J.Woolf, and J.E.Gaustad found that "the water-vapor bands in the spectrum of Mu Cephei are surprisingly strong. So far no satisfactory explanation has been found for this phenomenon, but it may be partly due to the large turbulent velocities in the atmosphere of this star".

According to some observers, the star varies in color as well as in light. It usually appears a deep orange-red but on occasion seems to take on a peculiar purple tint. Since human eyes vary in color sensitivity, and since color is affected by atmospheric and instrumental factors, it is still uncertain whether such changes are real. In their list of "The Finest Deep-Sky Objects", J.Mullaney and W. McCall (1966) find the star "almost red in a 3-inch (45X), deep orange in an 8-inch (70X) and yellow-orange in the Allegheny 13-inch refractor." According to the Arizona-Tonantzintla Catalog (1965) Mu Cephei has a color index (B-V) of +2.26 magnitudes; the visual magnitude was 4.13 at the time the measurement was made. In order to fully realize the peculiar tint of the "Garnet Star", the light should be compared with a white star such as Alpha Cephei at the time.

The annual proper motion of Mu is only 0.002"; the radial velocity (somewhat variable) is about $9\frac{1}{2}$ miles per second in recession. The ADS Catalog lists two faint companions to the star, probably optical attendants only:

Mag 12.3 at 19.4" in PA 262°
Mag 12.7 at 40.9" in PA 299°

XI   Mag 4.29; Spectra A3 & dF7. Position 22022n 6423. A fine double star, usually considered the most attractive in the constellation, with the possible exception of Delta itself. Xi Cephei is a physical pair, the components showing a common proper motion of 0.23" per year in PA 66°, but revealing only slight evidence of slow

orbital revolution. The PA is decreasing at about 7° per
century, and the separation has widened somewhat from the
first measurement of 5.6" made by F.G.W.Struve in 1831.
The individual magnitudes are 4.6 and 6.5; spectra A3 and
dF7; the actual luminosities are about 10 and 1½ times the
Sun. There is a slight color contrast in the pair, and the
fainter star seems ruddy or "tawny" to some observers. The
projected separation of the pair is about 185 AU.

Xi Cephei is about 80 light years distant, and has a
space motion which seems to class it as an outlying member
of the Taurus moving group associated with the Hyades star
cluster. The radial velocity of the system is about three
miles per second in approach.

A third faint component is listed in the ADS Catalog
at a distance of 97" in PA 200°. This star, magnitude 12.7,
is not a physical member of the system.

KRG 60   Krueger 60. Position 22262n5727. A noted
double star, one of the nearest of the visual
binaries. It is located near Delta Cephei, about 43' to
the south and 1^m preceding in RA. The main pair, Krueger
60 A and B, are about 2.5" apart, and form a rapid binary
system with an orbital period of 44.46 years. A third star
of the 10th magnitude, called "C", was 27" distant in 1890
but is not a true member of the system, and does not share
the large proper motion of the orbiting pair. The A to C
separation is now more than 60" and will increase continually. Incidentally, it was this wide pair which was discovered by A.Krueger at Helsinki. The much closer physical
companion was detected at Lick with the 36-inch refractor
by S.W.Burnham in 1890.

The components of Krueger 60 are both low-luminosity
red dwarf stars, and are separated by an actual distance
of 9.2 AU or about 850 million miles, comparable to the
separation of Saturn and the Sun. The semi-major axis of
the orbit is 2.38", the eccentricity is 0.42, and periastron is in 1970. Facts about the two stars are given in
the following short table.

	Mag	Spect.	Mass	Lum.	Diam.	Abs.Mag.
Krg 60A	9.8	dM3	0.26	.0013	0.34	+11.8
Krg 60B	11.4	dM4e	0.14	.0004	0.19	+13.4

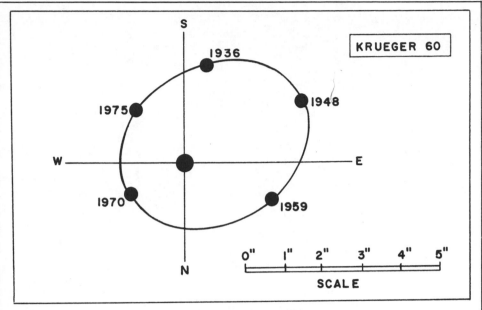

The Krueger 60 system is quite near us in space, at a distance of 13.1 light years. The annual proper motion is
0.86" in PA 246°, and the radial velocity is about 14
miles per second in approach.

Krueger 60B, the small component, is a star of special
interest, since it is one of the smallest stellar masses
known. Up to 1966, only two other stars are definitely
known to have smaller masses: Ross 614B in Monoceros and
the components of Luyten's Flare Star (L726-8) or UV Ceti
system in Cetus. The masses of the latter are at present
the smallest known, each at 4% the solar mass. For Ross
614B the figure is about 8%. Another famous star of abnormally small mass is the well known Proxima Centauri, or
Alpha Centauri C.

One of the peculiarities of Krueger 60B is shared by
UV Ceti and Proxima Centauri as well. All three objects
are "flare stars", variables which may show extremely sudden increases of light in a time of one or two minutes.
During a flare, an emission spectrum appears, superimposed
upon the normal features of a red dwarf. The cause of such
flares is somewhat controversial, but it seems probable
that the outbursts are similar in nature to the so-called
"solar flares" which occur on our own Sun and presumably

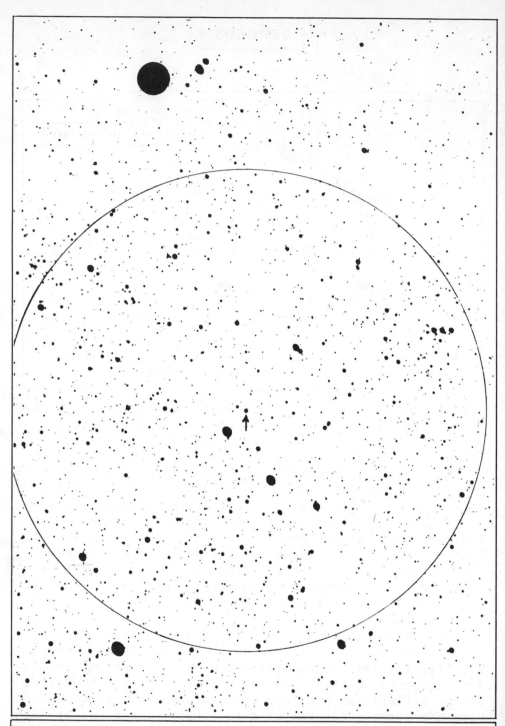

KRUEGER 60 Identification chart, from a 13-inch telescope
plate obtained at Lowell Observatory. Circle diameter = 1°
North is at the top; limiting magnitude about 15. Delta
Cephei is the bright object near the top of the chart.

on other normal stars as well. The total energy released
in a flare on Krueger 60B is about equal to that emitted
in such a solar flare. On the Sun, however, such a flare
represents only a small increase in the total brightness,
whereas on a faint red dwarf it more than doubles the total
radiation. Thus it is possible that all red dwarfs may be
potential flare stars. The frequency of flares is not
well known, but it appears that a number of hours of spot
checking are required before one is accidentally caught.
Possibly owing to this observational problem, stars of the
type are still regarded as rare; only eleven known examples
are listed in the Moscow General Catalog (1958).

Though not a brilliant object, Krueger 60 is of great
interest to any serious observer. A good 6-inch telescope
used with a fairly high power will usually resolve the
pair, and the change in position angle, due to the binary
motion, can be detected in an interval of only a few years.
The chance of detecting a flare adds to the interest in
observing this unusual system. (Krueger 60B also has a
variable star designation- DO Cephei).

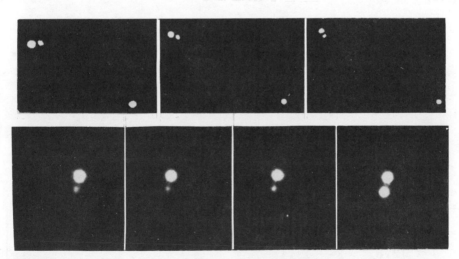

KRUEGER 60. (First row) The rotating double star is here
shown in 1908, 1915, and 1920. Photographed at Yerkes
Observatory by E.Barnard.
          (Second row) Four exposures of the system
made at Sproul Observatory, showing a flare on Krueger 60B.

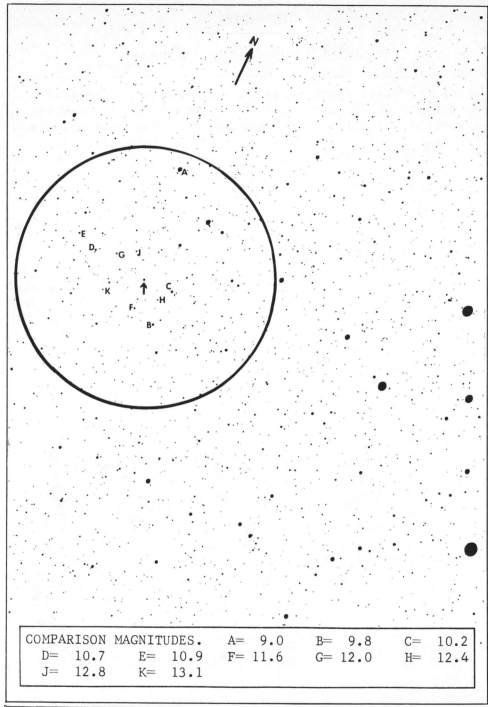

COMPARISON MAGNITUDES.   A= 9.0   B= 9.8   C= 10.2
   D= 10.7   E= 10.9   F= 11.6   G= 12.0   H= 12.4
   J= 12.8   K= 13.1

S CEPHEI. Finder chart made from a 13-inch telescope plate
at Lowell Observatory. Circle diameter = 1° with north at
the top; limiting magnitude about 15. Brightest star near
the right edge is magnitude 5.9.

DESCRIPTIVE   NOTES   (Cont'd)

S   Variable. Position 21359n7824. A famous long-
period pulsating variable star, located about
midway between Kappa and Gamma Cephei. The magnitude range
is from 7½ at maximum to less than 12½ at minimum, in a
cycle averaging about 487 days. S Cephei is noted as one
of the deepest colored stars available to observers with
small and moderate size telescopes. Its intense shade of
red makes it a vivid and conspicuous object in the field
of any good instrument, and renders a finding chart more
or less superfluous. The discovery of the star is credited
to Lalande in October 1789; he apparently did not observe
it long enough to detect the variability, but noted the
strong and unusual color. It was not until the later observations of Hencke (1855 to 1858) that the star was found
to be a variable.

S Cephei is one of the "Carbon stars", similar in
type to the celebrated "Crimson star" R Leporis. The spectral type is usually given as N8, but on the newer "carbon
star" classification it would be called C74 . These stars
are cooler even than the M-type red giants, and the unusually low temperature allows the bands of carbon compounds
to appear in the spectrum. The color index of S Cephei is
about 5½ magnitudes, one of the most extreme cases known.

The peak absolute magnitude of the star is believed
to be about -1.5 (luminosity = 330 suns) and the resulting
distance is close to 2000 light years. S Cephei shows an
annual proper motion of about 0.01" and a radial velocity
of 20 miles per second in approach.

Variables of this class are pulsating red giants
rather similar in type to Omicron Ceti and Chi Cygni.
While the exact place of these stars in the evolutionary

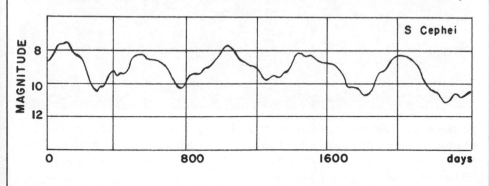

U CEPHEI. The eclipsing variable is shown in its normal state (top) on June 12, 1964; and at primary minimum the following evening (below). Lowell Observatory 13-inch telescope photographs.

picture is still uncertain, it is thought that the difference between M-type and N-type stars may be partly a matter of temperature, as well as a fundamental difference in
chemical constitution. (Refer also to Y Canum Venaticorum,
TX Piscium, and R Leporis)

U   Variable. Position 00577n8137. A fine eclipsing
           binary star, discovered by W.Ceraski in 1880. It
is a rapidly rotating double in which the bright primary
is occulted at periodic intervals by a larger but fainter
companion. U Cephei is one of the brightest and most easily
observed objects of its type, and is visible at any time
of the year because of its position in the northern sky
only $8\frac{1}{2}°$ from the Pole.
           The magnitude of U Cephei is normally 6.8; the
fall to minimum requires 4 hours and is followed by a 2-
hour total eclipse. During the total phase the magnitude
remains constant at 9.2. A slight secondary minimum, midway between the main eclipses, is caused by the partial
hiding of the fainter star by the bright one. The photographic range of the system is given in the Moscow General

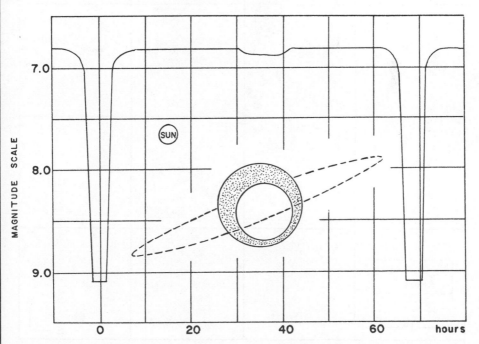

Catalog (1958) as 6.63 to 9.79. Relative sizes and orbital motion of the pair are shown in the diagram on page 605. The chief facts about the two components are given in the brief table below.

	Spect.	Diam.	Mass.	Lum.	Abs.Mag.
A	B8 V	2.9	4.7	130:	-0.5
B	gG8	4.7	1.9	25:	+1.5

The luminosities are derived from the spectral types, and the resulting distance of the system is about 1000 light years. The two stars are some 6½ million miles apart, and complete their orbital revolution in slightly under 2.5 days. The period has been increasing slowly during the

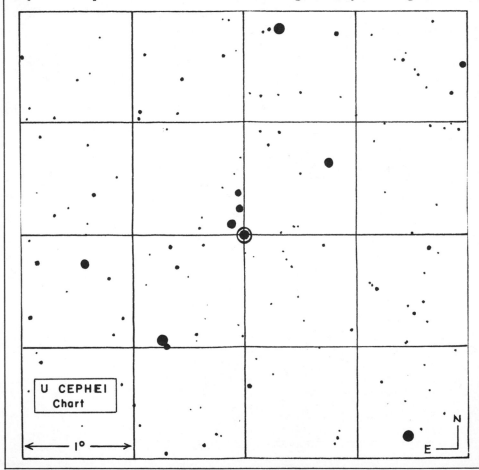

U CEPHEI
Chart

1°

N
E

last 80 years, and there are also small occasional changes
which remain unexplained. In the Harvard "Second Catalogue
of Variable Stars" of 1907, the exact period is given as
2d 11h 49m 44.55s, but by 1963 it was found to have changed
to 2d 11h 49m 51.08s. The difference may appear slight, but
after thousands of revolutions it is sufficient to alter
the predicted time of eclipse by several hours. This slow
increase in period appears to be due to a gradual change
in the relative mass of the components, the operating mechanism being a gaseous streamer between the stars. Matter
is moving from the expanded giant G-star toward the bright
B-type primary.

Another peculiarity of the system is probably due to
the same cause. It has been known for some time that the
radial velocity curve of the B-star is asymmetric, which
would appear to indicate that the orbit is fairly eccentric; however, the eclipse light curve shows that the orbit
is nearly circular! The discrepancy appears to be caused
by the moving gas streams, which distort the radial velocity measurements. It is now known that erroneous orbits
and spuriously large masses have been computed for binary
stars of this type, when the presence of gas streams was
not recognized. Beta Lyrae is a well known case.

In addition to the eclipsing components, there is
some evidence for the existence of a third star in the U
Cephei system; the computed period is about 30.7 years.
The annual proper motion of the system is about 0.025";
the radial velocity averages about 3 miles per second in
recession. (See also Beta Persei, U Sagittae, and Beta
Lyrae)

RZ   Variable. Position 22375n6436. One of the RR
        Lyrae or "Cluster variable" stars, pulsating
with the short period of 7h 24½m, discovered by Miss H.
Leavitt at Harvard in 1907. The photographic range is 9.2
to 9.8, with a spectral change of A0 to A3. The light curve
resembles those of the cepheids, with a rapid rise and a
slower decline. These variations are attributed to a periodic pulsation of the outer layers of the star; some of
the theories concerning such phenomena are briefly discussed under "Delta Cephei". RZ itself is chiefly noted for
the claim that it has the highest space velocity known for

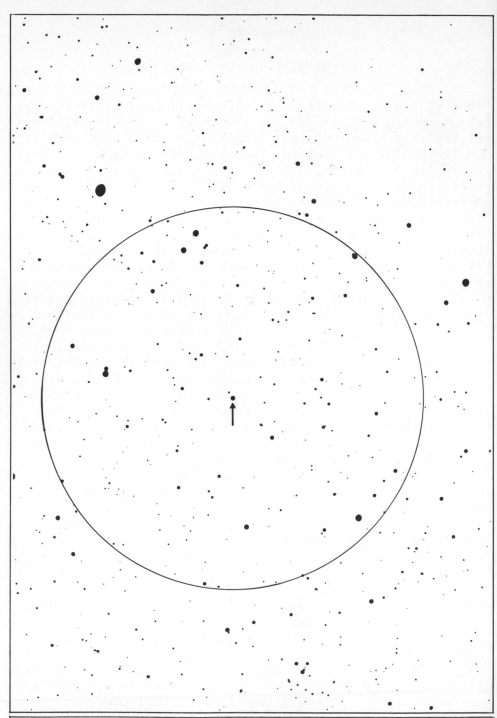

RZ CEPHEI. Identification field of the variable, from a Lowell Observatory 13-inch telescope plate. Circle diameter is 1°, with north at the top. Limiting magnitude about 14.

any star, but the exact figure is somewhat uncertain, and depends upon the value accepted for the distance. A figure of about 680 miles per second has been quoted by a number of writers, but is now definitely known to be in error due to incorrect calibration of the distance. The star is believed to have an absolute magnitude of about +0.5, and a distance of close to 2000 light years. The observed annual proper motion (0.20" in PA 26°) then indicates a velocity across the line of sight of about 400 miles per second. This figure is virtually identical with the true space motion, since the star is moving almost entirely "side-on" as seen from the Earth. The radial velocity is only 0.5 mile per second in approach.

RZ Cephei has on several occasions shown sudden and unexplained changes in the period. In 1898 the exact figure was 7h 24m 28.76s, which remained absolutely constant for some years. In August 1901 the period shortened by 3.98 seconds, in November 1916 it increased by 4.33 seconds, and in December 1923 it increased again by 1.84 seconds. In recent years it seems to have remained constant. (For a discussion of the "cluster variables" refer to RR Lyrae. See also Delta Cephei)

VV   (Mag 4.90 (variable). Spectrum M2ep Ia + B9. Position 21552n6323. A well known eclipsing variable star, often claimed to be one of the most colossal binary systems yet discovered. It is located about $1\frac{1}{4}°$ southwest of the double star Xi Cephei, and is identified in Norton's Atlas by the number 360, the underlining indicating the zone number in the catalog of Piazzi. VV Cephei consists of a red giant star and a smaller blue companion revolving in their orbits in the unusually long period of 20.34 years or 7430 days. Eclipses of the smaller star by the giant occur in 1936, 1956, 1977, etc. The first eclipse actually observed was that of 1936- 37, detected by D.B. McLaughlin. The disappearance of the bright lines of the B-star in that year immediately suggested eclipse by the red giant. Observations at Harvard soon showed that the brightness of the star had dropped by over $\frac{1}{2}$ magnitude. The eclipse is total, lasting for 15 months, and is preceded and followed by long partial phases lasting about 4 months each. During eclipse the magnitude drops from 6.7

DESCRIPTIVE NOTES (Cont'd)

to 7.4 (photographic) but the light during eclipse does not remain constant, as the accompanying light curve clearly shows. The fluctuations seem to be in the nature of a 350 day cycle with an amplitude of about 0.3 magnitude, probably indicating that the red star is itself a slowly pulsating variable. There are also more sudden variations which are attributed to the blue star, and possibly connected with a gaseous "shell" surrounding it. It appears virtually certain that both stars are intrinsically variable, a factor which complicates the interpretation of the eclipse light curve.

The spectral classes are M2 Ia and about B9, the bright lines of the B-star vanishing completely during an eclipse. VV Cephei is one of the eclipsing stars which shows direct evidence of a huge atmospheric corona around the red giant; the spectrum of the B-star does not regain its normal appearance until nearly 1½ years after the eclipse has ended. A similar effect is seen during the eclipses of Zeta Aurigae.

The red giant member of the pair is often said to be among the largest known stars, and a diameter of over 1200 times that of the Sun has been quoted in many texts. There is, however, considerable controversy concerning the true sizes and masses of the components. In a study of 260 spectrograms obtained at the University of Michigan Observatory, Dr.B.F.Peery found masses of 41 and 84 solar masses for the blue and red stars, respectively; and derived

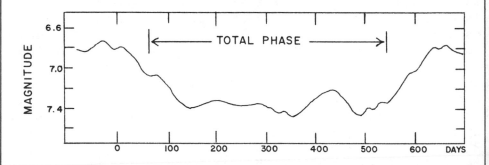

VV CEPHEI- A photographic light curve of the eclipse of 1956--1958, with magnitude scale at the left. First contact is at phase "0 days". Note the irregularities during total eclipse.

DESCRIPTIVE NOTES (Cont'd)

a diameter of 1620 times that of the Sun for the red giant. The spectroscopic orbit has an eccentricity of 0.25, with the red star about 1.2 billion miles from the center of gravity of the system. This interpretation requires the red giant to have an absolute magnitude of -5 or higher, and suggests a distance of at least 3000 light years for the system. According to these results, VV Cephei may be the largest star actually observed, though it might still be exceeded in size by the strange companion of Epsilon Aurigae, provided that mysterious object is accepted as a star.

In an analysis of the system, Dr.L.Fredrick at Sproul Observatory has obtained rather different results. More than 1000 plates of the star, taken with the 24-inch refractor, have been measured in an attempt to determine the parallax and orbital motion of the red star. The resulting distance is only about 650 light years, which gives the total luminosity of the system as about 250 times the Sun. According to this solution, the unusually large masses derived from the spectroscopic orbit cannot be accurate. The apparent orbit of the red star, derived from astrometric plate measurements, gives the system a total mass of about 20 suns, but it is uncertain which component is the more massive. The size of the red star may be as much as 600 times the Sun, though a value of 400 or so is thought to be more likely.

There is thus a considerable discrepancy between the spectroscopic and astrometric data concerning VV Cephei, but it is not certain what causes may be responsible. Dr.Fredrick suggests that the red star may not be a supergiant, though its sharp absorption lines seem to classify it as such. A strong magnetic field may act to produce the same effect, and the star has been classed as a magnetic variable by H.Babcock (1958). At any rate, the new astrometric information casts considerable doubt on the earlier picture of VV Cephei.

APPROXIMATE TIME-TABLE FOR VV CEPHEI		
First Contact	June 11, 1956	October 15, 1976
Totality begins	Aug 27, 1956	January 1, 1977
Totality ends	Dec 7, 1957	April 12, 1978
Last Contact	Feb 17, 1958	June 23, 1978

NGC 188. The unusually ancient galactic star cluster in Cepheus, photographed with the 13-inch telescope at Lowell Observatory.

NGC 188   Position 00394n8503. A galactic cluster of unusual interest, located some 4° from Polaris, and therefore observable throughout the year. The cluster is approximately 15' in diameter and contains some 150 stars, the majority of which are fainter than the 13th magnitude. The cluster can be detected with a 6-inch or 8-inch telescope using low power, and appears as a large but dimly luminous spot with only a few of the brighter members showing individually. The computed distance is slightly over 5000 light years, and the distance above the galactic plane is about 1800 light years. Spectroscopic observations show a radial velocity of about 30 miles per second in approach. Early type stars are completely lacking in this cluster; the 10 brightest members are yellow giants of luminosity class III whose spectra range from G8 to K4; their absolute magnitudes lie in the range of 0 to +2.

NGC 188 is famous as the oldest known galactic star cluster, and has received much attention from astronomers specializing in the problems of stellar evolution. In the article on M13 in Hercules, a brief account is given of the method of determining relative ages of star clusters by plotting the members on the H-R Diagram. The typical diagram for a globular star cluster turns out to be very different from that of a usual galactic star cluster. This is due to a difference in age; most galactic star clusters are relatively young, while the globulars are recognized as being extremely ancient. In the accompanying diagram (page 614) the plotted results for a number of galactic clusters are shown. The youngest cluster on the diagram is evidently NGC 2362, which contains highly luminous blue giant stars; only slightly older is the beautiful Double Cluster in Perseus, with a computed age of only a million years. Older star clusters contain no blue giants, since such stars are relatively short-lived, and are the first to evolve away from the main sequence stage of their lives. As the cluster grows older, the less luminous stars eventually begin their own evolution. The relative ages of star groups can thus be determined by comparing the highest points on the H-R diagram at which main sequence stars still exist. As an example, we may take the cluster M41 in Canis Major. The diagram shows that all stars brighter

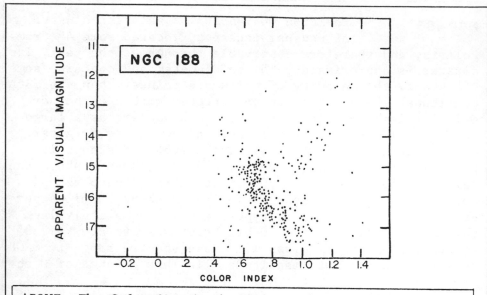

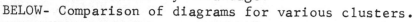

ABOVE- The Color-Magnitude Diagram for NGC 188, from
observations by A.Sandage.

BELOW- Comparison of diagrams for various clusters.

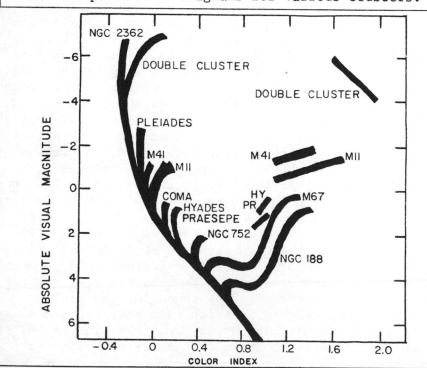

than absolute magnitude -1.5 have evolved away from the main sequence, whereas the "turn-off" point for the Hyades cluster is near +0.8. Clearly then, the Hyades group is older than M41, and a globular cluster such as M3 is vastly older than either.

There are, however, a few galactic clusters which have H-R diagrams resembling those of the globulars. M67 in Cancer and NGC 188 are the two chief examples.

While a comparison of the various H-R diagrams can be used to reveal the relative ages of star groups, the problem of reducing these to a definite age in years is not completely solved. For NGC 188 itself, A.R.Sandage originally estimated an age of some 24 billion years, older than anything else known in the Universe. However, new findings about stellar evolution have made revisions necessary. M41 may be about 60 million years old, the Hyades about 400 million, and M67 possibly close to 10 billion. In the case of NGC 188, all the stars brighter than absolute magnitude $+4\frac{1}{2}$ have evolved away from the main sequence and the currently accepted age is some 12 to 14 billion years. This exceeds the ages computed for many of the globular star clusters. (Refer also to M67 in Cancer, M13 in Hercules)

NGC 6946   Position 20339n5958. A large Sc-type spiral galaxy, lying on the Cepheus-Cygnus border about 2° southwest of Eta Cephei. The galactic star cluster NGC 6939 lies approximately 38' to the northwest, and the two objects may be viewed together in the field of a wide-angle eyepiece. NGC 6946 is about 11th magnitude visually, has a rather low surface brightness, and appears nearly circular to the eye, measuring about 8' in diameter. Photographs show that the system is actually oriented some 25° or 30° from the face-on position, the longest dimension lying northeast-southwest. S.van den Bergh, in his catalog "A Reclassification of the Northern Shapley-Ames Galaxies" (1960) gives the total integrated magnitude as 9.67 (pg) and the apparent dimensions as 9.0' x 7.5'. The integrated spectral class is F5.

Owing to the low surface brightness, the small central nucleus is the only detail which appears clearly to the visual observer. The outer haze reveals its counterclockwise spiral structure to the photographic plate, and the greatest telescopes resolve the arms into long chains of bright condensations - star clouds and masses of nebulosity. Multiple branching of the arms makes it difficult to trace the course of any one arm from its point of origin in the central mass to its gradual disappearance on the galaxy's outer rim. There are, however, at least four well defined arm segments, and several fainter branches or "sub arms". The most prominent arm, on the northeast side, can be traced out about 5' from the nucleus, and ends in several bright patches of nebulosity.

NGC 6946 has long been a controversial object from its possible membership in the Local Group of Galaxies. E.Hubble (1936) pointed out that "a total absorption of one magnitude would place this spiral outside the Local Group while two or three magnitudes would lead to a smaller distance and membership in the group". From recent studies (1963) it seems almost certain that this galaxy is not a member, but is apparently one of the nearest galaxies beyond. The radial velocity, after correction for the solar motion, is only 200 miles per second in recession, which implies a distance in the range of 10 - 20 million light years. The galaxy lies only 11° from the central plane of the Milky Way, and is heavily obscured by dust clouds in our own star system; the distance therefore remains indeterminate. A similar situation exists with regard to IC 342 in Camelopardus but a red shift only half that of NGC 6946 makes its membership in the Local Group almost a certainty.

NGC 6946 is noted for an unusual number of supernovae, having shown such outbursts in 1917, 1939, 1948, and 1968. The first of these, discovered by G.Ritchey at Mt.Wilson in July 1917, led directly to the studies which ultimately proved the "spiral nebulae" to be other galaxies. The exploding star was detected when well past its maximum, at magnitude 14.6. The actual peak brilliancy may have been about 12.9. Tentatively accepting a distance of about 10 million light years and an obscuration of about 1 magnitude, the actual luminosity of this star is found to be about equal to 100 million suns. NGC 6946 is also one of the nearby spirals which has been identified as a source of radio radiation by R.H.Brown and C.Hazard at Jodrell Bank in England. (See also IC 342 in Camelopardus)

NGC 6946. One of the nearby spiral galaxies, photographed with the 100-inch reflector at Mt. Wilson Observatory.

SPIRAL GALAXY NGC 6951 in CEPHEUS. This system is usually
classed as an early type barred spiral.
                     Palomar Observatory 200-inch telescope.

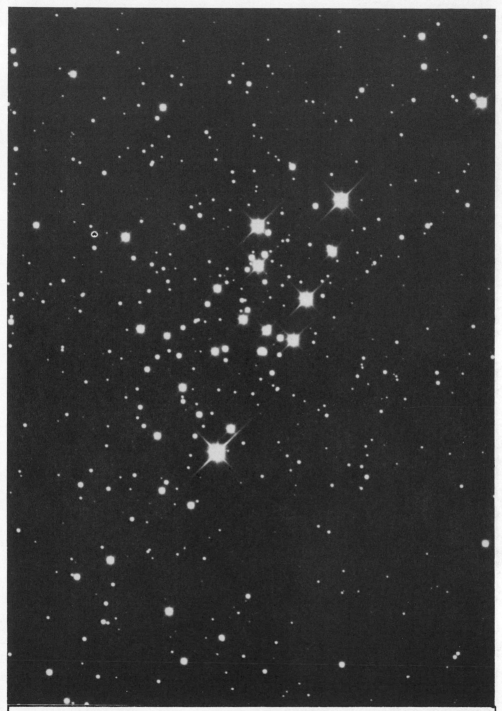

STAR CLUSTER NGC 7510 in CEPHEUS. This irregular group lies near the Cepheus-Cassiopeia border. Lick Observatory photograph with the 120-inch reflector.

DEEP-SKY OBJECTS IN CEPHEUS. Top: The diffuse nebula NGC 7023. Below: The open star cluster NGC 6939 and the spiral galaxy NGC 6946.

Lowell Observatory photographs

CETUS

LIST OF DOUBLE AND MULTIPLE STARS

NAME	DIST	PA	YR	MAGS	NOTES	RA & DEC
λ2	1.9	148	62	6 - 12	cpm pair, PA dec, spect A3	00052s2247
Σ3065	9.6	108	55	8½- 8½	relfix, spect G0	00054s1430
β486	3.0	4	58	6 - 12	relfix, cpm, spect gM4	00119s0804
β393	0.7	22	54	6 - 8	PA slow inc, spect A0	00158s2125
ι	63.4	17	10	3½- 12	(h1953) spect K2	00169s0906
h1957	6.1	24	52	7½- 8½	cpm. spect G0; B = 0.2" pair	00193s2317
A431	0.2	353	57	8½- 8½	Binary, 54 yrs; PA dec, spect G5	00246s0809
h1968	17.3	230	54	8- 10½	optical. AB dist	00251s1641
	88.2	123	21	- 11	inc, spect F8	
B1909	0.2	109	62	7 - 7	binary, 5.6 yrs; spect G0	00258s2036
12	10.7	197	61	6½- 11	(h322) optical, PA & dist inc, spect M0	00275s0414
β1158	79.2	87	29	7- 8½	(h1981) spect A3	00286s1022
β1158b	0.3	188	66	9 - 9	PA inc, spect G0	
A111	0.2	280	57	9 - 9	Binary, 10.8 yrs;	00299s0527
	2.5	199	58	- 12½	spect dG8; all cpm AC PA dec.	
Σ39	0.6	258	61	7½- 8	PA inc, spect G0	00319s0449
	19.7	45	53	- 8½		
13	0.2	240	61	5½- 6½	binary, 6.9 yrs;	00327s0352
	24.5	43	00	- 12½	Spect F8 (*)	
β395	0.7	106	61	6½- 6½	binary, 25 yrs; PA inc, spect G5	00347s2502
15	30.5	307	08	7½-11½	(Hd 31) spect K0	00355s0047
Σ49	8.1	321	62	6½- 10	optical, dist inc, spect G0	00382s0730
h323	64.3	289	22	6 - 8½	optical, spect G7	00382s0438
Mu1 1	2.5	195	55	7 - 9½	relfix, spect F0	00432s1642
β494	1.3	165	62	8½- 8½	relfix, spect F8	00444s0131
B13	8.8	167	26	6 - 13	spect gK2	00468s2425
β1160	1.3	114	33	6 - 12	cpm; spect gK5	00469s1350
β301	1.0	319	43	8½- 13	relfix, spect F5	00470s2140
	11.0	300	17	- 9½		

CETUS

LIST OF DOUBLE AND MULTIPLE STARS   (Cont'd)

NAME	DIST	PA	YR	MAGS	NOTES	RA & DEC
Stn 3	1.8	256	59	7 - 8	cpm; PA dec,	00498s2253
	32.6	196	08	- 12	spect G0	
β734	10.8	346	34	6 - 11	relfix, spect gK2	00502s2417
Stn 4	5.4	9	52	6½- 8½	relfix, spect dF6	00508s2503
h2004	3.4	240	43	7 - 10	spect A0	00552s1916
S390	6.4	216	52	8 - 8	slight dist dec,	00557s1557
					spect F5	
Σ80	25.4	331	61	7½- 8	dist inc, spect K0	00568n0031
A1902	0.2	293	58	8 - 8½	PA inc, spect F5	00568s0057
A1903	0.3	255	62	9 - 9½	binary, 150 yrs;	00570s0128
	5.8	30	58	- 13	PA inc, spect G5,	
					all cpm	
Σ81	17.4	68	11	7½-10½	spect F5	00575s0217
26	16.0	253	26	6½- 9	(Σ84) relfix,	01012n0106
					cpm; spect dF0, dG8	
Σ86	15.0	145	61	8 - 8½	PA dec, spect F2	01023s0544
Σ91	4.1	317	62	6½- 7½	relfix, spect F5	01046s0200
Σ101	21.0	341	15	8 - 10½	spect G5	01114s0753
37	49.7	331	31	5 - 7	wide cpm pair;	01119s0812
					spect dF2, dK0	
Σ106	4.7	307	53	8½- 8½	relfix, spect G0	01138s0725
Σ103	5.3	246	30	7½-10½	relfix, spect G5	01141s0131
Σ110	7.5	354	62	8 - 8½	relfix, spect F8	01155s1236
Σ111	20.7	330	21	8½- 10	relfix, spect F7	01155s0436
42	1.5	9	61	6 - 7	(Σ113) PA inc,	01172s0046
					spect G8, A7	
42b	0.1	99	61	7½- 7½	(φ 337) PA inc	
h2036	2.0	351	61	7 - 7	PA dec, dist inc,	01175s1604
					spect G0	
h2043	5.0	74	52	6½- 8½	relfix, cpm pair;	01201s1921
					spect F5	
Se 1	2.7	85	45	7 - 8½	relfix, spect A5	01212s2437
β1163	0.3	218	61	6 - 6	binary, 16 yrs;	01218s0710
					spect dF2	
Σ120	7.3	278	33	7 - 11	relfix, spect A0	01225s0612
Σ124	7.1	233	02	8 - 10	relfix, spect F5	01238s1409
β399	1.6	302	42	6½- 10	relfix, cpm pair,	01253s1110
					spect K0	
h3437	12.3	247	18	7½- 9½	relfix, spect F0	01257s1731
48	22.0	250	09	6- 13	(λ14) Spect A1	01271s2154

LIST OF DOUBLE AND MULTIPLE STARS   (Cont'd)

NAME	DIST	PA	YR	MAGS	NOTES	RA & DEC
Kui 7	0.1	344	58	7 - 7½	PA inc, spect F5	01351s0940
L726-8	2.4	29	60	12½- 13	Red dwarf binary, B = flare star UV Ceti (*)	01364s1813
h2067	33.9	92	19	7 - 11	spect F5. UV Ceti is 13' to SW	01369s1803
Σ147	2.0	91	61	6 - 7	cpm, dist dec, spect dF2, dF3	01393s1134
Σ150	36.2	196	51	8 - 9½	relfix, spect A	01409s0720
β36	2.1	167	62	6½- 9	relfix, spect G0	01422s0701
Σ171	32.5	163	61	8½- 8½	slight dist inc, both spectra G5	01462s0140
Σ171b	4.2	316	61	9 - 13	(β511) slight PA dec, spect G5	
I 450	0.5	213	59	8½- 8½	spect A5	01496s2325
Σ186	1.3	51	61	7 - 7	binary, 158 yrs; PA inc, spect dG0	01533n0136
58	2.8	16	33	7 - 12	(β7) slight PA inc, spect A0	01554s0218
Sh 24	8.4	304	59	8 - 9	(H58) relfix, spect F2	01566s2310
61	43.1	194	55	6- 10½	cpm pair, both spect G5, AC is optical	02012s0035
	83.0	326	09	- 12		
β516	0.6	305	61	8 - 8	PA inc, dist dec, spect A5	02027s0112
Σ218	4.9	248	51	7 - 8	relfix, spect F0	02062s0040
66	16.2	232	58	6 - 7½	relfix, easy cpm pair; spect dF9, dG4 (Σ231)	02102s0238
Hst 1	1.5	67	61	8 - 10	binary, about 170 yrs; PA inc, spect dK4	02134s1828
O	0.7	123	62	var-10	MIRA, classic long period variable; AB binary (*)	02168s0312
	118	78	25	- 9		
Σ265	12.1	136	31	8 - 8½	relfix, spect G0	02220s0159
β517	11.2	248	16	7½-12½	spect K0	02224s0407
	56.1	290	14	-11½		
Σ266	7.4	267	37	8 - 8½	relfix, spect F2	02224s0220

LIST OF DOUBLE AND MULTIPLE STARS   (Cont'd)

NAME	DIST	PA	YR	MAGS	NOTES	RA & DEC
H 80	12.1	295	62	6 - 9	cpm pair, relfix spect Ap	02236s1534
Kui 8	0.5	29	60	7 - 7½	PA inc, spect K0	02255n0144
β518	1.6	142	53	6½- 11	slight PA inc, spect K2	02269n0921
h3502	28.4	85	19	7 - 11	relfix, spect A5	02277s2254
Σ274	13.5	219	32	7 - 7½	cpm pair, relfix, spect A2	02289n0052
Σ276	2.2	265	53	9 - 9	slow PA inc, spect G5	02300n0607
Σ280	3.6	246	61	7½- 7½	relfix, spect K2	02316s0551
ν	8.0	83	31	5- 9½	(Σ281) (78 Ceti) cpm, relfix, spect G8, F7	02332n0523
h3511	14.8	98	53	7 - 9	relfix, spect G0	02337s2137
Kui 9	0.9	315	55	6½- 9½	PA inc, spect K0p	02360n0314
ε	0.1			5½ - 5½	(83 Ceti) binary 2.6 yrs, spect F5	02371s1205
A1928	0.2	70	66	8½- 8½	binary, 18 yrs; PA inc, spect G4	02373s0004
84	4.1	313	61	6 - 9	(Σ295) PA dec, spect dF6	02386s0054
A2337	2.2	252	55	7 - 13	PA inc, spect A0	02388n0413
	26.5	130	18	- 13		
h3524	19.4	152	34	7 - 7½	PA inc, spect G0	02406s2030
γ	2.7	295	61	3½- 6	(Σ299) slight PA inc, spect A2 (*)	02407n0302
Σ313	5.5	192	26	8½- 9	relfix, spect F8	02472n0844
Σ323	2.7	100	62	8 - 8	perhaps slight PA dec, spect B9	02500n0616
Σ330	8.8	192	30	7½- 9½	relfix, spect G5	02546s0047
A2413	0.4	33	61	8½- 9	binary, 150 yrs; PA inc, spect G0	02546n0141
Σ332	12.5	53	23	8½- 8½	relfix, spect G5	02553n0013
Σ334	1.1	311	66	7½- 8	cpm. slow PA dec, spect F0	02567n0627
ΣI 6	81.2	163	35	7 - 7	probably cpm; relfix, Spect G0, G0	03058n0727

CETUS

LIST OF DOUBLE AND MULTIPLE STARS   (Cont'd)

NAME	DIST	PA	YR	MAGS	NOTES	RA & DEC
94	3.2	233	62	5- 11½	(h663) PA & dist dec, spect F8	03102s0123
Σ 367	0.8	155	66	8 - 8	binary, about 870 yrs; PA dec, spect F8	03115n0033
95	1.0	239	66	6 - 10	binary, about 200 yrs; PA inc, Spect K1	03158s0107
Σ 3046	3.4	260	62	8 - 8½	cpm, PA inc, spect G5	23538s0946

LIST OF VARIABLE STARS

NAME	MagVar	PER	NOTES	RA & DEC
δ	4.04--4.07	.1611	Beta Canis Majoris type, spect B2	02369n0007
O	3----9.5	331	MIRA. Typical long period variable star (*)	02168s0312
R	7.4--14..	166	LPV. Spect M4e	02235s0024
S	7.7--14.7	320	LPV. Spect M3e--M4e	00215s0936
T	5.2-- 6.9	160:	Semi-reg; spect M5e	00192s2020
U	6.8--13.4	235	LPV. Spect M2e--M4e	02313s1322
V	8.5--14.6	260	LPV. Spect M3e	23553s0914
W	6.8--14.5	351	LPV. Spect S7e	23596s1458
X	8.4--13..	177	LPV. Spect M2e	03169s0115
Z	8.4--14..	185	LPV. Spect M1e--M4e	01042s0145
RR	8.7--9.5	.5530	Cl.Var; Spect A5--F0	01296n0105
RS	8.2---8.6	Irr	Spect G	02254n0000
RW	9.0--10.0	.9752	Ecl.Bin; spect A5	02129s1226
RY	9.2--12..	374	LPV. Spect M5e	02137s2045
SS	9.4--13.0	2.974	Ecl.bin; spect A0	02460n0133
SW	8.6---9.5	Irr	Spect M6	01359n0106

LIST OF VARIABLE STARS   (Cont'd)

NAME	MagVar	PER	NOTES	RA & DEC
TT	9.9--10.4	.4860	Ecl.Bin; W Ursae Majoris type, spect A	01445s1000
TV	8.7-- 9.1	9.103	Ecl.bin; spect F0	03120n0234
TW	8.8--9.6	.3169	Ecl.bin; W Ursae Majoris type; spect G5 + G5	01465s2108
TZ	8.0---8.5	95:	Semi-reg; spect M3	02203s1026
UV	7-----12	Irr	Flare star (L726-8b) spect dM6e (*)	01364s1813
UY	8.7--10.4	440	Semi-reg; spect M2	00246s0653
UZ	8.6---9.6	122	Semi-reg, spect M2	02036s1027
WW	9.3--16..	Irr	Z Cam. type	00089s1146
WX	10.5...18	---	Nova 1963	01142s1812
WY	9.6--10.4	1.939	Ecl.Bin; spect A2	01331s1212
XY	8.6---9.3	1.390	Ecl.Bin; spect A0	02570n0319
XZ	8.5---9.2	.451	Cl.var; spect A	01579s1635

LIST OF STAR CLUSTERS, NEBULAE, AND GALAXIES

NGC	OTH	TYPE	SUMMARY DESCRIPTION	RA & DEC
45		⊘	S; 12.1; 8.0' x 5.5' eF,L,lbM, 8^m star 4½' sp. loose scattered structure, spiral pattern dim.	00114s2327
151		⊘	Sb; 12.4; 3.1' x 1.1' pF,pL,lE, vglbM	00316s0958
157	3[2]	⊘	Sc; 11.1; 2.8' x 2.1' pB,L,E (*)	00323s0840
175		⊘	SBb; 12.8; 1.5' x 1.3' pB,pL,E,gbM, θ structure	00349s2012
178		⊘	SB/pec; 12.9; 1.2' x 0.7' F,S,mE,bM	00366s1427
210	452[2]	⊘	Sb; 12.0; 4.5' x 2.4' B,pS,E,psbM; faint outer arms form encircling ring	00380s1409
227	444[2]	⊘	E2; 13.5; 0.8' x 0.6' F,pL,lbM	00401s0148
237		⊘	Sc; 13.2; 1.2' x 0.8' vF,pS,lE,lbM	00409s0024

LIST OF STAR CLUSTERS, NEBULAE, AND GALAXIES   (Cont'd)

NGC	OTH	TYPE	SUMMARY DESCRIPTION	RA & DEC
245	445[2]	⬯	Sc; 12.9; 1.0' x 0.9' F,pS,1E	00437s0159
246	25[5]	◎	vF,L,1E, mag 8½, diam 4' x 3½', with 12^m 07 central star.. Galaxy NGC 255 lies 15' to NNE (*)	00446s1209
247	20[5]	⬯	Sc; 10.7; 18.0' x 5.0' F,eL,vmE,sN (*)	00446s2101
255	472[2]	⬯	Sb; 12.8; 1.5' x 1.5' F,pS,R,gbM; faint outer arms	00452s1145
268	463[3]	⬯	Sc; 13.2; 1.1' x 0.9' vF,pS,1E	00476s0528
274	429[3]	⬯	E1?; 13.0; 0.9' x 0.8' pB,pS,smbM. Forms 1' pair with NGC 275	00485s0720
275		⬯	SB; 13.0; 1.1' x 0.8' F,S,R. 1' sf NGC 274	00485s0720
309		⬯	Sc; 12.5; 2.4' x 2.1' pB,pL,R. fine spiral (*)	00540s1013
337	433[3]	⬯	SBc; 12.2; 2.0' x 1.5' pF,L,E,1bM; coarse spiral pattern	00573s0751
357	434[2]	⬯	SBa; 13.0; 1.6' x 1.1' F,S,1E,sbM	01008s0637
---	I.1613	⬯	I; 12.0; 11' x 9' vL,F, member of Local Group of Galaxies (*)	01025n0152
---	New 1	⬯	S ; 12.8; 3.5' x 3.5' F,R, outer arms dim	01026s0629
428	622[2]	⬯	Sc; 11.7; 3.9' x 3.5' F,L,R,bM; spiral pattern coarse & irregular	01104n0043
450	440[3]	⬯	Sc; 12.6; 2.5' x 2.0' vF,L,SBN. 12' nf 38 Ceti	01130s0107
521	461[2]	⬯	SB; 13.0; 2.0' x 2.0' R,F,pL, delicate multiple arms; NGC 533 foll 14'	01220n0128
533	462[2]	⬯	E2; 13.0; 0.9' x 0.7' pB,pL,1E, gbM	01229n0130

LIST OF STAR CLUSTERS, NEBULAE, AND GALAXIES   (Cont'd)

NGC	OTH	TYPE	SUMMARY DESCRIPTION	RA & DEC
578		⊖	Sc; 11.7; 4.5' x 2.5' B,pL,pmE,gmbM	01280s2256
584	100[1]	⊖	E4; 11.5; 1.7' x 1.0' vB,pL,R,mbM	01288s0707
596	4[2]	⊖	E2; 12.2; 1.0' x 0.9' pB,R,bM. 6^m star 12' foll	01303s0717
615	282[2]	⊖	Sb; 12.6; 2.7' x 0.8' pB,pL,1E (*)	01326s0753
636	283[2]	⊖	E1; 12.6; 0.7' x 0.7' pB,vS,R,mbM	01366s0745
681	481[2]	⊖	Sa/Sb; 12.9; 1.3' x 1.2' pF,cL,E,1bM. 16.7' NNW from Chi Ceti. Thin dust lane across equator, resembles "Sombrero" NGC 4594 in Virgo	01467s1040
701	62[1]	⊖	Sb; 12.7; 1.6' x 0.5' F, pL, E, g1bM	01486n0957
720	105[1]	⊖	E4; 11.7; 1.3' x 0.7' cB,pL,1E,psmbM	01506s1359
779	101[1]	⊖	Sb; 11.8; 3.0' x 0.5' cB,L,mE,mbM, nearly edge-on spiral	01572s0612
788	435[2]	⊖	Sa; 12.6; 1.2' x 0.9' pF,pS,1E,bM	01586s0703
864	457[3]	⊖	Sc; 12.0; 2.8' x 2.8' eF,cL,R,gbM, 12^m star sf nucleus 35"	02128n0545
895	438[2]	⊖	Sb; 12.2; 2.8' x 2.2' F,L,E	02191s0545
908	153[1]	⊖	Sc; 11.1; 4.0' x 1.3' cB,vL,E; thick spiral arms	02208s2127
936	23[4]	⊖	SBa; 11.3; 3.0' x 2.0' vB,vL,1E,mbMN. Has thick central bar, large faint outer halo. NGC 941 foll 12'	02251s0122
941	261[3]	⊖	Sc; 12.9; 1.9' x 1.3' vF,S,1E. 12' pair with NGC 936	02260s0122

LIST OF STAR CLUSTERS, NEBULAE, AND GALAXIES (Cont'd)

NGC	OTH	TYPE	SUMMARY DESCRIPTION	RA & DEC
955	278[2]	⊖	E7?; 13.1; 2.0' x 0.5' Possibly edge-on S0, lens shape. pB,S,mE,sbM. 25' west from 75 Ceti.	02280s0119
958	237[2]	⊖	Sb; 13.0; 1.8' x 0.6' pF,E,bM; nearly edge-on	02281s0309
991	434[3]	⊖	Sc; 12.7; 1.8' x 1.8' vF,cL,vlbM	02332s0722
1022	102[1]	⊖	Sb; 12.0; 1.0' x 0.6' cB,pL,E, compact spiral	02361s0653
1035	284[2]	⊖	Sb; 12.8; 2.0' x 0.5' L,pF,mE, nearly edge-on	02370s0820
1042		⊖	Sc; 12.5; 3.0' x 3.0' F,pS,R, spiral arms well defined, but narrow and faint. (Misidentified as NGC 1048 in Shapley-Ames Catalog & Skalnate Pleso)	02380s0840
1052	63[1]	⊖	E3; 11.6; 0.7' x 0.5' B,pL,R,mbM	02386s0828
1055	6[2]	⊖	Sb; 12.0; 5.0' x 1.0' pF,cL,E; edge-on; sombrero structure with equatorial dust lane. 11^m star 1' n	02392n0016
1068	M77	⊖	Sb; 10.0; 2.5' x 1.7' vB,pL,R,SBN. faint outer ring 6' diam. (*)	02401s0014
1073	455[3]	⊖	SBc; 12.0; 4.0' x 4.0' vF,L,lbM; spiral arms faint	02412n0110
1087	466[2]	⊖	Sc; 11.2; 2.3' x 1.3' pB,cL,lE,mbM. spiral arms coarse, but compact	02439s0042
1090	465[2]	⊖	Sb; 12.8; 4.0' x 1.5' vF,pL,E,bM. 15' n from NGC 1087	02440s0027

DESCRIPTIVE NOTES

ALPHA   Name- MENKAR or MENKAB. Mag 2.52; Spectrum
M2 III. Position 02597n0354. Menkar is an
orange giant star, located about 150 light years distant,
with an actual luminosity of about 175 suns (absolute magnitude -0.8). The radial velocity is 15½ miles per second
in approach; the annual proper motion is 0.07".

The 5th magnitude blue star 93 Ceti is located
15.8' distant in PA 5°, nearly due north. This wide pair
does not form a true double, but the contrasting colors of
the two stars makes an interesting sight in the low power
telescope. 93 Ceti is magnitude 5.6, spectrum B7 III; its
computed distance is about 500 light years, the actual
luminosity about 600 times that of the Sun.

Between the wide pair, on the east side, are two
11th magnitude stars with a separation of 1.7'; the southern member is a 10" pair, probably physically connected.

BETA   Name- DENEB KAITOS or DIPHDA. Mag 2.00; Spectrum K1 III, position 00411s1816. The dist-
ance is approximately 60 light years; the actual luminosity
about 40 times that of the Sun. Beta Ceti shows an annual
proper motion of 0.23" and a radial velocity of 8 miles
per second in recession.

The large dim spiral galaxy NGC 247 lies slightly
less than 3° distant toward the SSE (See page 649).

GAMMA   Position 02407n0302. A rather close but fine
double star, usually considered the most note-
worthy in the constellation, and discovered by F.G.W.
Struve in 1836. The components, separated by about 2.7",
are usually described as yellow and blue, but these colors
appear to be at least partly illusionary since the spectral
types are now known to be A2 and F3. The fainter star has
been described as "tawny" or "dusky" by some observers.
The chief facts about the two stars are given here:

	Mag.	Spect.	Lum.	Abs.Mag.
A	3.6	A2 V	13	+ 2.0
B	6.2	dF3	1.2	+ 4.6

Gamma Ceti is at a distance of about 70 light years, and
shows an annual proper motion of 0.21" in PA 225°.   The

two stars undoubtedly form a long-period binary, but the observed change in the PA has amounted to only 3° in the last century, suggesting that the period may be at least several thousand years. The projected separation is about 60 AU.

A third faint companion, sharing the proper motion of the bright pair, lies at 14' distance toward the NW, in PA 315°. This is LTT 10888 or BD+2°418, a 10th magnitude red dwarf of spectral type dM. The true distance from the two bright stars is at least 18,000 AU.

ETA   Mag 3.44; spectrum K3 III. Position 01061s1027.
            The computed distance is about 100 light years which leads to an actual luminosity of 35 times the Sun. The annual proper motion is 0.25" in PA 122°; the radial velocity is 7 miles per second in recession.

OMICRON Name- MIRA, "The Wonderful". The brightest and most famous of the long-period pulsating variable stars, and the standard object of its type. Position 02168s0312. It varies in brightness from 9th magnitude or less at minimum to 3rd or 4th at maximum, sometimes - but rarely - attaining 2nd magnitude. Once, in 1779 it rose to nearly 1st magnitude and was almost the equal of Alpha Tauri (Aldebaran). The period averages 331 days, but there are often considerable irregularities both in period and light range.

Mira was the first of the long-period variables to be discovered, by the Dutch astronomer David Fabricus, on August 13, 1596. He seems to have thought the star a nova and evidently did not look for its return. Thus the star was not seen again until 1603, when Bayer included it in

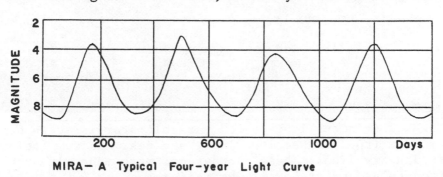

MIRA— A Typical Four-year Light Curve

MIRA. The famous long-period variable star is shown near minimum (top) in December 1961, and near maximum (below) in January 1965. These photographs were made with the 13-inch telescope at Lowell Observatory.

DESCRIPTIVE NOTES (Cont'd)

his famous atlas. Not aware of its variability, he catalogued it as a 4th magnitude star and assigned the identifying greek letter "Omicron". Sometime later it was found
that Omicron Ceti had mysteriously vanished, but in less
than a year it had reappeared and was shining once again
with its normal brightness. Continuing observations eventually revealed that the star was subject to a nearly regular cycle of variations, reaching naked-eye visibility
for only a few weeks out of each year. No other case of
stellar variability was then known, and as astronomers
became aware of the unusual fluctuations of Omicron Ceti
they honored the star with the name it now bears - Mira,
"the Wonderful". The name was first suggested by Hevelius.

The records of the variations of Mira go farther back
than those of any other variable star, every maximum since
1638 having been observed. A typical four-cycle light
curve appears in the graph on page 631, and an "idealized
light curve", made by integrating the observations over a
30-year interval, appears below.

CHARACTERISTICS OF THE MIRA-TYPE STARS. These stars form
the most numerous class of variables known in the Universe
at the present time, nearly 4000 having been catalogued.
Their leading characteristics are:

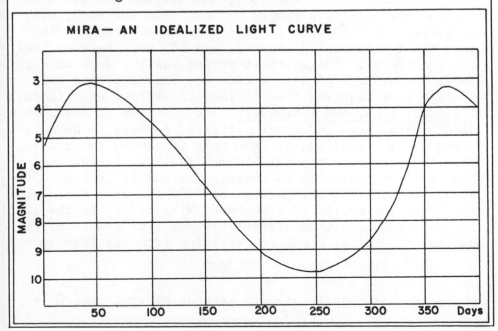

MIRA— AN IDEALIZED LIGHT CURVE

DESCRIPTIVE NOTES (Cont'd)

1. The light range is very great, averaging 5 or 6 magnitudes, and in a few stars exceeding 9 magnitudes. The
range of Chi Cygni has exceeded ten magnitudes on occasion.
2. The periods range from about 60 days to 700 days,
with a few stars exceeding these limits. Periods between
200 and 400 days are the most common. In general, it seems
that the stars of longer period have a greater range and
a deeper color, but not necessarily a higher actual luminosity.
3. The variations do not repeat themselves with absolute
regularity; there are often considerable changes from one
cycle to the next, both in period and amplitude.
4. Stars of the type are all red giants with absolute
magnitudes (maximum) generally lying in the range of -1
to about -3. About 90% of these stars fall into spectral
class M, classes N and S claim about 5% each, and a very
few belong to the rare class R.

LIGHT RANGE AND PERIOD OF MIRA. The irregularities of this
star are typical of the class; the following figures are
based on continuous records of the star over a 30-year
interval, from 1910 to 1940.

The highest maximum during this interval was magnitude 2.50, the lowest was 4.80, the average was 3.49. At
minimum the magnitude ranged between 8.60 and 9.60, with
an average of 9.30. The longest period recorded was 355
days between successive maxima, and 353 days between successive minima. The shortest period was 304 days in each
case. The average of all periods was 331 days, and the
average time required from minimum to maximum was 112 days.

DISTANCE, SIZE, AND LUMINOSITY. The distance of Mira, from
direct parallaxes and other criteria, appears to be quite
close to 220 light years. From this it can be calculated
that the star at a typical minimum is slightly fainter
than our Sun, while at an average maximum it is some 250
times brighter. At the maximum of 1779 the star must have
reached a luminosity of 1100 suns. Mira is one of the 10
largest stars measured directly by means of the interferometer, the actual diameter resulting from the formula:

$$\text{Diam} = \frac{d \times 93{,}000{,}000}{\pi}$$

where d = the angular size in seconds of arc, and π =

the parallax in seconds of arc. The result can be no more accurate than the accepted values of d and π , both being subject to a certain margin of error. But although we can not expect absolute exactness, the general order of size is clearly indicated. Interferometer observations give about 0.056" as the apparent diameter, and the parallax is about 0.015". The solution of the equation then gives us a diameter of about 350 million miles, or some 400 times the diameter of the Sun. During the course of its pulsations Mira probably undergoes some changes in size; according to some estimates it may attain a diameter of 500 times that of the Sun, when at maximum.

The tremendous size, however, is deceptive, for the mass of the star is probably no more than twice that of our Sun, and the resulting density is about 0.0000002 that of the Sun, a virtual vacuum by usual earthly standards! The great light range is also - in a sense - deceptive, since it is known that the actual increase in total radiation is only about $2\frac{1}{2}$ times. Much of the apparent loss of light at minimum is a temperature effect, the star radiating chiefly in the invisible infrared. At maximum the energy output shifts over into the visible spectrum. The alternate veiling and unveiling of the star by a cloud of low temperature compounds may also play its part, these substances tending to disperse as the temperature rises. The temperature, color, and spectral type all vary in the course of each cycle. At minimum Mira is one of the coolest stars known, with a spectral type of M9 and a temperature of about 1900°K. At maximum the temperature has risen to about 2500° and the spectral type has shifted to M6e. Due to the temperature change, the red tint of the star slowly deepens as the star fades. A peculiar fact about these changes is that the highest temperature is reached - not at maximum - but some days later when the star has already begun to fade.

Spectroscopically, Mira is a remarkable object with its strong dark bands of titanium oxide and its bright emission lines of hydrogen; these features are typical of the long-period variables. A recent discovery of great interest is the finding of water vapor (literally steam) in the atmosphere of Mira and in other red giants such as Mu Cephei and Y Canum Venaticorum.

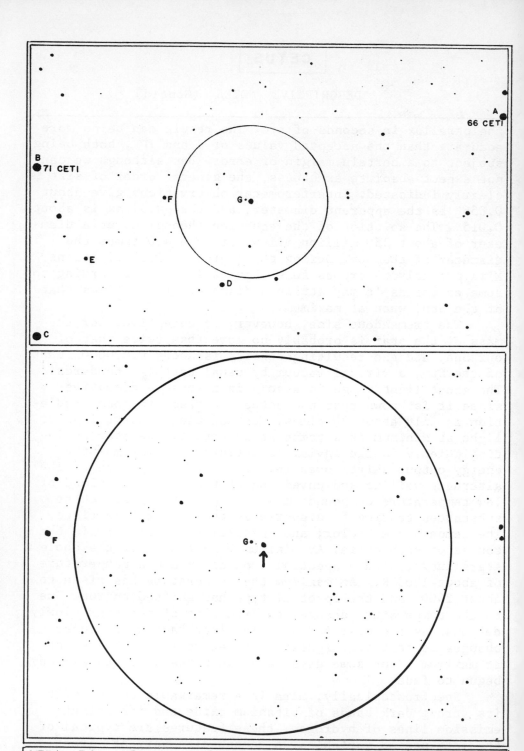

MIRA- Identification charts. Each circle = 1° diameter, with north at the top. Comparison magnitudes are: A = 5.7, B = 6.4, C = 7.1, D = 8.0, E = 8.6, F = 8.8, G = 9.2. Charts made from Lowell Observatory 13-inch telescope plate

DESCRIPTIVE NOTES (Cont'd)

THE THEORY OF PULSATION. All the physical characteristics
of Mira seem to change regularly in the course of each
cycle, suggesting that we are observing a periodic pulsation of the star, or at least of the outer layers. The
evidence for actual expansion and contraction, however, is
not as clear as in the case of the "cepheid" variables of
much shorter periods, and the reason for the pulsations
has long been a highly controversial question. When the
red giants were believed to be newly formed stars it was
suggested that the energy released by contraction might
equal the radiant energy output, and that a conflict between the two forces might result in pulsation. The best
evidence today seems to indicate that the red giants are
actually older stars which are nearing the point of hydrogen exhaustion and thus entering a critical phase in their
evolutionary history. By plotting the theoretical life
history track or "evolutionary track" of many stars on the
H-R diagram, it is found that the red variables lie near
the point where hydrogen has been consumed in the core of
the star and where the next reaction - helium burning - is
about to begin. Although the picture is still unclear and
the conclusion may be premature, it is tempting to attrib-
ute the pulsations to some conditions connected with the
onset of the helium reaction. It should also be remembered that the term "pulsation" is a convenient descriptive
phrase which should perhaps not be taken too literally.
The term "pulse" might be preferable, since it seems very
likely that the actual operating mechanism in these stars
is something in the nature of a shock wave or "front"
which pulses through the outer layers of the star after
having originated in some disturbance in the interior.
(Refer also to Delta Cephei and Alpha Orionis)

THE COMPANION TO MIRA. As early as 1918, A.H.Joy had detected peculiarities in the spectrum of Mira which indicated the existence of a B-type companion. In 1923 the companion was seen visually for the first time by R.G.Aitken
with the 36-inch refractor at Lick Observatory. The separation was then about 0.9" in PA 130° and the color was
noticeably bluish, agreeing with an estimated spectral
class of about B8. In the following years the star was
observed many times, though on occasion it was found to be

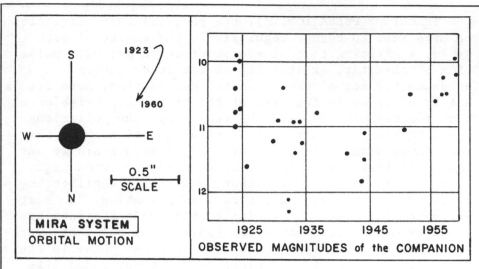

MIRA SYSTEM
ORBITAL MOTION

OBSERVED MAGNITUDES of the COMPANION

completely invisible although seeing conditions were quite favorable. There is little doubt that the companion is intrinsically variable by about 2 magnitudes, the reported range being from 10th to 12th. The star was apparently at maximum in 1923, but fainter than 12th magnitude in 1932. The observations from 1923 to 1958 are plotted above. No regular periodicity appears to be evident.

The star shares the proper motion of Mira itself (0.23" in PA 182°) and also shows the same radial velocity of about 38½ miles per second in recession. The system is thus known to be a physical one, but the orbital motion is rather slow. Both the distance and PA have decreased some- what since discovery and the observed arc seems to imply a highly eccentric orbit with a period of at least two cen- turies. Early in 1962 the companion was observed visually by Dr.F.Holden with the Lowell 24-inch refractor; and the author of this book was at that time enabled to make his own observation and measurement of this very unusual pair. Under very good seeing conditions, and with Mira itself near minimum, both stars were easily seen, and a measure- ment of 0.7" in PA 123° was obtained. The change since 1923 is thus not very great, and the possibility of a short per- iod orbit is definitely ruled out. The most recent studies indicate a probable period of about 260 years, with the separation at discovery (about 70 AU) near the maximum. The projected separation (1962) is about 50 AU.

DESCRIPTIVE NOTES (Cont'd)

The orbital elements of the Mira system are of great importance since they would reveal for the first time the actual mass of a red giant star, a quantity which has never been determined directly. Estimates of 15 or more solar masses are certainly not verified by observations of Mira; if we accept the period of 260 years the total mass of the system is found to be about 3.7 suns. Rather surprisingly, also, the blue companion appears to have about twice the mass of the red giant "primary"!

The blue star is a peculiar object, a hot sub-dwarf which seems to be intermediate in luminosity and density between the main sequence stars and the true white dwarfs. Its computed absolute magnitude is +5 to +7 and the estimated diameter is about 1/11 that of the Sun; this gives a density of about 3300 times that of the Sun. The star has a strong continuous spectrum with bright hydrogen lines showing dark centers; there are also bright lines of helium and ionized calcium. The spectral features are somewhat variable, and at times are reminiscent of the famous "permanent nova" P Cygni.

It is perhaps worthy of note that the combination of a red giant and a bluish subdwarf does not appear to be unique. Several peculiar variables show just such a combination spectrum, typical examples being Z Andromedae and R Aquarii. These are the mysterious "symbiotic stars" and it may be that in the Mira system we are observing for the first time the individual components of such a pair. The recurrent novae T Coronae and RS Ophiuchi also seem to be close binaries of this same type, differing only in that their much smaller separation allows an exhange of material between the components, with evidently violent results. (For additional information on other noted long-period variables, refer to: Chi Cygni, R Leonis, R Andromedae, R Hydrae, R Aquilae, R Centauri, R Leporis, & S Cephei. For red giant stars in general, evolutionary history, etc, refer to Alpha Orionis. Symbiotic stars are chiefly described under R Aquarii and Z Andromedae)

TAU (52 Ceti) Mag 3.50; spectrum G8 V. Position 01417s1612. Tau Ceti is one of the nearest of the naked-eye stars; according to the most reliable published data, it probably ranks 7th on the list, which is given here. Distances are in light years. For a more complete list of nearby stars, refer to the Index and Tables section of this Handbook.

1. Alpha Centauri	4.3	5. Procyon	11.3
2. Sirius	8.7	6. Epsilon Indi	11.4
3. Epsilon Eridani	10.8	7. Tau Ceti	11.8
4. 61 Cygni	11.1	8. 40 Eridani	16.3

Tau Ceti is a main sequence G star about 90% the diameter of our Sun, and about 45% the luminosity (absolute magnitude +5.7). The annual proper motion is 1.92" in PA 297°; the radial velocity is 9½ miles per second in approach.

As one of the nearest stars of the solar type, Tau Ceti is of special interest. It is among stars of this type that we should expect to find other planetary systems if such exist. A search for radio signals of artificial origin was begun in 1959 under the code-name "Project Ozma" with Tau Ceti as one of the most promising targets. Such a survey might eventually provide the only definite evidence of inhabited worlds beyond our own Solar System. The results, so far, have been completely negative.

The curious red dwarf flare star UV Ceti lies about 2½° distant from Tau, to the southwest. (See page 641)

13 (Ho 212) Mag 5.24; spectrum F8 V. Position 00327s0352. Double star of unusually short period, discovered by G.W.Hough in 1887 with the 18½-inch refractor at Dearborn Observatory. The two stars are normally too close to be resolved by amateur telescopes, but at widest separation (about 0.35") a good 12-inch glass will show both components, seeing conditions permitting. The period is only 6.91 years, with periastron in mid-1967.

13 CETI- STATISTICS
Semimajor axis = 0.24" = 4½ AU
Eccentricity = 0.73
Total luminosity = 2 X Sun
Periastron = 1.3 AU (1967.40)
Period = 6.91 yrs
Rotation = direct
Computed masses = 1.27 & 1.05

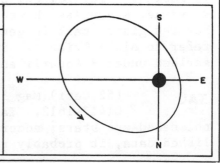

DESCRIPTIVE NOTES (Cont'd)

In addition to the visual pair, the primary star is itself
a spectroscopic binary with a period of 2.0819 days and
individual masses of about 0.95 and 0.32. The fainter star
of this pair is evidently a red dwarf. The distance of the
whole system is about 55 light years, the annual proper
motion is 0.41" in PA 93°, and the radial velocity is 5½
miles per second in recession.

A distant companion of the 12th magnitude was 37"
away at discovery in 1877. This star is not a physical
member of the system, and the separation is changing from
the proper motion of the orbiting pair. A minimum distance
of 19" was reached in 1959, and the separation is now once
again increasing.

L 726-8   (UV Ceti System). Position 01364s1813. This
                is a red dwarf binary system of unusual in-
terest, containing two of the smallest and faintest stars
yet identified. It appears to be the 6th nearest star to
the Solar System; recent parallax measurements give a distance of 9.0 light years, just a shade more distant than
Sirius. The annual proper motion of the star is uncommonly
large, amounting to 3.35" yearly in PA 80°. The star was
discovered by W.J.Luyten of the University of Minnesota
and was announced by Harvard Observatory in April 1949.

The red dwarf components of L726-8 are both of
exceptionally low mass and luminosity, the combined mass
of the pair being a mere 0.08 the mass of the Sun. Each
component, therefore, has a lower mass than any other visible star known. Ross 614b in Monoceros, previously the
star of smallest known mass, is twice as massive as either
component of UV Ceti. (A mass of about 0.03 has recently
been computed for the unseen fainter component of the WZ
Sagittae system).

The orbit of the UV Ceti system is still uncertain.
Luyten has obtained a period of about 54 years with an
average separation of 2.4" and an eccentricity of about
0.06. The computations of P.van de Kamp, however, lead to
a much more eccentric orbit with a period of about 200
years and a separation ranging from 1.5" to about 9". The
observed motion of the pair is shown by the solid portion
of the outlined orbit on page 643; the remaining part of
the curve is still subject to considerable revision.

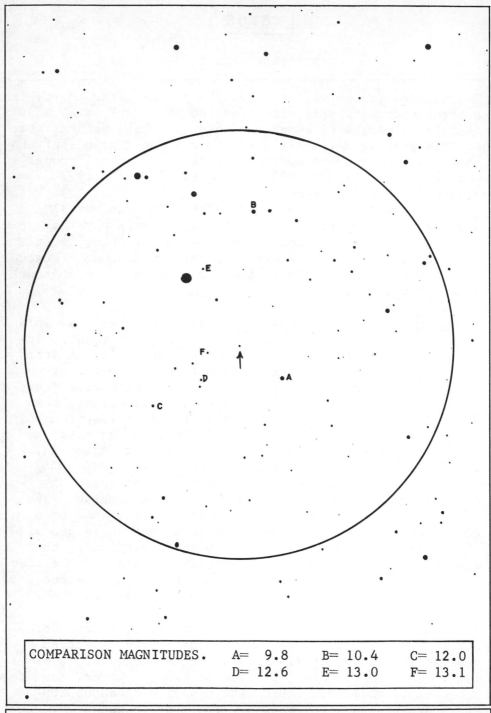

COMPARISON MAGNITUDES.	A= 9.8	B= 10.4	C= 12.0
	D= 12.6	E= 13.0	F= 13.1

L726-8. Identification chart for the UV Ceti System, made from a Lowell Observatory 13-inch telescope plate. Circle diameter = 1° with north at the top. Limiting magnitude about 15.

DESCRIPTIVE NOTES (Cont'd)

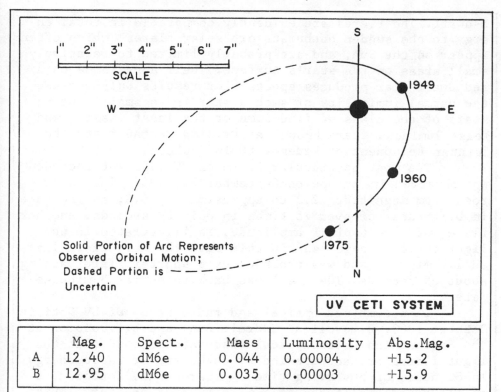

SCALE

1" 2" 3" 4" 5" 6" 7"

S

1949

W ———————————————— E

1960

1975

N

Solid Portion of Arc Represents
Observed Orbital Motion;
Dashed Portion is — — —
Uncertain

UV CETI SYSTEM

	Mag.	Spect.	Mass	Luminosity	Abs.Mag.
A	12.40	dM6e	0.044	0.00004	+15.2
B	12.95	dM6e	0.035	0.00003	+15.9

Strictly speaking, the variable star designation "UV"
should be applied only to the fainter member of the pair,
a remarkable object often called "Luyten's Flare Star".
It is the classic example of the type. The flares occur
with great suddenness and are always of very short duration, usually lasting only a few minutes. In a typical
flare the light of the star increases by one or two magnitudes in less than a minute. The fading is somewhat slower, but the light is usually back to normal in two or
three minutes. During the flare, a bright continuous spectrum appears, superimposed upon the normal spectrum of an

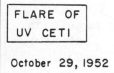

FLARE OF
UV CETI

October 29, 1952

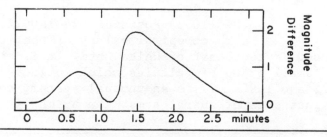

Magnitude
Difference

2

1

0

0   0.5   1.0   1.5   2.0   2.5   minutes

M-dwarf. The flares are evidently comparable in total energy to the sudden outbursts or "solar flares" which often
appear on the Sun, and are probably limited to relatively
small areas on the star's surface. Their appearance on a
red dwarf star produces spectacular results only because
the normal luminosity of such a star is so small. Among
stars of the class we find some of the least massive and
least luminous stars known, as Proxima Centauri and the
fainter component of Krueger 60 in Cepheus.

The most spectacular flare of UV Ceti yet recorded
was observed in Europe on September 24, 1952. The star
rose from magnitude 12.3 to approximately 6.8, an increase
in brightness of over 75 times in only 20 seconds. Another
flare, of more typical amplitude, is illustrated in the
light curve on page 643. In this case the main flare lasted 1.8 minutes and was preceded by a shorter flare lasting
about 55 seconds. The light was back to normal 3 minutes
later.

In a combined optical and radio study of UV Ceti in
October 1963, B.Lovell at Jodrell Bank and L.H.Solomon at
the Smithsonian Astrophysical Observatory found that the
light flares are accompanied by outbursts of radio energy.
In 82 hours of observing time 14 flares were detected, 6
of which had amplitudes of ½ magnitude or greater. This
study showed that the radio radiation tends to reach its
peak intensity several minutes after maximum light.

The radial velocity of the UV Ceti system is about
17½ miles per second in recession, and the computed space
motion closely matches that of the Hyades star cluster in
Taurus which is centered over 45° distant. If actually a
member, UV Ceti must be one of the most extreme outliers,
and possibly the nearest Hyades star to the Sun. The main
mass of the cluster is 120 light years distant. (For data
on other noted flare stars, refer to Proxima Centauri and
Krueger 60 in Cepheus.)

M77   (NGC 1068) Position 02401s0014. A bright and
                 compact spiral galaxy of the 10th magnitude,
located 1° southeast of Delta Ceti. It is the chief member
of a small group of galaxies which includes NGC 1055, 1073,
1087, and 1090. M77 is an unusual system, containing three
distinct sets of spiral arms. The bright inner spiral pat-

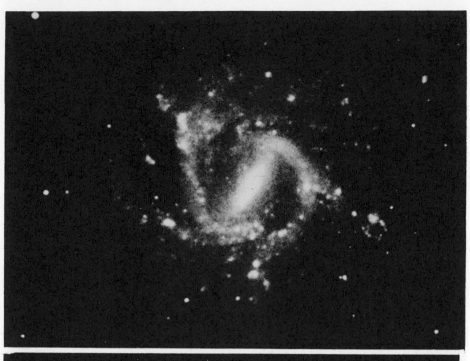

GALAXIES IN CETUS. Top: The barred spiral NGC 1073, one of the members of the M77 group. Below: The many-armed spiral NGC 157. Palomar Observatory 200-inch telescope.

tern measures about 40" x 20" and is resolved by large
telescopes into many luminous knots and condensations; in
instruments as small as 4-inch aperture some of this mottled effect may be detected on very fine nights. A second
fainter spiral pattern continues out to a radius of about
50". Finally, long exposures reveal very faint outer arms
of amorphous texture and very low surface brightness,
forming a 6' diameter elliptical ring around the whole
system.

This galaxy and the famous "Sombrero" (NGC 4594 in
Virgo) were the first two systems in which a very large
red shift was detected, thus introducing the astronomical
world to the mystery of the "expanding universe". Using
the 24-inch refractor at Lowell Observatory in November
1913, V.M.Slipher obtained spectra with exposures of over
$6\frac{1}{2}$ hours, and measured a recessional velocity of about 670
miles per second. Since the Milky Way system would appear
very nearly face-on as seen from M77, very little correction is required for the Sun's motion in our own galaxy.
The best of modern measurements give the true red shift as
about 620 miles per second. This suggests a distance of
somewhat over 60 million light years, the exact value not
well determined. The diameter of the main body is about
40,000 light years, while the outer ring measures some 100,
000 light years across. Rotation studies indicate a total
mass of about 100 billion suns, and the total luminosity
may lie in the range of 30 to 40 billion suns.

M77 is one of the peculiar "Seyfert galaxies" which
show very small bright nuclei whose spectra reveal strong
emission lines. Galaxies of the class, first studied by
C.K.Seyfert (1943) are now known to be moderately strong
radio sources; M77 appears in standard catalogs of radio
sources under the designation 3C71. Radial velocity measurements show that gas clouds are moving at velocities up
to 360 miles per second in the central area of the system,
presumably having been ejected from the nuclear region.
M.F.Walker at Lick Observatory (1966) estimates individual
masses of about 10 million suns for these clouds, and finds
linear diameters of 750 to 900 light years. "The entire
central region of the system is disrupted, perhaps as a
result of earlier generations of the type of explosion now
observed.... a new unknown source of energy in the nucleus
of M77 may be required to account for the observations."

GALAXIES IN CETUS. Top: The compact spiral galaxy M77, as photographed with the Mt.Wilson 100-inch reflector. Below: The fine spiral NGC 309, as recorded with the 200-inch reflector at Palomar.

GALAXIES IN CETUS. Top: The large dim spiral NGC 247.
Below: The faint irregular system IC 1613, a member of
the Local Group of Galaxies.   Palomar Observatory

DESCRIPTIVE NOTES (Cont'd)

According to D.E.Osterbrock and R.A.R.Parker (1965) "the
nuclei of Seyfert galaxies may perhaps be thought of as
miniature quasi-stellar radio sources located at the centers of otherwise normal galaxies". Similar outbursts
seem to be occurring in other unusual systems such as M82
in Ursa Major, M87 in Virgo, and NGC 5128 in Centaurus.
(For a discussion of quasi-stellar radio sources, refer to
3C273 in Virgo)

NGC 247   Position 00446s2101. A large but very dim
spiral galaxy located slightly less than 3°
SSE of the 2nd magnitude star Beta Ceti. In small telescopes, using low power wide-field oculars, it may be detected as a much-elongated smear of faint haze, oriented
nearly north and south, with an extreme length of about
18'. The total magnitude is close to 11. Photographs show
a very patchy and irregular distribution of star clouds in
which the actual spiral pattern is only faintly recognized.
A small bright central mass dominates the system. The
northern quarter of the galaxy is occupied by a large dark
oval area about $4\frac{1}{2}$' x $1\frac{1}{2}$', neatly enclosed by the loop of
the system's northernmost spiral arm. This feature may be
either an obscuring cloud of some sort or an actual vacant
area between star clouds.

NGC 247 is one of the larger members of a sparse
cluster of galaxies centered near the South Galactic Pole
and including a number of large, dim, loose-structured
spirals, chiefly NGC 45 in Cetus, and NGC 55, 253, 300,
and 7793 in Sculptor. Owing to the southern declination,
the group has not been adequately studied, but is probably
the nearest aggregation of galaxies beyond the Local Group.
NGC 247 itself shows a corrected radial velocity of very
nearly zero, and is estimated to be some 6 to 8 million
light years distant. The brightest member of the group is
NGC 253 which lies $4\frac{1}{2}$° to the south in Sculptor; it is
possibly even nearer than NGC 247, since the spectrum
shows a blue shift. G.de Vaucouleurs (1959) finds some
evidence that these galaxies form an expanding association
with a total computed mass of some 150 billion suns. The
absolute magnitude of NGC 247 itself is close to -20.
(Refer to NGC 253, NGC 55, and NGC 7793, all located in
Sculptor)

IC 1613   Position 01025n0152. A faint dwarfish
irregular galaxy, similiar in type and
structure to the better known Magellanic Clouds, but much
smaller and less luminous. It is one of the members of
the Local Group of Galaxies which includes our own Milky
Way system, and is therefore of great interest although it
can hardly be recommended as a subject for small telescopes. The surface brightness is extremely low, although
the total magnitude is about 11½. The galaxy has the form
of an irregular bar about 11' in length, with a detached
star cloud lying some 7' distant toward the northeast; the
maximum extent of the system is about 14'. With the large
modern reflectors, IC 1613 is well-resolved into stars
across its entire surface, and the stellar population is
evidently very similar to that of the Magellanic Clouds.
     Among the millions of stars composing this dwarf
galaxy, a rich population of cepheid variables has been
found. According to a study by W.Baade, other variables
identified in the system include 7 irregular types, one
nova, one eclipsing binary, and one long-period variable.
A study of the cepheids has established the distance of
the galaxy as about 1.8 million light years, slightly closer than the great Andromeda spiral M31. The actual diameter of IC 1613 is about 9000 light years, and the total
luminosity appears to be about 6 million times that of the
Sun. This is one of the least luminous galaxies known.
It is also one of the few galaxies that does not show a
red shift; the radial velocity is about 80 miles per second in approach.
     The most interesting portion of IC 1613 is the
group of bright stars in the northeast portion which form
a huge association about 1000 light years in diameter. The
group contains much dust and gas, and has been identified
by Baade as a region where star formation is still in
progress. The brightest stars in the association are blue
giants with absolute magnitudes up to -7, comparable to
Rigel. A second smaller association contains chiefly red
giant stars, plus one unusual variable which has been identified as a probable cepheid despite its abnormally long
period of 146 days. (Refer also to NGC 6822 in Sagittarius and the two Magellanic Clouds in Dorado and Tucana)

DEEP-SKY OBJECTS IN CETUS. Top: The planetary nebula NGC 246. Below: The spiral galaxy NGC 615. Palomar Observatory 200-inch telescope photographs.

CELESTIAL HANDBOOK— VOLUME ONE

CONSTELLATION INDEX and STAR ATLAS REFERENCE

CONSTELLATION	PAGE	NORTON'S ATLAS CHART	SKALNATE-PLESO ATLAS CHART		
ANDROMEDA	103	3	II	V	
ANTLIA	160	8, 10	XIII	VIII	
APUS	163	16	XVI	XIV	
AQUARIUS	165	4, 14	XI	XV	VI
AQUILA	197	13, 14	X	XI	V
ARA	235	12, 16	XIV	XV	XVI
ARIES	245	5	VI	II	
AURIGA	253	5, 7	II	III	
BOOTES	297	11	IV	IX	I
CAELUM	313	6	XII		
CAMELOPARDALIS	314	1, 2	I	II	III
CANCER	337	7	VIII	III	VII
CANES VENATICI	353	9	IV	III	I
CANIS MAJOR	381	8	VII	VIII	
CANIS MINOR	447	7	VII	VIII	
CAPRICORNUS	452	14	XI	XV	X
CARINA	458	8, 16	XIII	XVI	XII
CASSIOPEIA	476	2, 3	I	II	V
CENTAURUS	535	10, 16	XIV	XIII	XVI
CEPHEUS	573	2	I	V	II
CETUS	621	4, 5	VI	XII	XI XV